# Data, Network, and Internet Communications Technology

**Ata Elahi, Ph.D**

Southern Connecticut State University

**Mehran Elahi, Ph.D**

Elizabeth City State University

Australia • Canada • Mexico • Singapore • Spain • United Kingdom • United States

**Data, Network, and Internet Communications Technology**
Ata Elahi and Mehran Elahi

**Vice President, Technology and Trades SBU:**
Alar Elken

**Editorial Director:**
Sandy Clark

**Senior Acquisitions Editor:**
Stephen Helba

**Senior Development Editor:**
Michelle Ruelos Cannistraci

**Marketing Director:**
Dave Garza

**Senior Channel Manager:**
Dennis Williams

**Marketing Coordinator:**
Stacey Wiktorek

**Production Director:**
Mary Ellen Black

**Senior Production Manager:**
Larry Main

**Production Editor:**
Benj Gleeksman

**Art/Design Coordinator:**
Francis Hogan

**Senior Editorial Assistant:**
Dawn Daugherty



Printed in the United States of America
1 2 3 4 5 XX 07 06 05

For more information contact Thomson Delmar Learning
Executive Woods
5 Maxwell Drive, PO Box 8007,
Clifton Park, NY 12065-8007
Or find us on the World Wide Web at
www.delmarlearning.com



Library of Congress Cataloging-in-Publication Data:
Elahi, Ata
Data, network, and Internet communications technology / by Ata Elahi, Mehran Elahi.
p. cm.
Includes bibliographical references and index.
ISBN (invalid) 0-401-87269-7
1. Computer Networks—Textbooks. 2. Data transmission systems—Textbooks. I. Elahi, Mehran. II. Title.
TK5105.5.E388 2005
004.6–dc22 2005041005

ISBN: 1401872697

## NOTICE TO THE READER

# Contents

## 21 Voice Over IP Protocol (VoIP) 395



## 22 Asynchronous Transfer Mode (ATM) 417



## 23 Network Security 439

# Preface

This book is the result of teaching the course, Data Communications and Computer Networks, at Southern Connecticut State University since 1987. The book covers the technology aspect of networks, rather than the theory of networks.

The beta version of this textbook was tested in undergraduate-level computer networks courses at Southern Connecticut State University. The textbook covers networking using a *direct and practical approach* that explains the technology in simple terms. This book covers the latest topics in networking technology, including DSL, cable modem, ATM, fast ethernet, LAN switching, gigabit ethernet, frame relay, SONET, wireless LAN, wireless MAN, Voice over IP, and network security.

## Intended Audience

This book is written primarily as an introduction to networking for students majoring in computer science, electronics technology, or engineering. The content is clear and easy to understand for those in technical colleges, while the broad range of topics is appealing for those in higher-level courses. For students in business applications and CIS courses, the instructor may want to omit part of text.

## Organization

The material in this textbook is presented *practically* rather than taking a theoretical and mathematical approach. Therefore, no specialized background is required to understand the material; the first three chapters of the book form the foundation for the rest of the text.

We have opted to focus on the technology aspect of networks, so each technology is presented in a separate chapter. In addition, we offer a brief introduction to computer architecture, because networks are simply another part of the computer.

***Chapter 1*** is an introduction to computer networks, covering network topologies and types of networks.

***Chapter 2*** covers basic data communications, including analog signal, digital signal, binary numbers, serial and parallel transmission, communication modes, digital encoding, error detection methods, and error correction used in networking. This chapter gives students the basic knowledge required for the rest of the textbook.

***Chapter 3*** presents an overview of computer architecture. The network is part of the computer; therefore, the reader should have a basic knowledge of microcomputer architecture. This chapter covers the basic components of a microcomputer: the CPU, types of memory, and computer buses.

***Chapter 4*** covers data communications media such as twisted-pair cable, coaxial cable, fiber-optic cable, and wireless communications. It also covers channel bandwidth, channel capacity, and throughput.

***Chapter 5*** covers the types of multiplexers (TDM, FDM, SPM, CDM, and WDM), T1 link architecture, and switching concepts.

***Chapter 6*** explains the function of standards organizations, and lists some of the computer protocols, followed by a summary of the open system model, which explains the function of each layer. IEEE 802 committee standards are also presented.

***Chapter 7*** presents modem technology, modulation methods, digital subscriber line (DSL) technology, and cable modem technology.

***Chapter 8*** covers the Ethernet network, from operation to technical specifications and cabling. It serves as the foundation for the material on Fast Ethernet and Gigabit Ethernet.

***Chapter 9*** explains the operation of the Token Ring network, Token Ring technical specifications, and Token Bus operation.

***Chapter 10*** presents Fast Ethernet technology, Fast Ethernet repeaters, and different types of media used for Fast Ethernet.

***Chapter 11*** presents networking interconnection devices such as repeaters, bridges, routers, gateways, and switches. It also explains switch operations and applications for VLAN.

***Chapter 12*** explains Gigabit Ethernet technology and the types of media used for Gigabit Ethernet, followed by an explanation of the applications of Gigabit Ethernet. It also covers 10 Gigabit Ethernet technology and the type of media used for 10 Gigabit Ethernet.

***Chapter 13*** covers Fiber Distributed Data Interface (FDDI) technology and its applications.

***Chapter 14*** covers Frame Relay technology, operations, and components.

***Chapter 15*** presents SONET components and architecture.

***Chapter 16*** covers Internet architecture, IP addressing, Transmission Control Protocol (TCP), Internet protocol, IPv6, and Internet II.

***Chapter 17*** explains Dynamic Host Configuration Protocol (DHCP), Internet Message Protocol (ICMP), Point-to-Point Protocol (PPP), Virtual Private Network (VPN), IP security, and routing tables.

***Chapter 18*** presents wireless LAN (IEEE 802.11 a, b, and g), WLAN topologies, WLAN Medium Access Control (MAC), Wi-FI certification, and WLAN security.

***Chapter 19*** covers Bluetooth topology, Bluetooth protocol architecture, and Bluetooth Physical link.

***Chapter 20*** explains wireless MAN (IEEE 802.16) topology, wireless MAN applications, and IEEE 802.16 protocol architecture.

***Chapter 21*** covers application of Voice over IP Protocol (VoIP), components of VoIP, Voice over IP protocols and standards, H.323, RTP, and Session Initiation Protocol (SIP).

***Chapter 22*** presents ATM network components, ATM switch architecture, and ATM Adaptation layer.

***Chapter 23*** covers elements of network security, cryptography, digital signatures, Kerberos, certificates, Secure Socket Layer protocol (SSL), IEEE 802.1x extensible authentication protocol (EAP), and firewalls.

***Chapter 24*** covers architecture of Universal Serial Bus (USB), pact formats, USB operations, USB transactions, and PCI Express Bus architecture.

### Features

- Each chapter begins with a quick look of topics to be covered in the chapter Outline and Objectives. Within the text, key terms are highlighted in bold repeated for reinforcement at the end of the chapter. A bulleted list of important points is listed in the chapter Summary. At the end of each chapter is a set of multiple-choice questions for a quick review of chapter concepts, followed by a set of more challenging short answer questions.
- Each chapter focuses on a separate networking technology, making the text flexible for readers to pick and choose selected topics.

- The book covers the latest emerging topics such as Bluetooth technology, which is covered in Chapter 19, Chapter 20 explores wireless MANs and Chapter 21 introduces Voice over IP Protocol (VoIP) and Chapter 23 explores security topics, such as cryptography and encryption.
- Chapters 16 and 17 are both devoted to internet topics and TCP/IP.
- Numerous examples and illustrations are clear and help make it easy to learn some of the more complicated concepts.
- Appendix A provides a list of computers and networks connectors.
- Appendix B contains answers to odd-numbered questions.
- Appendix C and D includes a list of acronyms and glossary terms.

**Supplements**
***e.resource* CD**
The following instructional aids are available all on one CD.

- **Instructor's Manual** contains answers to all end-of-chapter review questions.
- **Exam View Testbank** includes over 700 review questions, including true/false, multiple-choice and short answer.
- **PowerPoint Presentation Slides** highlight key concepts from each chapter and can be used as handouts or as a springboard for lecture and discussion.
- **Image Library** includes all images from the textbook so instructors can customize handouts, tests, and power point presentations.

## Acknowledgments

Many people contributed to the development of this book. We would like to express our deep appreciation to Professor Greg Simmons of DeVry University for his in-depth review of the manuscript, and valuable suggestions and comments that enabled us to improve the quality of this textbook. We would like to thank Thomas Sadowski and Megan Daman. Many thanks to the staff of Thomson Delmar Learning particularly our Acquistions Editor, Steve Helba, Development Editor Michelle Ruelos Cannistraci, Production Editor Benj Gleeksman and Marketing Director Dave Garza.

We would like to thank Dr. Edward Harris Dean of the School of Communication, Information and Library Science and Professor Winnie

Yu, Chairperson of Computer Science at Southern Connecticut State University; Dr. Carolyn Mahoney (Provost and Vice Chancellor for Academic Affairs), Dr. Ronald Blackmon (Dean of School of Math, Science & Technology), and particularly Dr. Akbar Eslami (Chairman of the Department of Technology) at Elizabeth City State University for their encouragement and support.

The author would like to thank the following reviewers who reviewed the earlier version of Network Communications Technology.

Mike Awwad, DeVry University, North Brunswick, NJ

Omar Ba-Rukab, Florida Institute of Technology, Melbourne, FL

Robert Diffenderfer, DeVry University, Kansas City, MO

Enrique Garcia, Laredo Community College, Laredo, TX

Dr. Rafiqul Islam, DeVry University, Alberta, Canada

Lynette Garetz, Heald College, Hayward, CA

Anu Gokhale, Illinois State University, Normal, IL

Nebojsa Jansic, DeVry University, Columbus, OH

Steve Kuchler, IVY Tech, Indianapolis, IN

Dr. William Lin, PhD, Indiana University—Purdue University, Indianapolis, IN

Clifford Present, DeVry University, Pomona, CA

Glenn Rettig, Owens Community College, Findlay, OH

Richard Rouse, DeVry University, Kansas city, MO

Mohammed Mehdi Shahsavari, Florida Institute of Technology, Melbourne, FL

Lowell Tawney, DeVry University, Kansas City, MO

Julius Willis, Heald College, Hayward, CA

The Authors and Thomson Delmar Learning would like to thank our reviewers for the final version of *Data, Network and Internet Communications Technology.*

Silver Dasgupta, Wentworth Institute of Technology

Phil Dumas, DeVry University, Orlando, FL

Bob Gill, British Columbia Institute of Technology, Canada

Bob Gruber, Norhtern Virginia Community College, Manassas, VA

William Lin, Indiana University / Purdue University at Indianapolis, IN

John Lyon, Onondaga Community College, Syracuse, NY

William Routt, Wake Technical Community College, Raleigh, NC

Gregory Simmons, DeVry, Columbus, OH

Randy Winzer, Pittsburgh State University, Pittsburgh, KS

*Dedication*

This book is dedicated to

our parents, Rahim and Shayesteh.
—Ata & Mehran

my children Shabnam and Aria.
—Ata

CHAPTER 1

# Introduction to Communications Networks

**OUTLINE**

1.1 Computer Networks
1.2 Network Models
1.3 Network Components
1.4 Network Topology
1.5 Types of Networks

**OBJECTIVES**

After completing this chapter, you should be able to:

- Explain the components of a data communication system
- Explain the advantages of computer networks
- Describe the components of a network
- Discuss the function of a client/server model
- Explain various networking topologies
- Describe different types of networks in terms of their advantages and disadvantages

**INTRODUCTION**

In order to transfer information from a source to a destination, some hardware components are required. Figure 1.1 shows the components of a data communication model for transmitting information from source to destination.

**Source:** The source station can be a computer and its function is to pass information to a transmitter for subsequent transmission.

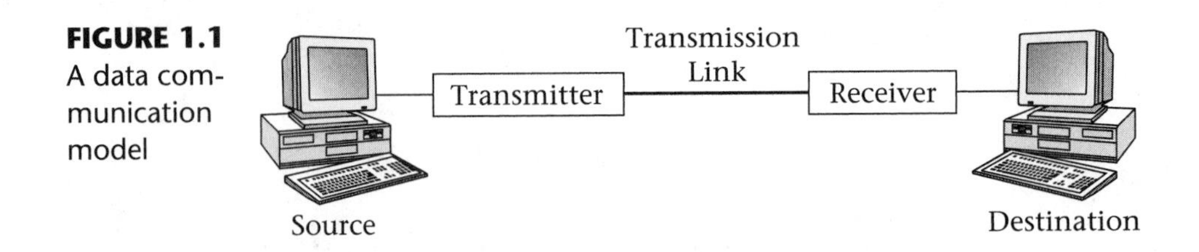


**FIGURE 1.1** A data communication model

**Transmitter:** The function of a transmitter is to accept information from the source and change the information such that it is compatible with the transmission link. Information is then transmitted over the communication link. Modem and network cards are examples of transmitters.

**Transmission:** The function of a transmission link is to carry information from a transmitter to a receiver. The transmission link can be a conductor, a fiber optic cable, or wireless media (air).

**Receiver:** The receiver accepts information from the transmission link and information is then converted to the proper form so that it is acceptable to the destination.

**Destination:** The receiver passes information to the destination for processing.

In networking technology both the receiver and transmitter come in one unit and are usually installed inside the computer, such as network cards or modems.

## 1.1 Computer Networks

Computer networking is a business tool for companies: for example, a bank can transfer funds between branches by using a network. People can access their bank accounts using automatic teller machines via a network. Travel agencies are using networks to make airline reservations. We can even make banking transactions and shop online using a network. Networking technology is growing fast and will enable students to access the Internet in any campus location using their laptop computers.

Networking is a generic term. Several computers connected together are called a **computer network.** A network is a system of computers and related equipment connected by communication links in order to share data. The related equipment may be printers, fax machines, modems, copiers, and so forth. The following are some of the benefits of using computer networks:

**Resource sharing:** Computers in a network can share resources such as data, printers, disk drives, and scanners.

**Reliability:** Since computers in a network can share data, if one of the computers on the network crashes, a copy of its resources might be found on other computers in the network.

**Cost:** Microcomputers are much less expensive than mainframes. Instead of using several mainframes, a network can use one mainframe as a server, with several microcomputers connected to the server as clients. This creates a client/sever relationship.

**Communication:** Users can exchange messages via electronic mail or other messaging systems, and they can transfer files.

## 1.2 Network Models

A computer in a network can be either a server, a client, or a peer. A **server** is a computer on the network that holds shared files and the network operating system (NOS) that manages the network operations. In order for a **client** computer to access the resources on the server, the client computer must request information from the server. The server will then transmit the information requested to the client.

Three models are used, based on the type of network operation required:

1. Peer-to-peer network (work group)
2. Server-based network
3. Client/server network

### Peer-to-Peer Model (Work Group)

In a **peer-to-peer model** there is no special station that holds shared files and a network operating system. Each station can access the resources of the other stations in the network. Individual stations can act as a server and/or as a client. In this model, each user is responsible for administrating and upgrading the software of his or her station. Since there is no centralized station to manage network operations, this model is typically used for a network of fewer than ten stations. Figure 1.2 shows a peer-to-peer network model.

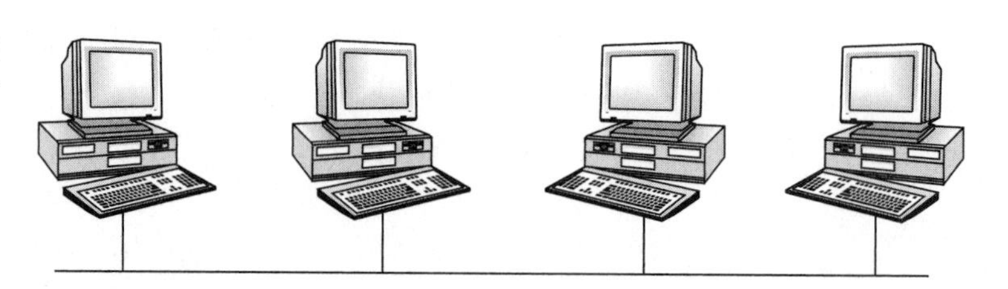

**FIGURE 1.2** Peer-to-peer network

### Server-Based Network

In a **file server model,** a server stores all of the network's shared files and applications such as word processor documents, compilers, database applications, and the network operating system that manages network operations. A user can access the file server and transfer shared files to his or her station.

Figure 1.3 shows a network with one file server and three users, or clients. Each client can access the resources on the server as well as the resources of other clients. Clients that are connected on a network may be able to freely exchange information with one another. The following are widely used servers:

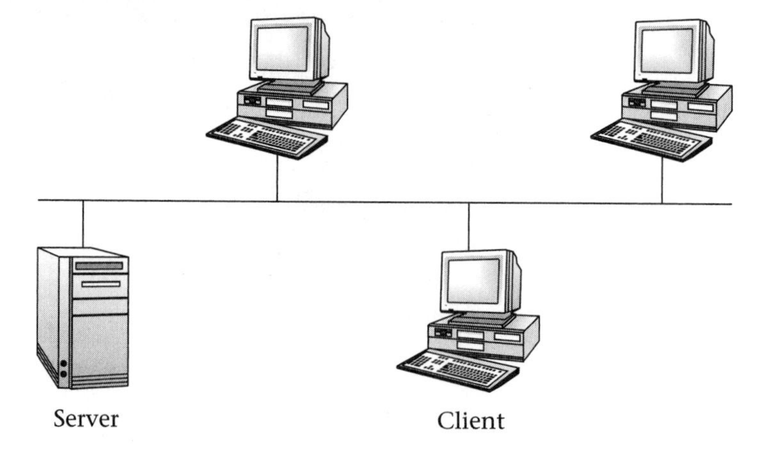


**FIGURE 1.3** Network with one server and three clients

**Mail server:** A mail server stores all the clients' electronic mail. The client can access the server and transfer incoming mail to its station. The client can also use the mail server to transfer outgoing mail to the network.

**Print server:** Clients can submit files to the server for printing.

**Communication server:** The server is used by clients to communicate with other networks via communication links.

### Client/Server Model

In the **client/server model,** a client submits its task to the server; the server then executes the client's task and returns the results to the requesting client station. This method of information sharing is called the *client/server model* and is depicted in Figure 1.4. In a client/server model, less information travels through the network compared to the server-based model, making more efficient use of the network.

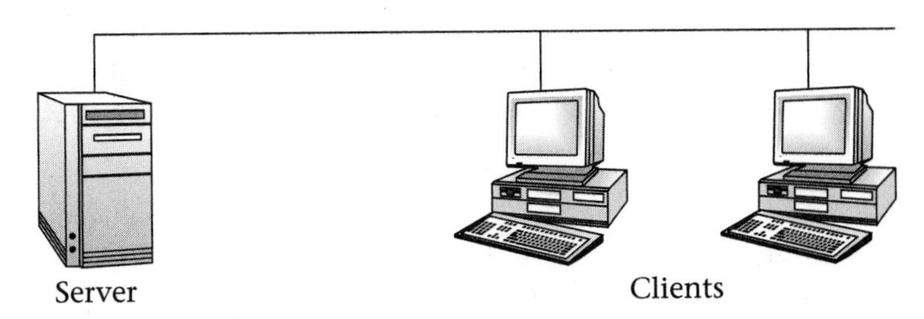


**FIGURE 1.4** Client/server model

## 1.3 Network Components

A computer network has several basic components:

1. **Network Interface Card (NIC):** Each computer in a network requires a network interface card, which allows the stations on the network to communicate with each other.
2. **Transmission Medium:** The transmission medium connects the computers together and provides a communication link between the computers on the network. Some of the more common types are twisted pair cable, coaxial cable, fiber optic cable, and wireless.
3. **Network Operating System (NOS):** The NOS runs on the server and provides services to the client such as login, password, print file, network administration, and file sharing. Most modern computer operating systems have NOS functionality.

## 1.4 Network Topology

The **topology** of a network describes the way computers are connected together. Topology is a major design consideration for cost and reliability. The following is a list of common topologies found in computer networking:

- Star
- Ring
- Bus
- Mesh
- Tree
- Hybrid

## Star Topology

In a **star topology,** all stations are connected to a central controller or hub, as shown in Figure 1.5. For any station to communicate with another station, the source must send information to the hub, and then the hub must transmit that information to the destination station. If station #1 wants to send information to station #3, it must send information to the hub; the hub must pass the information to station #3.

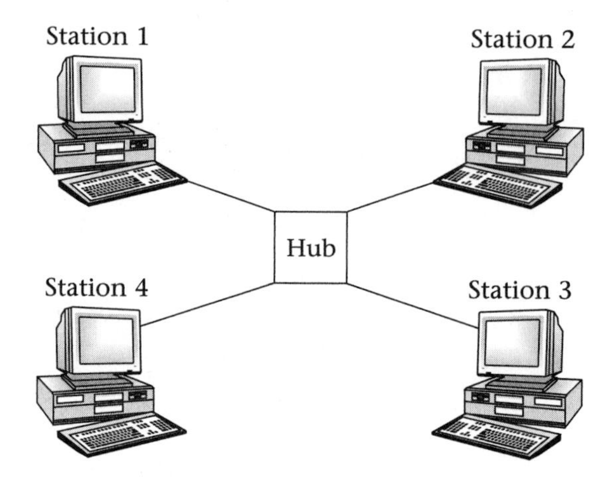


**FIGURE 1.5**
Star topology

The disadvantage of star topology is that the operation of the entire network depends on the hub. If the hub breaks down, the entire network is disabled. The advantages of star topology are as follows:

- It is easy to set up the network.
- It is easy to expand the network.
- If one link to the hub breaks, only the station using that link is affected.

It is possible for a network to have one topology electrically but another topology physically. For example, Ethernet with UTP cable uses star topology physically but the stations are connected using a bus topology.

## Ring Topology

IBM invented ring topology, which is well known as *IBM Token Ring*. In a **ring topology** all the stations are connected in cascading order to make a ring, as shown in Figure 1.6. The source station transfers information to the next station on the ring, which checks the address of the information.

If the address of the information matches the station's address, the station copies the information and passes it to the next station; the next station repeats the process and passes the information on to the next station, and so on, until the information reaches the source station. The source then removes the information from the ring. The arrows in Figure 1.6 indicate the direction in which the information flows.

**FIGURE 1.6**
Ring topology

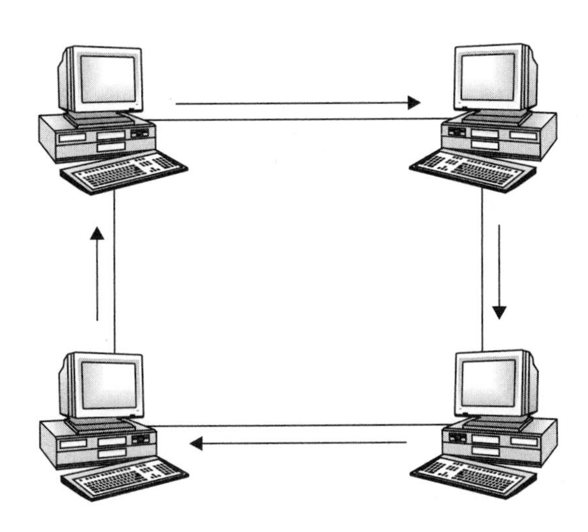

The disadvantages of ring topology are as follows:

- If a link or a station breaks down, the entire network is disabled.
- Complex hardware is required (the network interface card is expensive).
- Adding a new client disrupts the entire network.

The advantages of ring topology are that it is easy to install and easy to expand.

### Bus Topology

A **bus network** is a multi-point connection in which stations are connected to a single cable called a *bus*. The bus topology is depicted in Figure 1.7. In the bus topology, all stations share one medium. Bus topology is widely used in LAN networking. Ethernet is one of the most popular LANs that use bus topology.

The advantages of bus topology are simplicity, low cost, and easy expansion of the network. The disadvantage of bus topology is that a breakdown in the bus cable brings the entire network down.

**FIGURE 1.7**
Bus topology

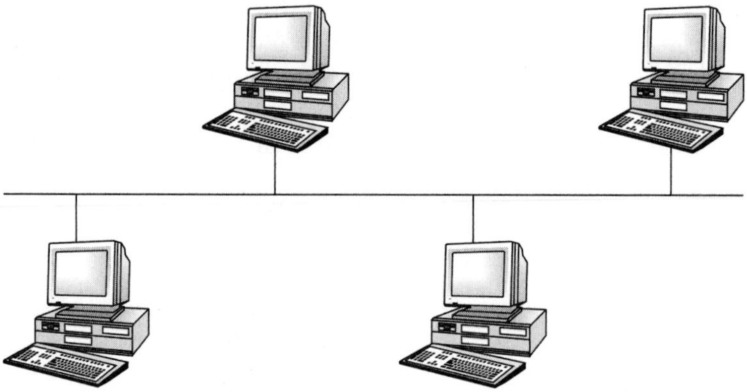

## Mesh Topology

**Mesh topology** can be full or partial. In a full mesh topology each station is directly connected to every other station in the network, as shown in Figure 1.8.

The advantage of a fully connected topology is that each station has a dedicated connection to every other station; therefore this topology offers the highest reliability and security. If one link in the mesh topology breaks, the network remains active.

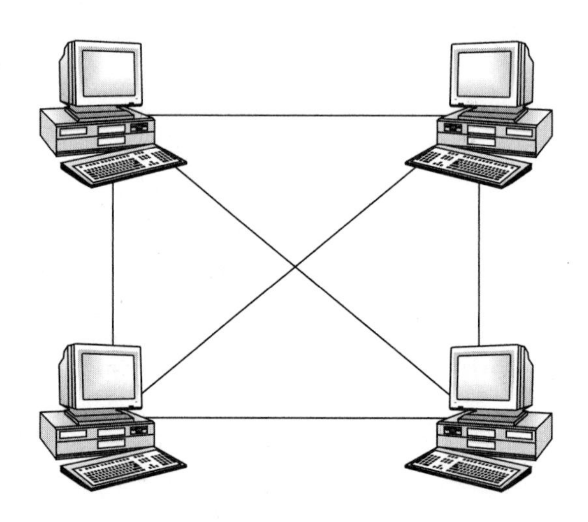

**FIGURE 1.8**
Full mesh topology

A major disadvantage of a fully connected topology is that it uses many connections and therefore requires a great deal of wiring, especially when the number of stations increases. Consider, for example, a fully connected network with 100 workstations. Workstation #1 would require 99 network connections to connect it to workstations 2 through 99. The number of connections is determined by $N(N - 1)/2$, where $N$ is the number of stations in the network. This type of topology is seldom used because it is not cost effective.

In partial mesh topology, some of the stations are connected to other stations, but some of the stations are connected only to those stations with which they exchange the most data. Figure 1.9 shows partial mesh topology.

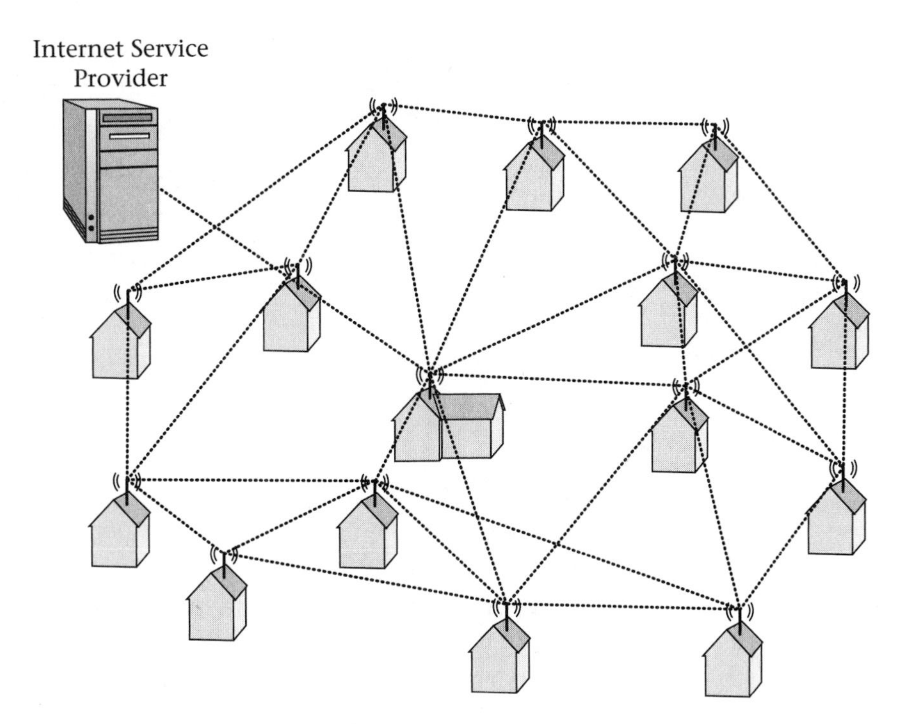


**FIGURE 1.9**
Partial mesh topology

## Tree Topology

The **tree topology** uses an active hub or repeater to connect stations together. The **hub** is one of the most important elements of a network because it links stations in the network together. The function of the hub is to accept information from one station and repeat the information to other stations on the hub, as shown in Figure 1.10.

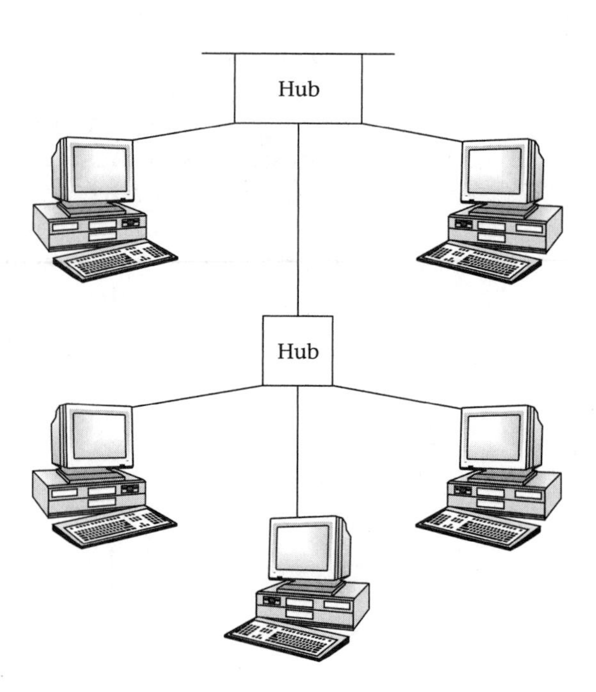


**FIGURE 1.10**
Tree topology

The advantage of this topology is that when one hub breaks, only stations connected to the broken hub will be affected. There are several types of hubs:

**Manageable hub:** Intelligent hubs are defined as *manageable hubs;* that is, each of the ports on the hub can be enabled or disabled by the network administrator using software.

**Standalone hub:** A standalone hub is used for workgroups of computers that are separate from the rest of the network. They cannot be linked together logically to represent a larger hub.

**Modular hub:** A modular hub comes with a chassis or card cage and the number of ports can be extended by adding extra cards.

**Stackable hub:** A stackable hub looks like a standalone hub, but several of them can be stacked or connected together in order to increase the number of ports.

## Hybrid Topology

**Hybrid topology** is a combination of different topologies connected together by a backbone cable, as shown in Figure 1.11. Each network is connected to the backbone cable by a device called a *bridge*.

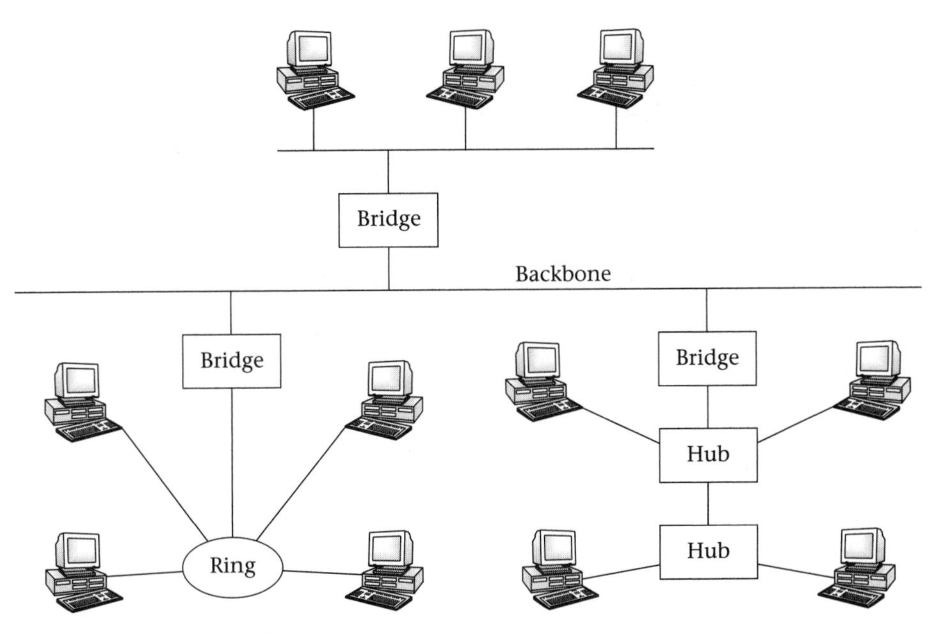


**FIGURE 1.11**
Hybrid topology

## 1.5 Types of Networks

The distance between computers that are connected as a network determines the type of network, such as a Local Area Network (LAN), Metropolitan Area Network (MAN), and Wide Area Network (WAN).

**Local Area Network (LAN)**

A **Local Area Network (LAN)** is a high-speed network designed to link computers and other data communication systems together within a small geographic area such as an office, a department, or a single floor of a multi-story building. Several LANs can be connected together in a building or campus to extend the connectivity. A LAN is considered a private network. The most popular LANs in use today are Ethernet, Token Ring, and Gigabit Ethernet.

**Metropolitan Area Network (MAN)**

A **Metropolitan Area Network (MAN)** can cover approximately 30 to 100 miles, connecting multiple networks that are located in different locations of a city or town. The communication links in a MAN are generally owned by a network service provider. Figure 1.12 shows a Metropolitan Area Network.

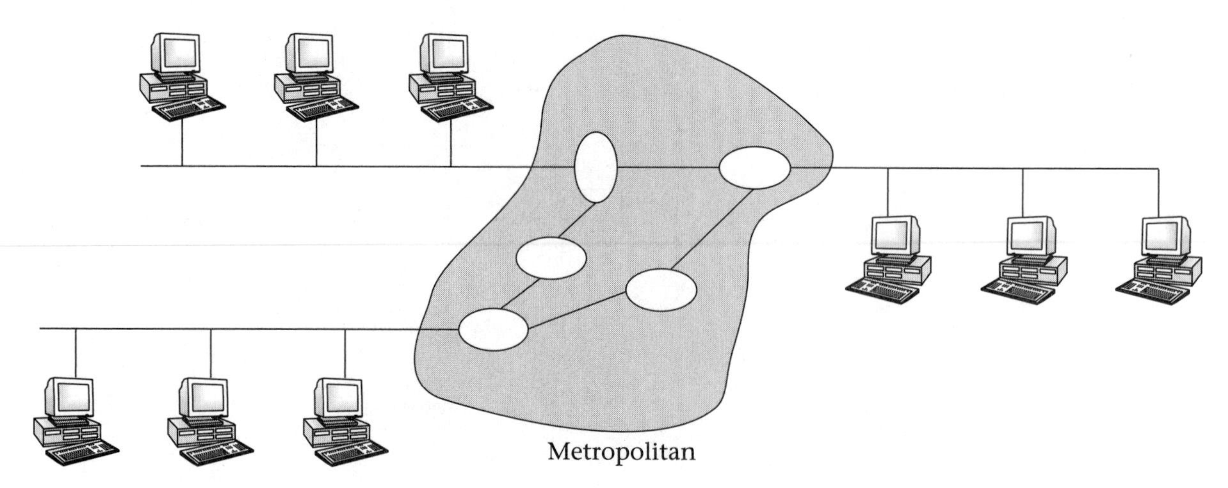


**FIGURE 1.12**
Metropolitan Area Network (MAN)

**Wide Area Network (WAN)**

A **Wide Area Network (WAN)** is used for long-distance transmission of information. WANs cover a large geographical area, such as an entire country or continent. WANs may use leased lines from telephone companies, Public Switch Data Networks (PSDN), or satellites for communication links.

**Internet**

The **Internet** is a collection of networks located all around the world that are connected by gateways, as shown in Figure 1.13. Each gateway has a routing table containing information about the networks that are connected to the gateway. Several networks may be connected to one gateway. The gateway accepts information from the source network and checks its routing table to see if the destination is in a network connected to the gateway. If the destination station is in a network connected to the gateway, it transmits the information to the destination network. Otherwise, it passes the information to the next gateway, which performs the same operation; this process continues until the information reaches its destination.

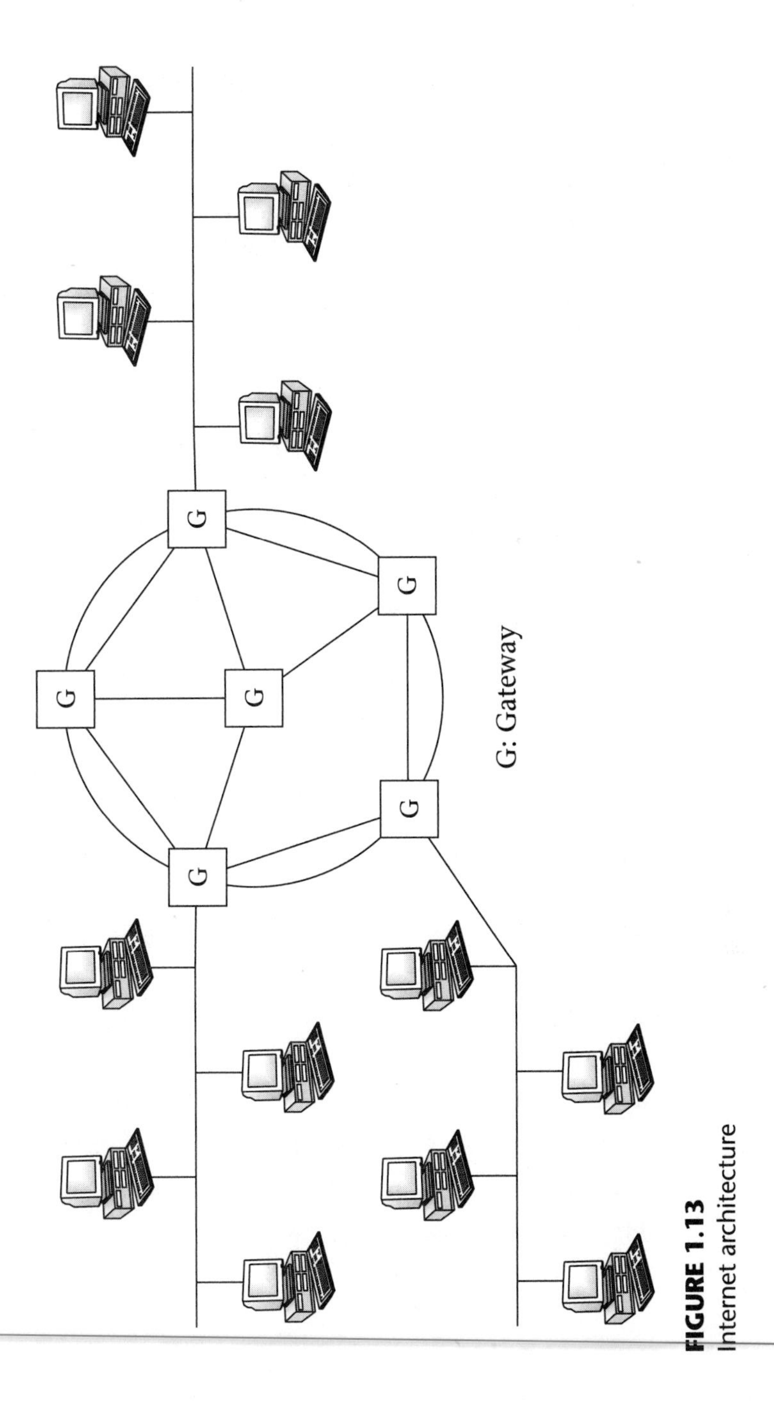


**FIGURE 1.13**
Internet architecture

## Summary

- A group of several computers connected by communication media is called a *computer network*.
- Some of the applications of computer networks are file sharing, reliability, hardware sharing, and electronic mail.
- In the client/server model of networking, the client submits the information to the server, which processes the information and returns the results to the client station.
- The components of a network are the Network Interface Card (NIC), the Network Operating System (NOS), and the communication link (transmission medium).
- Computers can be connected in the form of star, ring, bus, mesh, tree, and hybrid topologies.
- The types of networks are LAN, MAN, WAN, and the Internet.

## Key Terms

Bus Network
Client
File Server Model
Hub
Hybrid Topology
Internet
Local Area Network (LAN)
Mesh Topology
Metropolitan Area Network (MAN)
Network Interface Card (NIC)
Network Operating System (NOS)
Client/Server Model
Computer Network
Peer-to-Peer Model
Ring Topology
Server
Star Topology
Topology
Transmission Medium
Tree Topology
Wide Area Network (WAN)

## Review Questions

### • Multiple Choice Questions

1. Several computers connected together are called a ______.
   a. computer network
   b. client
   c. server
   d. hub
2. In a ______ network, the client submits a task to the server, and then the server executes and returns the result to the requesting client station.
   a. peer-to-peer
   b. file server-based
   c. client/server
   d. all of the above
3. A computer in a network can function as a ______ or as a ______.
   a. client, server
   b. client, user
   c. a and b
   d. all of the above
4. A ______ stores all of the client's electronic mail.
   a. file server
   b. print server
   c. communication server
   d. mail server
5. A ______ uses a modem or other type of communication link to enable clients to communicate with other networks.
   a. mail server
   b. communication server
   c. a and b
   d. none of the above
6. In a ______ topology, all stations are connected to a central controller or hub.
   a. star
   b. ring
   c. bus
   d. mesh
7. In a ______ topology, all stations are connected in cascade.
   a. star
   b. ring
   c. tree
   d. bus
8. A ______ topology is a combination of different topologies connected together by a backbone cable.
   a. star
   b. ring
   c. bus
   d. hybrid
9. The network topology that uses a hub is ______.
   a. ring
   b. bus
   c. star
   d. mesh
10. The type of topology that uses multi-point connections is a ______.
    a. bus
    b. star
    c. ring
    d. full mesh

11. A fully connected network with five stations requires ______.
    a. 5
    b. 10
    c. 20
    d. 15
12. The network used for office buildings is ______.
    a. LAN
    b. MAN
    c. WAN
    d. Internet
13. The topology used for Ethernet is a ______.
    a. bus
    b. star
    c. ring
    d. a and b
14. The Internet is a collection of LANs connected together by ______.
    a. routers
    b. switches
    c. gateways
    d. repeaters
15. Computers on a campus are connected by ______.
    a. LAN
    b. WAN
    c. MAN
    d. Internet

## • Short Answer Questions

1. What are the components of a communication model?
2. Explain the function of a server.
3. What is the function of the client in a file server model?
4. Explain the term client/server model.
5. What are the advantages of a client/server model?
6. What are the network components?
7. A network operating system runs on the ______.
8. List the networking topologies.
9. What is the disadvantage of a fully connected topology?
10. What is a hub?
11. What are the types of networks?
12. What does MAN stand for?
13. Explain the Internet operation.
14. What does WAN stand for?
15. What are the advantages of using bus topology?

CHAPTER 2

# Data Communications

OUTLINE

2.1 Analog Signals
2.2 Digital Signals
2.3 Binary Numbers
2.4 Coding Schemes
2.5 Transmission Modes
2.6 Transmission Methods
2.7 Communication Modes
2.8 Signal Transmission
2.9 Digital Signal Encoding
2.10 Error Detection Methods
2.11 Error Correction

OBJECTIVES

After completing this chapter, you should be able to:

- Distinguish between analog and digital signals
- Distinguish between periodic and non-periodic signals
- Convert decimal numbers to binary, binary to hexadecimal, and vice versa
- Represent characters and decimal numbers in the 7-bit ASCII code
- Compare serial, parallel, asynchronous, and synchronous transmission
- List the communication modes

- Explain the different types of digital encoding methods
- Calculate a Block Check Character (BCC)
- Calculate a Frame Check Sequence (FCS), which is used for error detection in networking
- Learn different error detection methods

## INTRODUCTION

In order to understand network technology it is important to know how information is represented for transmission from one computer to another. Information can be transferred between computers in one of two ways: by an analog signal or a digital signal.

## 2.1 Analog Signals

An **analog signal** is a signal whose amplitude is a function of time and changes gradually as time changes. Analog signals can be classified as non-periodic or periodic signals.

***Non-periodic signal.*** In a non-periodic signal there is no repeated pattern in the signal, as shown in Figure 2.1.

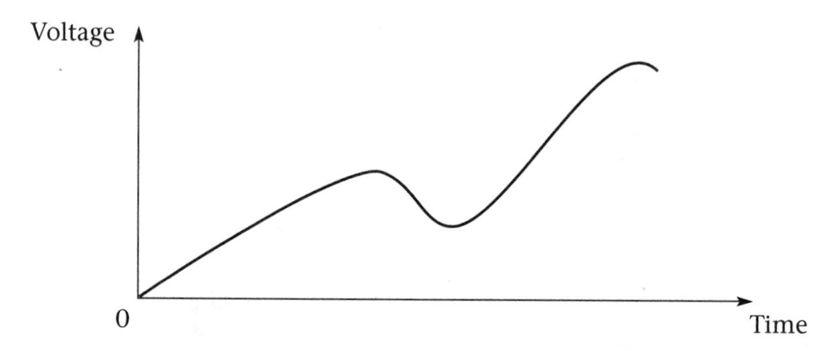


**FIGURE 2.1**
Representation of a non-periodic analog signal

***Periodic signal.*** A signal that repeats a pattern within a measurable time period is called a *periodic signal* and completion of a full pattern is called a *cycle*. The simplest periodic signal is a sine wave, which is shown in Figure 2.2. In the time domain, sine wave amplitude $a(t)$ can be represented mathematically as $a(t) = A\,sin(\omega t + \theta)$ where $A$ is the maximum amplitude, $\omega$ is the angular frequency, and $\theta$ is the phase angle.

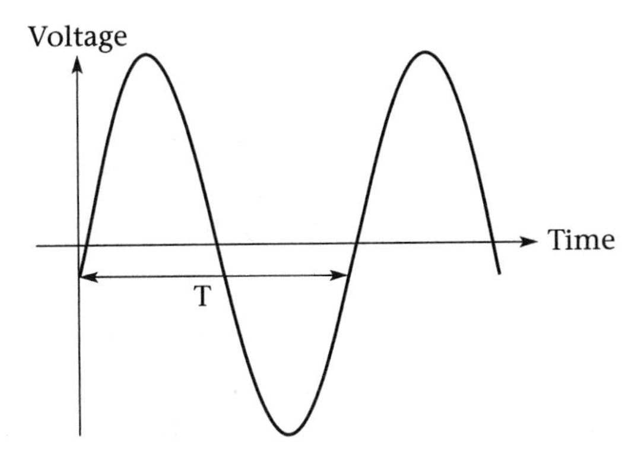


**FIGURE 2.2**
Time domain representation of a sine wave

A periodic signal can also be represented in the frequency domain where the horizontal axis is the frequency and the vertical axis is the amplitude of signal. Figure 2.3 shows the frequency domain representation of a sine wave signal.

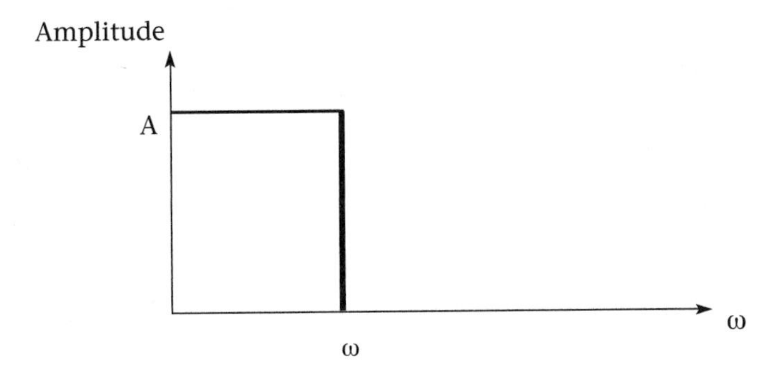


**FIGURE 2.3**
Frequency representation of a sine wave

Usually an electrical signal representing voice, temperature, or a musical sound is made of multiple waveforms. These signals have one fundamental frequency and multiple frequencies that are called *harmonics*.

## Characteristics of an Analog Signal

The characteristics of a periodic analog signal are frequency, amplitude, and phase.

***Frequency.*** Frequency ($F$) is the number of cycles in one second; $F = \frac{1}{T}$, represented in *Hz* (Hertz). If each cycle of an analog signal is repeated every one second, the frequency of the signal is one Hz. If each cycle of an analog signal is repeated 1000 times every second (once every millisecond) the frequency is:

$$f = \frac{1}{T} = \frac{1}{10^{-3}} = 1000 \text{ Hz} = 1 \text{ kHz}$$

Table 2.1 shows different values for frequency and their corresponding periods.

**TABLE 2.1** Typical Units of Frequency and Period

| Units of Frequency | Numerical Value | Units of Period | Numerical Value |
|---|---|---|---|
| Hertz ( Hz) | 1 Hz | Second (s) | 1 s |
| Kilo Hertz (kHz) | $10^3$ Hz | Millisecond (ms) | $10^{-3}$ s |
| Mega Hertz (MHz) | $10^6$ Hz | Micro Second (μs) | $10^{-6}$ s |
| Giga Hertz (GHz) | $10^9$ Hz | Nanosecond (ns) | $10^{-9}$ s |
| Tera Hertz (THz) | $10^{12}$ Hz | Pico Second (ps) | $10^{-12}$ s |

***Amplitude.*** The amplitude of an analog signal is a function of time (as shown in Figure 2.4) and may be represented in volts (unit of voltage). In other words, the amplitude is the signal's voltage value at any given time. At the time of $t_1$, the amplitude of signal is $V_1$.

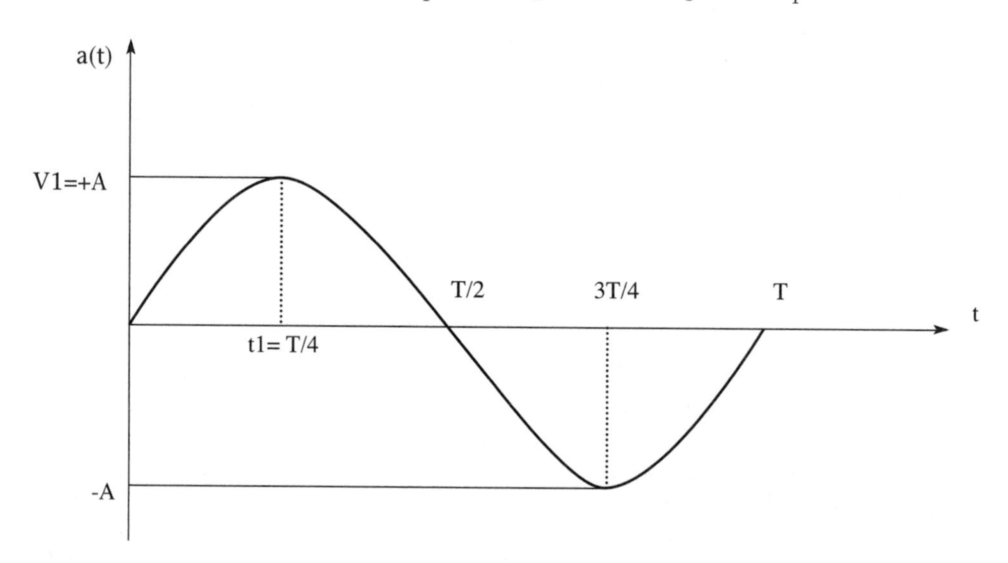


**FIGURE 2.4**
A sine wave signal over one cycle

***Phase.*** Two signals with the same frequency can differ in phase. This means that one of the signals starts at a different time from the other one. This difference can be represented by degrees, from 0 to 360 degrees or by radians where $360° = 2\pi$ radians. A sine wave signal can be represented by equation the $a(t) = A\,sin(\omega t + \theta)$ where $A$ is the peak amplitude; $\omega$ (omega) is frequency in radians per second; $t$ is time in seconds; and $\theta$ is the phase angle. Cyclic frequency $f$ can be expressed in terms of $\omega$ according to $f = \frac{\omega}{2\pi}$. A phase angle of zero means the sine wave starts at time $t = 0$ and a phase angle of 90 degrees mean the signal starts at 90 degrees as shown in Figure 2.5.

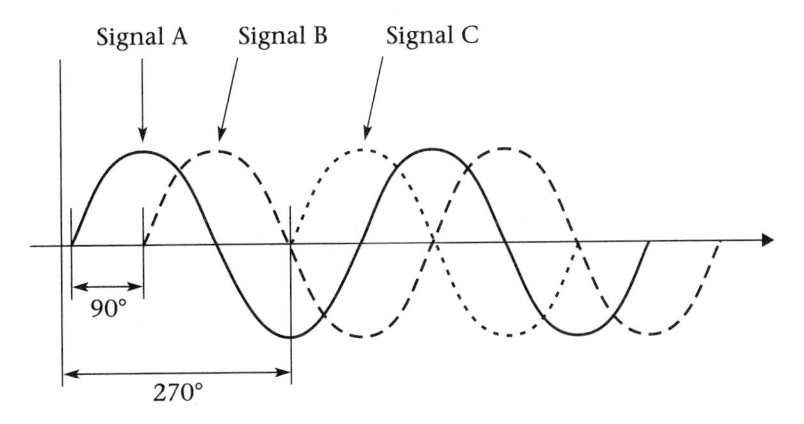


**FIGURE 2.5**
Three sine waves with different phases

**Example 2.1.** Find the equation for a sine wave signal with frequency of 10 $Hz$, maximum amplitude of 20 volts and phase angle of zero.

$$\omega = 2\pi f = 2 \times 3.1416 \times 10 = 62.83\ \frac{\text{rad}}{\text{sec}}$$

$$a(t) = 20\sin(62.83t)$$

## 2.2 Digital Signals

Modern computers communicate by using digital signals. **Digital signals** are represented by two voltages: one voltage represents the number 0 in binary and the other voltage represents the number 1 in binary. An example of a digital signal is shown in Figure 2.6, where 0 volts represents 0 in binary and +5 volts represents 1.

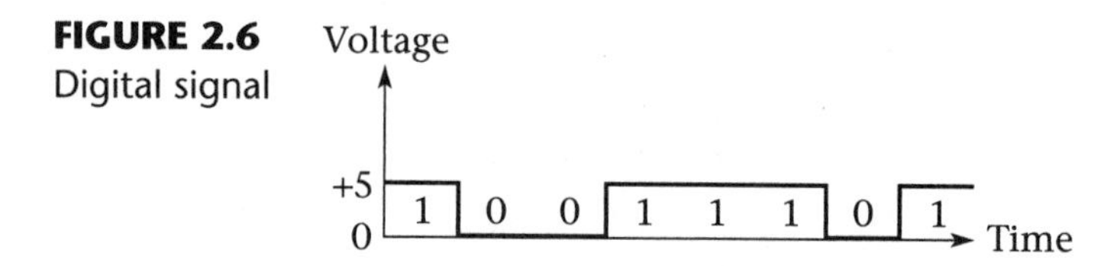


**FIGURE 2.6** Digital signal

## 2.3 Binary Numbers

**Binary,** or base-2, numbers are represented by 0 and 1. A binary digit, 0 or 1, is called a **bit.** Eight bits are equal to one **byte**. Two or more than two bytes is called a **word**. The hexadecimal number system has a base of 16, and therefore has 16 symbols (0 through 9, and A through F). Table 2.2 shows the decimal numbers, their binary values from 0 to 15, and their hexadecimal equivalents.

**TABLE 2.2** Decimal Numbers with Binary and Hexadecimal Equivalents

| Decimal | Binary (Base 2) | Hexadecimal (Base 16) or HEX |
|---|---|---|
| 0 | 0000 | 0 |
| 1 | 0001 | 1 |
| 2 | 0010 | 2 |
| 3 | 0011 | 3 |
| 4 | 0100 | 4 |
| 5 | 0101 | 5 |
| 6 | 0110 | 6 |
| 7 | 0111 | 7 |
| 8 | 1000 | 8 |
| 9 | 1001 | 9 |
| 10 | 1010 | A |
| 11 | 1011 | B |
| 12 | 1100 | C |
| 13 | 1101 | D |
| 14 | 1110 | E |
| 15 | 1111 | F |

**Converting from Hex to Binary**

Table 2.2 can also be used to convert a number from hexadecimal to binary, and from binary to hexadecimal.

**Example 2.2.** Convert the binary number 001010011010 to hexadecimal. Each 4 bits are grouped from right to left. By using Table 2.2, each 4-bit group can be converted to its hexadecimal equivalent.

| 0010 | 1001 | 1010 |
|---|---|---|
| **2** | **9** | **A** |

**Example 2.3.** Convert $(3D5)_{16}$ to binary. By using Table 2.2, the result in binary is

| 3 | D | 5 |
|---|---|---|
| **0011** | **1101** | **0101** |

The resulting binary number is: 001111010101.

**Example 2.4.** Convert 6DB from hexadecimal to binary. By using Table 2.2, the result in binary is

| 6 | D | B |
|---|---|---|
| 0110 | 1101 | 1011 |

The resulting binary number is: 011011011011.

**Converting from Binary to Decimal**

In general, any binary number can be represented by equation 2.1.

$$(a_5\,a_4\,a_3\,a_2\,a_1\,a_0\,.\,a_{-1}\,a_{-2}\,a_{-3})_2 \tag{2.1}$$

where
$a_i$ is a binary digit or bit (either 0 or 1).

Equation 2.1 can be converted to decimal number by using equation 2.2.

$$(\underbrace{a_5\,a_4\,a_3\,a_2\,a_1\,a_0}_{\text{Integer}}\,.\,\underbrace{a_{-1}\,a_{-2}\,a_{-3}}_{\text{Fraction}})_2 = a_0 \times 2^0 + a_1 \times 2^1 + a_2 \times 2^2 + a_3 \times 2^3 + \cdots\cdots + a_1 \times 2^{-1} + a_{-2} \times 2^{-2} + \cdots\cdots \tag{2.2}$$

**Example 2.5.** To convert $(110111.101)_2$ to decimal:

$$\begin{aligned}(110111.101)_2 &= 1{*}2^0 + 1{*}2^1 + 1{*}2^2 + 0{*}2^3 + 1{*}2^4 \\ &\quad + 1{*}2^{5}{*} + 1{*}2^{-1} + 0{*}2^{-2} + 1{*}2^{-3} \\ &= 55.625\end{aligned}$$

## 2.4 Coding Schemes

Since computers can understand only binary numbers (0 or 1), all information (such as numbers, letters, and symbols) must be represented as binary data. One commonly used code to represent printable and nonprintable characters is the American Standard Code for Information Interchange (ASCII).

**ASCII Code**

Each character in ASCII code is represented by 8 bits; the most significant bit is used for parity bit. Table 2.3 shows the **ASCII code** and its hexadecimal equivalent.

Characters from hexadecimal 00 to 1F and 7F are control characters, which are nonprintable characters, such as NUL, SOH, STX, ETX, ESC, and DLE (data link escape).

**Example 2.6.** Convert the word "Network" to binary and show the result in hexadecimal. By using Table 2.3 each character is represented by seven bits and results in:

| 1001110 | 1100101 | 1110100 | 1110111 | 1101111 | 1110010 | 1101011 |
|---|---|---|---|---|---|---|
| N | e | t | w | o | r | k |

or in hexadecimal

| **4E** | **65** | **74** | **77** | **6F** | **72** | **6B** |
|---|---|---|---|---|---|---|

**Universal Code or Unicode**

Unicode is a new 16-bit character encoding standard for representing characters and numbers in languages such as Greek, Arabic, Chinese, and Japanese. The ASCII code uses eight bits to represent each character in Latin, and it can represent 256 characters. The ASCII code does not support mathematical symbols and scientific symbols. **Unicode** uses 16 bits, which can represent 65536 characters or symbols. A character in unicode is represented by16-bit binary, equivalent to four digits in hexadecimal. For example, the character B in unicode is U0042H (U represents unicode). The ASCII code is represented between $(00)_{16}$ to $(FF)_{16}$. For converting ASCII code to unicode, two zeros are added to the left side of ASCII code; therefore, the unicode to represent ASCII characters is between $(0000)_{16}$ to $(00FF)_{16}$. Table 2.4 shows some of the unicode for Latin and Greek characters. Unicode is divided into blocks of code, with each block assigned to a specific language. Table 2.5 shows each block of unicode for some different languages.

**TABLE 2.3** American Standard Code for Information Interchange (ASCII)

| Binary | Hex | Char | Binary | Hex | Char | Binary | Hex | Char | Binary | Hex | Char |
|---|---|---|---|---|---|---|---|---|---|---|---|
| 0000000 | 00 | **NUL** | 0100000 | 20 | **SP** | 1000000 | 40 | @ | 1100000 | 60 | ’ |
| 0000001 | 01 | **SOH** | 0100001 | 21 | **!** | 1000001 | 41 | **A** | 1100001 | 61 | **a** |
| 0000010 | 02 | **STX** | 0100010 | 22 | “ | 1000010 | 42 | **B** | 1100010 | 62 | **b** |
| 0000011 | 03 | **ETX** | 0100011 | 23 | **#** | 1000011 | 43 | **C** | 1100011 | 63 | **c** |
| 0000100 | 04 | **EOT** | 0100100 | 24 | **$** | 1000100 | 44 | **D** | 1100100 | 64 | **d** |
| 0000101 | 05 | **ENQ** | 0100101 | 25 | **%** | 1000101 | 45 | **E** | 1100101 | 65 | **e** |
| 0000110 | 06 | **ACK** | 0100110 | 26 | **&** | 1000110 | 46 | **F** | 1100110 | 66 | **f** |
| 0000111 | 07 | **BEL** | 0100111 | 27 | ‘ | 1000111 | 47 | **G** | 1100111 | 67 | **g** |
| 0001000 | 08 | **BS** | 0101000 | 28 | **(** | 1001000 | 48 | **H** | 1101000 | 68 | **h** |
| 0001001 | 09 | **HT** | 0101001 | 29 | **)** | 1001001 | 49 | **I** | 1101001 | 69 | **i** |
| 0001010 | 0A | **LF** | 0101010 | 2A | * | 1001010 | 4A | **J** | 1101010 | 6A | **j** |
| 0001011 | 0B | **VT** | 0101011 | 2B | + | 1001011 | 4B | **K** | 1101011 | 6B | **k** |
| 0001100 | 0C | **FF** | 0101100 | 2C | , | 1001100 | 4C | **L** | 1101100 | 6C | **l** |
| 0001101 | 0D | **CR** | 0101101 | 2D | – | 1001101 | 4D | **M** | 1101101 | 6D | **m** |
| 0001110 | 0E | **SO** | 0101110 | 2E | . | 1001110 | 4E | **N** | 1101110 | 6E | **n** |
| 0001111 | 0F | **SI** | 0101111 | 2F | / | 1001111 | 4F | **O** | 1101111 | 6F | **o** |
| 0010000 | 10 | **DLE** | 0110000 | 30 | **0** | 1010000 | 50 | **P** | 1110000 | 70 | **p** |
| 0010001 | 11 | **DC1** | 0110001 | 31 | **1** | 1010001 | 51 | **Q** | 1110001 | 71 | **q** |
| 0010010 | 12 | **DC2** | 0110010 | 32 | **2** | 1010010 | 52 | **R** | 1110010 | 72 | **r** |
| 0010011 | 13 | **DC3** | 0110011 | 33 | **3** | 1010011 | 53 | **S** | 1110011 | 73 | **s** |
| 0010100 | 14 | **DC4** | 0110100 | 34 | **4** | 1010100 | 54 | **T** | 1110100 | 74 | **t** |
| 0010101 | 15 | **NACK** | 0110101 | 35 | **5** | 1010101 | 55 | **U** | 1110101 | 75 | **u** |
| 0010110 | 16 | **SYN** | 0110110 | 36 | **6** | 1010110 | 56 | **V** | 1110110 | 76 | **v** |
| 0010111 | 17 | **ETB** | 0110111 | 37 | **7** | 1010111 | 57 | **W** | 1110111 | 77 | **w** |
| 0011000 | 18 | **CAN** | 0111000 | 38 | **8** | 1011000 | 58 | **X** | 1111000 | 78 | **x** |
| 0011001 | 19 | **EM** | 0111001 | 39 | **9** | 1011001 | 59 | **Y** | 1111001 | 79 | **y** |
| 0011010 | 1A | **SUB** | 0111010 | 3A | : | 1011010 | 5A | **Z** | 1111010 | 7A | **z** |
| 0011011 | 1B | **ESC** | 0111011 | 3B | ; | 1011011 | 5B | [ | 1111011 | 7B | { |
| 0011100 | 1C | **FS** | 0111100 | 3C | < | 1011100 | 5C | \ | 1111100 | 7C | \ |
| 0011101 | 1D | **GS** | 0111101 | 3D | = | 1011101 | 5D | ] | 1111101 | 7D | } |
| 0011110 | 1E | **RS** | 0111110 | 3E | < | 1011110 | 5E | ^ | 1111110 | 7E | ~ |
| 0011111 | 1F | **US** | 0111111 | 3F | **?** | 1011111 | 5F | – | 1111111 | 7F | **DEL** |

**TABLE 2.4** Unicode Values for some Latin and Greek Characters

| Latin | | Greek | |
|---|---|---|---|
| **Character** | **Code (Hex)** | **Character** | **Code (Hex)** |
| A | U0041 | φ | U03C6 |
| B | U0042 | α | U03B1 |
| C | U0043 | γ | U03B3 |
| 0 | U0030 | μ | U03BC |
| 8 | U0038 | β | U03B2 |

**TABLE 2.5** Unicode Block Allocations

| Start Code (Hex) | End Code (Hex) | Block Name |
|---|---|---|
| U0000 | U007F | Basic Latin |
| U0080 | U00FF | Latin supplement |
| U0370 | U03FF | Greek |
| U0530 | U058F | Armenian |
| U0590 | U05FF | Hebrew |
| U0600 | U06FF | Arabic |
| U01A0 | U10FF | Georgian |

## 2.4 Transmission Modes

When data is transferred from one computer to another by digital signals, the receiving computer has to distinguish the size of each signal to determine when a signal ends and when the next one begins. For example, when a computer sends a signal as shown in Figure 2.7, the receiving computer has to recognize how many ones and zeros are in the signal. Synchronization methods between source and destination devices are generally grouped into two categories: asynchronous and synchronous.

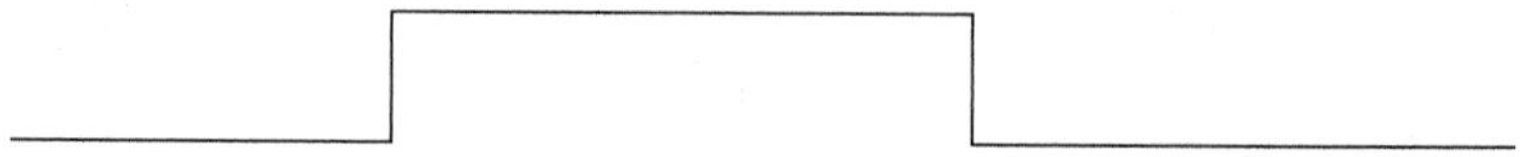

**FIGURE 2.7**
Digital signal

## Asynchronous Transmission

**Asynchronous transmission** occurs character by character and is used for serial communication, such as by a modem or serial printer. In asynchronous transmission each data character has a start bit that identifies the start of the character, and one or two bits that identify the end of the character, as shown in Figure 2.8. The data character is 7 bits. Following the data bits may be a parity bit, which is used by the receiver for error detection. After the parity bit is sent, the signal must return to high (logic one) for at least one bit time to identify the end of the character. The new start bit serves as an indicator to the receiving device that a data character is coming and allows the receiving side to synchronize its clock. Since the receiver and transmitter clock are not synchronized continuously, the transmitter uses the start bit to reset the receiver clock so that it matches the transmitter clock. Also, the receiver is already programmed for the number of bits in each character sent by the transmitter.

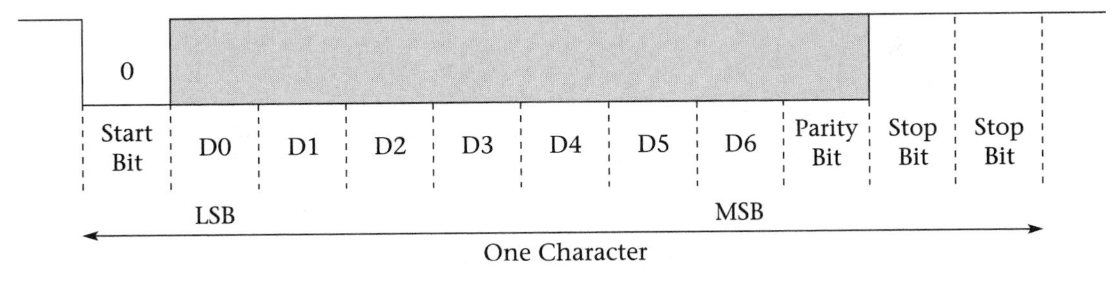


**FIGURE 2.8**
Asynchronous transmission

## Synchronous Transmission

Some applications require transferring large blocks of data, such as a file from disk or a print job from a computer to a printer. **Synchronous transmission** is an efficient method of transferring large blocks of data by using time intervals for synchronization.

One method of synchronizing the transmitter and receiver is through the use of an external connection that carries a clock pulse. The clock pulse represents the data rate of the signal, as shown in Figure 2.9, and is used to determine the speed of data transmission. The receiver of Figure 2.9 reads the data as 01101.

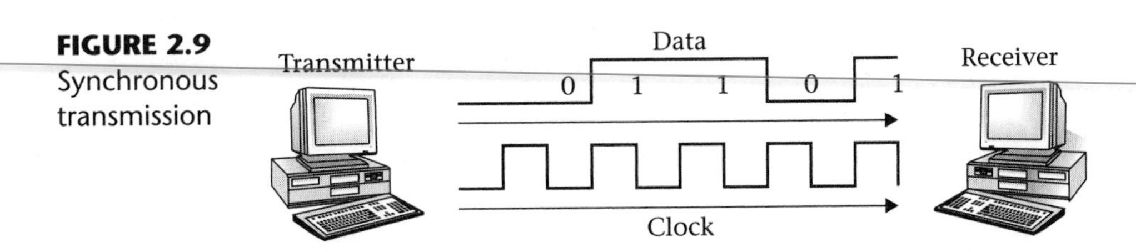


**FIGURE 2.9**
Synchronous transmission

Figure 2.9 shows that an extra connection is required to carry the clock pulse for synchronous transmission. In networking, one medium is used for transmission of both information and the clock pulse. The two signals are encoded in such a way that the synchronization signal is embedded into the data. This can be done with Manchester encoding or Differential Manchester encoding, which are discussed later in the chapter.

## 2.6 Transmission Methods

There are two types of transmission methods used for sending digital signals from one station to another across a communication channel: serial transmission and parallel transmission.

### Serial Transmission

In **serial transmission,** information is transmitted one bit at a time over one wire, as shown in Figure 2.10.

**FIGURE 2.10**
Serial transmission

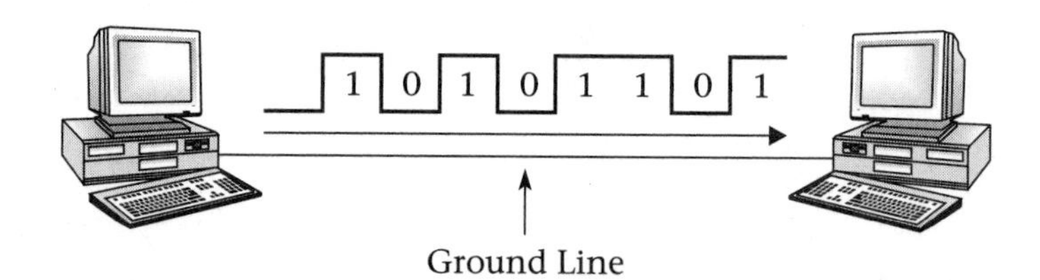


### Parallel Transmission

In **parallel transmission,** multiple bits are sent simultaneously, one byte or more at a time, instead of bit by bit as in serial transmission. Figure 2.11 shows how computer A sends eight bits of information to computer B at

**FIGURE 2.11**
Parallel transmission

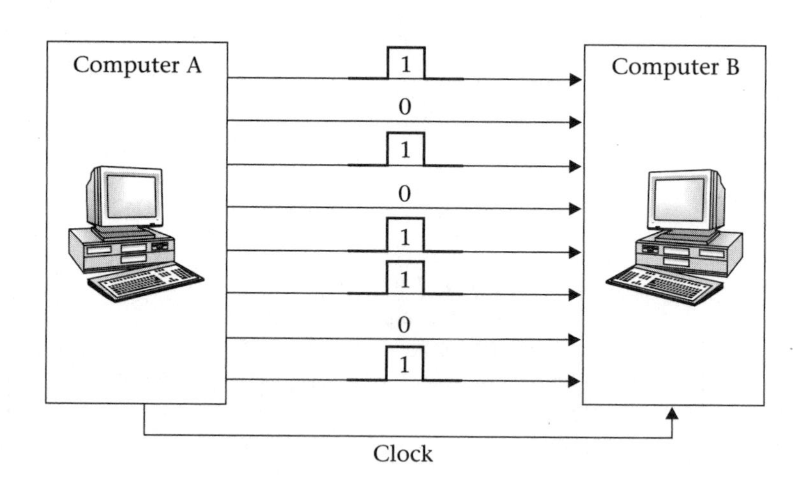

the same time by using eight different wires. Parallel transmission is faster than serial transmission, at the same clock speed.

## 2.7 Communication Modes

A communication mode specifies the capability of a device to send and receive data by determining the direction of the signal between two connections. There are three types of communication modes: simplex, half-duplex, and full-duplex.

**Simplex Mode** In **simplex mode,** transmission of data goes in one direction only, as shown in Figure 2.12. A common analogy is a commercial radio or TV broadcast; the sending device never requires a response from the receiving device.

**FIGURE 2.12**
Simplex transmission

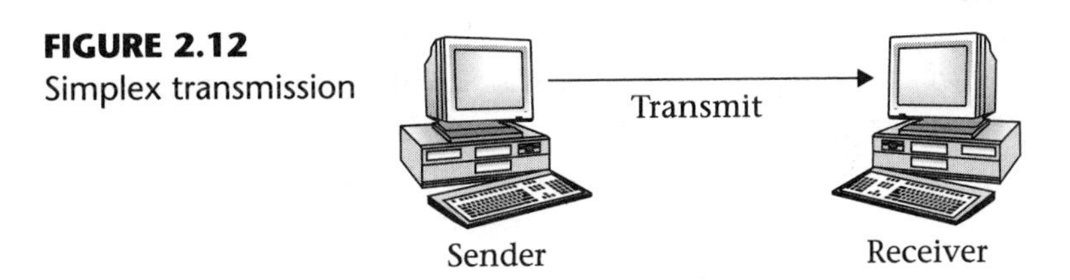


**Half-Duplex Mode** In **half-duplex mode,** two devices exchange information, as shown in Figure 2.13; however, information can be transmitted across the channel only one direction at a time. A common example is Citizen Band radio (CB) or ham radio; a user can either talk or listen, but both parties cannot talk at the same time.

**FIGURE 2.13**
Half-duplex transmission

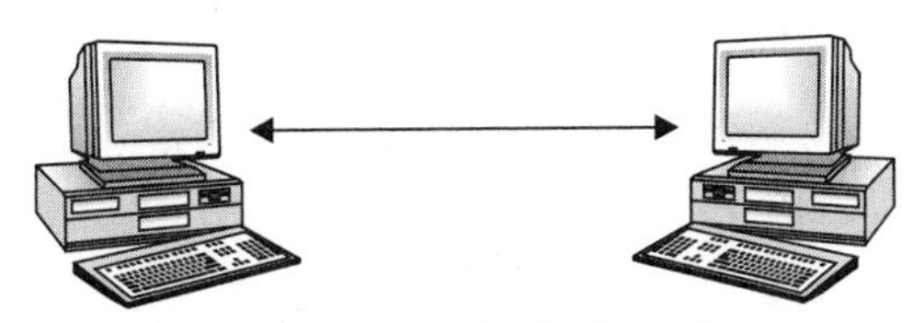


**Full-Duplex Mode** In **full-duplex mode,** both computers can send and receive information simultaneously, as shown in Figure 2.14. An example of full-duplex is our modern telephone system, in which both users may talk and listen at the same time, with their voices carried two ways simultaneously over the phone lines.

**FIGURE 2.14**
Full-duplex transmission

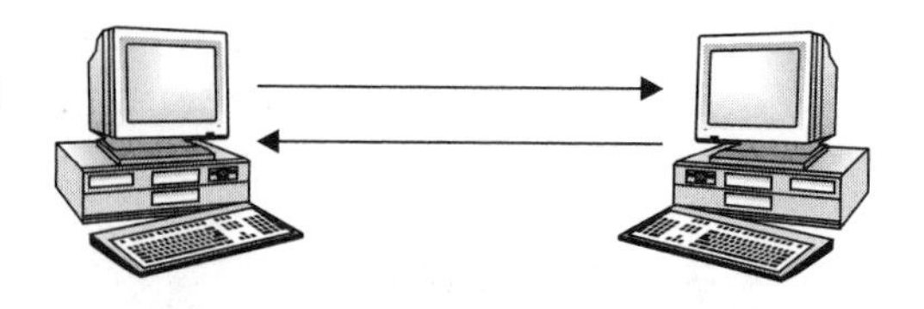

## 2.8 Signal Transmission

There are two methods used to transfer information over media: baseband and broadband transmission.

**Baseband Transmission Mode** When the entire bandwidth of a cable is used to carry only one signal, the cable operates in **baseband mode**. Many digital signals use baseband transmission. An example is Local Area Network.

**Broadband Transmission Mode** When the bandwidth of a cable is used to carry several signals simultaneously, the cable operates in **broadband mode**. For example, cable TV transmission works in broadband mode because it carries multiple channels using multiple signals over the cable. Broadband mode frequently uses analog signals.

## 2.9 Digital Signal Encoding

Digital signal encoding is used to represent binary values in the form of digital signals. The receiver of the digital signal must know the timing of each signal, such as the start and end of each bit. The following are some methods used to represent digital signals:

- Unipolar Encoding
- Polar Encoding
- Bipolar Encoding
- Non-Return to Zero (NRZ)
- Non-Return to Zero Inverted (NRZ-I)
- Manchester and Differential Manchester Encoding

Manchester and Differential Manchester, and Non-Return to Zero Inverted (NRZ-I) encoding schemes are used in LANs, and Non-Return to Zero is used in WANs. Each encoding technique is described next.

**Unipolar Encoding**

In **unipolar encoding** only positive voltage or negative voltage is used to represent binary 0 and 1. For example, +5 volts represents binary 1 and zero volts represents 0, as shown in Figure 2.15.

**FIGURE 2.15**
Unipolar encoding signals

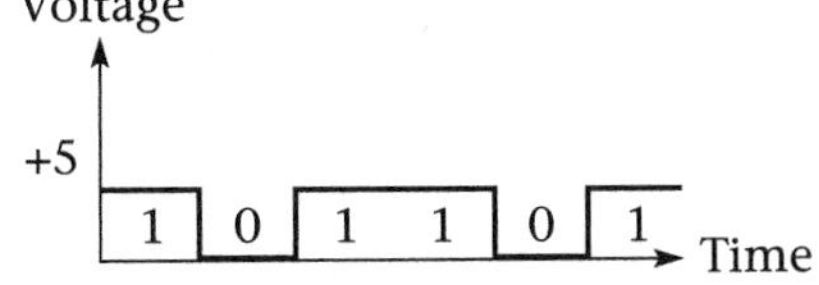


**Polar Encoding**

In **polar encoding** positive and negative voltages are used to represent binary zero and one. For example +5 volts represents binary one and −5 volts represent binary zero, as shown in Figure 2.16.

**FIGURE 2.16**
Polar encoding signals

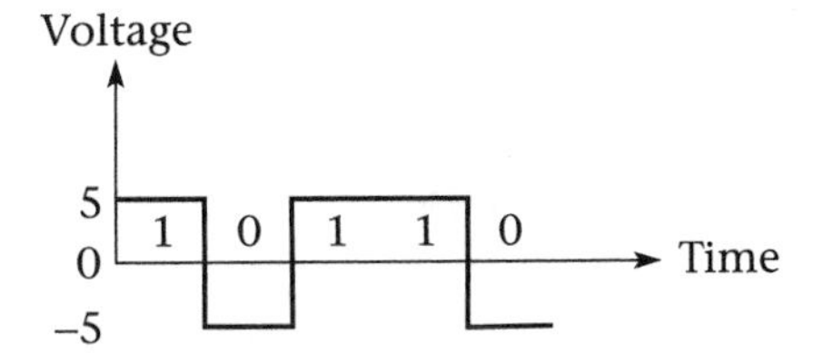


**Bipolar Encoding**

In **bipolar encoding,** signal voltage varies in three levels: positive, zero, and negative voltage. One of the most popular bipolar encoding methods is alternate mark inversion (AMI).

In AMI encoding, binary 0 is represented by zero volts and binary 1 is represented by alternates between positive and negative voltages, as shown in Figure 2.17.

**FIGURE 2.17**
Bipolar encoding signals

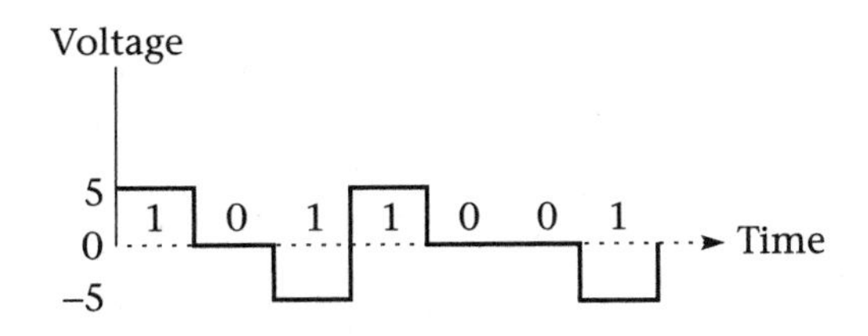


**Non-Return to Zero Encoding (NRZ)**

NRZ is a simple form of polar encoding, using two voltage levels for representing 1 and 0, with binary 0 represented by positive voltage and binary 1 represented by negative voltage, as shown in Figure 2.18.

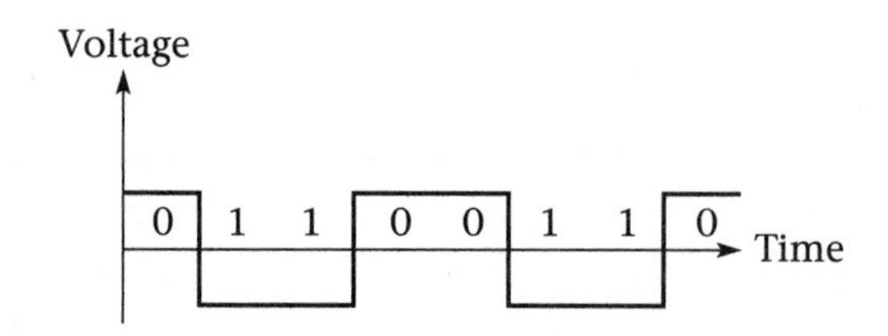


**FIGURE 2.18**
NRZ encoding signals

**Non-Return to Zero Inverted Encoding (NRZ-I)**

In **NRZ-I** there is a transition at the start of logic 1 (low to high or high to low) and no transition at the start of 0, as shown in Figure 2.19.

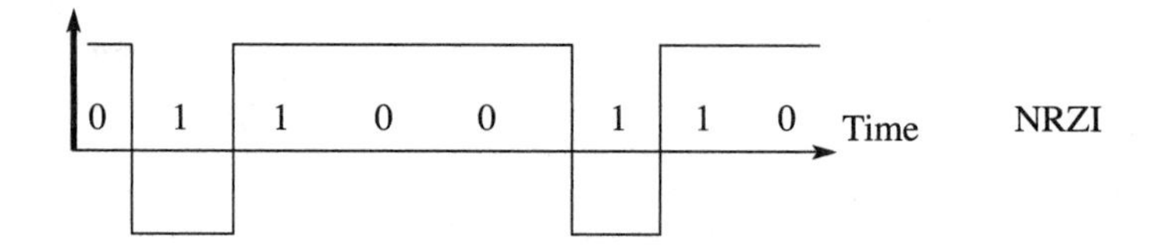


**FIGURE 2.19**
NRZ-I encoding signals

**Manchester and Differential Manchester Encoding**

In **Manchester** and **Differential Manchester encoding,** the clock pulse is embedded into the signal. Therefore, the receiver does not require any additional signal to represent the clock pulse. Table 2.6 shows how to convert digital signals to Manchester encoding and Differential Manchester encoding, and Figure 2.20 shows Manchester and Differential Manchester encoding.

**TABLE 2.6** Conversion Methods of Digital Signal to Manchester and Differential Manchester

| Digital Signal | Manchester Encoding | Differential Manchester Encoding |
|---|---|---|
| Logic 1 | Transition from high to low at the middle of the signal | Transition only in the middle of the signal |
| Logic 0 | Transition from low to high at the middle of signal | Transition at the start of zero and also middle of zero (original signal) |

**FIGURE 2.20**
Manchester and Differential Manchester encoding

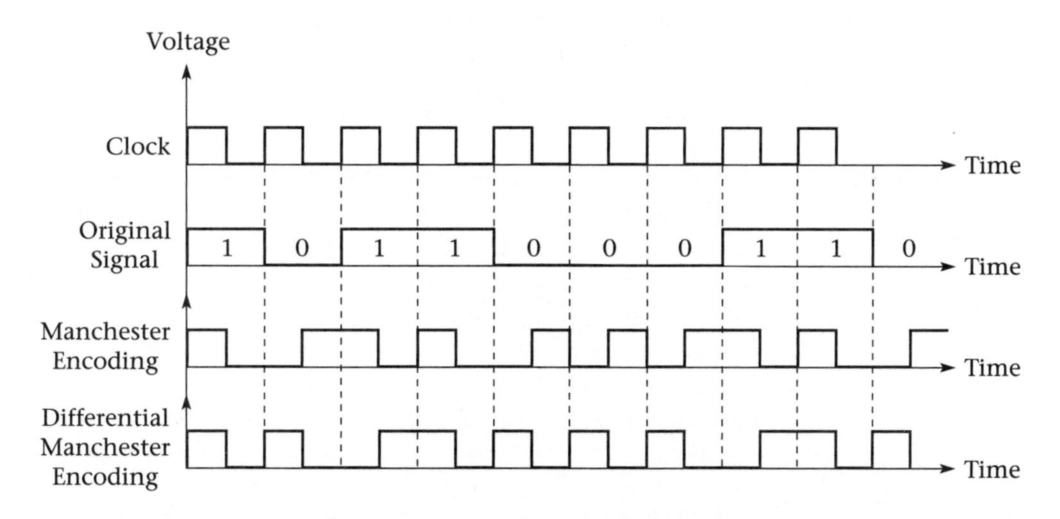


## 2.10 Error Detection Methods

When the transmitter sends a frame to the receiver, the frame can become corrupted due to external and internal noise. The receiver must first check the integrity of the frame. The following are some possible sources of error:

**Impulse noise:** A non-continuous pulse for a short duration is called an **impulse noise**. It may be caused by a lightning discharge or a spike generated by a power switch being turned off and on.

**Crosstalk:** This type of noise can be generated when a transmission line carrying a strong signal is coupled with a transmission line carrying a weak signal. The transmission line with the strong signal will produce noise **(crosstalk)** on the transmission line with the weak signal.

**Attenuation:** When a signal travels on a transmission line, the strength of the signal is reduced over distance. This reduction is called **attenuation**. A weak signal is more affected by noise than a strong signal.

**White noise or thermal noise:** This type of noise exists in all electrical devices and is generated by moving electrons in the conductor.

The following methods can be used to detect an error or errors:

- Parity Check
- Block Check Character (BCC)
- One's Complement of the Sum
- Cyclic Redundancy Check (CRC)

## Parity Check

The simplest error detection method is the **parity check**. The parity check method can detect one error, and is used in both the asynchronous transmission method and the character-oriented synchronous transmission method. A parity bit is an extra bit that the transmitter adds to the information before transmitting to the receiver. The value of the parity bit selected by the transmitter determines whether the data is given an even number of ones (even parity) or an odd number of ones (odd parity). For example, if a transmitter uses even parity to transmit the ASCII character 1000011 (uppercase e), the transmitter adds parity bit 1 to the character so that the number of ones in the character becomes an even **1**1000011. The transmitter would then transmit 11000011 to the receiver. The receiver checks the number of the ones in the character. If the number of ones is even, there is no error detected in the character. Otherwise the character contains an error. Parity error detection is used in serial communications. Figure 2.21 shows the logic diagram for a parity bit generator using Exclusive-OR gates.

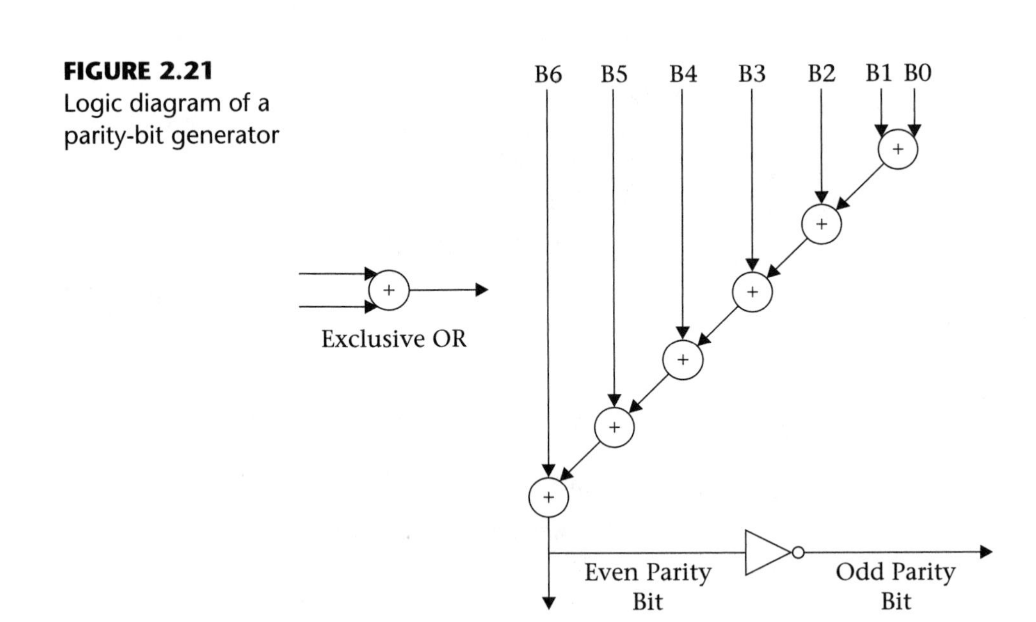


**FIGURE 2.21** Logic diagram of a parity-bit generator

**Block Check Character (BCC)**

**Block Check Character (BCC)** uses vertical and horizontal parity bits in order to detect double errors. Remember, the parity check is limited to detection of only one error and is used only for transmitting single characters. When a block of characters is transmitted, BCC can be used to correct one error and detect two errors. A parity bit is added to each character (row parity) in a block of characters; then column parity is computed. As illustrated in Table 2.7, the result of the column parity bits is called *Block Check Character (BCC)*. Table 2.7 shows that odd parity use for the rows and even parity use for columns results in a BCC of 01011110. For example, if two bits are changed in row one, such as B5 from 0 to 1 and B2 from 1 to 0, the row parity does not change; but the BCC will change, indicating the detection of two errors in one row.

**TABLE 2.7** BCC Calculation for the Word "NETWORK"

| Parity | B6 | B5 | B4 | B3 | B2 | B1 | B0 | |
|---|---|---|---|---|---|---|---|---|
| 1 | 1 | 0/**1** | 0 | 1 | 1/**0** | 1 | 0 | N |
| 0 | 1 | 0 | 0 | 0 | 1 | 0 | 1 | E |
| 0 | 1 | 0 | 1 | 0 | 1 | 0 | 0 | T |
| 0 | 1 | 0 | 1 | 0 | 1 | 1 | 1 | W |
| 0 | 1 | 0 | 0 | 1 | 1 | 1 | 1 | O |
| 0 | 1 | 0 | 1 | 0 | 0 | 1 | 0 | R |
| 1 | 1 | 0 | 0 | 1 | 0 | 1 | 1 | K |
| 0 | 1 | 0/**1** | 1 | 1 | 1/**0** | 1 | 0 | BCC |

**One's Complement of the Sum**

**One's complement of the sum** method is used for error detection of the Transmission Control Protocol (TCP) header and Internet Protocol (IP) header. At the transmitter side, the 16-bit one's complement sum of the header is calculated. The result of this calculation is transmitted with the information to the receiver. At the receiver side, the 16-bit one's complement of the header is calculated and compared to the result with the one's complement of the transmitter. If the two results are equal, no error is detected. Otherwise there is an error in the information. Figure 2.22 shows the one's complement of the sum for a four-byte header.

**Cyclic Redundancy Check (CRC)**

The parity bit and BCC can detect single and double errors. The **cyclic redundancy check (CRC)** method is used for detection of a single error, more than a single error, and burst error (when two or more consecutive bits in a frame have changed).

**FIGURE 2.22**
One's complement of the sum

| Transmitter side | | Receiver side | |
|---|---|---|---|
| 1000010 | | 1000010 | |
| 1111101 | | 1111101 | |
| 0000001 | | 0000001 | |
| + 0111100 | | + 0001100 | |
| --------- | | --------- | |
| 1111100 | Carry is discarded | 1001100 | Carry is discarded |
| Therefore: One's Complement is 0000011 | | Therefore: One's Complement is 1110011 | |

The CRC uses modulo-2 addition to compute the Frame Check Sequence (FCS). In modulo-2 addition:

$$1 + 1 = 0, \quad 1 + 0 = 1, \quad \text{and} \quad 0 + 0 = 0.$$

The following procedure is used to calculate FCS: At the transmitter side: Frame $M$ is $k$ bits. $P$ is a divisor of $n + 1$ bits. FCS is $n$ bits, and it is the remainder of $2^n * M/P$ using modulo-2 division.

At the transmitting side, the FCS, which is the remainder from dividing $2^n * M$ by $P$, is calculated. The transmitter will transmit frame $T = 2^n * M + \text{FCS}$ to the receiver, where $T$ is $k + n$ bits.

At the receiving side, the receiver divides $T$ by $P$ using modulo-2 division. If the result of this division generates a remainder of zero, no error is detected in the frame. Otherwise the frame contains one or more errors.

**Example 2.7.** Find the Frame Check Sequence (FCS) for the following message. The divisor is given.

$$\text{Message } M = 111010, K = 6 \text{ bits}$$

$$\text{Divisor } P = 1101 \; n + 1 = 4 \text{ bits}$$

Therefore $2^n * M = 111010000$. By dividing $2^n * M$ by $P$ using modulo-2 division, FCS = 010 as shown in Figure 2.23.

At the transmitter side the FCS is added to $2^n * M$, and the transmitter transmits frame $T = 111010010$ to the receiver.

At the receiver side, the receiver divides $T$ by $P$, and if the result has a remainder of zero, there is no error in the frame. Otherwise the message contains an error. Since this division takes time, special hardware is designed to generate FCS.

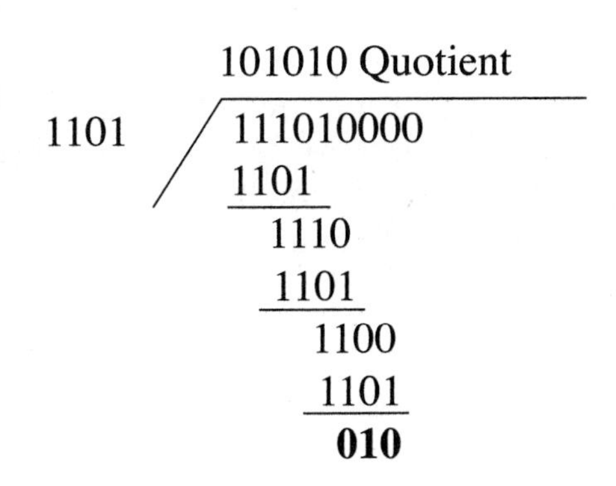


**FIGURE 2.23**
FCS calculation

***CRC polynomial and architecture.*** A binary number represented by $b_5 b_4\ b_3 b_2\ b_1 b_0$ where $b_i$ represents each $b$ it that can be represented by a polynomial:

$$b_5X^5 + b_4X^4 + b_3X^3 + b_2X^2 + b_1X + b_0$$

**Example 2.8.** Represent $P$ = 1101101 by polynomial.

$$\begin{aligned} P(X) &= 1.X^6 + 1.X^5 + 0.X^4 + 1.X^3 + 1.X^2 + 0.X + 1 \\ &= X^6 + X^5 + X^3 + X^2 + 1 \end{aligned}$$

The following CRC polynomials are IEEE and ITU standards:

$$\text{CRC-12} = X^{12} + X^{11} + X^3 + X^1 + 1$$

$$\text{CRC-16} = X^{16} + X^{15} + X^2 + 1$$

$$\text{CRC-ITU} = X^{16} + X^{12} + X^5 + 1$$

$$\begin{aligned} \text{CRC-32} = {} & X^{32} + X^{26} + X^{23} + X^{22} + X^{16} + X^{11} + X^{10} \\ & + X^8 + X^7 + X^5 + X^4 + X^2 + X + 1 \end{aligned}$$

The CRC method uses a special integrated circuit (IC) to generate the FCS. The design of this IC is based on the CRC polynomial. In general, a CRC polynomial can be represented by:

$$P(X) = X^n + \cdots + a_4X^4 + a_3X^3 + a_2X^2 + a_1X + 1$$

Figure 2.24 shows the general architecture of a CRC integrated circuit. $C_i$ is a one-bit shift register and the output of each register is connected to the input of Exclusive-OR gate; $a_i$ is the coefficient of a CRC polynomial.

In Figure 2.24, if $a_i$ equals zero, then there is no connection between the feedback line and the XOR gate. In order to find the FCS, the initial value for $C_i$ is set to zero, and the message $2^n * M$ is shifted $k + n$ times through the CRC circuit. The final contents of $C_{n-1}, \ldots, C_4, C_3, C_2, C_1, C_0$ is the Frame Check Sequence (FCS).

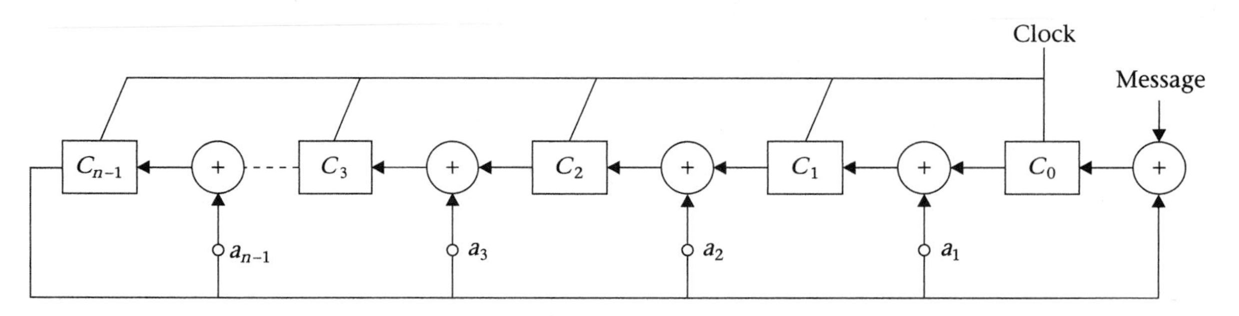


**FIGURE 2.24**
General architecture CRC polynomial

**Example 2.9.** Show the CRC circuit for the polynomial:

$$P(X) == X^5 + X^4 + X^2 + 1$$

In this polynomial the value for $a_1$, $a_3$ is zero and Figure 2.25 shows the CRC circuit for the polynomial.

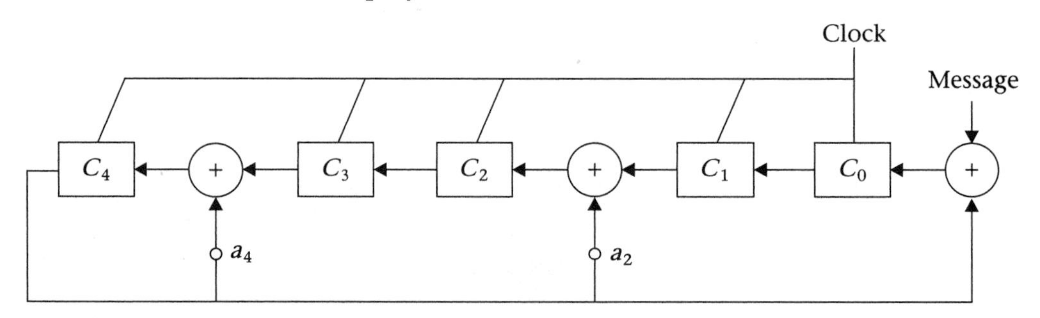


**FIGURE 2.25**
CRC circuit for polynomial $P(X) == X^5 + X^4 + X^2 + 1$

**Example 2.10.** Find FCS for Message $M$=111010; assume $P$= 1101:

$$P(X) = X^3 + X^2 + 1$$

The circuit for $P(X)$ is shown in Figure 2.26, where $a_1 = 0$, $a_2 = 1$, and $a_3 = 1$. Table 2.8 shows the contents of each register after shifting one bit at the time. After shifting 9 ($k + n$) times, the contents of the registers is FCS.

**FIGURE 2.26**
CRC circuit for $P = 1101$

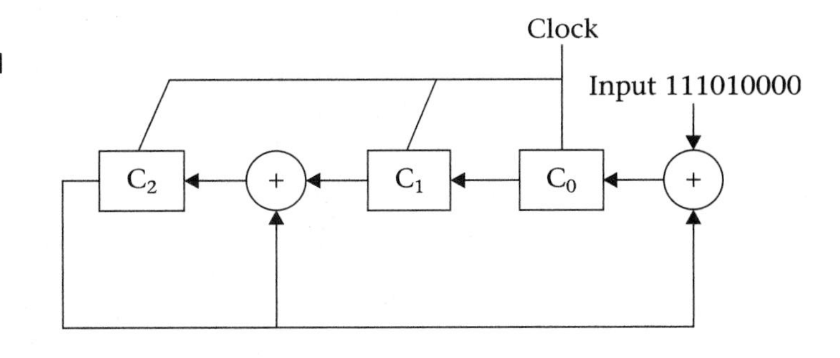


**TABLE 2.8** FCS Value for Message $M = 111010$ and $P = 1101$

| C2 | C1 | C0 | Input |
|---|---|---|---|
| 0 | 0 | 0 | Initial Value |
| 0 | 0 | 1 | 1 |
| 0 | 1 | 1 | 1 |
| 1 | 1 | 1 | 1 |
| 0 | 1 | 1 | 0 |
| 1 | 1 | 1 | 1 |
| 0 | 1 | 1 | 0 |
| 1 | 1 | 0 | 0 |
| 0 | 0 | 1 | 0 |
| 0 | 1 | 0 | 0 |

010 is Frame Check Sequence (FCS)

## 2.12 Error Correction

Error correction can be used in certain situations. One case is when a receiver detects a corrupted packet and the transmitter is not able to transmit a copy of the packet to the receiver. Another example is when the transmitter is too far from the receiver and it takes too long to request a copy of the packet. An example of the first case is a satellite orbiting Saturn. A ground station sends information to the satellite and the satellite detects an error in the information. To send and process the information all over again is a very time-consuming process. Another application of error correction is in memory. Suppose a CPU stores a 4-bit data, 1101, in a memory location (RAM). The CPU then reads the same memory location as 1001. Note that one of the data bits did not store correctly and this may be caused by a faulty memory cell. For error correction the transmitter attaches some extra bits, called *check bits*, to the data and the receiver uses these extra bits for error correction. This method is called Forward Error Correction (FEC); a simple version of FEC is the **hamming code.**

**Hamming Code** The hamming code is used to detect and correct one error. Figure 2.27 shows a block diagram for hamming code. In Figure 2.27, the transmitter side uses the $M$-bit data to generate $H$ check bits. The transmitter then transmits $M + H$ bits to the receiver. At the receiver side, the receiver uses $M$-bit data to generate new $H$ check bits. The check bits generated by the receiver are compared with check bits sent by the transmitter. The result of this comparison is called the *syndrome* and might have the following values:

1. If all bits of the syndrome are zeros, then there is no error detected.
2. If only one bit of syndrome is set to one, then the error is in check bits.
3. If more than one bit of the syndrome is set to one, then the value of the syndrome determines the position of the data bit in error. For correction, complement the bit that is in error.

**FIGURE 2.27**
Block diagram of hamming code

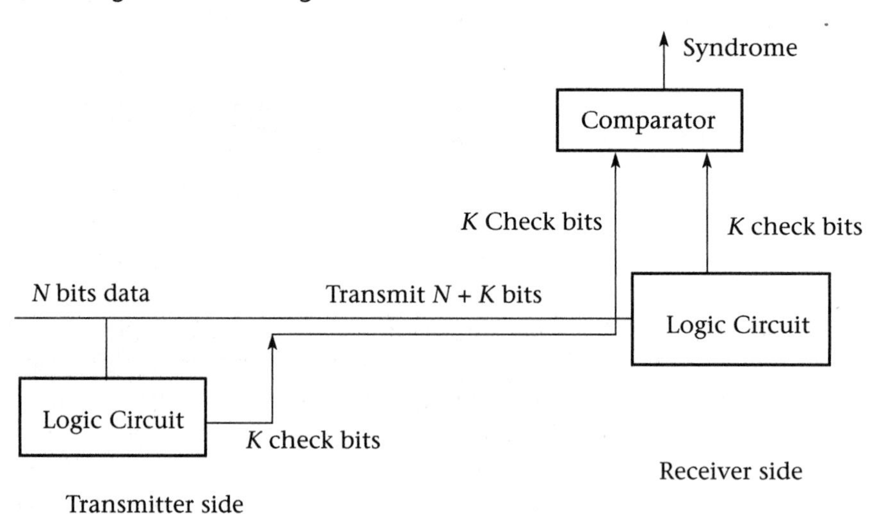

The relation between $H$ check bits and $M$ data bits is given by equation 2.3:

$$2^H > M + H - 1 \quad \text{or} \quad 2^H - H + 1 > M \tag{2.3}$$

Where $M$ is the number of bits in the message and $H$ is number of the check bits, Table 2.9 shows different values for $M$ and its corresponding check bits. For example, if $M = 8$, then $H = 4$.

**TABLE 2.9** *M* and Corresponding Value of *H*

| Number of Data Bits *M* | Number of Check bits *H* |
|---|---|
| 1 | 2 |
| 2 | 3 |
| 3 | 3 |
| 4 | 3 |
| 5 | 4 |
| 8 | 4 |
| 12 | 4 |
| 16 | 5 |

**Generating Check Bits**

An eight-bit message M8 M7 M6 M5 M4 M2 M1 requires 4 check bits and it is represented by C8, C4, C2, and C1. The check bits inserted into the message are shown in Table 2.10, which shows the locations of the check bits and data bits. The locations of check bits are to the power of 2.

The following equations are used to find the value of check bits:

$$\text{C1} = \text{M1 } xor \text{ M2 } xor \text{ M4 } xor \text{ M5 } xor \text{ M7}$$

$$\text{C2} = \text{M1 } xor \text{ M3 } xor \text{ M4 } xor \text{ M6 } xor \text{ M7}$$

$$\text{C4} = \text{M2 } xor \text{ M3 } xor \text{ M4 } xor \text{ M8}$$

$$\text{C8} = \text{M5 } xor \text{ M6 } xor \text{ M7 } xor \text{ M8}$$

**Example 2.11.** Find the check bits for 11001101 at the transmitter side:

$$\text{C1} = 1 \; xor \; 0 \; xor \; 1 \; xor \; 0 \; xor \; 1 = 1$$

$$\text{C2} = 1 \; xor \; 1 \; xor \; 1 \; xor \; 0 \; xor \; 1 = 0$$

$$\text{C4} = 0 \; xor \; 1 \; xor \; 1 \; xor \; 1 = 1$$

$$\text{C8} = 0 \; xor \; 0 \; xor \; 1 \; xor \; 1 = 0$$

**TABLE 2.10** Location of Check Bits

| Bit Position | Data Bit and Check Bits Position |
|---|---|
| 0001 | C1 |
| 0010 | C2 |
| 0011 | M1 |
| 0100 | C4 |
| 0101 | M2 |
| 0110 | M3 |
| 0111 | M4 |
| 1000 | C8 |
| 1001 | M5 |
| 1010 | M6 |
| 1011 | M7 |
| 1100 | M8 |

The transmitter transmits 11000 1101101 to the receiver side and the receiver receives 11100 1101101. The receiver then generates check bits as follows:

$$C1 = 1 \textit{ xor } 0 \textit{ xor } 1 \textit{ xor } 0 \textit{ xor } 1 = 1$$

$$C2 = 1 \textit{ xor } 1 \textit{ xor } 1 \textit{ xor } 1 \textit{ xor } 1 = 1$$

$$C4 = 0 \textit{ xor } 1 \textit{ xor } 1 \textit{ xor } 1 = 1$$

$$C8 = 0 \textit{ xor } 1 \textit{ xor } 1 \textit{ xor } 1 = 1$$

At the receiver side: C8 C4 C2 C1 = 0101

*xor*

At transmitter side: C8 C4 C2 C1 = 1111

---

1010 Syndrome

This number shows that the bit located at position 1010 is in error and must be complemented.

## Summary

- Information transfer between two computers occurs in one of two types of signals: digital or analog.
- Modern computers work with digital signals.
- A digital signal is represented by two voltages.
- Binary is the representation of a number in base-2.
- One digit in binary is called a *bit* and 8 bits are equal to one byte. More than one byte is called a *word*.
- Information is represented inside the computer by binary or base-2.
- Binary coded decimal (BCD) is used for representing decimal numbers from 0 to 9.
- Information is represented by ASCII code inside the computer; ASCII code is made of 7 bits.
- There are two methods used for transmission of data: synchronous and asynchronous transmission.
- Parallel transmission is a method by which data is transmitted byte by byte.
- Serial transmission is a method by which data is transmitted one bit at the time over a single transmission media.
- Asynchronous transmission adds extra bits (start bit and stop bit) to the character for synchronization.
- Synchronous transmission clock pulses are used for synchronization.
- There are three types of communication modes: simplex transmission, half-duplex transmission, and full-duplex transmission.
- Baseband mode uses the bandwidth of a transmission media to carry only one signal.
- Broadband mode uses the bandwidth of a transmission media to carry multiple signals.
- Digital information can be represented by several forms of digital signal, such as non-return to zero (NRZ), non-return to zero inverted (NRZ-I), Manchester encoding, Differential Manchester encoding, and bipolar encoding.
- Some sources of error for digital communications are impulse noise, crosstalk, attenuation, and white noise.

- Parity check, block check character (BCC), one's complement of the sum, and cyclic redundancy checks (CRC) are used for error detection in networking.
- Hamming code can detect and correct one error.

## Key Terms

Alternate Mark Inversion (AMI)
Analog Signal
ASCII Code
Asynchronous Transmission
Attenuation
Baseband Mode
Binary
Bipolar Encoding
Bit
Block Check Character (BCC)
Broadband Mode
Byte
Crosstalk
Cyclic Redundancy Check
Differential Manchester Encoding
Digital Signal
Full-Duplex Mode
Half-Duplex Mode
Hamming Code
Impluse Noise
Manchester Encoding
Non-Return to Zero (NRZ)
Non-Return to Zero Inverted (NRZ-I)
One's Complement of the Sum
Parallel Transmission
Parity Check
Polar Encoding
Serial Transmission
Simplex Mode
Synchronous Transmission
Thermal Noise
Unicode
Unipolar Encoding
White Noise
Word

## Review Questions

### • Multiple Choice Questions

1. Frequency (F) is the number of cycles in one second, and can be represented as ______.
   a. F = 1/T
   b. F = T
   c. F = −1/T
   d. F = −T
2. Modern computers work with ______ signals.
   a. digital
   b. analog
   c. a and b
   d. none of the above
3. Unicode is a new ______ bit character encoding standard code.
   a. 16-
   b. 18-
   c. 8-
   d. 12-
4. ______ transmission transmits data character by character.
   a. Asynchronous
   b. Synchronous
   c. Full-duplex
   d. Half-duplex
5. __________ transmission uses asynchronous transmission.
   a. Serial
   b. Parallel
   c. Broadband
   d. Full-duplex
6. In ______ mode, transmission of data goes in only one direction.
   a. simplex
   b. half-duplex
   c. full-duplex
   d. serial
7. In ______ mode, both computers can send and receive information simultaneously.
   a. simplex
   b. half-duplex
   c. full-duplex
   d. serial
8. The ______ of a communication signal is the range of frequencies that the signal occupies.
   a. data rate
   b. bandwidth
   c. baud rate
   d. broadband
9. The bandwidth of each computer for an Ethernet LAN with 20 computers is ______.
   a. 1 Mbps
   b. 10 Mbps
   c. 500 Kbps
   d. 2 Mbps
10. Cyclic redundancy check can ______.
    a. detect a single error and correct it
    b. detect double errors and correct them
    c. detect one or more
    d. correct more than one error

11. Of the following digital encodings, ______ carries the clock pulse.
    a. Manchester encoding
    b. NRZ
    c. RZ
    d. RS-232

12. The decimal value for $(111101)_2$ is ______.
    a. 44
    b. 63
    c. 61
    d. 52

13. The hexadecimal value for $(111110111)_2$ is ______.
    a. 1F6
    b. 1F7
    c. FB1

14. The binary value for 45 is ______.
    a. 101011
    b. 101101
    c. 101111
    d. 011111

15. A range of frequencies carried by a medium is called ______.
    a. a broadband signal
    b. a based-band signal
    c. an analog signal
    d. a digital signal

16. Asynchronous communication uses __________.
    a. stop and start bits to indicate the start of the character and the end of the character
    b. a start bit to synchronize transmission
    c. start and stop bits for clocking
    d. none of the above

17. What is the efficiency of serial connection using asynchronous transmission with 1 start bit, 2 stop bits, and 7 data bits? ______
    a. 70%
    b. 75%
    c. 80%
    d. 65%

## • Short Answer Questions

1. Sketch an analog signal.
2. What is frequency?
3. What is the unit of frequency?
4. What is the frequency of an analog signal that is repeated every .02 ms?
5. Explain the amplitude of an analog signal.
6. Sketch a digital signal.
7. What is a bit?
8. What is a byte?
9. What is a word?

10. Convert the following binary number to hex:

$$(111000111001)_2 = (\_\_\_\_\_)_{16}$$

11. Convert the following binary number to decimal:

$$(11111111)_2 = (\_\_\_\_\_)_{10}$$

$$(10110001)_2 = (\_\_\_\_\_)_{10}$$

12. Convert the following number to binary:

$$(FDE6)_{16} = (\_\_\_\_\_)_2$$

13. Convert the word DIGITAL to binary using the ASCII table.
14. Convert the word NETWORK to hexadecimal.
15. Write your name in binary ASCII; then change the result to hexadecimal.
16. What is serial transmission?
17. What is parallel transmission?
18. What is the advantage of parallel transmission over serial transmission?
19. Explain the following terms:
    a. simplex
    b. half-duplex
    c. full-duplex
20. What is a synchronous transmission?
21. Why is a clock pulse needed for transmission of a digital signal?
22. Show the format of asynchronous transmission.
23. Sketch a clock pulse.
24. List two types of digital encoding methods in which the clock is embedded to the data signal.
25. List methods of error detection.
26. List sources of errors in networking.
27. Represent binary 110101 with a polynomial:

$$X^5 + X^4 + X^2 + 1$$

28. Find the BCC for the word "ETHERNET."
29. Show the CRC circuit for 1011.
30. Find the FCS for message 10110110 using the circuit in question 29.

31. Find the one's complement of the sum for the word "NETWORK."
32. Show the digital wave form for 0101011110.
33. Draw Manchester encoding and Differential Manchester encoding for binary 010110110.
34. Calculate the frequency of a signal repeated every 0.0005 seconds.
35. Find the FCS for data unit 111011 with divisor 1011.
36. What is a burst error?
37. Find the check bits for 10001101 using the hamming code.

# CHAPTER 3

# Introduction to Computer Architecture

**OUTLINE**



**OBJECTIVES**

After completing this chapter, you should be able to:

- List the components of a microcomputer and their functions
- List the components of a CPU
- Distinguish between a CPU and a microprocessor
- Compare a RISC processor and a CISC processor
- Discuss different types of memory
- Differentiate between various types of computer buses

**INTRODUCTION**

Just as the architecture of a building defines its overall design and functions, so does computer architecture define the design and functionality of a computer system. The components of a microcomputer are designed to interact with one another, and this interaction plays an important role in the overall system operation.

## 3.1 Components of a Microcomputer

A standard microcomputer consists of a microprocessor (central processing unit or CPU), buses, memory, parallel input/output, serial input/output, programmable I/O interrupt, and direct memory access DMA. Figure 3.1 shows the components of a microcomputer.

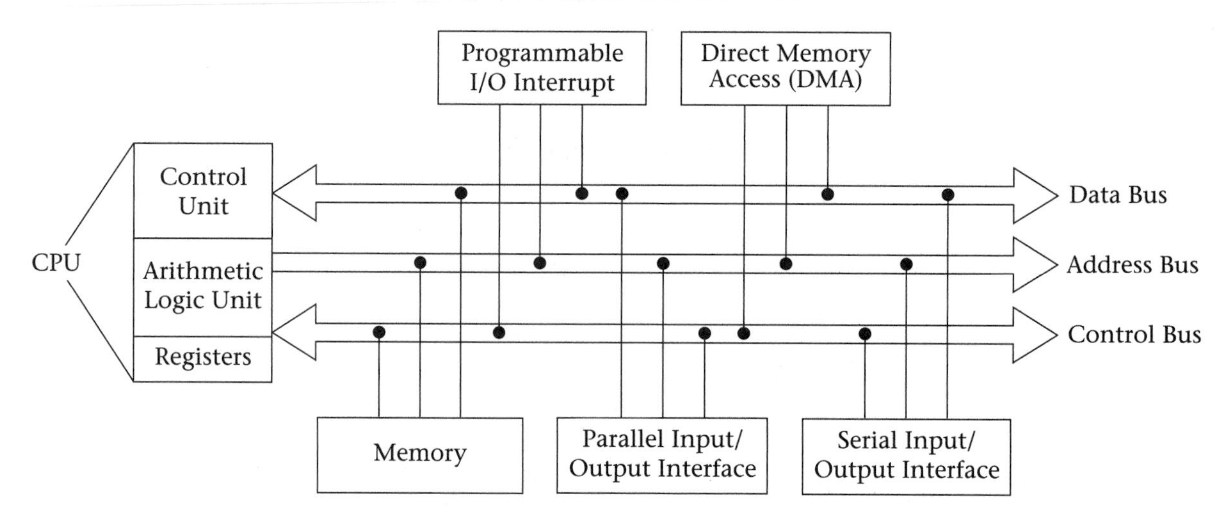


**FIGURE 3.1**
Components of a microcomputer

### Central Processing Unit (CPU)

The **central processing unit (CPU)** is the "brain" of the computer and is responsible for accepting data from input devices, processing the data into information, and transferring the information to memory and output devices. The CPU is organized into the following three major sections:

1. Arithmetic Logic Unit (ALU)
2. Control Unit
3. Registers

The function of the **arithmetic logic unit (ALU)** is to perform arithmetic operations such as addition, subtraction, division, and multiplication, and logic operations such as AND, OR, and NOT.

The function of the **control unit** is to control input/output devices, generate control signals to the other components of the computer such as read and write signals, and perform instruction execution. Information is moved from memory to the registers; the registers then pass the information to the ALU for logic and arithmetic operations.

The function of the microprocessor and the CPU are the same. If the control unit, registers, and ALU are packaged into one integrated circuit (IC), then the unit is called a *microprocessor*; otherwise the unit is called a *CPU*. The difference in packaging is shown in Figure 3.2.

**FIGURE 3.2**
Block diagram of microprocessor and CPU

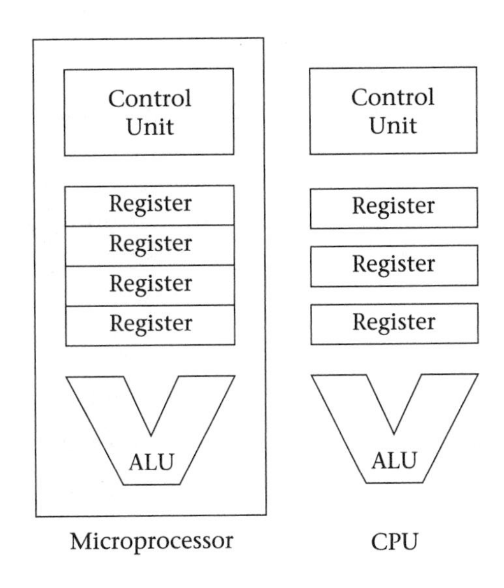


There are two types of technology used to design a CPU: **Reduced Instruction Set Computer (RISC)** and **Complex Instruction Set Computer (CISC)**.

***CISC architecture.*** In 1978, Intel developed the 8086 microprocessor chip. The 8086 was designed to process a 16-bit data word; it had no instruction for floating-point operations. At the present time, the Pentium processes 32-bit and 64-bit words and it can process floating-point instructions. Intel designed the Pentium processor in such a way that it can execute programs written for earlier 8086 processors.

The characteristics of 8086 are called *Complex Instruction Set Computers (CISC)*, which include instructions for earlier Intel processors. Another CISC processor is VAX 11/780, which can execute programs for the PDP-11 computer. The CISC processor contains many instructions with different addressing modes; for example, the VAX 11/780 has more than 300 instructions with 16 different address modes.

The major characteristics of CISC processors are as follows:

1. A large number of instructions
2. Many addressing modes

3. Variable length of instructions
4. Most instructions can manipulate operands in the memory
5. Control unit is microprogrammed

***RISC architecture.*** Until the mid-1990s, computer manufacturers were designing complex CPUs with large sets of instructions. At that time, a number of computer manufacturers decided to design CPUs capable of executing only a very limited set of instructions.

One advantage of a reduced instruction set computer is that it can execute its instructions very fast because the instructions are simple. In addition, the RISC chip requires fewer transistors than the CISC chip. Some of the RISC processors are the PowerPC, MIPS processor, IBM RISC System/6000, ARM, and SPARC.

The major characteristics of RISC processors are as follows:

1. Require few instructions
2. All instructions are the same length (they can be easily decoded)
3. Most instructions are executed in one machine clock cycle
4. Control unit is hardwired
5. Few address modes
6. A large number of registers

## Computer Bus

When more than one wire carries the same type of information, it is called a *bus*. The most common buses inside a microcomputer are the address bus, the data bus, and the control bus.

***Address bus.*** The address bus defines the number of addressable locations in a memory IC by using the $2^n$ formula, where $n$ represents the number of address lines. If the address bus is made up of three lines then there are $2^3 = 8$ addressable memory locations, as shown in Figure 3.3. The size of the address bus directly determines the maximum number of memory locations that can be accessed by the CPU.

***Data bus.*** The data bus is used to carry data to and from the memory and represents the size of each location in memory. In Figure 3.3 each location can hold only four bits. If a memory IC has eight data lines, then each location can hold eight bits. The size of a memory IC is represented by $\mathbf{2^n \times m}$ where $n$ is the number of address lines and $m$ is the size of each location. In Figure 3.3, where $n = 3$ and $m = 4$, the size of the memory is:

$$2^3 * 4 = 32 \text{ bits}$$

**FIGURE 3.3**
A memory with three address lines and four data lines

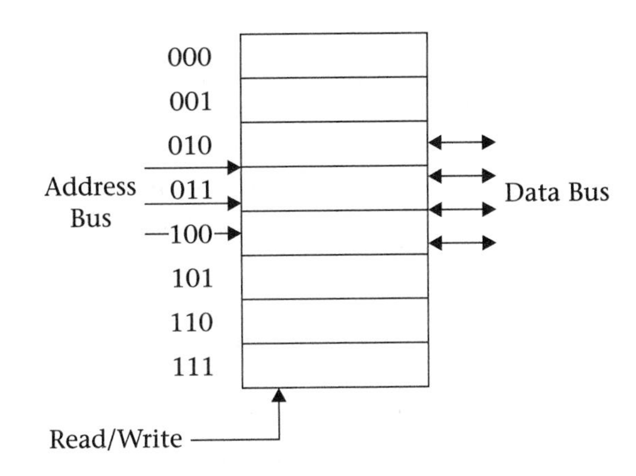


***Control bus.*** The control bus carries control signals from the control unit to the computer components in order to control the operation of each component. In addition, the control unit receives control signals from computer components. Some of the control signals are as follows:

**Read signal:** The read line is used to read information from memory or input/output (I/O) devices

**Write signal:** The write line is used to write data into the memory

**Interrupt:** Indicates an interrupt request

**Bus request:** The device is requesting to use the computer bus

**Bus grant:** Gives permission to the requesting device to use the computer bus

**I/O read and write:** I/O read and write is used to read from or write to I/O devices

## Memory

In general, memory can hold information either temporarily or permanently. The following are some types of memory:

- Semiconductor Memory or Memory IC
- Floppy Disk and Hard Disk
- Tape
- CD ROM (Compact Disk-Read Only Memory)
- Flash Memory

## Semiconductor Memory

There are two types of semiconductor memory: Random Access Memory (RAM) and Read Only Memory (ROM).

Data can be read from or written into **Random Access Memory (RAM)**. The RAM can hold the data as long as power is supplied to it. Figure 3.4 shows a general block diagram of RAM consisting of a data bus, address bus, and read/write signals. The data bus carries information out of or into the RAM. The address bus is used to select a memory location. The read signal becomes active when reading data from RAM and the write line becomes active when writing to the RAM. Remember, RAM can hold information only when it has power.

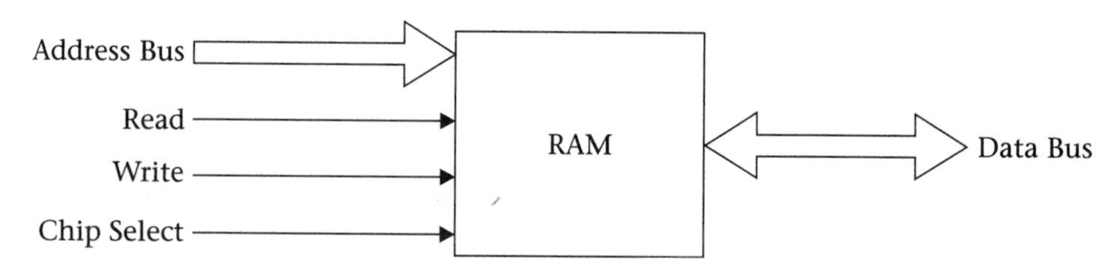


**FIGURE 3.4**
RAM block diagram

There are many types of RAM, including **Dynamic RAM (DRAM), Synchronous DRAM (SDRAM),** EDORAM, DDR SDRAM, RDRAM, and **Static RAM (SRAM).**

***Dynamic RAM (DRAM)*** is used in main memory. It needs to be refreshed (recharged) about every 1 ms. The CPU cannot read from or write to memory while the DRAM is being refreshed—this makes DRAM the slowest running memory. A DRAM comes in different types of packaging, such as the SIMM (Single In-Line Memory Module) and the DIMM (Dual In-Line Memory Module). The SIMM is a small circuit board that holds several chips and has a 32-bit data bus. The DIMM is a circuit board that holds several memory chips and has a 64-bit data bus.

***Synchronous DRAM (SDRAM)*** technology uses DRAM and adds a special interface for synchronization. It can run at much higher clock speeds than DRAM. SDRAM uses two independent memory banks. While one bank is recharging, the CPU can read and write to the other bank. Figure 3.5 shows a block diagram of SDRAM. Table 3.1 compares the throughput of SRAM and DRAM.

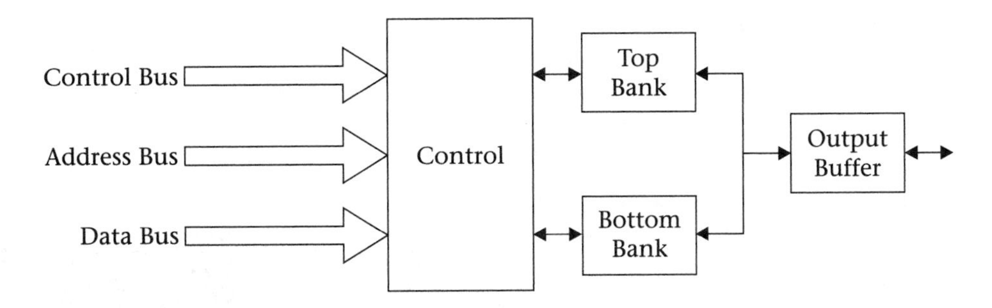


**FIGURE 3.5**
Block diagram of SDRAM

**TABLE 3.1** Throughput of SDRAM Compared to DRAM

| Bits Transfer | Two Banks of SDRAM |
|---|---|
| 1 bit | 1.8 times DRAM |
| 2 bits | 2.4 times DRAM |
| 8 bits | 4 times DRAM |
| 16 bits | 4.4 times DRAM |

***Extended Data Out RAM (EDORAM)*** transfers blocks of data to or from memory.

***Double Data Rate SDRAM (DDR SDRAM)*** is a type of SDRAM that transfers data at both the rising edge and the falling edge of the clock.

***Rambus DRAM (RDRAM)*** was developed by Rambus Corporation. It uses multiple DRAM banks with a new interface that enables DRAM banks to transfer multiple words and also transfer data at the rising edge and the falling edge of clock. The RDRAM refreshing is done by the interface. The second generation of RDAM is called *DRDRAM (Direct RDRAM)* and it can transfer data at a rate of 1.6 Gbps. Figure 3.6 shows an RDRAM module.

**FIGURE 3.6**
Rambus memory module
(Courtesy Samsung Corp.)

**DRAM Packaging**

DRAM comes in different types of packaging such as **SIMM (Single In-Line Memory Module)** and **DIMM (Dual In-Line Memory Module).**

Figure 3.7 shows SIMM, which is a small circuit board that holds several chips and has a 32-bit data bus.

**FIGURE 3.7**
DRAM SIMM

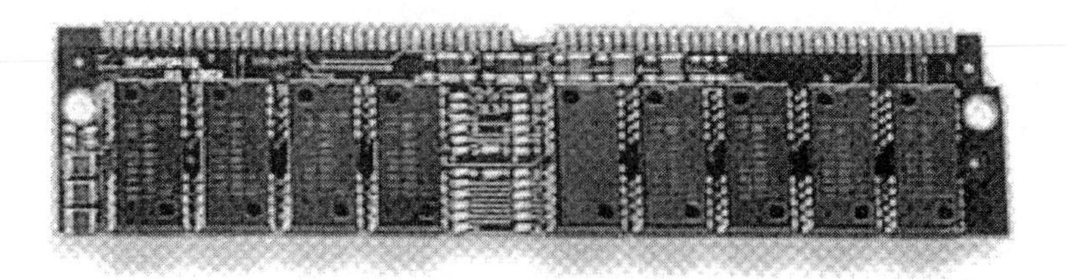

DIMM is a circuit board that also holds several memory chips and has a 64-bit data bus.

**Static RAM (SRAM)** is used in cache memory. SRAM is almost twenty times faster than DRAM and is also much more expensive.

As its name suggests, information can only be read from **Read Only Memory (ROM)**. ROM holds information permanently, even while there is no power to the ROM. Two types of ROM are listed next.

**Erasable Programmable Read Only Memory (EPROM):** EPROM can be erased with ultraviolet light and reprogrammed with a device called an *EPROM programmer*. Flash ROM is a type of EEPROM.

**Electrically Erasable PROM (EEPROM):** EEPROM can be erased by applying specific voltage to one of the pins and can be reprogrammed with an EPROM programmer.

**Flash Memory:** Flash memory is a type of EEPROM that allows multiple memory locations to be written or erased in one operation but only one memory location at a time can be erased or written.

**Parallel Input/Output Interface**

The parallel I/O interface is used to connect parallel devices, such as printers and scanners, to the computer. Figure 3.8 shows Centronics parallel connectors.

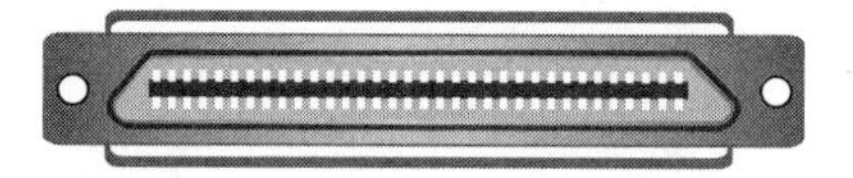

**FIGURE 3.8**
Centronics male and female connectors

### Serial Input/Output Interface

The serial I/O interface is used to connect serial I/O devices, such as a serial printer and modem, to the computer. The most common serial connector is RS-232D. RS-232 is commonly used either with a 25 pin-connector, DB-25, or a 9-pin connector, DB-9, as shown in Figure 3.9. For pin assignments refer to Appendix A.

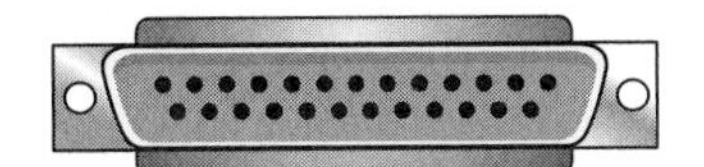
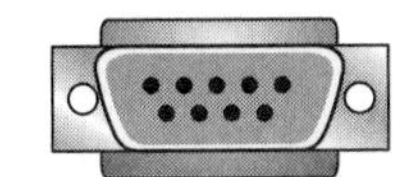

**FIGURE 3.9**
DB-25 and DB-9 serial connectors

### Direct Memory Access

**Direct Memory Access (DMA)** allows for the transfer of blocks of data from memory to an I/O device or vice versa. Without DMA, the CPU reads data from memory and writes it to an I/O device. Transferring blocks of data from memory to an I/O device requires the CPU to do one read and one write for each operation. This method of data transfer takes a lot of time. The function of DMA is to transfer data from memory to an I/O device directly, without using the CPU, so that the CPU is free to perform other functions.

The DMA performs the following functions in order to use the computer bus:

- The DMA sends a request signal to the CPU.
- The CPU responds to the DMA with a grant request, permitting the DMA to use the bus.
- The DMA controls the bus and the I/O device is able to read or write directly to or from memory.
- The DMA is able to load a file off an external disk into main memory when large blocks of data need to be transferred to a sequential range of memory. DMA is much faster and more efficient than a CPU.

### Programmable I/O Interrupt

When multiple I/O devices such as external drives, hard disks, printers, monitors, and modems are connected to a computer, as shown in Figure 3.10, a mechanism is necessary to synchronize all device requests. The function of a programmable interrupt is to check the status of each device and inform the CPU of the status of each; for example, the printer is not ready, a disk is write protected, this is an unformatted disk, there is a missing connection to a modem. Each device sends a signal to the programmable I/O interrupt controller in order to update its status. Figure 3.10 shows the programmable I/O interrupt controller.

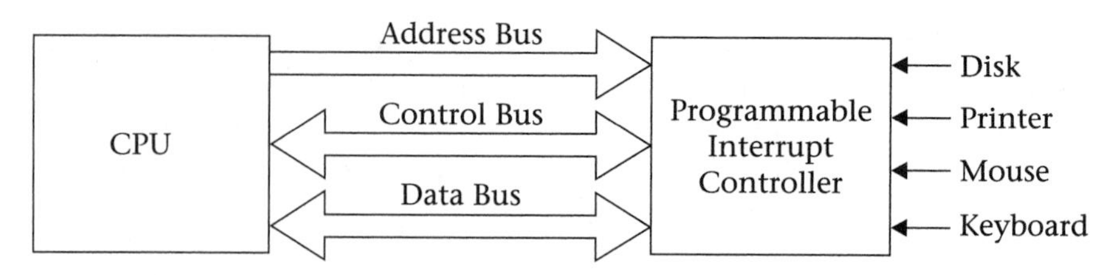


**FIGURE 3.10**
Programmable interrupt controller

## 3.2 Memory Hierarchy

Computers come with three types of memory, which are arranged in a hierarchical fashion, as shown in Figure 3.11.

1. **Cache memory** is the fastest type of memory and is most often used inside the microprocessor IC and on the motherboard. It is about twenty times faster than main memory and therefore, more expensive than main memory. Cache memory uses SRAM. Programs that reside in main memory are divided into blocks of data, with some of the blocks moved to the cache; the CPU accesses the cache to get the data.
2. **Main memory** uses DRAM and SDRAM. The program to be executed moves from secondary memory (disk or tape) into main memory.
3. **Second memory** refers to memory available on media such as a hard disk, tape, floppy disk, Zip drive, Jaz drive, and CD-ROM.

**FIGURE 3.11**
Memory hierarchy of a microcomputer

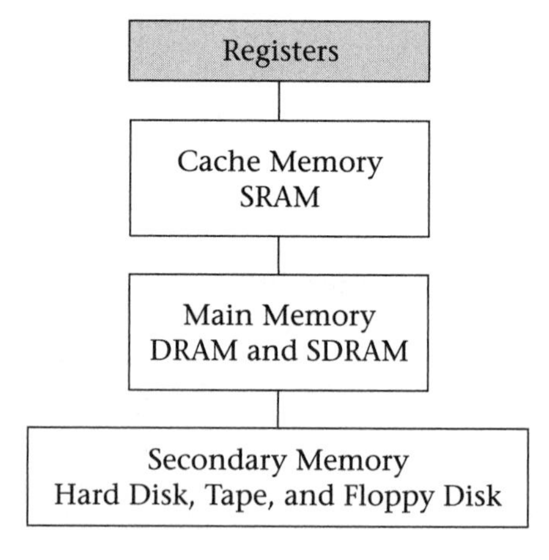

## 3.3 Disk Controller

The disk controller moves the disk drive head, reads, and/or writes data. Today there are two types of disk controllers: **IDE (Integrated Disk Electronics)** and **SCSI (Small Computer System Interface)**.

**Integrated Disk Electronics (IDE)**

An IDE disk drive is connected to the ISA bus with a flat ribbon cable. The IDE disk controller supports two hard disks, each with a 528-megabyte capacity. In 1994, hard disk drive vendors introduced EIDE (Extended IDE) which supports four devices, such as hard disks, tape drives, CD-ROM devices, and larger hard disk drives. The EIDE has two connectors. Each cable is connected to the EIDE controller and can support two hard disk drives with a capacity of up to 250 gigabytes. EIDE is used in IBM-compatible computers.

**Small Computer System Interface (SCSI)**

The Small Computer System Interface (SCSI) standard is defined by the American National Standards Institute (ANSI) for connecting daisy-chaining multiple I/O devices, such as scanners, hard disks, and CD-ROMs to the microcomputer, as shown in Figure 3.12.

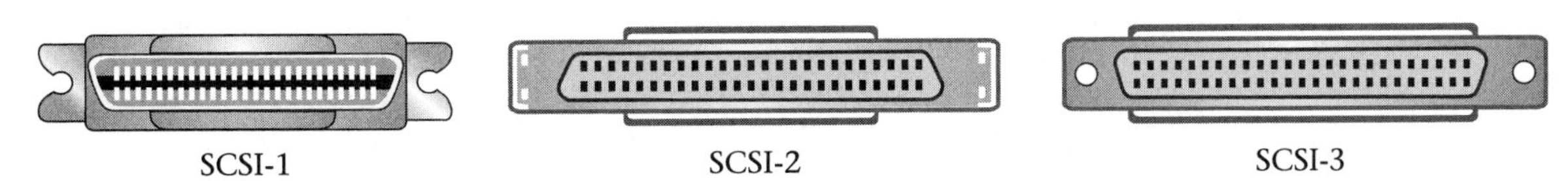


**FIGURE 3.12**
SCSI bus

SCSI is a standard interface for all types of microcomputers. It is used in Macintosh, RISC workstations, and minicomputers, as well as in higher-end IBM-compatible computers. The SCSI bus comes with different types of controllers. Table 3.2 shows the characteristics of several

**TABLE 3.2** Characteristics of Several Types of SCSI Controllers

| | Bandwidth | Data Rate MB/s* |
|---|---|---|
| SCSI-1 | 8 bits | 5 |
| SCSI-2 | 16 bits | 10–20 |
| Ultra SCSI | 8 bits | 20 |
| SCSI-3 | 16 bits | 40 |

*MB/s millions/bytes per second

types of SCSI controllers, and Figure 3.13 shows SCSI-1, SCSI-2, and SCSI-3 connectors.

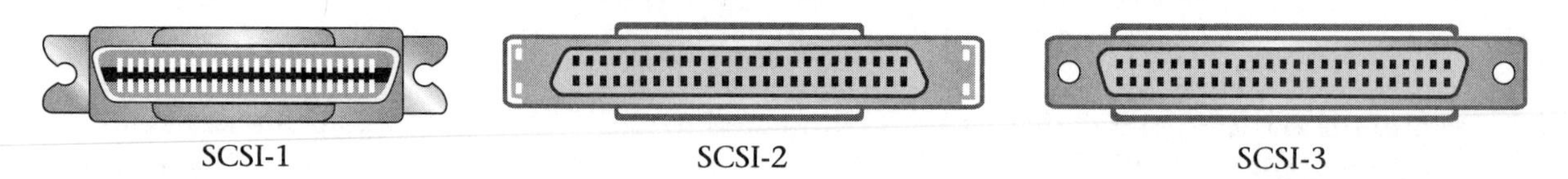


**FIGURE 3.13**
SCSI-1, SCSI-2, and SCSI-3 connectors

## 3.4 Microcomputer Bus

There are currently a number of different computer buses on the market that are designed for microcomputers, including ISA, MCA, EISA, VESA PCI, FireWire, USB, and PCI Express. Universal Serial Bus (USB) and PC Express are covered in more detail in Chapter 24.

### ISA Bus

The **Industry Standard Architecture (ISA) bus** was introduced by IBM for the IBM PC using an 8088 microprocessor. The ISA bus has an 8-bit data bus, and 20 address lines at a clock speed of 8 MHz. The PC AT type uses the 80286 processor, which has a 16-bit data bus and 24-bit address lines and is compatible with the PC.

### Microchannel Architecture Bus

The **Microchannel Architecture (MCA) bus** was introduced by IBM in 1987 for its PS/2 microcomputer. The MCA bus is a 32-bit bus that can transfer four bytes of data at a time and runs at a 10-MHz clock speed. It also supports 16-bit data transfer and has 32-bit address lines. Microchannel architecture was so expensive that the non-IBM vendors developed a comparable but less expensive solution called the *EISA bus*.

### EISA Bus

The **Extended ISA (EISA) bus** is a 32-bit bus that also supports 8- and 16-bit data transfer bus architectures. EISA runs at 8-MHz clock speeds and has 32-bit address lines.

### VESA Bus

The **Video Electronics Standards Association (VESA) bus,** which is also called a *Video Local bus (VL-BUS),* is a standard interface between the computer and its expansion. As applications became more graphically intensive, the VESA bus was introduced to maximize throughput of video graphics memory. The VESA bus provides fast data flow between stations and can transfer up to 132 Mbps.

### PCI Bus

The **Peripheral Component Interconnect (PCI) bus** was developed by Intel Corporation. PCI bus technology includes a 32/64-bit bus that runs at a 33/66-MHz clock speed. PCI offers many advantages for connections to hubs, routers, and Network Interface Cards (NIC). In particular, PCI provides more bandwidth: up to one gigabit per second as needed by these hardware components.

The PCI bus was designed to improve the bandwidth and decrease latency in computer systems. Current versions of the PCI bus support data rates of 1056 Mbps and can be upgraded to 4224 Mbps. The PCI bus can support up to 16 slots or devices in the motherboard. Most suppliers of ATM (Asynchronous Transfer Mode) and 100BaseT NICs offer a PCI interface for their products. The PCI bus can be expanded to support a 64-bit data bus. Table 3.3 compares different bus architectures showing characteristics of ISA, EISA, MCA, VESA, and PCI Buses. Figure 3.14 shows the PCI bus.

**TABLE 3.3** Characteristics of Various Buses

| Bus Type | ISA | EISA | MCA | VESA | PCI | PCI-64 |
|---|---|---|---|---|---|---|
| Speed (MHz) | 8 | 8.3 | 10 | 33 | 33 | 64 |
| Data Bus Bandwidth (bits) | 16 | 32 | 32 | 32 | 32 | 66 |
| Max. Data Rate (MB/s)* | 8 | 32 | 40 | 132 | 132 | 508 |
| Plug & Play capable | no | no | yes | yes | yes | yes |

*MB/s millions/bytes per second

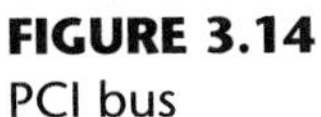
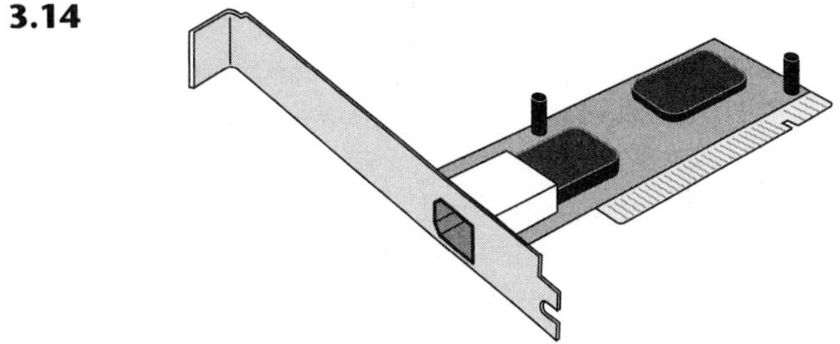

**FIGURE 3.14**
PCI bus

### PC Card

In the early 1990s, the rapid increase in the demand for mobile computing caused the development of smaller and more portable processing devices such as laptop computers. One of these developments was PC card technology. The almost spontaneous world-wide adaptation of the PC card was due in large part to the standard specification of the PC card by the Personal Computing Memory Card International Association.

The PC card standard provides standards for three types of cards: Type I, Type II, and Type III. The Type I card is used for memory devices such as RAM, Flash Memory, and SRAM. Type II is used for I/O devices such as a fax and modem, and Type III is used for rotating mass storage devices.

The PC card adapter is covered by metal casing and can be inserted and removed from the laptop computer at any time. The PC card adapter is show in Figure 3.15.

**FIGURE 3.15**
Inserting a PC card

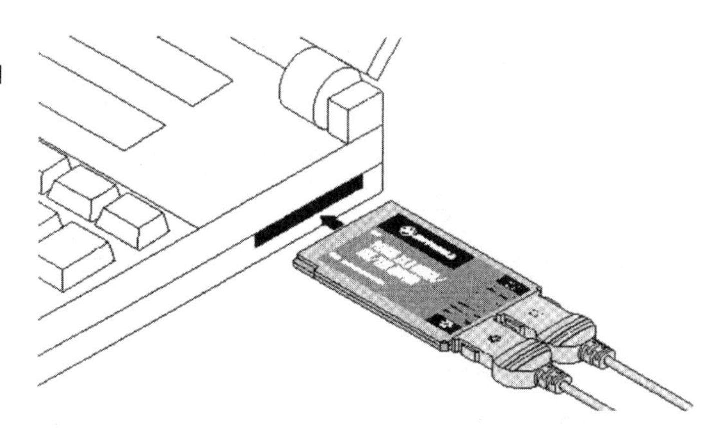

## 3.5 Plug-and-Play

When adding a new adapter to the computer, the Interrupt Request (IRQ), I/O address, and some DMA addresses must be set. Computers with an ISA bus must have these settings assigned manually by the user. The current EISA, MCA, and PCI bus technology can automatically assign the IRQ and I/O addresses. This automation process is called **plug-and-play (PnP).** Most new computers have the ability for plug-and-play; however, the adapter card itself must also have the capability to support plug-and-play. The adapter reports to the computer what IRQ and I/O addresses it is able to use, and the computer attempts to allocate the requested resources for the adapter. Operating systems such as Windows 98, Windows NT, Windows 2000, Windows 2003, Linux, and Macintosh OS support plug-and-play technology.

## 3.6 FireWire

**FireWire** or IEEE 1394 is a high-speed serial bus used for connecting digital devices, such as a digital video or camcorder. The bus is able to transfer data at the rate of 100, 200, or 400 Mpbs. The IEEE 1394 cable consists of six copper wires; two of the wires carry power and four of the wires carry signals, as illustrated in Table 3.4. Some FireWire connectors come with four pins, without having power pins. Figure 3.16 shows FireWire male and female connectors.

**TABLE 3.4** IEEE 1394 Pins

| Pin | Signal Name | Description |
|---|---|---|
| 1 | Power | unregulated DC; 17–24 V no load |
| 2 | Ground | Ground return for power and inner cable shield |
| 3 | TPB– | Twisted-pair B, differential signals |
| 4 | TPB+ | Twisted-pair B, differential signals |
| 5 | TPA– | Twisted-pair A, differential signals |
| 6 | TPA+ | Twisted-pair A, differential signals |

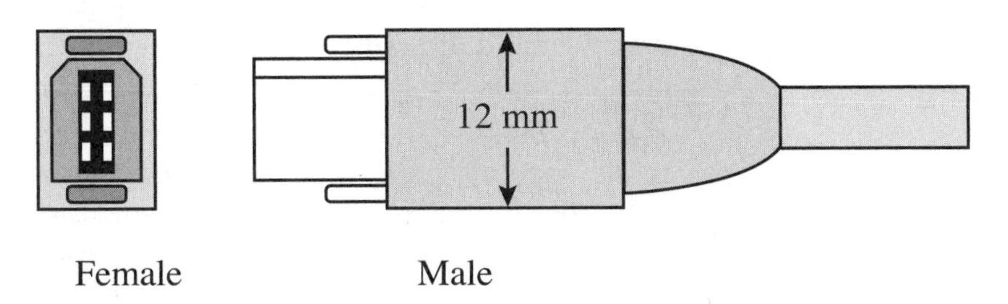


**FIGURE 3.16**
FireWire male and female connectors

## 3.7 Intel Microprocessor Family

Intel designs and manufactures microprocessors for IBM-compatible microcomputers. Each processor has a number or name that is used by the computer designer to access the information provided by the manufacturer of the processor.

Intel microprocessor IC numbers and names are 8088, 80286, 80386, 80486, Pentium, Pentium II, Pentium III, and Pentium IV. Recently Intel and HP developed a new processor called *Itanium*, which is described later in this chapter. The Pentium III offers Single Instruction Multiple Data (SIMD) for floating-point operations. The following is a list of microprocessor characteristics:

**Register size:** Registers are used to store information inside the processor. Register sizes can vary from 8- to 16- to 32- to 64-bits.

**Number of registers:** A processor with several registers can store more information in the CPU for processing.

**Data bus size:** The size of the data bus determines how many bits of data can be transferred in parallel to or from memory or input/output ports.

**Address bus size:** The typical address sizes are 16, 32, and 64 bits. The size of the address bus determines the number of memory locations that may be accessed by the microprocessor.

**Clock speed:** The speed of the clock determines the speed at which the processor executes instructions.

**Math coprocessor:** The math coprocessor is a special processor that performs complex mathematical operations.

**Real mode:** Real mode allows for software compatibility with older software. It enables the processor to emulate the lowest Intel 8088 processor and use only the first 1 MB of memory.

**Protected mode:** Protected mode is a type of memory usage available on 80286 (and later) model microprocessors. In protected mode, each program can be allocated a certain section of memory and other programs cannot use this memory. Protected mode also enables a single program to access more than 1 MB of memory.

**Cache size:** Cache memory is a small amount of high-speed memory used for temporary data storage based between the processor and main memory. The size of the cache can help to speed up the execution time of a program.

**MMX technology:** Intel's **MMX technology** is designed to speed up multimedia and communications applications, such as video, animation, and 3D graphics. The technology includes Single Instruction Multiple Data (SIMD) technique (meaning that with one instruction, the computer can perform multiple operations), 57 new instructions, eight 64-bit MMX register, and four new data types.

Table 3.5 provides a quick reference to this list of characteristics.

**TABLE 3.5** Characteristics of Intel Microprocessor

| | 80486dx | Pentium | Pentium Pro | Pentium Pro II | Pentium II |
|---|---|---|---|---|---|
| Register Size | 32 bit | 32 bit | 32 bit | 32/64 bit | 32/64 bit |
| Data Bus Size | 32 bit | 64 bit | 64 bit | 64 bit | 64 bit |
| Address Size | 32 bit | 32 bit | 32 bit | 32 bit | 32 bit |
| Max Memory | 4 GB | 4 GB | 4 GB | 4 GB | 4 GB |
| Clock Speed | 25, 33 MHz | 60, 166 MHz | 150, 200 MHz | 233, 340, 400 MHz | 450, 500 MHz |
| Math Processor | Built in | Built in | Built in | Built in | Built in |
| L1 Cache* | 8 KB, 16 KB | 8 KB Instruction 8 KB Data | 8 KB Instruction 8 KB Data | 16 KB Instruction 16 KB Data | 16 KB Instruction 16 KB Data |
| L2 Cache** | No | No | 256 KB or 512 KB | 512 KB | 512 KB |
| MMX Technology | No | No | Yes | Yes | Yes |

*L1 Cache is the cache memory built inside the microprocessor.

**L2 Cache is not part of the microprocessor; it is in a separate IC.

## 3.8 Itanium Architecture

Intel and Hewlett-Packard developed the Itanium processor. The Itanium processor is also known as *IA-64 (Intel Architecture 64-bit processor)*. The IA-64 uses 64-bit registers and performs 64-bit arithmetic and logical operations. The Itanium architecture also provides full compatibility with Intel's 32-bit architecture (IA-32). Figure 3.17 shows the internal architecture of IA-64.

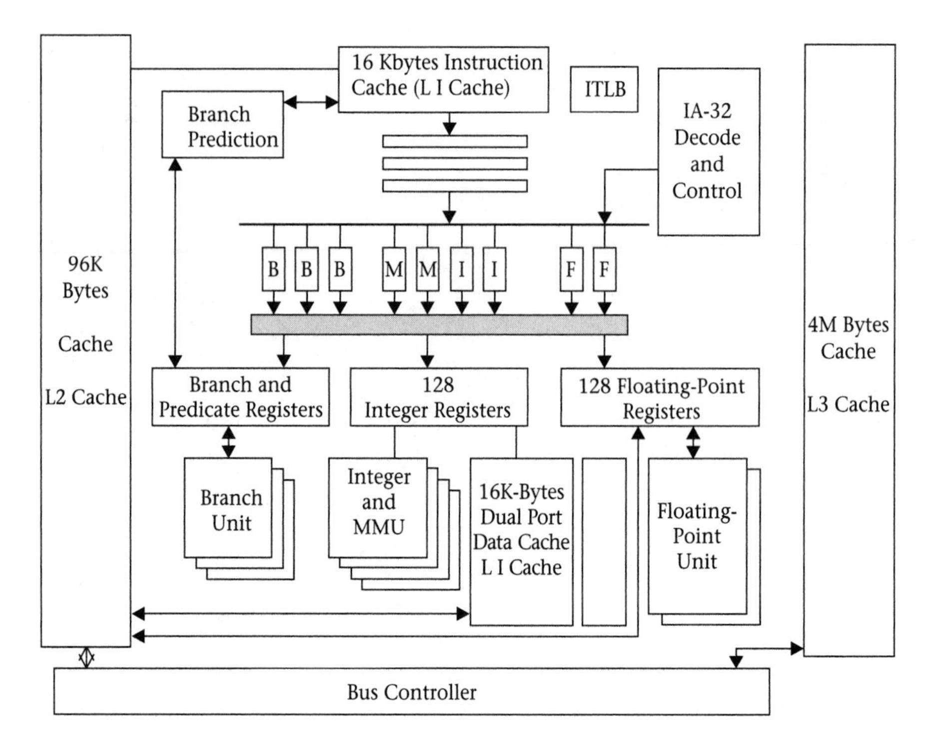


**FIGURE 3.17**
Itanium architecture

The following are characteristics of the Itanium processor:

- 64-bit integer registers
- 128 82-bit floating registers
- 64 1-bit predicate registers
- 8 64-bit branch registers
- 4 arithmetic logic units
- 3 branch processing units
- 2 floating-point units
- L1, L2, and L3 caches

## Summary

- The components of a computer are the CPU, memory, parallel I/O, serial I/O, programmable interrupt, and DMA.
- The function of the CPU is to process information using the arithmetic logic unit (ALU).
- The components of the CPU are the arithmetic logic unit (ALU), the control unit, and registers.
- Most computers use three types of memory: cache memory (SRAM), main memory (DRAM or SDRAM), and secondary memory (hard disk, tape drive, and floppy disk).
- Semiconductor memory types are DRAM, SDRAM, EDORAM, DDR, SDRAM, RDRAM, ROM, and EPROM.
- SRAM is used in cache memory; DRAM and SDRAM are used in main memory.
- SCSI-1, SCSI-2, and SCSI-3 are computer peripheral controllers.
- ISA bus, EISA, MCA, and PCI are microcomputer buses.
- FireWire is a high-speed serial bus with a data rate up to 400 Mbps.

## Key Terms

Arithmetic Logic Unit (ALU)

Cache Memory

Central Processing Unit (CPU)

Complex Instruction Set Computer (CISC)

Control Unit

Direct Memory Access (DMA)

Dual In-Line Memory Module (DIMM)

Double Data Rate SDRAM (DDR SDRAM)

Dynamic RAM (DRAM)

Erasable Programmable Read Only Memory (EPROM)

Extended Data Out RAM (EDORAM)

Extended ISA (EISA) Bus

FireWire

Industry Standard Architecture (ISA) Bus

Intel Microprocessor Family

Integrated Disk Electronics (IDE)

Main Memory

Microchannel Architecture (MCA) Bus

MMX Technology

Peripheral Component Interconnect (PCI) Bus

Plug-and-Play (PnP)

Rambus DRAM (RDRAM)

Random Access Memory (RAM)

Read Only Memory (ROM)

Reduced Instruction Set Computer (RISC)

Secondary Memory

Single In-Line Memory Module (SIMM)

Small Computer System Interface (SCSI)

Static RAM (SRAM)

Synchronous DRAM (SDRAM)

Video Electronics Standards Association (VESA)

## Review Questions

### • Multiple Choice Questions

1. The function of the ______ is to perform arithmetic operations.
   a. bus
   b. serial port
   c. ALU
   d. control unit
2. When you compare the functions of a CPU and a microprocessor, ______.
   a. they are the same.
   b. they are not the same.
   c. the CPU is faster than the microprocessor.
   d. the microprocessor is faster than the CPU.
3. RISC processors use ______.
   a. complex instruction sets
   b. reduced instruction sets
   c. a and b
   d. none of the above
4. The CISC processor control unit is ______.
   a. hardware
   b. microcode
   c. a and b
   d. none of the above
5. ______ memory types are used for main memory.
   a. ROM and SDRAM
   b. SRAM and DRAM
   c. SDRAM and DRAM
   d. DRAM and EPROM

6. ______ holds information permanently, even when there is no power.
   a. ROM
   b. DRAM
   c. RAM
   d. SRAM

7. Direct memory access allows for the transfer of blocks of data from memory to an I/O device (or vice versa) without using the ______.
   a. CPU
   b. data bus
   c. control bus
   d. DMA controller

8. ______ is the fastest type of memory.
   a. Cache memory
   b. Main memory
   c. Secondary memory
   d. Hard disk

9. Of the following buses, ______ is/are 32-bit.
   a. ISA
   b. PCI and EISA
   c. EISA and ISA
   d. MCA and ISA

10. Of the following operating systems, ______ support plug-and-play.
    a. Windows NT and Windows 95
    b. Windows 98 and Windows NT
    c. Windows XP and Windows 2000
    d. DOS and Windows NT

- **Short Answer Questions**

1. List the components of a microcomputer.
2. Explain the functions of a CPU.
3. List the functions of an ALU.
4. What is the function of a control unit?
5. Distinguish between a CPU and a microprocessor.
6. What does RAM stand for?
7. What is SRAM? Discuss its applications.
8. Define DRAM and SDRAM and explain their applications.
9. Explain the function of an address bus and a data bus.
10. What does IC stand for?

11. What is the capacity of a memory IC with 10 address lines and 8 data buses?
12. What is ROM?
13. What does EEPROM stand for, and what is its application?
14. What does RDRAM stand for?
15. What is SIMM?
16. Explain the function of cache memory and give its location.
17. List types of memory used for main memory.
18. List types of memory used for secondary memory.
19. Explain the function of DMA.
20. What is the application of a parallel port?
21. What is the application of a serial port?
22. What is an interrupt?
23. Explain plug-and-play.
24. List operating systems that support plug-and-play.
25. What is the application of FireWire?
26. What does SCSI stand for?
27. What does IDE stand for?
28. List some of the computer buses.
29. Explain the difference between CISC processors and RISC processors.

CHAPTER 4

# Communications Channels and Media

**OUTLINE**

4.1 Conductive Media

4.2 Fiber-Optic Cable

4.3 Wireless Transmission

4.4 Transmission Impairment

4.5 Bandwidth, Latency, Throughput, and Channel Capacity

**OBJECTIVES**

After completing this chapter, you should be able to:

- List the types of communication media currently in use
- Distinguish among the different types of unshielded twisted-pair (UTP) cables
- List the different types of coaxial cables and their applications
- Discuss the different types of fiber-optic cables and their usage
- Explain the operation of wireless transmission
- Explain signal attenuation and channel bandwidth

**INTRODUCTION**

A transmission medium is a path between the transmitter and the receiver in a transmission system. The type of transmission medium is defined by the various characteristics of the digital signal, including the signal rate, data rate, and the bandwidth of the channel. The bandwidth of a channel determines the range of frequencies that the channel can transmit.

There are three types of communications media currently in use:

- Conductive, such as twisted-pair wire and coaxial cable
- Fiber-optic cable
- Wireless

## 4.1 Conductive Media

The most popular **conductive media** used in networking are unshielded twisted-pair (UTP) cable, shielded twisted-pair (STP) cable, and coaxial cable.

### Twisted-Pair Cable

**Unshielded twisted-pair (UTP) cable** is the least expensive transmission medium and is typically used for LANs. Electrical interference, such as external electromagnetic noise generated by nearby cables, can have a devastating effect on the performance of UTP cable. One way of improving the effect of noise on UTP cable is to shield the cable with metallic braid. **Shielded twisted-pair (STP) cable** provides better performance but is more difficult to work with. Figures 4.1(a) and 4.1(b) show examples of UTP and STP cable.

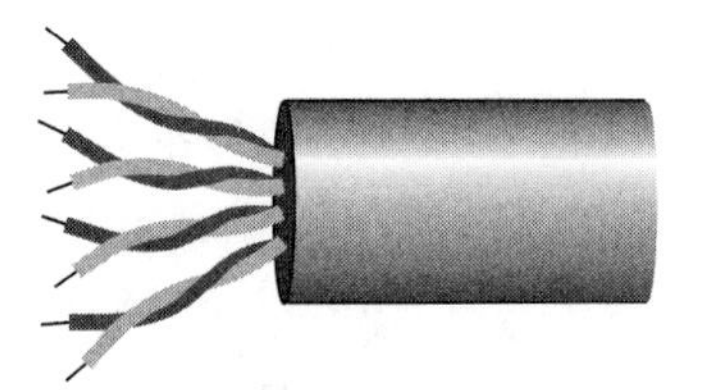

**FIGURE 4.1(a)**
Unshielded twisted-pair (UTP) cable

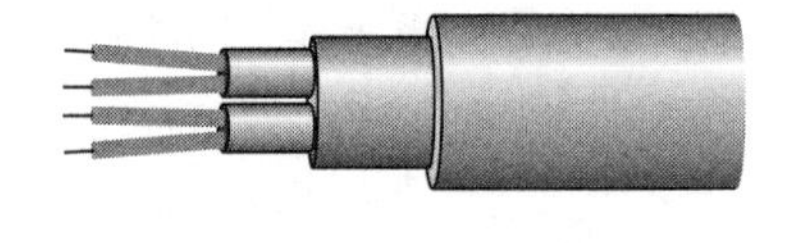

**FIGURE 4.1(b)**
Shielded twisted-pair (STP) cable

Unshielded twisted-pair cable is divided into categories CAT-1 through CAT-6 by the Electronic Industries Association (EIA). There are also proprietary enhancements to the CAT-5 specification that allow for even better performance over longer distances.

The EIA provides specifications for UTP cable, as shown in Table 4.1. These standards apply to four-pair UTP. UTP uses **RJ-45 and RJ-11 connectors**, as shown in Figure 4.2.

**TABLE 4.1** UTP Specifications

| Type of UTP | Performance | Application |
|---|---|---|
| CAT-1 | none | none |
| CAT-2 | 1 MHz | Telephone wiring |
| CAT-3 | 16 MHz | 10Base-T, Token Ring 4 Mbps, ISDN low speed |
| CAT-4 | 20 MHz | Token Ring 16 |
| CAT-5/5e | 100 MHz | 100Base-T, 100 VG-any LAN, 20 Mbps Token Ring |
| CAT-6 | 250 MHz | Gigabit Ethernet |

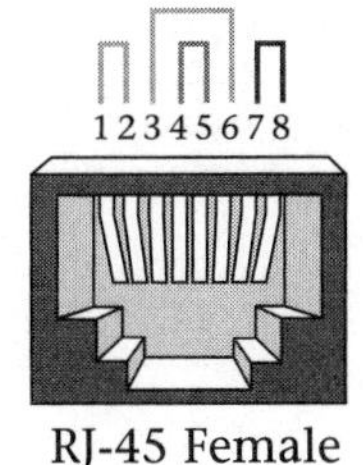
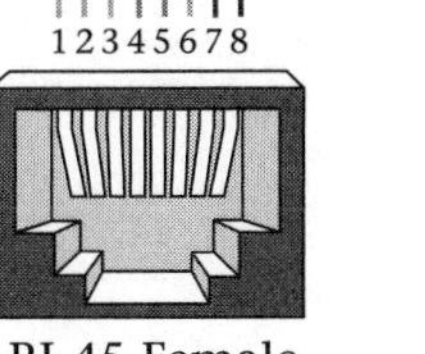

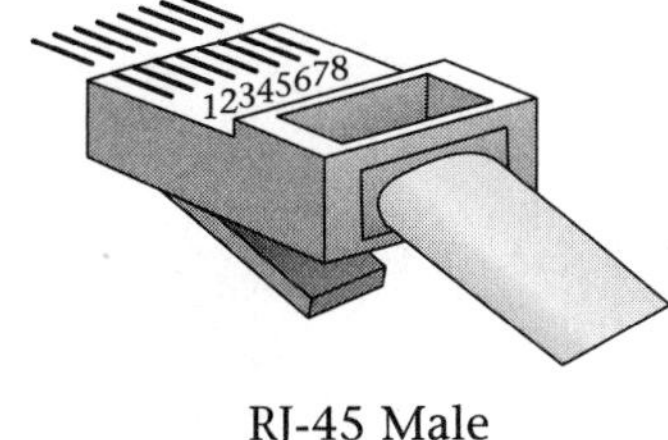

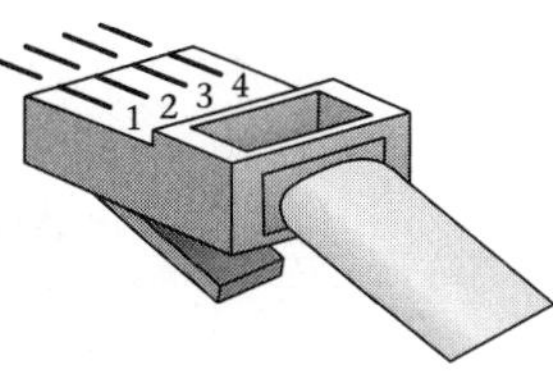


**FIGURE 4.2**
RJ-45 and RJ-11 connectors

## Coaxial Cable

**Coaxial cable** is used to transmit high-speed digital and analog signals over long distances. Figure 4.3 shows a coaxial cable that has an outer insulating cover made of polyvinyl chloride (PVC) or Teflon protecting the coaxial cable. Under the outer cover is a wire mesh shield, which provides excellent protection from external electrical noise. This shield is made of mesh wire or foil, or both. Under the shield is a plastic insulator, which isolates the center conductor from the shield.

The center conductor is a solid copper or aluminum wire that is shielded from external interference signals. Table 4.2 shows different types of coaxial cable, categorized by Radio Government (RG) rating. RG represents a set of specifications for cable such as the conductor diameter, thickness, and type of insulator. Coaxial cable uses a BNC connector, as shown in Figure 4.4.

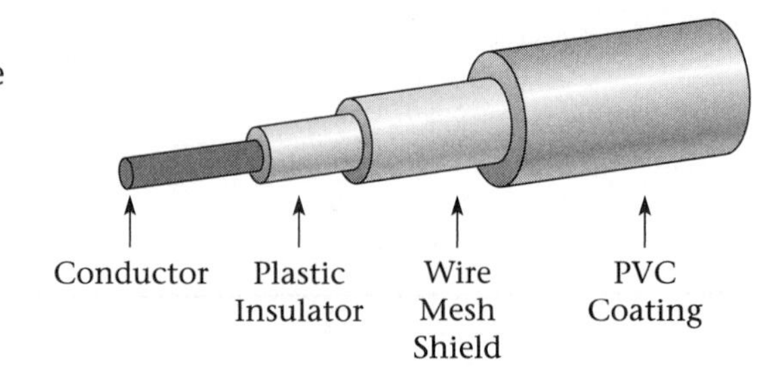


**FIGURE 4.3**
Coaxial cable

**TABLE 4.2** Various Types of Coaxial Cable and Their Applications

| Cable Type | Impedance | Application |
|---|---|---|
| RG-59 | 75 ohms | Cable TV |
| RG-58 | 50 ohms, 5 mm in diameter | 10 BASE2 or ThinNet |
| RG-11 and RG-8 | 50 ohms, 10 mm in diameter | 10 BASE5 or ThickNet |

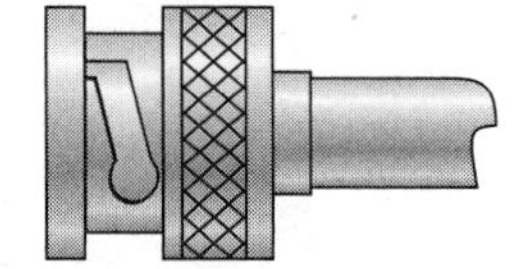

**FIGURE 4.4**
BNC connector

## 4.2 Fiber-Optic Cable

A **fiber-optic cable** is made of fiber that is covered by a buffer and a jacket. The fiber is composed of a core of thin glass or plastic covered by cladding, which may also be glass or plastic. This fiber (the core and the cladding) is then covered by a buffer to strengthen it. The buffer is finally covered by a plastic outer layer called the *jacket,* which acts as a protective coating or shield. Figure 4.5 illustrates the structure of a fiber-optic cable.

To transmit information using optical fiber, the digital information is converted to light pulses by **light-emitting diodes (LED)** or **injected-laser diodes (ILD)** and sent through the fiber-optic cable. An LED is a diode that generates a low-power light. At the receiving end, a photodiode or a photo transistor is used to convert the light pulse signals back into electrical signals.

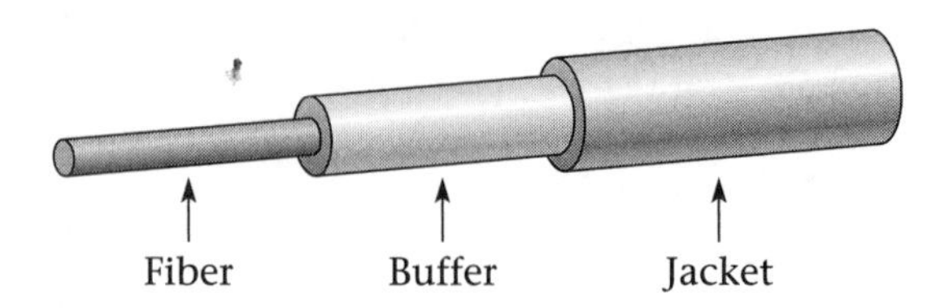


**FIGURE 4.5**
Fiber-optic cable

The advantages of fiber-optic cable include the following:

- Longer distance transmission due to reduced signal loss (attenuation)
- Greater bandwidth up to the gigahertz range
- Immunity from any kind of noise or external interference such as electromagnetic signals
- Smaller size
- Secure media

Some disadvantages of fiber-optic cables are as follows:

- Network interface cards and cabling can be expensive
- Connection to the network is more difficult

## Characteristics of Light

The source of signals for fiber-optic cable is light. The characteristics of light are propagation speed, wavelength, and attenuation.

***Propagation speed.*** Light propagates through a vacuum at the speed of $3.0*10^8$ m/s.

***Wavelength.*** The length of a wave is measured in meters and is represented by λ (lambda). The **wavelength** is the distance between two successive peaks of a wave or the distance traveled by one cycle of a wave, as shown in Figure 4.6.

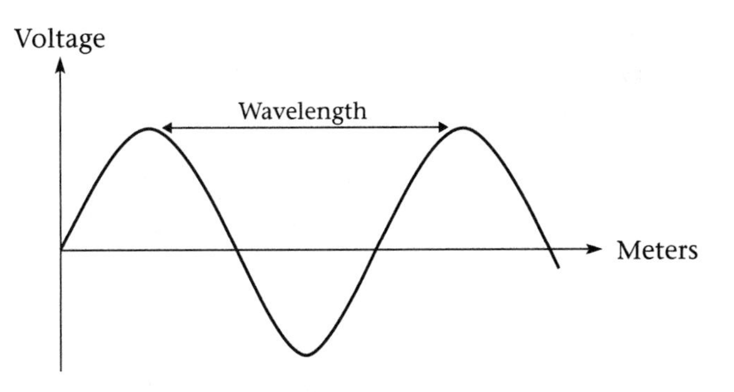


**FIGURE 4.6** Wavelength of a sine wave

Equation 4.1 describes the wavelength in terms of the speed of light and the frequency of a signal.

$$\lambda = C/f \qquad (4.1)$$

where:

C = the speed of light ($3*10^8$ m/s)
$f$ = frequency of the signal

***Attenuation of light.*** **Attenuation** is the reduction of the strength of a signal. When light travels through a fiber it loses energy. The greatest loss of energy is caused by absorption. Absorption is caused by fiber materials, and the optical power is converted to another form of energy such as heat. Attenuation is defined in Equation 4.2.

$$A = 10 \log_{10} \frac{Pt}{Pr} \qquad (4.2)$$

where:

A = Attenuation in decibel
Pt = Power of light at the transmitter side
Pr = Power of light at the receiver side (after transmission)

The attenuation of a fiber cable is specified by the manufacturer. Figure 4.7 shows the attenuation of a 1-Km fiber-optic cable with different signal wavelengths. Figure 4.7 shows two windows with the least attenuation (1300 nm and 1550 nm). The 850-nm wavelength window offers the most economical solution (use less expensive emitting diode). Fiber-optic systems operate at the wavelength defined by one of these three windows:

The first window is centered at the wavelength of 850 nm (nanometers = $10^{-9}$ meters).

The second window is centered at 1300 nm.

The third window is centered at 1550 nm.

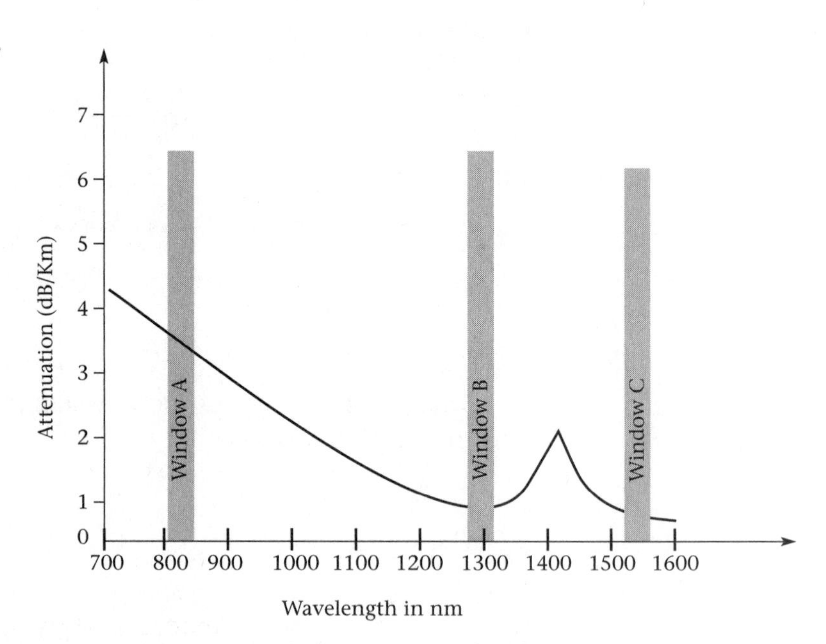


**FIGURE 4.7** Attenuation of fiber cable versus wavelength

### Types of Fiber-Optic Cable

In a fiber-optic cable, the angle of light reflection is directly dependent upon the diameter of the fiber. As the diameter increases, the light is reflected more and it takes more time to travel a given distance. There are two types of fiber-optic cable: Single-Mode Fiber (SMF) and Multimode Fiber (MMF).

***Single-Mode Fiber (SMF).*** In **single-mode fiber**, only one light ray propagates through the fiber, as shown in Figure 4.8. The core diameter of single-mode fiber is between 7 and 10 microns or micrometers (1 micron = $10^{-6}$ meters) and the cladding diameter of single-mode fiber is 125 microns. The light wavelengths that are used for SMF are 1300 and 1550 nanometers. Manufacturers of fiber-optic cable represent the fiber cable by ratio of core over cladding diameters, for example, 8/125 for single mode.

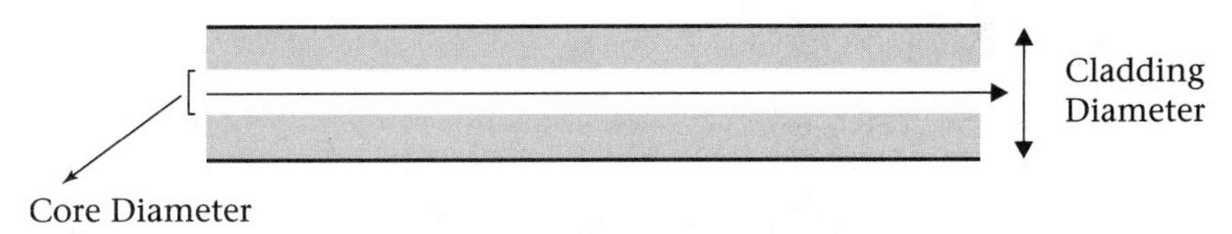


**FIGURE 4.8**
Light propagation in single-mode fiber (SMF)

***Multimode Fiber (MMF).*** In **multimode fiber**, more than one light ray can propagate through the fiber, since each light ray propagates at different wavelengths, as shown in Figure 4.9. Multimode fiber has a core diameter larger than the wavelength of the light source being used. For multimode fiber, the core diameter ranges from 50 micrometers to 1000 micrometers, and the wavelength of the light is about 1 micrometer. This means light can propagate through the fiber in many different ray paths or modes. A single-mode fiber cable has a smaller diameter than a multimode fiber cable.

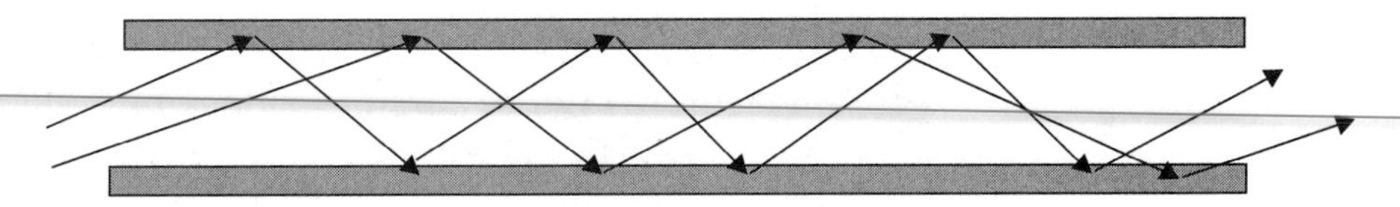

**FIGURE 4.9**
Light propagation in multimode step index fiber

There are two types of multimode fibers: multimode step index fiber and multimode graded index fiber.

***Multimode step index fiber.*** This is a simple type of multimode fiber in which the index of refraction (the ability of the material to bend light) is the same all across the core of the fiber. Therefore, rays of the light can propagate, as shown in Figure 4.9. For step index fiber the bandwidth is typically 20 to 30 MHz over a distance of one kilometer.

***Multimode graded index fiber.*** In multimode graded index fiber, the index of refraction across the core is gradually changed from the maximum at the center to a minimum near the edges. This type of fiber causes the light to travel faster in the low index of refraction material than in the high-refraction material. Typical bandwidth for graded index fibers range from 100 MHz*Km to 1 GHz*Km. Figure 4.10 shows multimode graded index fiber-optic cable.

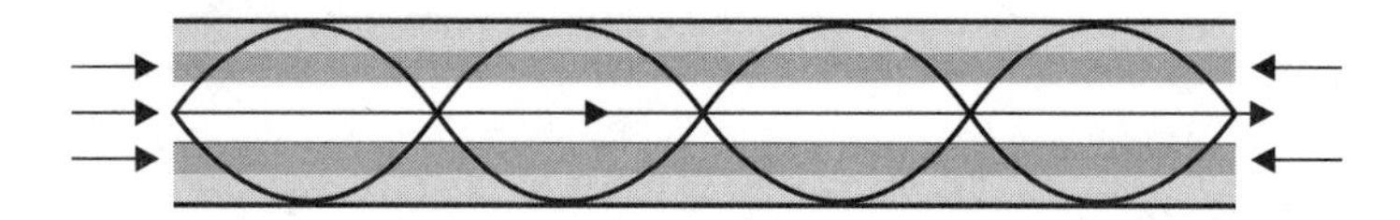

**FIGURE 4.10**
Light propagation in multimode graded index fiber

## Modal Bandwidth

**Modal bandwidth** is specified in units of MHz*Km. The modal bandwidth indicates the amount of bandwidth supported by a fiber cable for a 1-Km (0.625 miles) distance. The modal bandwidth is given by the manufacturer of the optical cable. For example, a cable with a modal bandwidth of 500 MHz*Km can support end-to-end bandwidth of 250 MHz at the maximum 2-Km (1.25 miles) distance.

## Fiber-Optic Connectors

There are three common types of fiber-optic connectors used for networking:

1. **Subscriber Channel Connector (SC):** The SC connector, shown in Figure 4.11, uses a push-pull locking system. SC connectors are used for CATV, telephone connections, and network.
2. **Straight Tip Connector (ST):** ST connectors use bayonet locking and are valued for their high reliability. The ST connector is also shown in Figure 4.11.

3. **MT-RJ Connector:** The MT-RJ is a duplex connector, as shown in Figure 4.12. The size of a MT-RJ connector is equal to that of an RJ-45 connector.

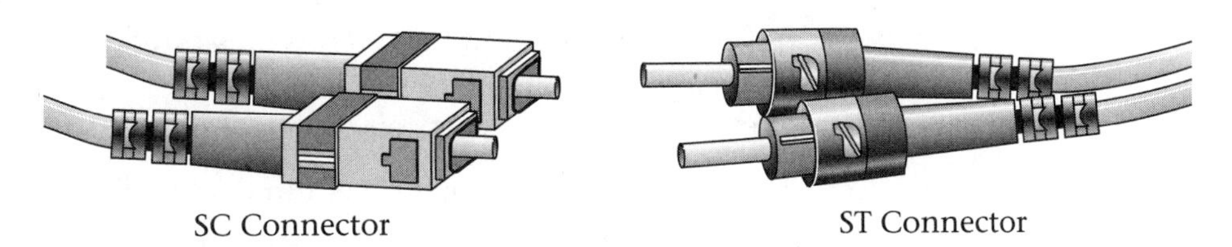


**FIGURE 4.11**
Fiber-optic ST and SC connectors

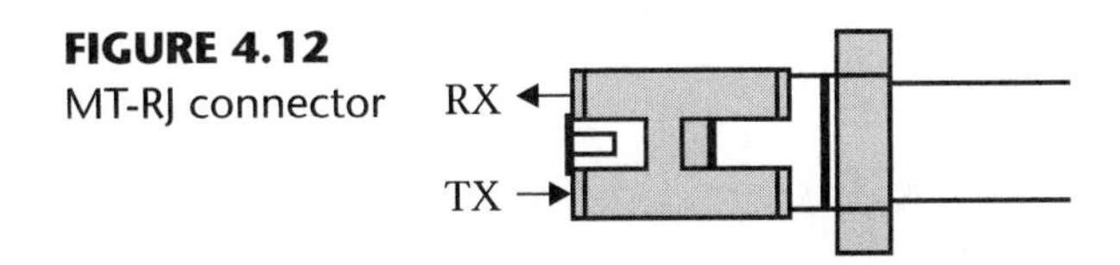


**FIGURE 4.12**
MT-RJ connector

## 4.3 Wireless Transmission

**Wireless transmission** does not use any transmission media, such as a conductor or optical cable, to transmit and receive information. Microwave, radio, infrared light, and laser are forms of wireless communication.

When electrons accelerate, they generate electromagnetic waves. Wireless transmission uses these electromagnetic waves. Table 4.3 shows the electromagnetic wave spectrum and its applications.

**TABLE 4.3** The Electromagnetic Frequency Spectrum and Its Applications

| Frequency Range | Name | Application |
|---|---|---|
| 3–30 KHz | Very Low Frequency (VLF) | Telephone |
| 30–300 KHz | Low Frequency (LF) | Radio Frequency for Navigation |
| 300–3000 KHz | Medium Frequency (MF) | AM Radio Frequency |
| 3–30 MHz | High Frequency (HF) | CB Radio, Shortwave Radio |
| 30–300 MHz | Very High Frequency (VHF) | TV and FM Radio |
| 300–3000 MHz | Ultra High Frequency (UHF) | TV, Military |
| 3–30 GHz | Super High Frequency (SHF) | Terrestrial and Satellite Microwave |
| 30–300 GHz | Extreme High Frequency (EHF) | Experimental |
| > 300 GHz | Infrared Light, Lasers | TV Remote Control and Laser Surgery |

## Microwave Transmission

**Microwave transmission** uses electromagnetic waves to transmit information through the atmosphere. The range of microwave frequency is between 2 GHz and 40 GHz. Since any electromagnetic wave with a frequency of approximately more than 100 MHz travels in a straight line, the type of transmission that microwaves provide is therefore referred to as **line-of-sight transmission.**

There are two types of microwave communication:

1. **Terrestrial:** A typical system for transmitting and receiving microwave signals is a parabolic antenna or dish, as shown in Figure 4.13. Terrestrial microwave is typically used for long distance telecommunications, and between buildings when wiring is impossible.

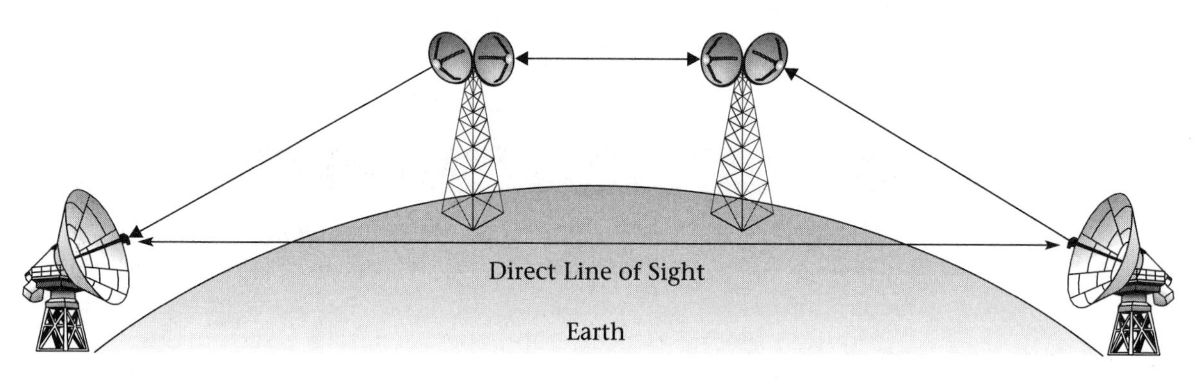


**FIGURE 4.13**
Terrestrial-based microwave system

2. **Satellite System:** A satellite system is used for communication between states and countries where it is not possible to use ground-based line of sight, as shown in Figure 4.14. The satellite is used to link two or more ground stations together (earth station). The satellite receives the transmitted signal from the earth station (up link) and retransmits the signal to another earth station (down link). Separate frequencies are used for up link and down link transmissions.

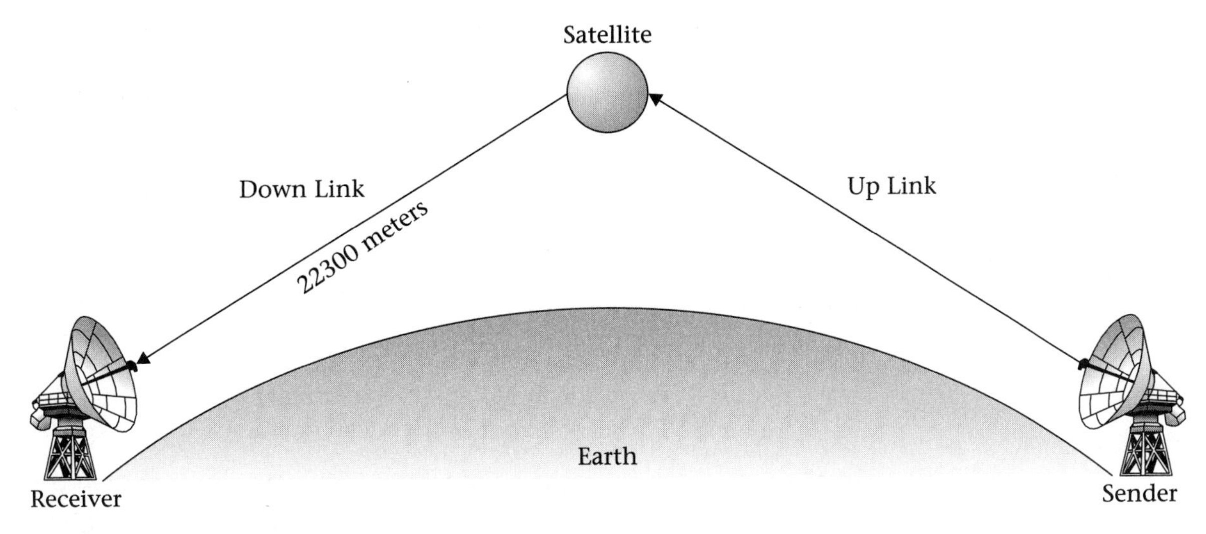


**FIGURE 4.14**
Satellite-based transmission system

## 4.4 Transmission Impairment

In a communication system, the transmitter sends information in the form of signals such as optical, electrical, or radio frequency to the receiver. The signals are sent on a communication channel. The receiver side of the signal has a different shape than the transmitter side. This difference can be caused by attenuation, distortion, or noise.

**Attenuation** When a signal travels from source to destination, the signal loses energy. The amount of lost energy depends on the type of the channel, the length of the channel, and the frequency of the signal, as shown in Figure 4.15. A longer channel will result in a higher loss of energy than a

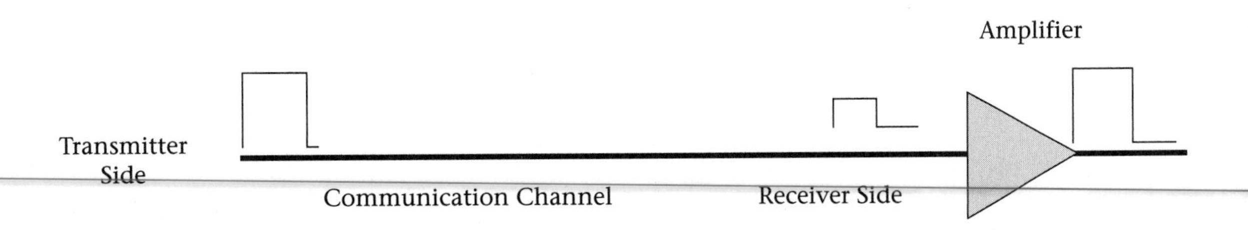


**FIGURE 4.15**
Attenuation and amplification of a signal as it travels through a communication channel

shorter channel. If a signal attenuates too much, then it becomes undetectable by the receiver.

To increase the length of a communication channel one must use an amplifier to strengthen the signal. The attenuation is measured in decibels (dB) and is defined in Equation 4.3.

$$Ap = 10\log_{10}\frac{Pt}{Pr} \tag{4.3}$$

where:

Ap = Power attenuation
Pt = Power of signal at the transmitter side
Pr = Power of signal at the receiver side

**Example 4.1.** A signal with the power of 500 mW is transmitted over a communication channel. At the receiver side the power of the signal is 50 mW. Calculate the power attenuation of the signal.

$$Ap = 10\log_{10}\frac{500}{50} = 10 \text{ dB}$$

Electrical signals are used for transmission of information over a conductive medium that acts as a communication channel. The electrical signal travels through the conductor. As a result of the resistance of the conductor the signal loses some voltage. This loss of voltage is called a *voltage drop*. Attenuation of electrical signal is calculated by Equation 4.4.

$$Pt = Vt * I, \quad \text{where } I = Vt/R \text{ then}$$
$$Pt = (Vt)^2/R$$
$$Pr = (Vr)^2/R$$

Substituting Pt with $(Vt)^2/R$ and Pr with $(Vr)^2/R$ then results in Equation 4.4.

$$Av = 10\log_{10}\frac{V_t^2}{V_r^2} \tag{4.4}$$

where:

Av = Voltage attenuation
Vt = Voltage of the signal at the transmitter side
Vr = Voltage of the signal at the receiver side (after transmission)

The attenuation of a cable is published by the cable manufacturer. Table 4.4 shows the attenuation of UTP Cat-5 and Cat-6 cables at different frequencies. It is notable that attenuation increases as frequency increases.

**TABLE 4.4** Attenuation of Cat-5 and Cat-6 Cables for Various Frequencies

| Frequency (MHz) | Cat-5 Attenuation (dB/100 m) | Cat-6 Attenuation (dB/100 m) |
|---|---|---|
| 1 MHz | 2 | 1.9 |
| 10 | 6.5 | 5.6 |
| 20 | 9.3 | 8.0 |
| 100 | 22 | 18.7 |
| 250 | NA | 31.0 |

The network designer uses attenuation data to find the maximum cable length that can be used without using a repeater.

**Example 4.2.** Find the maximum length of a Cat-6 cable that transmits a signal with 250 MHz, assuming the voltage of the signal at the transmitter side is 5000 mV and at the receiver side is 200 mV.

$$Av = 10\log_{10}\frac{(5000)^2}{(200)^2} = 27.8\,\text{dB}$$

From Table 4.4, attenuation for 100 meters of Cat-6 at frequency 250 MHz is 31 dB; thus the maximum cable length is 89.6 meters. If the transmitter transmits information at 20 MHz using Cat-6 , then the maximum length of the cable would be 347.4 meters.

## 4.5 Bandwidth, Latency, Throughput, and Channel Capacity

**Bandwidth**

In general, **bandwidth** is the maximum rate of data transfer over a communication link. It is categorized by the type of signal over the communication link.

***Analog bandwidth.*** Analog bandwidth is the difference between the highest and the lowest frequencies in a communication channel. For example, the highest frequency of the human voice is 3300 Hz and the lowest frequency of the human voice is 300 Hz. Therefore, the bandwidth of the human voice is simply:

$$3300\text{ Hz} - 300\text{ Hz} = 3000\text{ Hz}$$

***Digital bandwidth.*** The bandwidth of a digital link is the maximum number of bits per second that can be transmitted over the communication link. For example, the bandwidth of a T1 link is 1.54 Mbps, meaning that it can transfer up to 1.54 million bits per second. Ethernet's bandwidth is 10 Mbps meaning that each bit takes 0.1 μs to get transmitted.

**Latency (Delay)**

**Latency** defines the time it takes to transmit one packet (unit of information) from source to destination. Latency delay consists of propagation delay, transmission time, and buffering time. Latency is defined in Equation 4.5. A two-way latency is called round trip time (RTT).

$$\text{Latency} = T_x + T_p + T_b \tag{4.5}$$

where:

$T_x$ = Transmission time
$T_p$ = Propagation delay
$T_b$ = Buffering time

***Transmission time.*** Transmission time is the time that it takes to transmit one bit on a transmission channel. If the data rate of a link is 1000 bits per second, then each bit takes 0.001 seconds to transmit. Transmission time is defined by Equation 4.6:

$$\text{Transmission Time } T_x = \frac{\text{Packet Size (bits)}}{\text{Bandwidth (bps)}} \tag{4.6}$$

**Example 4.3.** Find the transmission time for transferring 1500 bytes using a communication link with a data rate of 10 Mbps.

$$T_x = \frac{1500 \text{ byte} \times \left(8 \frac{\text{bit}}{\text{byte}}\right)}{10 \times 10^6 \frac{\text{bit}}{\text{sec}}} = 0.0012 \text{ sec}$$

***Propagation delay.*** Propagation delay (or propagation time) is the time that it takes the signal to travel from source to destination. Propagation delay is defined by Equation 4.7:

$$T_p = \frac{\text{Length of the Communication Link (meters)}}{\text{Speed of Light in the Medium}} \tag{4.7}$$

where:

Speed of light = $3 \times 10^8$ m/s in vacuum and less in wire and fiber medium

Electrical and optical signals travel considerably less than almost at the speed of light and are generally taken as $2.3 \times 10^8$ *m/s* and $2 \times 10^8$ *m/s*, respectively.

**Example 4.4.** Find the propagation time for transferring 100 bytes over 200 Km of fiber-optic cable.

$$T_p = \frac{2 \times 10^5\, m}{2 \times 10^8 \left(\frac{m}{s}\right)} = 0.001 \text{ sec}$$

***Buffering time.*** A transmitted packet might be stored in several locations before reaching its final destination. The time a packet spends in a temporary location (known as a *buffer*) is called *buffering time* or *queue time*.

**Throughput** The throughput of a communication channel is defined by the number of bits transmitted over a communication link and is shown by Equation 4.8:

$$\text{Throughput} = \frac{\text{Transfer Size}}{\text{Latency}} \qquad (4.8)$$

**Example 4.5.** Calculate the transmission time and the throughput of a communication link for a user to download 1500 bytes of information from a server. The user's computer is connected to the server by a modem with the data rate of 50 Kbps, and the distance between the two computers is 4000 Km. Assume there is no buffering delay.

$$T_x = \frac{1500 \text{ byte} \times \left(8 \frac{\text{bit}}{\text{byte}}\right)}{50{,}000 \text{ bps}} = 0.24 \text{ sec}$$

$$T_p = \frac{4 \times 10^6 \text{ m}}{2.3 \times 10^8 \left(\frac{\text{m}}{\text{s}}\right)} = 0.018 \text{ sec}$$

$$\text{Latency} = 0.24 + 0.018 = 0.258 \text{ sec}$$

$$\text{Throughput} = \frac{1500 \text{ byte} \times \left(8 \frac{\text{bit}}{\text{byte}}\right)}{0.258 \text{ sec}} = 46.51 \text{ kbps}$$

**Channel Capacity** The bandwidth of a channel **(channel capacity)** is defined as the range of frequencies that pass through the channel. **Nyquist's theorem** is defined the maximum data rate ($\frac{\text{bits}}{\text{sec}}$) in a noiseless channel and is represented mathematically by Equation 4.9:

$$\text{Max Data Rate (MDR)} = 2W*\log_2 N \text{ bps} \tag{4.9}$$

where:

W = Bandwidth of channel
N = Number of signal levels or voltage levels

**Example 4.6.** Find the maximum data rate of a channel with a bandwidth of 4000 Hz transmitting two voltages (e.g., binary = two levels, 0 and 1).

$$\text{MDR} = 2*4000*\log_2 2$$
$$\text{MDR} = 8000 \text{ bps}$$

Equation 4.9 is valid only when using a noiseless channel. Noise affects the data rate of a channel. Figure 4.16 shows transmission of a digital signal through a noisy channel.

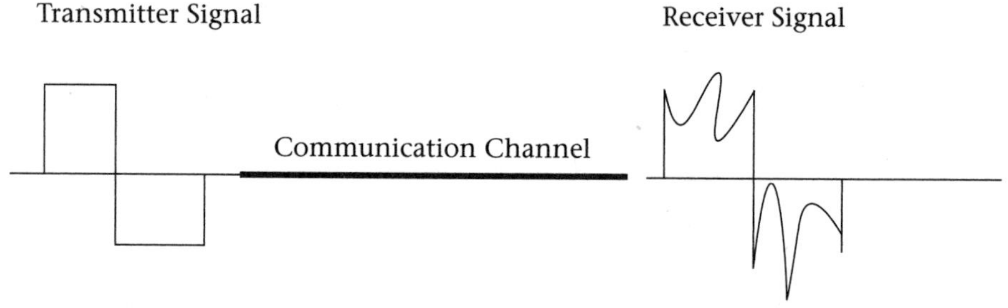


**FIGURE 4.16**
Transmission of a digital signal through a noisy channel

Figure 4.16 indicates that the presence of noise in a communication channel may cause distortion of the incoming signal such that the output signal is no longer a square wave. When the amplitude of the noise is larger than expected, total destruction of the original signal occurs, which then affects the data rate of the communication channel. The capacity of a channel may be obtained using Shannon's theorem and is represented mathematically by Equation 4.10:

$$\text{Max Data Rate (MDR)} = W*\log_2\left(1 + \frac{S}{N}\right) \text{ bps} \tag{4.10}$$

where:

$\frac{S}{N}$ = the signal-to-noise ratio (SNR), ratio of average signal power to average noise power at the receiver, which is usually given in decibels (dB) as shown in Equation 4.11.

**Decibel** is defined as:

$$\text{Decibels} = 10*\log_{10}\left(\frac{S}{N}\right) \tag{4.11}$$

Example: If $\left(\frac{S}{N}\right) = 10$, then (SNR) in decibels is:

$$\begin{aligned} \text{SNR} &= 10*\log_{10} 10 \\ &= 10 \text{ dB} \end{aligned}$$

**Example 4.7.** Find the maximum data rate of a channel with a bandwidth of 4000 Hz and a signal-to-noise ratio of 30 *dB*.

$$30 \text{ dB} = 10*\log_{10}\frac{S}{N}$$

Therefore,

$$\frac{S}{N} = 10^3$$

Subsequently,

$$\begin{aligned} \text{MDR} &= \text{W}*\log_2 (1 + 10^3) \\ &= 4000*9.967 \text{ bps} \end{aligned}$$

## Summary

- Transmission media are used to connect computers.
- The types of communication media are conductors, optical cable, and wireless.
- The types of conductors used for networking are, unshielded twisted-pair (UTP) cable, shielded twisted-pair (STP) cable, and coaxial cable.
- UTP cable contains several pairs of wires and is divided by EIA into categories such as Cat-1, Cat-5, and Cat-6.
- Unshielded-twisted pair (UTP) uses an RJ-45 connector.
- Coaxial cable is used for transmitting high-speed information over relatively long distances.
- Fiber-optic cable transfers information in the form of light.
- Light-emitting diodes (LED) and laser diodes are used to convert a digital signal to an optical signal for transmitting information over optical cable.
- There are two types of optical cable: single-mode fiber (SMF) and multimode fiber (MMF).

- Single-mode fiber (SMF) uses only one ray of the light source.
- In MMF step index fiber, the index of refraction is the same across the core of the fiber and a ray of light makes a sharp refraction at the core cladding boundary.
- In MMF grade index fiber, the index of refraction changes from maximum at the center to minimum near the edge of the core, causing the light to bend in a curved shape.
- Wireless transmission uses microwave, radio, or infrared light signals to transmit information.
- Very high frequency (VHF) and ultra high frequency (UHF) use line-of-sight to transmit information.
- Microwave uses radio waves in the range of 1-GHz to 23-GHz frequency.
- There are two types of microwave communications: terrestrial and satellite systems.
- Channel bandwidth is defined as the range of frequencies that passes through the channel.
- Analog bandwidth is the difference between the highest and the lowest frequencies in a communication channel.
- Digital bandwidth is the maximum number of bits per second that can be transmitted over a communication link.
- Latency (delay) defines the time it takes to transmit one packet (unit of information) from source to destination.
- Attenuation is the loss of energy in a signal after transmission.

## Key Terms

Attenuation
Bandwidth
Channel Capacity
Coaxial Cable
Conductive Media
Decibel
Fiber-Optic Cable
Injected-Laser Diode
Latency
Light-Emitting Diode (LED)
Line-of-Sight Transmission
Microwave Transmission
Model Bandwidth
Multimode Fiber (MMF)
Multimode Graded Index Fiber
Multimode Step Index Fiber
MT-RJ Connector
Nyquist's Theorem
Propagation Delay
RJ-45 and RJ-11 Connectors

Shielded Twisted-Pair (STP) Cable
Single-Mode Fiber (SMF)
Terrestrial
Transmission Time
Unshielded Twisted-Pair (UTP) Cable
Wireless Transmission
Wavelength

## Review Questions

### • Multiple Choice Questions

1. A ______ cable is the least expensive transmission media.
   a. UTP
   b. STP
   c. fiber-optic
   d. coaxial
2. ______ cables are used to transmit high-speed and analog signals.
   a. UTP and coaxial
   b. STP and UTP
   c. Coaxial and fiber-optic
   d. Fiber-optic and STP
3. ______ connector(s) is/are used in fiber-optic cable.
   a. SC and BNC
   b. ST and SC
   c. RJ-11 and ST
   d. BNC and SC
4. ______ transmission does not use any transmission medium.
   a. WAN
   b. LAN
   c. Wireless
   d. Internet
5. Wireless transmission uses ______ waves.
   a. optical true
   b. electrical
   c. electromagnetic
   d. digital
6. Of following UTP cables, ______ is suitable for a data rate of 100 Mbps.
   a. Cat-2
   b. Cat-4
   c. Cat-3
   d. Cat-5
7. Of the following transmission media, ______ are used for high-speed transmission.
   a. coaxial cable and fiber-optic cable
   b. fiber-optic cable and UTP CAT-2 cable
   c. microwave and fiber-optic cable
   d. UTP CAT-2 cable and microwave cable
8. The type of fiber-optic cable used for long distance transmission is ______.
   a. multimode graded index
   b. single-mode
   c. multimode step index
   d. STP

9. The speed of the light in a vacuum is ______.
   a. $3*10^8$ m/s
   b. $5*10^8$ m/s
   c. $3*10^2$ m/s
   d. $6*10^{20}$ m/s
10. The maximum length of a fiber cable with modal bandwidth of 1000 MHz*Km in order to transmit information with a 200-MHz speed is ______.
    a. 1 Km
    b. 10 Km
    c. 5 Km
    d. 20 Km

## • Short Answer Questions

1. List the major communications media.
2. What does UTP stand for?
3. What does STP stand for?
4. Which organization defines standards for cables?
5. What is the performance of Cat-5 UTP cable?
6. What type of light source is used for fiber-optic cable?
7. What are the advantages of fiber-optic cable over conductive cable?
8. What are the types of fiber-optic cables?
9. What does SMF stand for?
10. What does MMF stand for?
11. What is the application of single-mode fiber?
12. What is the application of multimode fiber?
13. What are the types of microwave communication?
14. What is the range of microwave frequencies?
15. What are the signal sources for optical cable?
16. What are the advantages of optical cable over coaxial cable?
17. What are the advantages of STP cable over UTP cable?
18. Explain digital bandwidth.
19. Explain latency and the causes of latency.
20. Find the transmission time of a packet of 1500 bytes transmitted over a 100-Mbps channel.

21. For the following case, what should the transmitter voltage be for 100 meters of Cat-5 cable with a receiver voltage of 500 mV?
    a. transmitted signal at 20 MHz
    b. transmitted signal at 100 MHz
22. 2000 bytes of data are to be transferred between a server and a host computer, which are connected via a 1000-meter Cat-5 cable with a transmission rate of 10 Mbps. Calculate the following:
    a. transmission time
    b. propagation delay
    c. round trip time
23. A packet of 100 bytes is transmitted over a 100-Km cable with bandwidth of 100 Mbps. Calculate the following:
    a. propagation delay of the link
    b. transmission time
    c. latency of the packet
    d. RTT
24. What is the bandwidth of a 20-Km link for transmitting 500 bytes of information such that the propagation delay is equal to transmission delay?
25. Find the time that it takes to transmit 1000-kByte file from a server that is located 4000 Km away from a host computer. Assume you are using a modem with a data rate of 52 Kbps and the size of each packet is 1000 bytes.
26. Calculate the latency for transmitting 1500 bytes of data over the following links:
    a. 100-meter copper with a bandwidth of 10 Mbps
    b. 4000-meter optical fiber with a bandwidth of 10 Mbps
27. 500 bytes of data are transmitted over 200 Km of fiber-optic cable.
    a. Find the data rate such that the transmission time becomes equal to the propagation time.
    b. What is the throughput of this communication link?
28. Find the maximum data rate of a communication link with a bandwith of 3000 Hz using eight signal levels.
29. Find the bandwidth of a communication channel in order to transfer the data at a rate of 100 Mbps; assume $\frac{S}{N}$ ratio is (50 dB).

CHAPTER 5

# Multiplexer and Switching Concepts

OUTLINE



OBJECTIVES

After completing this chapter, you should be able to:

- Explain the operation of multiplexers and demultiplexers
- List the types of multiplexers and their operations
- Discuss how a telephone system operates
- Tell how pulse code modulation converts voice to digital signals
- Explain T1 link technology, including how to calculate its data rate
- Discuss switching concepts
- List the types of switching methods

INTRODUCTION

Because long-distance transmission lines are expensive, a method to allow several devices to share one transmission line is necessary to defray the cost of wiring. The multiplexer provides a solution to this problem. Figure 5.1 shows four terminals sharing one transmission line to send information to the host computer by using a multiplexer instead of four transmission lines.

**FIGURE 5.1**
Application of multiplexer

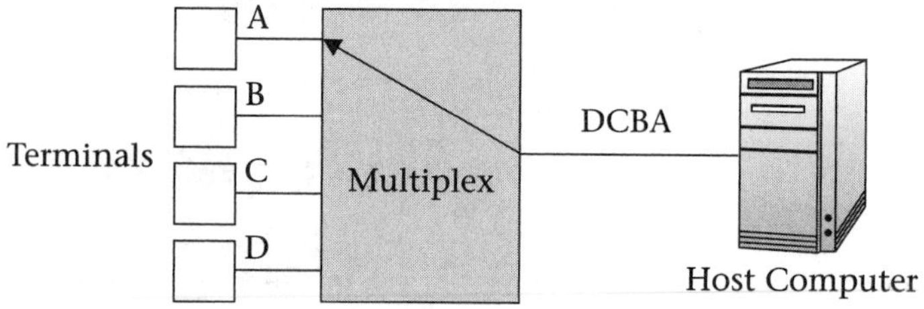

Terminals
A
B
C
D
Multiplex
DCBA
Host Computer

A **multiplexer** is a device that combines several low-speed data channels and transmits all of the data on a single high-speed channel. A common application of multiplexing is long-distance communication using high-speed point-to-point links for transferring large quantities of voice signals and data between users. Figure 5.2 shows the basic architecture of a multiplexer. A multiplexer that has N inputs and one output is called an N-to-1 multiplexer. Figure 5.2 shows a 4-to-1 multiplexer. The internal switch selects one input line at a time and transfers the input to the output. When the switch is in position A, it transfers input A to the output, and then the switch moves to position B and transfers input B to the output. This method continues until the switch moves to position D and transfers input D to the output. After this function is completed, the switch starts over from A input.

**FIGURE 5.2**
4-to-1 multiplexer

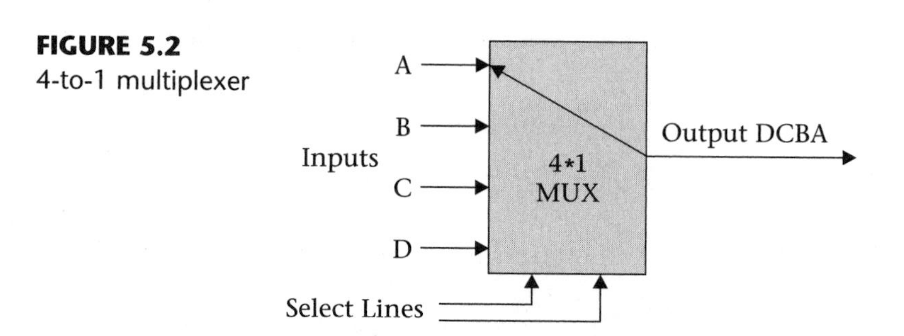


The opposite of a multiplexer is a **demultiplexer (DMUX)**, shown in Figure 5.3. The switch moves to send each input to the appropriate output. A DMUX has one input and N outputs and is called a *1-to-N demultiplexer.* When the switch is in position 0, it transfers A to output port 0, and then moves to output port 1 and transfers B to this port. This process continues until the switch moves to output port 3 and transfers D to port 3. One cycle is complete and the transfer of data starts over from port 0.

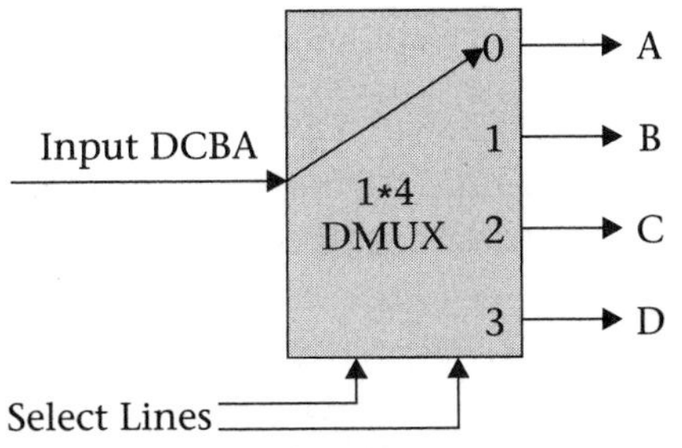

**FIGURE 5.3**
1-to-4 demultiplexer

## 5.1 Types of Multiplexers

Multiplexers can be categorized into the following types and each type has a specific application:

1. Time Division Multiplexing (TDM)
2. Frequency Division Multiplexing (FDM)
3. Statistical Packet Multiplexing (SPM)
4. Fast Packet Multiplexing (FPM)
5. Code Division Multiplexing (CDM)
6. Wavelength Division Multiplexing (WDM)

**Time Division Multiplexing (TDM)**

In **Time Division Multiplexing (TDM)**, multiple digital signals can be carried on a single transmission path by interleaving each input of the multiplexer. A TDM operates with a preassigned equal time slot to each input. It divides the bandwidth of the multiplexer's output into fixed segments. Each input to the MUX is given a fixed unit of time. First, information from input one is transmitted, then from input number two, and so on in a regular sequence, as shown in Figure 5.4.

One disadvantage of TDM is that the bandwidth of any one input is not available to other inputs when that input to the TDM is inactive. In Figure 5.4, the inputs to B and D are inactive at the times $t_1$ and $t_2$, and the outputs in frames #2 and #3 have idle times. Another disadvantage of

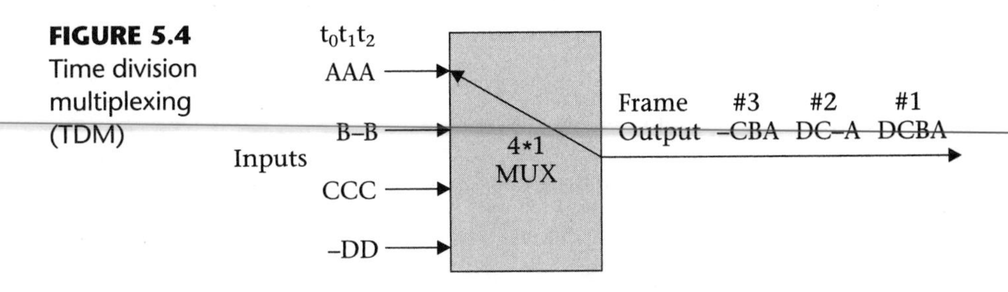


**FIGURE 5.4**
Time division multiplexing (TDM)

TDM is that it is not able to change the bandwidth of the input dynamically, and therefore cannot transport a combination of voice, fax, and data.

**Frequency Division Multiplexing (FDM)**

**Frequency Division Multiplexing (FDM)** divides the bandwidth of a transmission line into channels, and each channel can transmit specific information. Figure 5.5 shows the multiplexing of several TV channels using FDM. The bandwidth of coaxial cable is about 500 MHz and it can carry 80 TV channels. Each TV channel is assigned a different frequency, each using 6 MHz of bandwidth. Therefore, FDM combines several signals for transmission on a single transmission line.

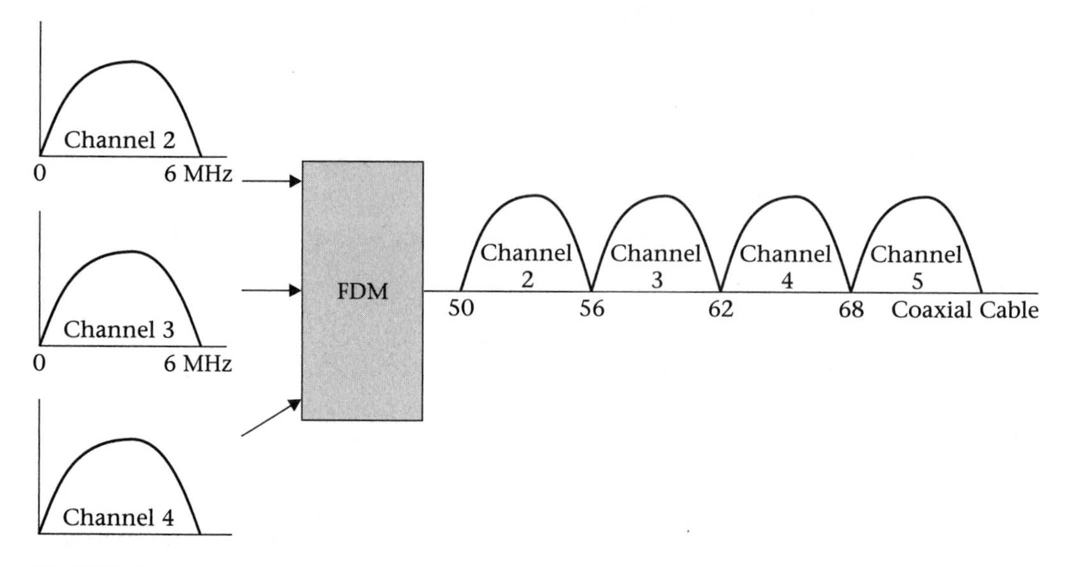


**FIGURE 5.5**
Frequency division multiplexing (FDM)

**Statistical Packet Multiplexing (SPM)**

**Statistical Packet Multiplexing (SPM)** dynamically allocates bandwidth to the active input channels, resulting in very efficient bandwidth utilization. In SPM, an idle channel does not receive any time allocation, as shown in Figure 5.6. SPM uses a store-and-forward mechanism in order to detect and correct any error from incoming packets.

**FIGURE 5.6**
Statistical packet multiplexing (SPM)

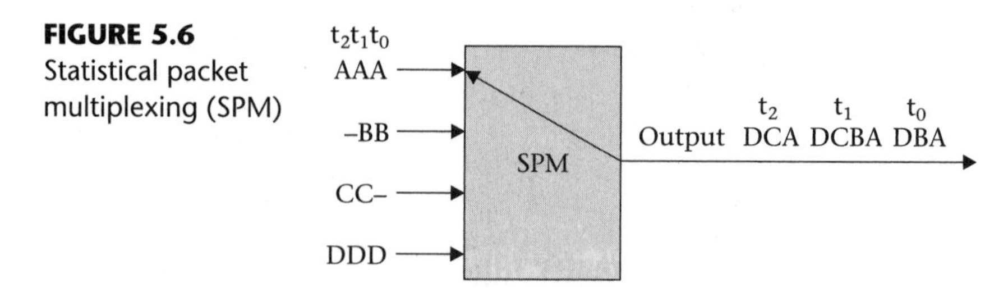

**Fast Packet Multiplexing (FPM)**

**Fast Packet Multiplexing (FPM)** uses the same method as SPM and has the ability to assign maximum bandwidth to any input needed. FPM does not use a store-and-forward mechanism, and therefore cannot perform error detection and correction. FPM will forward a packet before it has been completely received by the multiplexer.

**Code Division Multiplexing (CDM)**

In Time Division Multiplexing (TDM) each end user is allocated a time slot for transmission. For instance, if 10 users are connected to a TDM and bandwidth of transmission link is 10 Mbps, then each user is capable of transmitting at the rate of 1 Mbps only. One disadvantage of TDM is that each user has to wait for its turn to transmit its information. **Code Division Multiplexing (CDM)** is similar to TDM but allows all users to transmit simultaneously.

***CDM operation.*** In CDM each bit time is divided into multiple bits that are called *chip bits*. This is done by multiplying logic 1 with chip sequence or assigning a chip bit to each node to represent logic 1. Table 5.1 shows the chip bits that are assigned to each node to represent logic 1 and the complement of the chip bits represents logic zero. Chip bits can be represented by a bipolar value so that +1 represents logic 1 and −1 represents logic zero.

**TABLE 5.1** Chip Bits for Node A and B

| End Node | Chip Bits for Logic 1 | Bipolar Representation of Chip Bits |
|---|---|---|
| A | 0101 | −1+1−1+1 |
| B | 1100 | +1+1−1−1 |

***Characteristics of chip bits.*** In general, the chip bits for $A$ can be represented by $A = (A_4\ A_3\ A_2\ A_1)$ and the chip sequence for $B$ can be represented by $B = B_4\ B_3\ B_2\ B_1$. One property of chip bits is that the inner product of two different chip bits is zero and inner product of the two identical chip bits is one.

The inner product $A$ and $B$ is represented by $A \cdot B$, and is defined by Equation 5.1:

$$A \cdot B = \frac{1}{m}\sum_{i=1}^{4} A_i B_i = \frac{1}{4}(A_1B_1 + A_2B_2 + A_3B_3 + A_4B_4) \tag{5.1}$$

Therefore,

$$A \cdot B = \frac{1}{4}(-1 + 1 - 1 + 1)(+1 + 1 - 1 - 1)$$
$$= \frac{1}{4}(-1 + 1 + 1 - 1) = 0.$$

And the inner product $A$ with itself $A \cdot A$ is:

$$A \cdot A = \frac{1}{4}(-1 + 1 - 1 + 1)(-1 + 1 - 1 + 1)$$
$$= \frac{1}{4}(+1 + 1 + 1 + 1) = 1$$

***CDM architecture.*** Figure 5.7 shows the general architecture of CDM, with three inputs and one output. The nodes $A$, $B$, and $C$ are the inputs and the chip bits for each input are 4 bits. The chip bits of the input are added and the result is transmitted over the communication link. At the receiver side, the inner product of the sum of the chip bits and the chip sequence of the input node result in the data bits of the input node.

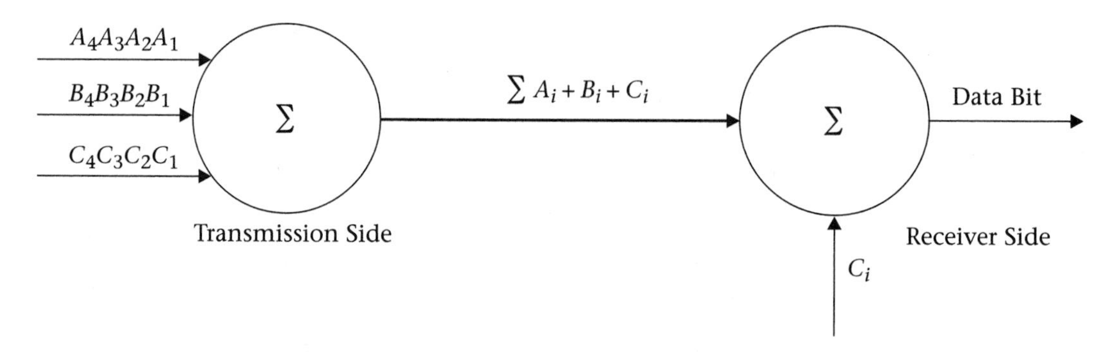


**FIGURE 5.7**
General architecture of code division multiplexing (CDM)

**Example 5.1.** Table 5.2 shows data to be transmitted by the nodes A, B, and C with their chip sequence.

a. Find the output of CDM.
b. Find the data bit that is transmitted by the node A at the receiver side.

**TABLE 5.2** Chip Bits and Data Bits for Nodes A, B, and C

| Node | Chip Sequence | Data to Be Transmitted |
|---|---|---|
| A | −1−1−1−1 | 101 |
| B | −1+1−1+1 | 110 |
| C | +1+1−1−1 | 001 |

The chip sequence for each data node is represented by Table 5.3.

**TABLE 5.3** Chip Bits for Data of Nodes A, B, and C

| | | | |
|---|---|---|---|
| Node A | −1−1−1−1 = (1) | +1+1+1+1 = (0) | −1−1−1−1 = (1) |
| Node B | −1+1−1+1 = (1) | −1+1−1+1 = (1) | +1−1+1−1 = (0) |
| Node C | −1−1+1−+1 = (0) | −1−1+1+1 = (0) | +1+1−1−1 = (1) |
| Sum | −3−1−−1+1 | −1+1+1+3 | +1−1−1−3 |

The sum of data is then transmitted to the receiver side. At the receiver side, the receiver uses the chip sequence of a specific node to recover the original data by using the inner product. For example, in order to recover user A's data, the inner product of A and sum is shown in Table 5.4.

**TABLE 5.4** Inner Product of Sum and A Input

| | | | |
|---|---|---|---|
| Sum | −3−1−−1+1 | −1+1+1+3 | +1−1−1−3 |
| Node A | −1−1−1−1 | −1−1−1−1 | −1−1−1−1 |
| Inner Product | (+3−1+1−1)/4 = +1 | (+1−1−1−3)/4 = −1 | (−1+1+1+3)/4 = +1 |

In Table 5.4, +1 represents 1 and −1 represents 0.

**Wavelength Division Multiplexing (WDM)**

**Wavelength Division Multiplexing (WDM)** is used to transmit multiple rays of light with different wavelengths in one optical cable in order to increase the capacity of the optical cable, rather than using multiple optical cables. The concept of wave division multiplexing is similar to frequency division multiplexing. Figure 5.8 shows two optical rays with different wavelengths multiplexed and transmitted over an optical fiber cable.

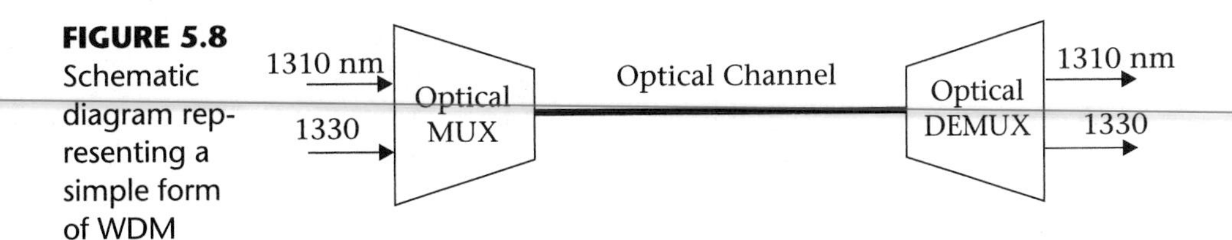


**FIGURE 5.8** Schematic diagram representing a simple form of WDM

***Components of a WDM.*** The components of a WDM are the optical transponder, optical multiplexer, optical amplifier, and optical demultiplexer.

***Optical transponder.*** The function of the optical **transponder** is to change the incoming ray's wavelength to another wavelength. Figure 5.9 shows the block diagram of an optical transponder.

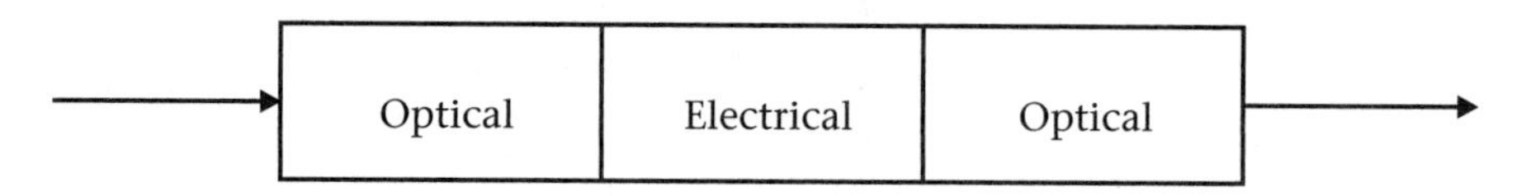


**FIGURE 5.9**
Block diagram of an optical transponder

***Optical multiplexer.*** An optical multiplexer is a device that combines multiple optical wavelengths (having different wavelength) for transmission over a single fiber optics cable.

***Optical amplifier.*** An optical amplifier is a device that amplifies an optical signal without converting it to an electrical signal.

***Optical demultiplexer.*** An optical demultiplexer is a device that separates multiple optical rays from fiber-optic cable.

There are two types of WDM: Dense Wavelength Division Multiplexing (DWDM) and Coarse Wavelength Division Multiplexing (CWDM)

***Dense Wavelength Division Multiplexing (DWDM).*** In DWDM, the wavelengths of the optical signals are close together. Current DWDM can transmit 60 to 80 wavelengths per channel with the wavelength spacing about 0.8 nm.

***Coarse Wavelength Division Multiplexing (CWDM).*** CWDM can transmit 4 to 8 wavelengths of optical signals, and wavelengths are spaced 20 nm apart from each other. The ITU specifies 18 channels using wavelength from 1270–1610 nm for CWDM, but some channels are not usable due to high signal attenuation. Figure 5.10 shows three optical rays having the same wavelengths connected to three optical transponders. The optical transponders change the wavelength of the incoming signals and are connected to the MUX. The outputs of the DMUX are connected to three optical transponders in order to convert the signal wavelength back to their original values.

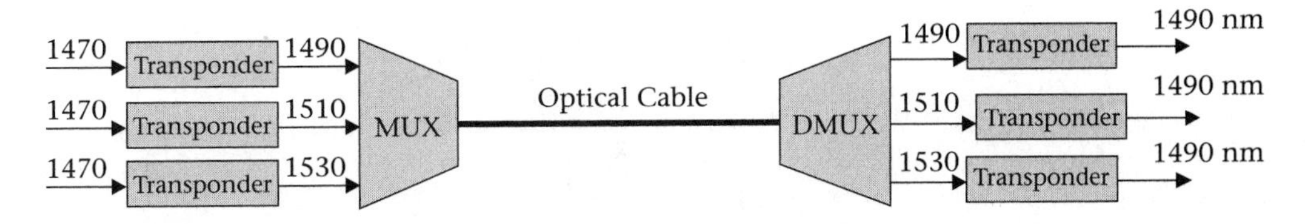


**FIGURE 5.10**
Coarse wavelength division multiplexing (CWDM)

## 5.2 Telephone System Operation

The wired telephone system transmits information in analog form from telephone set to the central office (CO). At the CO, the analog signal is converted to a digital signal, which is transferred to the next central office, as shown in Figure 5.11. This digital signal is then converted to an analog signal and transmitted to the user. The method of conversion from analog to digital is called **Pulse Code Modulation (PCM)**.

**FIGURE 5.11**
Telephone system architecture

## 5.3 Digitizing Voice

Voice is an analog signal. In the central office of the telephone company it is digitized by a device called a **codec** (coder-decoder). The function of a codec is to digitize the voice signal and convert an already digitized signal to analog. According to the Nyquist theorem, in order to convert an analog signal into a digital signal, the analog signal must be sampled at least the rate of two times the highest frequency of the analog signal. The voice signal is sampled at 8000 samples per second because human speech is below 4000 Hz as shown in Figure 5.12. This method is called **Pulse Amplitude Modulation (PAM)**.

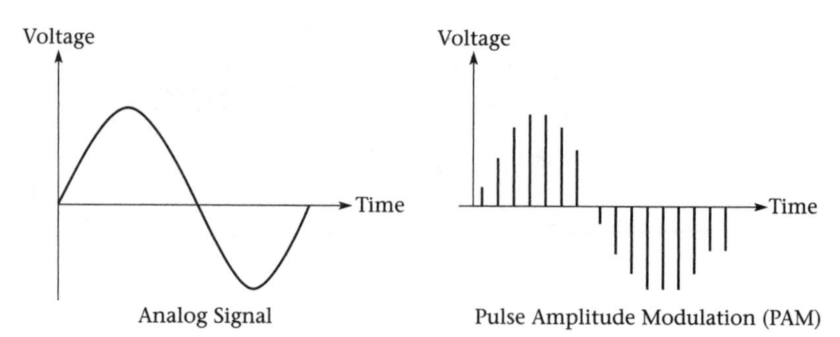


**FIGURE 5.12**
Analog signal and pulse amplitude modulation (PAM)

Each PAM sample is represented by eight bits. In Figure 5.13 it is represented by four bits. Remember, this method of converting voice to digital signal is called *pulse code modulation (PCM)*. Since voice is digitized at the rate of 8000 samples per second and each sample is represented by 8 bits, the data rate of the human voice is 8000 * 8 = 64 kbps.

**FIGURE 5.13**
Binary value for each PAM

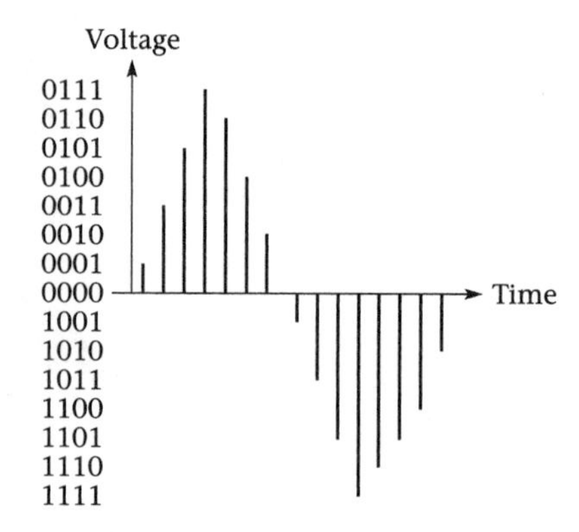


## 5.4 T1 Links

Long-distance carriers use TDM to transmit voice signals over high-speed links. One of the applications of TDM is the T1 link. A **T1 link** carries a level-1 digital signal (DS-1). A DS-1 is generated by multiplexing 24 voice digital signals (digital signal level-0 or DS-0), as shown in Figure 5.14. Pulse code modulation (PCM) is used to convert each analog signal to a digital signal. Each frame is made of 24 * 8 bits = 192 bits, with one extra bit added to separate each frame, making each frame 193 bits. Each frame represents 1/8000th of a second. Therefore, the data rate of a T1 link is 193 * 8000 = 1.544 Mbps.

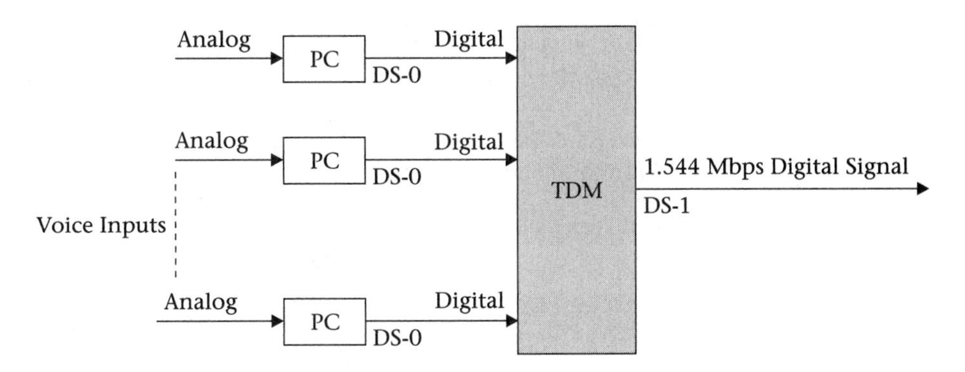


**FIGURE 5.14**
Architecture of T1 link

Table 5.5 shows TDM carrier standards for North America. Look at the table and you will see that aq DS-2 can carry 96 voice channels with 168 Kbps overhead; therefore the data rate for DS-2 is 6.312 Mbps (96 * 64 Kbps + 168 Kbps overhead). Figure 5.15 shows the DS-1 frame format; the 1-bit gap is used to separate each frame.

**TABLE 5.5** TDM Carrier Standards for North America

| Frame Format | Line | Number of Voice Channels | Data Rates (Mbps) |
|---|---|---|---|
| DS-1 | T1 | 24 | 1.544 |
| DS-1C | T-1C | 48 | 3.152 |
| DS-2 | T2 | 96 | 6.312 |
| DS-3 | T3 | 672 | 44.736 |
| DS-4 | T4 | 4032 | 274.176 |

| 1 bit | Byte #24 | Byte #23 | | Byte #2 | Byte #1 |
|---|---|---|---|---|---|

**FIGURE 5.15**
DS-1 frame format

## 5.5 Switching Concepts

A communication network that has more than two computers must establish links between computers in order for them to be able to communicate with each other. One way to connect these computers is via a fully connected network or (mesh), as shown in Figure 5.16.

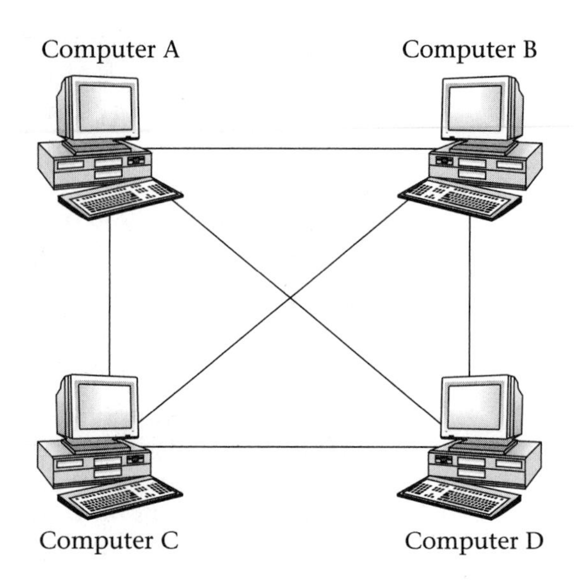


**FIGURE 5.16**
Fully connected network

The advantage of this method is that all stations can communicate with each other. The disadvantage is the large number of connections required, when the number of stations is greater than four. To overcome this disadvantage, a device called a *switch* is used to connect stations, as shown in Figure 5.17.

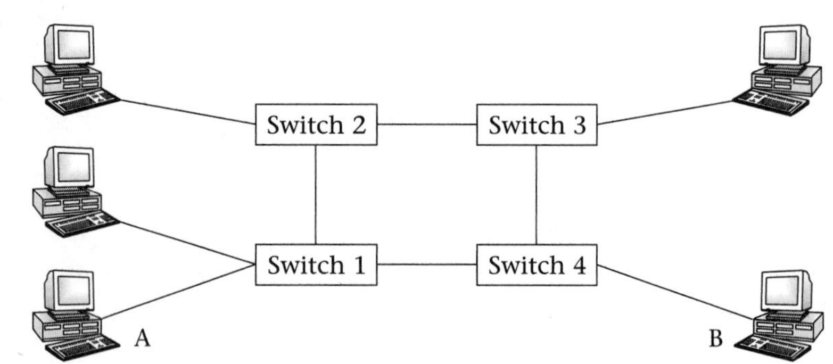


**FIGURE 5.17**
Stations connected by switches

The following types of switching are used in networking:

- circuit switching
- message switching
- packet switching
- cell switching (covered in Chapter 22)

**Circuit Switching**

In **circuit switching,** also called a *connection-oriented circuit,* a physical connection must be established for the duration of the transmission (such as in a telephone system). The application of circuit switching is for real-time communications. By dialing a telephone system, a connection is established and then communication begins, disconnecting at the end of the communication. Figure 5.18 shows circuit switching with multiple stations.

**FIGURE 5.18** Circuit switching

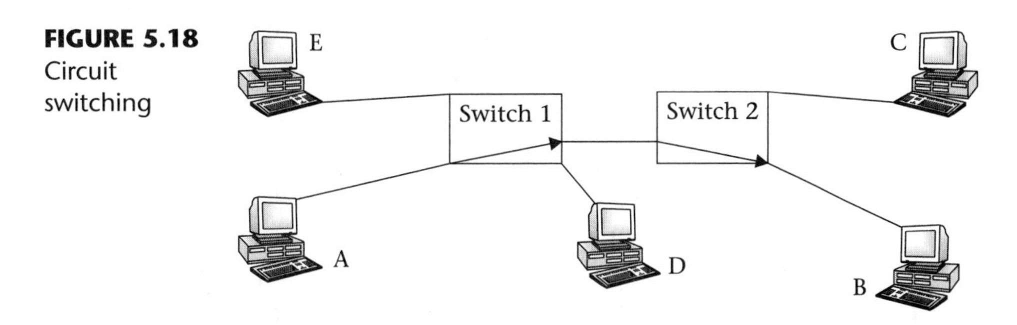


***Advantages of circuit switching.*** Circuit switching is used for real-time communication. There is no delay or congestion in the communication link because a physical connection exits between the source and the destination.

***Disadvantages of circuit switching.*** In circuit switching, only two stations can use the communication link at the same time. Therefore, it is not cost effective. For example, in Figure 5.18, if stations A and B are communicating with each other, station D cannot communicate with station C. Station D must wait until A and B have finished their communication, and then D may start communicating with C. In addition, if two stations such as A and B want to make a connection with C at the same time, a contention will occur and both stations must wait.

**Message Switching**

In message switching, station A sends its message to the switch, the switch stores that message, and then forwards it to the destination. The disadvantage of message switching is that the switch needs to have a large buffer to store incoming messages from other links.

**Packet Switching**

Figure 5.19 shows a network with several switches. Source A has a message and wants to transfer it to destination B. Source A divides the message into packets and sends each packet, possibly by a different route. This process is known as **packet switching.** Each packet goes to the switch, which stores the packet and looks at the routing table inside the switch to find the next switch or destination. The packets may take different routes and be received at the destination out of order. To prevent mistakes in reassembling the packets, each packet is given a sequence number that will be used by the destination to put the packets back in order.

In Figure 5.19 the source divides the message into three packets: A, B, and C. The source transmits packet A to switch #1. Switch #1 stores packet A and looks at the congestion on all outgoing links, and finds that the link to switch #2 is less congested. Switch #1 then sends packet A to switch #2. Switch #2 stores the packet, finds out from its routing table that packet A must go to switch #4, and then switch #4 forwards packet A to switch #5. Packets B and C take different routes; therefore, the packets might be received out of order. The destination uses the sequence numbers of the packets to put them in the proper order. The Internet uses packet switching. This type of service is also called a *connectionless-oriented circuit.*

## Virtual Circuit

A **Virtual circuit** is a type of packet switching that operates on the same concept as packet switching, but the routing of the packets is specified before transmission. As seen in Figure 5.19, the source specifies the route, which is represented by dotted lines. Therefore all the packets from source A go via the dotted line. By using this method, all packets will be received at the destination in the proper order.

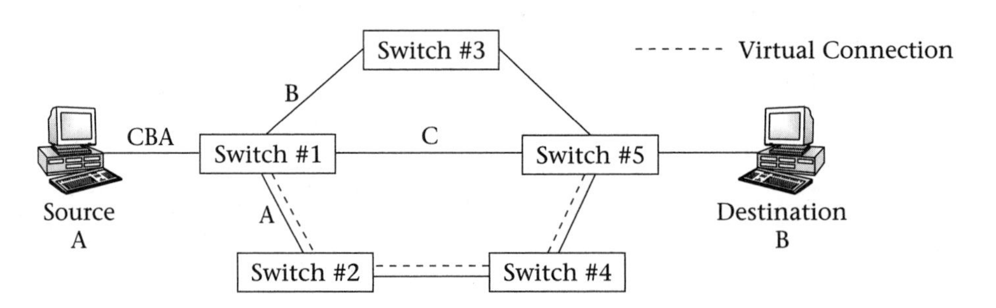


**FIGURE 5.19**
Packet switching and virtual circuit

## Summary

- A multiplexer is used to share media among several users.
- In Time Division Multiplexing (TDM), the users are assigned equal time to use the digital channel.
- Frequency Division Multiplexing (FDM) is used for analog transmission, with the bandwidth of the analog channel divided into smaller channels.
- In statistical TDM the bandwidth is dynamically allocated to active users.
- Code Division Multiplexing (CDM) allows all users to transmit simultaneously.
- Wavelength Division Multiplexing (WDM) is used for transmitting multiple rays of light with different wavelengths over one optical cable.
- The Pulse Code Modulation (PCM) method is used in the central switch to convert the human voice to a digital signal.
- The bandwidth of the human voice is below 4000 Hz and it is digitized at the rate of 64 kbps.
- A T1 link is a special digital transmission line that has 24 inputs (each input is 64 kbps on the voice channel) and one output, with a data rate of 1.544 Mbps.
- There are four types of switching used in networking: circuit switching, packet switching, virtual circuit, and cell switching.
- A message is divided into pieces. Each piece is called a *packet.*
- Packet switching treats each packet of a message separately.
- In circuit switching a physical connection must be established between the source and the destination before transmitting information.
- A virtual circuit is a type of packet switching. In a virtual circuit all packets of a message are transmitted in a specific path called a *virtual path.*

## Key Terms

Circuit Switching
Coarse Wavelength Division Multiplexing (CWDM)
Code Division Multiplexing (CDM)
Codec
Demultiplexer (DMUX)
Dense Wavelength Division Multiplexing (DWDM)
Fast Packet Multiplexing (FPM)
Frequency Division Multiplexing (FDM)
Multiplexer

Packet Switching
Pulse Amplitude Modulation (PAM)
Pulse Code Modulation (PCM)
Statistical Packet Multiplexing (SPM)
T1 Link
Time Division Multiplexing (TDM)
Transponder
Virtual Circuit
Wavelength Division Multiplexing (WDM)

## Review Questions

### • Multiple Choice Questions

1. Several devices can share one transmission line by using a ______.
   a. multiplexer
   b. demultiplexer
   c. BUS
   d. CPU
2. ______ divides the bandwidth of a transmission line into channels.
   a. TMD
   b. FMD
   c. SPM
   d. FSPM
3. Code division multiplexing allows the users to transmit ______.
   a. in an assigned time slot
   b. one at a time
   c. simultaneously
   d. none of the above
4. Wave division multiplexing is used for ______.
   a. optical signals
   b. analog signals
   c. digital signals
   d. radio frequency signals
5. The function of an optical transponder is to ______.
   a. change optical signals to electrical signals
   b. change the power of an optical signal
   c. change the wavelength of an optical signal
   d. change the electrical signal to optical
6. ______ dynamically allocates bandwidth to active inputs.
   a. TDM to FDM
   b. SPM to FPM
   c. FPM to TDM
   d. FDM to SPM
7. The channel bandwidth of a telephone system for human voice is ______.
   a. 4000 Hz
   b. 400 Hz
   c. 40 Hz
   d. 4 Hz
8. One of the applications of TDM is to provide a ______ link.
   a. DSL
   b. T1
   c. cable modem
   d. LAN

9. Virtual circuit is a type of ______.
   a. circuit switching
   b. packet switching
   c. a and b
   d. message switching

10. Pulse code modulation is used to convert ______.
   a. digital to analog
   b. analog to digital
   c. digital to digital
   d. analog to analog

11. Of the following switching methods, ______ delivers packets in order.
   a. packet switching
   b. virtual circuits
   c. circuit switching
   d. b and c

12. The bandwidth of a telephone system is ______.
   a. 3 Khz
   b. 4 Khz
   c. 8 Khz
   d. 40 Khz

## • Short Answer Questions

1. Show an 8-to-1 MUX and a 1-to-8 DMUX.
2. List the types of multiplexers.
3. Explain TDM operation.
4. Describe statistical packet multiplexer.
5. What is the difference between CDM and TDM?
6. What is an application of WDM?
7. What are the types of WDM?
8. What is the function of an optical transponder?
9. What is the type of signal used between two central switches of a telephone system?
10. What is the function of a codec?
11. What does PCM stand for and what is its application?
12. Why must the human voice be sampled at the rate of 8000 samples per second?
13. How many voice channels does a T1 link carry?
14. What type of multiplexer is used in T1 link?
15. What is the data rate of the human voice and why?
16. What is the data rate of a T1 link?
17. What is difference between a DS-1 and a T1 link?
18. What is the data rate of a T3 link?

19. How many voice channels can be carried by a T3 Link?
20. Explain the following switching operations:
    a. circuit switching
    b. message switching
    c. packet switching
    d. virtual circuit
21. Show the frame format of a T1 link.
22. Why is the data rate of a T1 link 1.54 Mbps?
23. The following inputs are connected to a 4 * 1 statistical multiplexer; show the outputs of the multiplexer:
    Input #1 A-A-A
    Input #2 BBB-B
    Input #3 - CC- -
    Input #4 DD-D
24. What is the sampling rate of a signal with the highest frequency of 1000 Hz?
25. Use the chip sequence in Table 5.3 to find data transmitted for node C at the receiver side; assume the nodes A, B, and C transmit the following data.

| Node A | 111 |
|---|---|
| Node B | 010 |
| Node C | 001 |

CHAPTER 6

# Standards Organizations and the OSI Model

**OUTLINE**

6.1 Communication Protocols
6.2 The Open System Interconnection Model
6.3 Error and Flow Control
6.4 Frame Transmission Methods
6.5 IEEE 802 Standards Committee

**OBJECTIVES**

After completing this chapter, you should able to:

- List some of the standards organizations responsible for developing standards for networks and data communication
- Discuss the concepts of communication protocols
- Explain the OSI model for networking, and the function of each layer
- Comprehend frame transmission methods
- Demonstrate your understanding of error and flow control
- List some of the IEEE 802 committee standards
- Draw the Logical Link Control (LLC) frame format

**INTRODUCTION**

There are several organizations that are constantly working toward developing standards for computers and communication equipment. The development of standards for computers enables hardware and software products made by different vendors to be compatible. Standardization allows products from different manufacturers to work together in creating customized systems. Without standards, only hardware and software made by the same manufacturer can work properly together. The following is a list of **standards organizations:**

*IEEE:* The Institute for Electrical and Electronics Engineers (IEEE) is the largest technical organization in the world. The mission of IEEE is to advance the field of electronics, computer science, and computer engineering. The IEEE also develops standards for computers, electronics, and local area networks (in particular, the IEEE 802 standards for the LAN).

*ITU:* The International Telecommunications Union (ITU) was founded in 1864 and became a United Nations Agency with the purpose of defining standards for telecommunications, Wide Area Networks (WAN), Asynchronous Transfer Mode (ATM), and Integrated Services Digital Networks (ISDN).

*EIA:* The Electrical Industry Association (EIA) is a trade association representing high technology manufacturers in the United States. The EIA develops standards for connectors and transmission media. Some of the well-known EIA standards are RS-232 and RJ-45.

*ANSI:* The American National Standards Institute (ANSI) was founded in 1918. ANSI is composed of 1300 members representing computer companies, with the purpose of developing standards for the computer industry. ANSI is the United States representative in the International Organization for Standardization (ISO). Some of the well-known ANSI standards are optical cable, programming language (ANSI C), and the fiber distributed data interface (FDDI).

*ISO:* The International Standards Organization (ISO) is an international organization comprised of national standards bodies of 75 countries. The ISO develops standards for a wide range of products, including the model for networks called the *Open System Interconnection (OSI)* model.

*IETF:* The Internet Engineering Task Force (IETF) develops standards for the Internet, such as Internet Protocol version 6 (IPv6). The IETF is composed of international network designers, network industries, and researchers.

## 6.1 Communication Protocols

A **communication protocol** is a set of rules used by computers that allows them to communicate with each other. Computers must follow these rules in order to communicate with each other. Some of the rules that define a protocol include the following:

***Size of information.*** Both computers must agree on the minimum and maximum size of information.

***How to represent information.*** Information may be Unicode, ASCII, or encrypted.

***Error detection.*** The method used by the receiver to check the integrity of information.

***Receipt of information.*** The transmitter must know that information has been received at the destination.

***Non–receipt of information.*** Both computers must know what to do if information sent is not received or if it is received but has been corrupted.

Some of the common network protocols include the following:

**TCP/IP:** Transmission Control Protocol/Internet Protocol; used in the Internet and many LANs.

**NetBEUI:** NetBIOS Extended User Interface is a small and fast protocol used for small LANs.

**X.25:** X.25 is a set of protocols used in packet-switching networks.

**IPX/SPX:** Novell NetWare uses Internet Packet Exchange/Sequenced Packet Exchange.

**NWlink:** NWlink is a Microsoft version of IPX/SPX.

## 6.2 The Open System Interconnection Model

The **Open System Interconnection (OSI) model** was developed by the **International Standards Organization (ISO)** for interoperability between equipment designed for networks. The ISO developed this open system reference model for networking. Any device that meets the OSI standards can be easily connected to any other device that adheres to the OSI model. An open system is a set of protocols that allows two computers to communicate with each other regardless of their design, manufacturer, or CPU type. The OSI model divides network communications into seven layers, with each layer performing specific tasks, as shown in Figure 6.1.

| | |
|---|---|
| Layer 7 | **Application Layer**<br>Performs information processing such as file transfer, e-mail, and Telnet. |
| Layer 6 | **Presentation Layer**<br>Defines the format of data to be sent:<br>ASCII, data encryption, data compression, and EBCDIC. |
| Layer 5 | **Session Layer**<br>Sets up a session between two applications by determining the type of communication such as duplex, half-duplex, synchronization, etc. |
| Layer 4 | **Transport Layer**<br>Ensures data gets to the destination. Manages error control, flow control, and quality of the service. |
| Layer 3 | **Network Layer**<br>Sets up connection, disconnects connection, provides routing and multiplexing. |
| Layer 2 | **Data Link Layer**<br>Manages framing, error detection, and retransmission of message. |
| Layer 1 | **Physical Layer**<br>Electrical Interface (type of signal), Mechanical interface (type of connector).<br>Converts electrical signals to bits, transmits and receives electrical signals. |

**FIGURE 6.1**
OSI model

**Layer 1: Physical Layer**

The **Physical layer** defines the type of signal, and type of connectors (such as RS-232 or RJ-45) to be used for the Network Interface Card (NIC). It defines cable types (such as coaxial cable, twisted-pair, or fiber-optic cable) to be used for the transmission media. It accepts the signal from the media and converts it to the bits, and also converts the bits to the signals for the transmission over media.

**Layer 2: Data Link Layer**

The **Data Link layer** has to define the frame format, such as the start of the frame, end of the frame, size of the frame, and type of transmission. The Data Link layer performs the following functions:

**On the transmitting side:** The Data Link layer accepts information from the Network layer and breaks the information into frames. It then adds the destination MAC address, source address MAC, and Frame Check Sequence (FCS) field, and passes each frame to the Physical layer for transmission.

**On the receiving side:** The Data Link layer accepts the bits from the Physical layer and forms them into a frame, performing error detection. If the frame is free of error, the Data Link layer passes the frame up to the Network layer.

**Frame synchronization:** It identifies the beginning and end of each frame.

**Flow control:** It distinguishes between control frames and information frames.

**Link management:** It coordinates transmission between transmitter and receiver.

**Determine contention method:** It defines an access method in which two or more network devices compete for permission to transmit information across the same communication media, such as token passing and Carrier Sense Multiple Access with Collision Detection (CSMA/CD).

There are several existing protocols for the Data Link layer, including the following:

- **Synchronous Data Link Control (SDLC):** SDLC was developed by IBM as link access for System Network Architecture (SNA).
- **High Level Data Link Control (HDLC):** HDLC is a version of SDLC modified by the ISO for use in the OSI model.
- **Link Access Procedure Balanced (LAPB):** HDLC was modified by ITU and it is called *LAPB* when used in ISDN.

***High Level Data Link Control (HDLC).*** High Level Data Link Control (HDLC) is designed to work with any type of station such as primary, secondary, and combined station (a combination of primary and secondary). The function of a **primary station** is to control the network and thus be able to transmit information at any time.

A **secondary station** can transmit information only when requested by a primary station. Figure 6.2(a) shows a point-to-point link using primary and secondary stations and Figure 6.2(b) shows multi-point links between primary and secondary stations.

A **combined station** operates as both a primary and secondary station in one unit. It can request data from other stations and can also respond to the requests of other stations. Figure 6.2(c) shows a point-to-point link between two combined stations.

**FIGURE 6.2(a)**
Point-to-point connection between primary and secondary stations

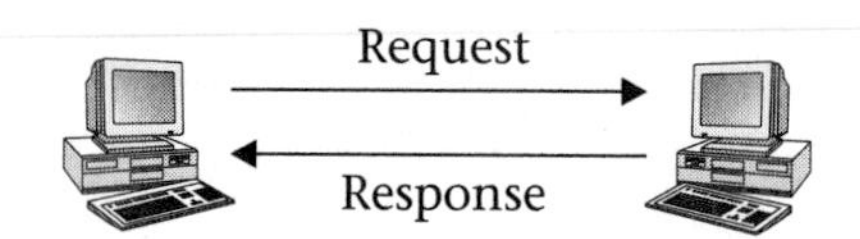


**FIGURE 6.2(b)**
Multi-point connection using one primary and several secondary stations

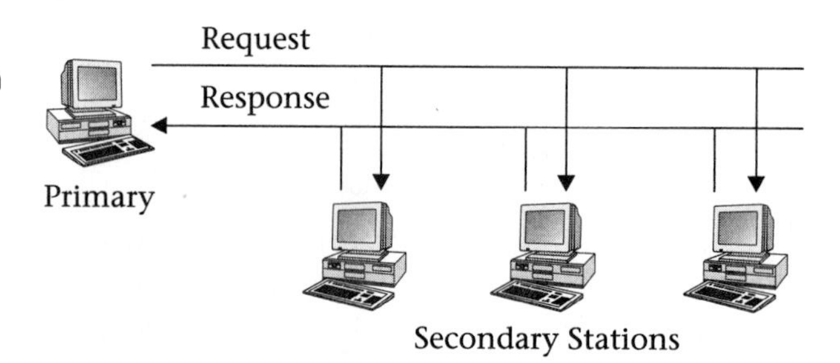


**FIGURE 6.2(c)**
Point-to-point connection for combined stations

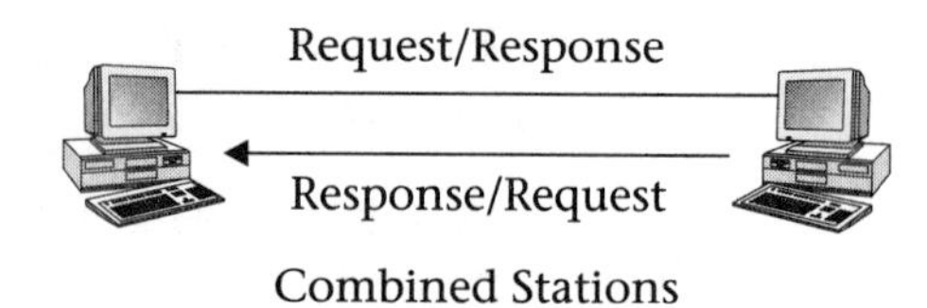


***HDLC frame format.*** Figure 6.3 shows a general frame format of the HDLC, where the flag field is 01010101. The address field gives the address of destination and the control field determines the type of information in the information field, such as Information Frame (I-Frame), Supervisory Frame (S-Frame), and Unnumbered Frame (U-Frame).

**FIGURE 6.3**
HDLC frame format

| 1 byte | 1 or more bytes | 1 or 2 bytes | variable | 2 or 4 bytes | 1 byte |
|---|---|---|---|---|---|
| Flag | Address | Control | Information | FCS | Flag |

***Information Frame (I-Frame).*** Figure 6.4(a) shows the control field for the information frame (I-Frame) where:

N(S) is sequence number of transmitted framė,

N(R) is sequence number of the next frame expected to receive,

P/F (Poll/Final) = 0 implies this frame is the last frame, and

P/F = 1 indicates primary station is requesting secondary station to send its frame.

**FIGURE 6.4(a)**
Control field for I-Frame

| 0 | N(S) | P/F | N(R) |
|---|---|---|---|

***Supervisory Frame (S-Frame).*** An S-frame is used for flow and error control such as receiver ready, receiver not ready, and receiver reject. The code field in Figure 6.4(b) defines the type of Supervisory frame.

**FIGURE 6.4(b)**
Control field for S-Frame

| 1 | 0 | CODE | P/F | N(R) |
|---|---|---|---|---|

***Unnumbered Frame (U-Frame).*** Figure 6.4(c) shows the control field for a U-Frame. The U-Frame is used for setting the mode of operation between source and destination, disconnecting a logical link, resetting a connection, and testing a connection. The code field defines the type of U-Frame.

**FIGURE 6.4(c)**
Control field for U-Frame

| 1 | 1 | CODE | P/F | CODE |
|---|---|---|---|---|

## 6.3 Error and Flow Control

Functions of the Data Link layer include error detection, error control, and flow control. During the transmission of a frame from a source to its destination, the frame may get corrupted or lost. It is the function of the Data Link layer of the destination to check for error in the frame and inform the source about the status of the frame. This function must be performed in order for the source to retransmit the frame. One of the methods used is **Automatic Repeat Request (ARQ)** in which positive or negative acknowledgment is used to establish reliable communication between the source and the destination. Automatic repeat request is carried out in two ways: stop-and-wait ARQ and continuous ARQ.

### Stop-and-Wait ARQ

In **Stop-and-Wait ARQ** the source transmits a frame and waits for a specific time for acknowledgment from the destination. If the source does not receive acknowledgment during this time, it retransmits the frame. This method is used for networks with a half-duplex connection.

***Case 1.*** The source station transmits a frame to the destination station. The destination station checks the frame for any errors. If there is no error in the frame, the destination station responds to the source station with a Positive Acknowledgment Frame ACK(N), where N is the sequence number of the frame. The source station then transmits the next frame, as shown in Figure 6.5.

**FIGURE 6.5** Positive acknowledgment

Source
Destination
I(N)
ACK(N)
I(N+1)

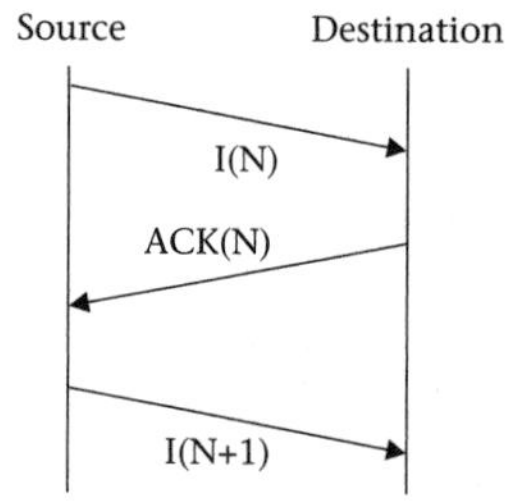

***Case 2.*** The source station transmits a frame to the destination station. The destination station checks the frame for error. If there is an error in the frame, the destination station responds to the source station with a Negative Acknowledgment Frame NACK(N), where N is the sequence number of the corrupted frame. Then the source retransmits the frame, as shown in Figure 6.6.

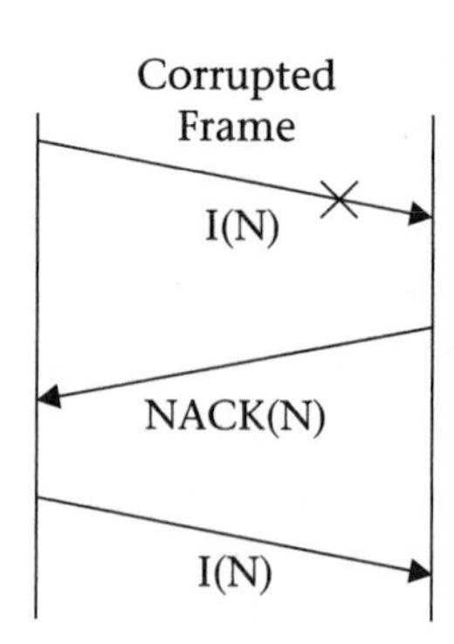


**FIGURE 6.6** Response to corrupted frame by destination

***Case 3.*** The source station transmits a frame to the destination station. The source does not receive any acknowledgment due to the loss of the frame or loss of acknowledgment from the destination. When the source starts to transmit a frame to the destination, it sets a timer and waits for acknowledgment. If the source does not receive an acknowledgment from the destination during that period of time, the frame is retransmitted, as shown in Figure 6.7.

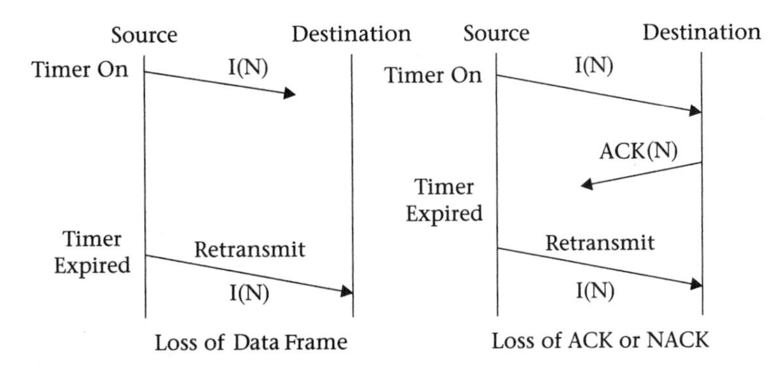


**FIGURE 6.7**
Loss of ACK or NACK and I-Frame

**Continuous ARQ**

In **continuous ARQ** the transmitter continuously transmits frames to the destination. The destination sends ACK or NACK on different channels. The continuous ARQ is used in a packet-switching network and full-duplex connection. There are two types of ARQ: Go-Back-N ARQ and Selective Reject ARQ.

***Go-Back-N ARQ.*** In the Go-Back-N ARQ method the transmitter continuously transmits and the receiver acknowledges each frame in a different channel, as shown in Figure 6.8.

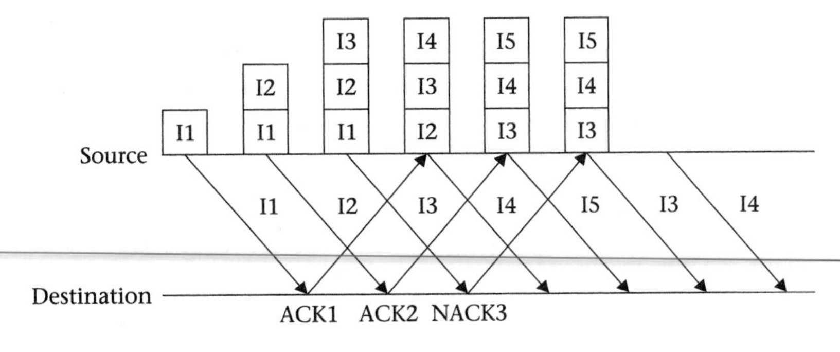


**FIGURE 6.8**
Go-Back-N ARQ

Figure 6.8 shows that the source transmitted frame I5 and received NACK from I3; the source then retransmits frames I3, I4, and I5. In Go-Back-N, the source should hold a copy of those frames not receiving acknowledgment. When the source receives acknowledgment for a frame, it can remove the frame from its buffer.

***Selective Reject ARQ.*** In selective ARQ the source will retransmit only those frames for which the destination had sent a negative acknowledgment. Figure 6.9 shows the source transmitted frame I3 and received NACK1, which indicated that frame I1 was corrupted. The source retransmits only frame I1. In this method, the destination must have the capability to reorder frames that are out of order.

**FIGURE 6.9**
Selective reject ARQ

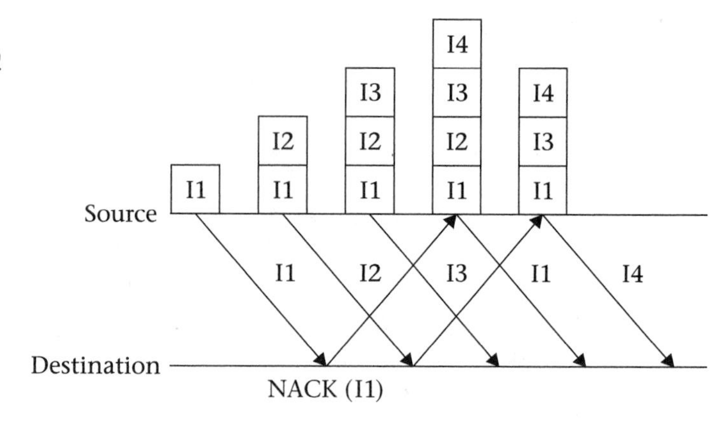


## Sliding Window Method

In continuous ARQ the source keeps a copy of transmitted frames in its buffer until it receives acknowledgment for a frame; it then removes the frame from its buffer. The continuous ARQ has the following deficiencies:

1. The destination may not have enough memory to store incoming frames.
2. The source may transmit frames faster than the destination can process them.
3. The source must hold a copy of all unacknowledged transmitted frames in its buffer; therefore, the source requires a large buffer.
4. The file to be transmitted is divided into packets. Each packet has a sequence number; if the sequence number becomes large, it decreases network efficiency.

The **Sliding Window Method** limits the number of frames waiting for acknowledgment in the source. For example, a source with a window of seven means the source can hold only seven unacknowledged frames in its buffer. The source will stop transmitting after having seven frames in its buffer and wait for an acknowledgment frame. When the source receives acknowledgment for a frame, it removes that frame from its buffer and transmits the next frame.

In order to prevent the need for a large sequence of numbers, most of the networking protocols use the following formula for assigning sequence numbers to each frame:

Sequence Number = Frame Number Modulo K

where:

K is the window size of sending station

For example: What is sequence number for frame number 25? Assume window size is 7:

Sequence Number = 25 Modulo 7 = 4

**Layer 3: Network Layer**

The function of the **Network layer** is to perform routing. Routing determines the route or pathway for moving information (in a network with multiple LANs). The Network layer checks the logical address of each frame and forwards the frame to the next router based on a routing table. The Network layer is responsible for translating each logical address (name address) to a physical address (MAC address). An example of Network layer protocol is Internet Protocol (IP).

The Network layer provides two types of services: connectionless and connection-oriented services. In connection-oriented services, the Network layer makes a connection between source and destination; then transmission starts. In connectionless service, there is no connection between source and destination. The source transmits information regardless of whether the destination is ready or not. A common example of this is e-mail.

**Layer 4: Transport Layer**

The **Transport layer** provides for the reliable transmission of data in order to ensure that each frame reaches its destination. If, after a certain period of time, the Transport layer does not receive an acknowledgment from the destination, it retransmits the frame and again waits for acknowledgment from the destination. An example of a Transport layer is Transmission Control Protocol (TCP).

**Layer 5: Session Layer**

The **Session layer** establishes a logical connection between the applications of two computers that are communicating with each other. It allows two applications on two different computers to establish and terminate a session. When a workstation connects to a server, the server performs the login process, requesting a username and password. This is an example of establishing a session.

**Layer 6: Presentation Layer**

The **Presentation layer** receives information from the Application layer and converts it to a form acceptable by the destination. The Presentation layer converts information to ASCII, or Unicode, or encrypts or decrypts the information.

**Layer 7: Application Layer**

The **Application layer** enables users to access the network with applications such as e-mail, FTP, and Telnet.

## 6.4 Frame Transmission Methods

The Data Link layer offers two types of frame transmission: asynchronous transmission and synchronous transmission.

### Asynchronous Transmission

In **asynchronous transmission** each character is transmitted separately, as shown in Figure 6.10. One disadvantage of this method is that extra bits must be added to each group of character bits, such as a start bit and stop bit, which represent the beginning and end of each character, and parity bit, which is used for error detection. This method is inefficient for transferring a large volume of information. The modem is the primary application that uses asynchronous protocol.

| Start Bit | Character Bits | Parity Bit | Stop Bits |
|---|---|---|---|

| Start Bit | Character Bits | Parity Bit | Stop Bits |
|---|---|---|---|

**FIGURE 6.10**
Asynchronous transmission

### Synchronous Transmission

**Synchronous transmission** is more efficient than asynchronous transmission for transmitting large blocks of data. Synchronous transmission can be character-oriented or bit-oriented.

***Character-oriented Synchronization.*** **Character-oriented synchronization** is used for character-oriented transmissions in which blocks of data, such as text files in ASCII format, are transferred over a network. Figure 6.11 shows the frame format for character-oriented transmission.

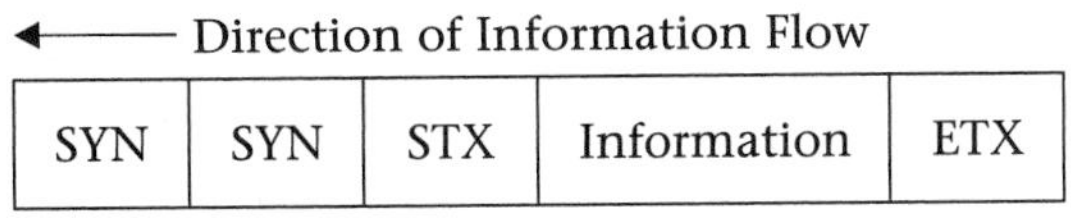

**FIGURE 6.11**
Character synchronization

← Direction of Information Flow

| SYN | SYN | STX | Information | ETX |
|---|---|---|---|---|

The following are the functions of each field in character synchronization frame format:

**SYN:** The **S**ynchronization Character is used by the receiving device to synchronize its timing. The SYN character is 16 in Hex, with two SYNs in each frame.

**STX:** The Start of Text (STX) field is used by the receiver in order to recognize the start of data. STX is 02 in Hex.

**ETX:** The End of Text (ETX) is used to inform the receiver that the end of the data has been reached. ETX is 03 in Hex.

Data in the information field is in the form of printable characters. Since the information field might contain STX or ETX characters, the transmitter precedes the STX or ETX with a DLE (Data Link Escape) character, as shown in Figure 6.12. Therefore the receiver will interpret a DLE STX as the start of the frame and a DLE ETX as the end of the frame. If the information field contains a DLE character, the transmitter sends a DLE DLE. When it sees a DLE DLE sequence, the receiver knows that this DLE does not belong to any of the start-of-text sequence or end-of-text sequence. The receiver then strips off one of the DLE characters within the information block.

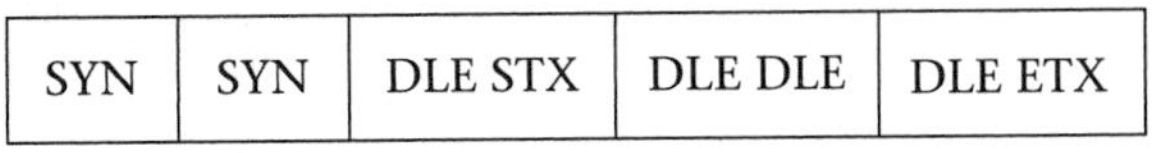

**FIGURE 6.12**
Character synchronization with DLE characters

| SYN | SYN | DLE STX | DLE DLE | DLE ETX |
|---|---|---|---|---|

***Bit-Oriented Synchronization.*** Since the character-oriented synchronization technique used in synchronous protocol needs several control characters, it is less efficient than bit-oriented synchronization.

**Bit-oriented synchronization** is used for bit-oriented transmission, when information is transmitted bit by bit. Figure 6.13 shows the frame format of a bit synchronization frame. The start of the frame and the end of the frame are represented by eight bits in the form of 01111110. If the information field happens to contain the binary value 01111110, which normally indicates the end of a frame, an adjustment must be made by the transmitter. The transmitter inserts an extra zero any time five ones are repeated in the information field. The receiver will discard this extra zero. This technique is called *bit insertion*, as shown in Figure 6.14.

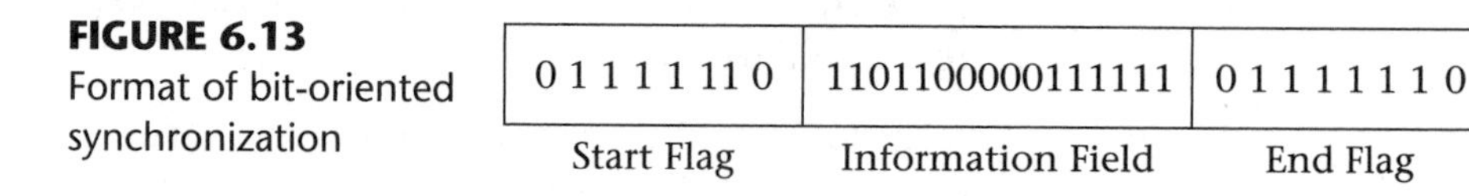


**FIGURE 6.13**
Format of bit-oriented synchronization

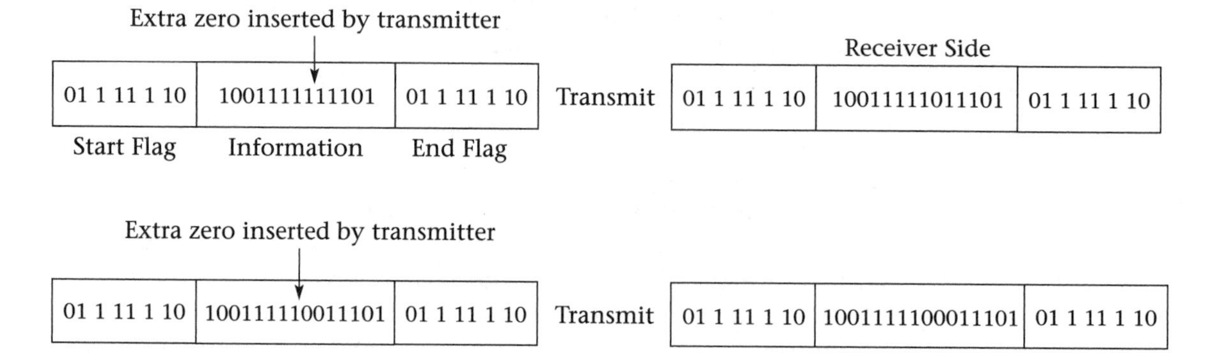


**FIGURE 6.14**
Bit insertion in information field

## 6.5 IEEE 802 Standards Committee

The **IEEE 802 committee** defined standards for the Physical layer and the Data Link layer in February of 1980, and called it *IEEE 802*, with "80" representing 1980, and "2" representing the month of February. Figure 6.15 shows the IEEE 802 standard and OSI model. The IEEE standard divides the Data Link layer of the OSI model into two sub-layers, Logical Link Control (LLC), and Media Access Control (MAC).

**FIGURE 6.15**
IEEE standard and OSI model

| OSI Model | IEEE 802 Model |
|---|---|
| Layer 4–7 | Layer 4–7 |
| Network Layer | Network Layer |
| Data Link Layer | Logical Link Control |
| | Media Access Control |
| Physical Layer | Physical Layer |

**Media Access Control (MAC)**

The **Media Access Control (MAC)** layer defines the method that stations use to access the network, such as:

- Carrier Sense Multiple Access Collision Detection (CSMA/CD)—used for Ethernet
- Control Token used in Token Ring Network and Token Bus Network

**Logical Link Control (LLC)**

The **Logical Link Control (LLC)** defines the format of the frame. It is independent of network topology, transmission media, and media access control.

Figure 6.16 shows different MAC layers for several IEEE 802 networks. All networks that are listed use the same logical link control. Figure 6.17 shows the frame format of the LLC, which is used by all IEEE 802.X projects.

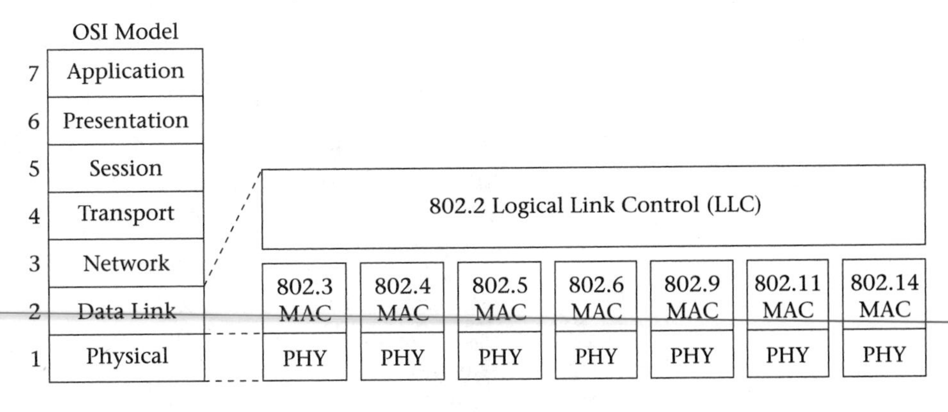


**FIGURE 6.16**
IEEE 802 reference model

**FIGURE 6.17** Logical link control frame format

| 1 byte | 1 byte | 1 or 2 bytes | |
|---|---|---|---|
| DSAP | SSAP | Control | Information |

The following are the functions of each field of the LLC frame format:

***Destination Service Access Point (DSAP).*** Since the destination station might run several network protocols such as Novell Netware, NetBIOS, Windows NT, and TCP/IP, the DSAP has to show the address of the protocol for the destination. Table 6.1 shows the most common values for the service access point (SSAP and DSAP).

**TABLE 6.1** DSAP and SSAP Values

| Protocol | SSAP and DSAP Values in Hex |
|---|---|
| IBM SNA | 04 |
| IP | 06 |
| 3Com | 80 |
| Novell | E0 |
| Banyan | BC |
| Net BIOS | F8 |
| LAN Manager | F4 |

***Source Service Access Point (SSAP).*** SSAP is a value of the source protocol and indicates the protocol that was used by the transmitter to send the packet.

***Control Field.*** The control field determines what type of information is stored in the information field, such as information frame, supervisory frame, and unnumbered frame. The supervisory frames are receiver ready, receiver not ready, and reject. Some of the unnumbered frames are reset, frame reject, disconnect, and set asynchronous respond mode.

The following is a partial list of standards developed for networking by the IEEE 802 committee:

802.1 Internetworking
802.2 Logical Link Control (LCC)
802.3 Ethernet
802.4 Token Bus LAN
802.5 Token Ring LAN

802.6 MAN
802.7 Broadband Technical Advisory Group
802.8 Fiber Optics Technical Advisory Group
802.9 Integrated Services Data Networks (ISDN)
802.10 Network Security
802.11 Wireless Networks
802.12 100 VG-any LAN
802.14 Cable Modem
802.16 Wireless MAN

## Summary

- Some of the standards organizations that develop standards for networks and data communications are the IEEE, ITU, EIA, ISO, IETF, and ANSI.
- A communication protocol is a set of rules used by two computers in order to communicate with each other.
- The most popular communication protocols are Transmission Control Protocol/Internet Protocol (TCP/IP), NetBEUI, IPX/SPX, NWlink, and DecNet.
- The International Standards Organization (ISO) developed a model for networks called the *Open System Interconnection (OSI) model.*
- The OSI model consists of seven layers, from top to bottom: Application layer, Presentation layer, Session layer, Transport layer, Network layer, Data Link layer, and Physical layer.
- The Application layer enables the user to access the network applications.
- The Presentation layer is responsible for representation of information such as ASCII, encryption, and decryption.
- The function of the Session layer is to establish a session between source and destination applications, and to disconnect a session between two applications.
- The function of the Transport layer is to ensure that data gets to the destination, to perform error control and flow control, and to assure quality of service.
- The function of the Network layer is to deliver information from source to destination and route the information.
- The function of the Data Link layer is framing, error detection, and retransmission.

- The functions of the Physical layer are electrical interface, type of signal, convert bits to signals (electrical or optical or wireless), and vise versa.
- There are two types of frame transmission: asynchronous transmission and synchronous transmission.
- Asynchronous transmission is character-oriented, and each character has start and stop bits.
- There are two types of synchronous transmission: character-oriented synchronization and bit-oriented synchronization.
- In character-oriented transmission, a block of data is transferred over the network in the form of ASCII code. Character-oriented transmission uses the SYN character for synchronization and information is transmitted character by character.
- Bit-oriented synchronization uses bit-oriented transmission. Information is transmitted bit by bit.
- The IEEE 802 committee developed the standard for the Physical and Data Link layers.
- The IEEE 802 divides the Data Link layer into two sub-layers: Logical Link Controls (LLC) and Media Access Control (MAC).
- All IEEE 802.X use the LLC frame format.
- The IEEE 802.2 is the standard for LLC.
- The IEEE 802.3 is the standard for Ethernet.
- The IEEE 802.4 is the standard for Token Bus.
- The IEEE 802.5 is the standard for the Token Ring.
- The IEEE 802.6 is the standard for MAN.
- The IEEE 802.11, IEEE 802.11a, IEEE 802.11b, and IEEE 802.11 are the standards for wireless communications.
- The IEEE 802.12 is the standard for 100 VG-any LAN.
- The IEEE 802.14 is the standard for cable modem.
- The IEEE 802.16 is the standard for wireless MAN.

## Key Terms

Application Layer
Asynchronous Transmission
Automatic Repeat Request (ARQ)
Bit-Oriented Synchronization
Character-Oriented Synchronization
Combined Station
Communication Protocol
Continuous ARQ
Data Link Layer
IEEE 802 Committee

International Standards Organization (ISO)
Logical Link Control
Media Access Control (MAC)
Network Layer
Open System Interconnection (OSI) Model
Physical Layer
Presentation Layer
Primary Station
Secondary Station
Session Layer
Sliding Window Method
Standard Organizations
Stop-and-Wait ARQ
Synchronous Transmission
Transport Layer

## Review Questions

### • Multiple Choice Questions

1. The IEEE developed the ______ standard for LAN.
   a. IEEE 802
   b. RS232
   c. OSI model
   d. All of the above
2. The ______ defines standards for telecommunications.
   a. IEEE
   b. ITU
   c. EIA
   d. ISO
3. The ______ defines standards for programming languages.
   a. IEEE
   b. ISO
   c. ANSI
   d. IETF
4. The ______ protocol is used on the Internet.
   a. TCP/IP
   b. X.25
   c. IPX/SPX
   d. NWLink
5. Microsoft's version of IPX/SPX is called ______.
   a. Net BEUI
   b. TCP/IP
   c. NWLink
   d. X.25

6. The OSI model contains ______ layers.
   a. 4
   b. 3
   c. 7
   d. 6
7. The ______ layer establishes a connection.
   a. Network
   b. Physical
   c. Data Link
   d. Application
8. In character-oriented transmission, information is transmitted ______.
   a. bit by bit
   b. character by character
   c. byte by byte
   d. word by word
9. The ______ layer defines the format of the frame.
   a. Transport layer
   b. MAC layer
   c. Network layer
   d. Physical layer
10. The layer of the OSI model responsible for forming a frame is ______.
    a. Data Link
    b. Transport
    c. Session
    d. Physical
11. The layer of the OSI model that performs encryption is the ______.
    a. Session layer
    b. Presentation layer
    c. Data Link layer
    d. Transport layer
12. The function of the Network layer is ______.
    a. error detection
    b. routing
    c. to set up a session
    d. encryption
13. The ______ layer of the OSI model converts electrical signals to bits.
    a. Physical
    b. Data Link
    c. Network
    d. Application

14. The ______ layer determines the route for packets transmitted from source to destination.
    a. Data Link
    b. Network
    c. Transport
15. Of the following standards, ______ applies to Logical Link Control.
    a. IEEE 802.3
    b. IEEE 802.4
    c. IEEE 802.2
    d. IEEE 802.5
16. Of the following standards, ______ applies to Token Ring.
    a. IEEE 802.3
    b. IEEE 802.5
    c. IEEE 802.2
    d. IEEE 802.4
17. A ______ protocol is HDLC.
    a. character-oriented
    b. bit-oriented
    c. character and bit-oriented
    d. word-oriented

## • Short Answer Questions

1. Explain why it is necessary to have standards for networking products.
2. List five standards organizations that are developing standards for networks and computers.
3. Define an open system.
4. Show the OSI model for a network.
5. List the function of the Application layer.
6. Explain the functions of the Presentation layer.
7. Define the functions of the Session layer.
8. Explain the functions of the Transport layer.
9. List the functions of the Network layer.
10. Explain the functions of the Data Link layer.
11. Define the functions of the Physical layer.
12. IEEE 802 subdivides the Data Link layer into ______ sub-layers, ______ and ______.

13. Show the frame format of IEEE 802.2.
14. Explain the functions of MAC layer.
15. What is IEEE 802.3?
16. What is IEEE 802.5?
17. What are the frame transmission methods?
18. Explain asynchronous transmission methods.
19. What does SDLC stand for?
20. List synchronous transmission methods.
21. What is an application of HDLC?
22. Explain character-oriented synchronization.
23. Explain bit insertion in bit-oriented synchronization.
24. Explain the function of synchronization bits.
25. Name an organization that makes standards for cable and connectors.
26. What is the function of SYN, STX, and ETX characters in character-oriented transmission?
27. List four communication protocols.
28. Explain communication protocols.
29. Which layer performs error detection?
30. Which layer converts electrical signals to bits?
31. Which layer is responsible for physical connection between source and destination?
32. How many layers are in the OSI model?
33. Which layer performs login?
34. Which layer performs information routing?
35. Show the frame format of bit-oriented synchronization.
36. Show the frame format for character-oriented synchronization.
37. Show the binary value of SYN, STX, and ETX characters.
38. List three types of HDLC frames.
39. Show the transmitted frame after bit insertion for the following frame:
    01111111000000011111011111110

CHAPTER

# Modem, DSL, Cable Modem, and ISDN

OUTLINE



OBJECTIVES

After completing this chapter, you should be able to:

- Discuss modem operation and the methods of signal modulation
- Explain the operation of a 56-K modem
- Understand the technology of Digital Subscriber Line (DSL) and xDSL
- Explain cable modem technology
- Have a clear understanding of modem technology

INTRODUCTION

In order for two computers to communicate with each other, a link between them is required. If these computers are some distance from each other, it is not cost effective to use a cable and link them together. A cheaper alternative is to use a telephone or cable TV line to provide the link. The device that enables users to establish a link between computers using a telephone line is called a *modem*. The types of modems that use telephone lines are dial-up, the Integrated Services Digital Network (ISDN), and Digital Subscriber Line (DSL).

Currently, about 63 million households have cable TV services. The same wire that brings TV signals to your house is a cable that can also provide Internet access with a speed 100 times faster than a dial-up modem. The device that enables computers to access the Internet by cable TV lines is called a *cable modem*.

## 7.1 Modem

To link computers for communication over traditional telephone wires, a modem must be used, as shown in Figure 7.1. A traditional telephone network operates with analog signals, whereas computers work with digital signals. Therefore a device is required to convert the computer's digital signal to an analog signal, compatible with the phone line **(modulation)**. This device must also convert the incoming analog signal from the phone line to a digital signal **(demodulation)**. Such a device is called a **modem**; its name derived from this process of modulation/demodulation.

**FIGURE 7.1**
Connection of two computers using a modem

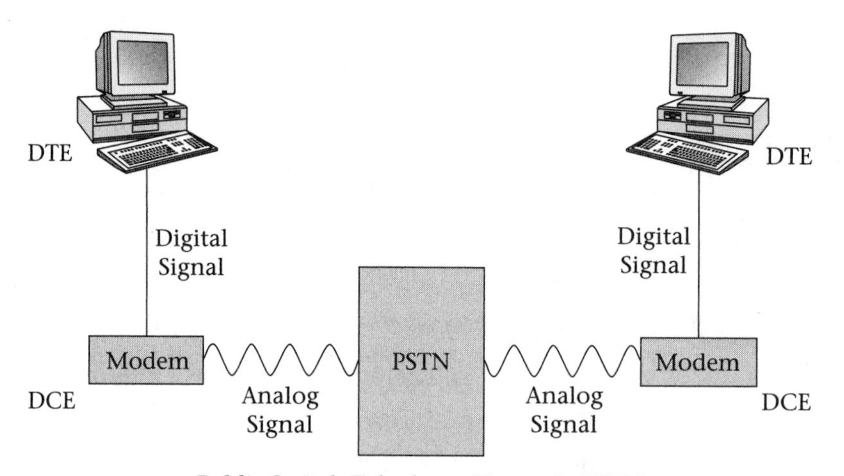


A modem is also known as *Data Communication Equipment (DCE)*, which is used to connect a computer or data terminal to a network. Logically, a PC or data terminal is also called *Data Terminal Equipment (DTE)*.

There are three types of modems. An **internal modem** is an expansion card that plugs into an ISA or PCI bus inside the computer. It connects to a phone line with an RJ-11 connector. There is also an **external modem** available. Its circuitry is housed in a separate casing and it typically uses a DB9 connector to attach to one of the computer's serial ports. The third type is used in laptop and notebook computers and consists of a PCMCIA card that houses the entire circuitry for the modem.

A modem's transmission speed can be represented by either a data rate or baud rate. The **data rate** is the number of bits that a modem can transmit in one second. The **baud rate** is the number of signals that a modem can transmit in one second.

## Modulation Methods

The carrier signal on a telephone line has a bandwidth of 4000 Hz. Figure 7.2 shows one cycle of a telephone carrier signal. The following types of modulation are used to convert digital signals to analog signals:

- Amplitude Shift Keying (ASK)
- Frequency Shift Keying (FSK)
- Phase Shift Keying (PSK)
- Quadrature Amplitude Modulation (QAM)

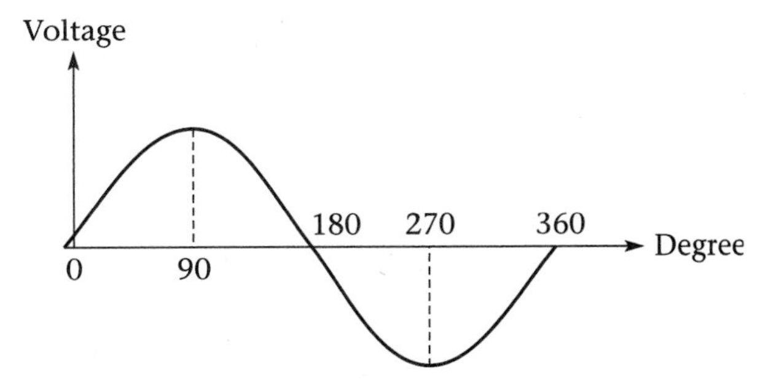


**FIGURE 7.2** Telephone carrier signal

***Amplitude Shift Keying (ASK).*** In **Amplitude Shift Keying (ASK)**, the amplitude of the signal changes. This also referred to as *amplitude modulation (AM)*. The receiver recognizes these modulation changes as voltage changes, as shown in Figure 7.3. The smaller amplitude is represented by *zero* and the larger amplitude is represented by *one*. Each cycle is represented by one bit, with the maximum bits per second determined by the speed of the carrier signal. In this case, the baud rate is equal to the number of bits per second.

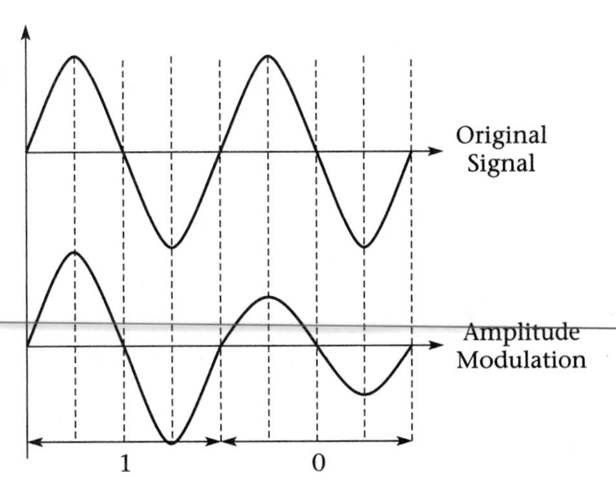


**FIGURE 7.3** Amplitude shift keying (ASK)

***Frequency Shift Keying.*** With **Frequency Shift Keying (FSK),** a *zero* is represented by no change to the frequency of the original signal, and a *one* is represented by a change to the frequency of original signal. This is shown in Figure 7.4. Frequency modulation is a term used in place of FSK.

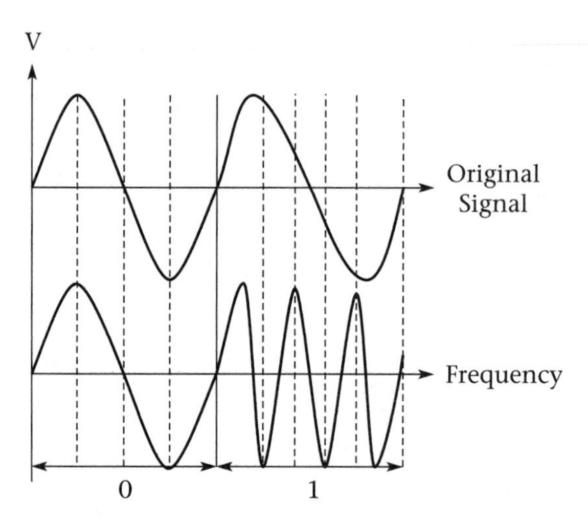


**FIGURE 7.4**
Frequency shift keying (FSK)

***Phase Shift Keying (PSK).*** Using the **Phase Shift Keying (PSK)** modulation method, the phase of the signal is changed to represent *ones* and *zeros*. Figure 7.5 shows a 90-degree phase shift. Figures 7.6(a), (b), and (c) show the original signals with a 90-degree shift, a 180-degree shift, and a 270-degree shift, respectively. In Figure 7.6, note that the original signal can be represented with four different signals: no shift, 90-degree shift, 180-degree shift, and 270-degree shift. Therefore, each signal can be represented by a two-bit binary number, as shown in Table 7.1.

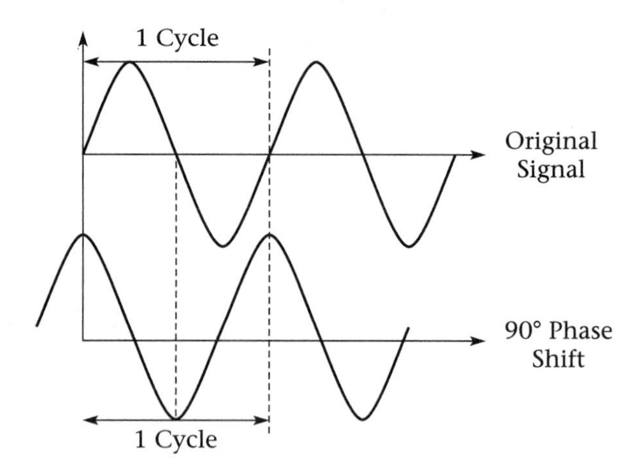


**FIGURE 7.5**
90-degree phase shift

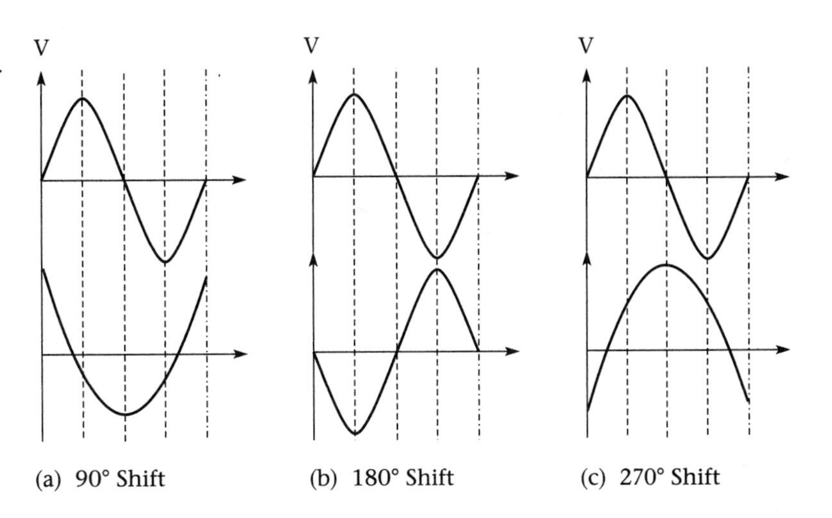


**FIGURE 7.6** Phase shift for 90, 180, and 270 degrees

The modem's speed using a 90-degree phase shift is 2*4000, which is equal to 8 Kbps. To increase the speed of the modem, the original signal can be shifted 45 degrees to generate eight distinct signals. Each signal can be represented by three bits. Therefore, the speed of the modem is increased to 3*4000, which is equal to 12 Kbps.

The relation between phase and the binary representation of each phase can be plotted on a coordinate system called a **constellation diagram.** Figure 7.7 is a constellation diagram showing the four distinct signals of

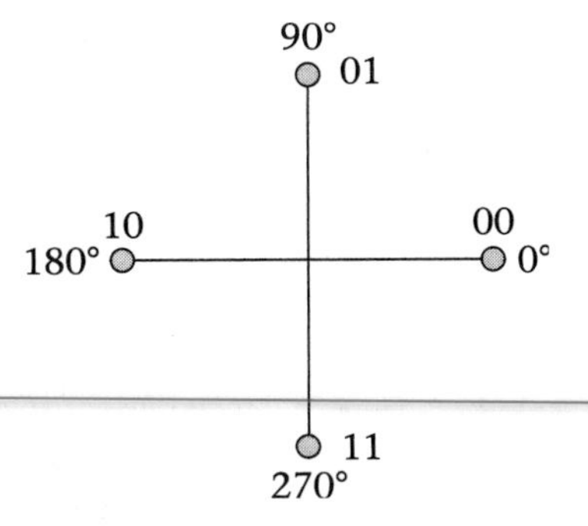


**FIGURE 7.7** Constellation diagram for Table 7.1

**TABLE 7.1** Phase Shift and Binary Value

| Phase Shift | Binary Value |
|---|---|
| No Shift | 00 |
| 90° | 01 |
| 180° | 10 |
| 270° | 11 |

a 90-degree shift, with each signal represented by two bits. Figure 7.8 shows a constellation diagram using a 45-degree shift and 3-bit representation (8-PSK).

**FIGURE 7.8**
Constellation diagram for 8-PSK

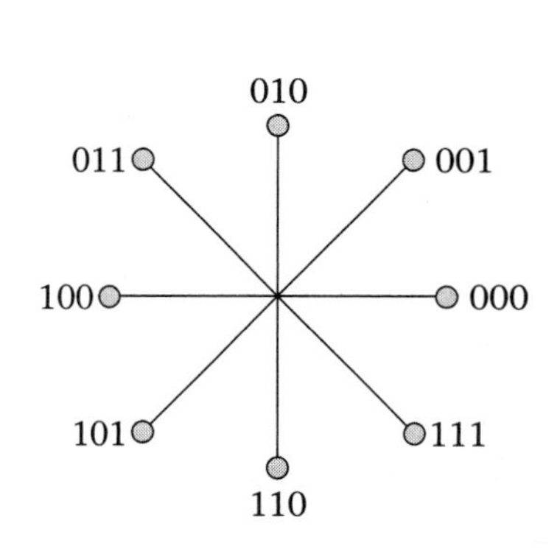


| Bits | Phase Shift |
|---|---|
| 000 | 0° |
| 001 | 45° |
| 010 | 90° |
| 011 | 135° |
| 100 | 180° |
| 101 | 225° |
| 110 | 270° |
| 111 | 315° |

***Quadrature Amplitude Modulation (QAM).*** One method to increase the transmission speed of a modem is to combine PSK and ASK modulation. This hybrid modulation technique is called **Quadrature Amplitude Modulation (QAM)** and is shown in Figure 7.9. Here we see the combination

**FIGURE 7.9**
8-QAM modulation

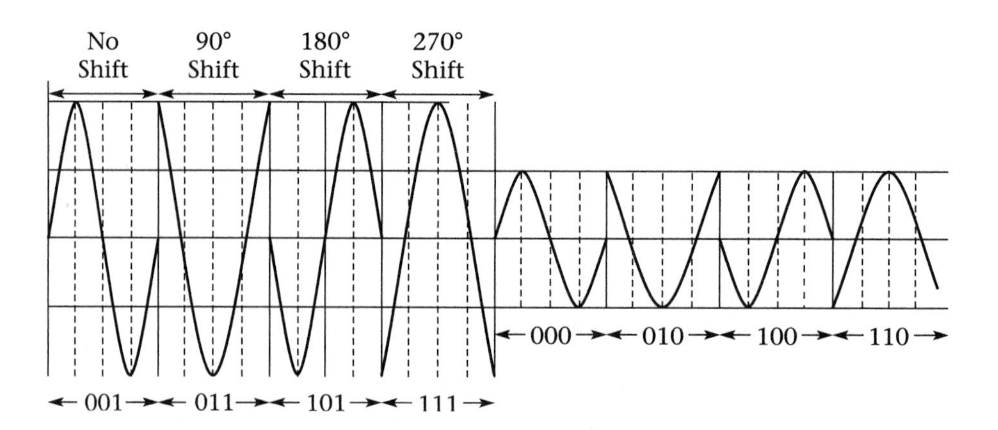

of four phases and two amplitudes, which generates eight different signals called *8-QAM*. Table 7.2 shows the binary value of each signal and provides a constellation diagram for 8-QAM (illustrated in Figure 7.10). The data rate of this modem is 3 bits*4K = 12 Kbps.

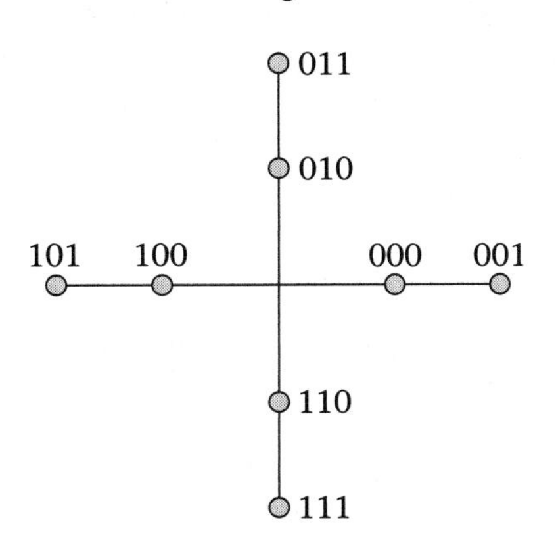


**FIGURE 7.10**
Constellation diagram for 8-QAM

**TABLE 7.2** Binary Value for 8-QAM

| Shift | Amplitude | Binary Value |
|---|---|---|
| No | A1 | 000 |
| No | A2 | 001 |
| 90° | A1 | 010 |
| 90° | A2 | 011 |
| 180° | A1 | 100 |
| 180° | A2 | 101 |
| 270° | A1 | 110 |
| 270° | A2 | 111 |

**Modem Standards** The International Telecommunications Union (ITU) is responsible for developing standards for modems. Currently, most modem manufacturers produce modems with a transmission speed of up to 56 Kbps. The following are some modem standards and their respective speeds:

| Name | Speed |
|---|---|
| V.90 or X2 | 56 Kbps receiving<br>33.6 Kbps transmitting |
| V.36 | 48 Kbps |
| V.34 | 28.8 Kbps |
| V.33 | 14.4 Kbps |
| V.32 | 14.4 Kbps |
| V.26 bis* | 1200, 2400 bps |

*bis means the modem has a switch and works with two different data rates.

**V.90 (56 Kbps) Modem** The maximum theoretical data rate for a modem, as set by the ITU, is 33.6 Kbps. Several manufacturers, however, have developed 56-Kbps modems. The 56-K modem was designed for one-end digital connection from the server to the Public Switch Telephone Network (PSTN); the subscriber side (the line that connects to the actual modem) remains analog. Figure 7.11 shows an application of a 56-K modem in which the server is connected to the PSTN without any modem and transmits information as a digital signal to the PSTN. At the central switch, digital information is converted to an analog signal and transmitted to the subscriber side. Only one side uses the modem, which will reduce the signal/noise ratio caused by conversion, and allows the server to transmit data at a rate of 56 Kbps. At the user side, the information is converted from digital to analog and transmitted to the central switch. These conversions produce noise and reduce the speed of the modem to 33.6 Kbps.

**FIGURE 7.11**
56-K modem connection

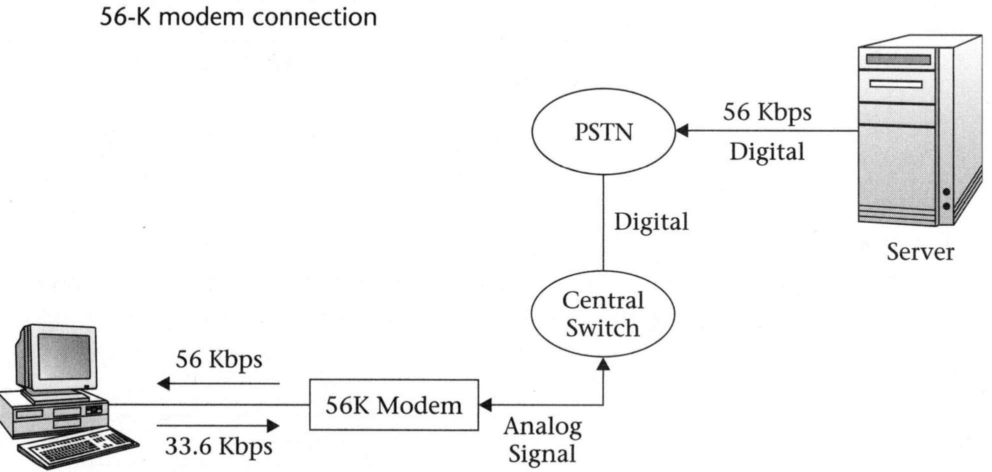


## 7.2 Digital Subscriber Line (DSL)

The **Digital Subscriber Line (DSL)** is the latest modem technology, using twisted-pair wires to deliver data and voice at speeds ranging from 64 Kbps to 50 Mbps. DSL uses current telephone wire (UTP) to transfer information at higher data rates than a modem. Currently, a modem transfers data at 56 Kbps, and networking technology transfers data at the rate of 10 to 1000 Mbps. Therefore, modems are becoming far too slow for transferring information across the Internet. DSL uses standard phone twisted-pair cable to transfer analog signals with Plain Old Telephone Service (POTS) and digital signals for data. DSL is implemented using several different technologies called *xDSL*.

***Asymmetrical DSL.*** ADSL supports voice and data simultaneously. The data rate from service provider to the user is 6 Mbps and is 786 Kbps from the user to the service provider (telephone switch).

***High Bit Rate DSL.*** HDSL supports data or voice, but not simultaneously, with a data rate of 768 Kbps.

***Symmetrical DSL.*** SDSL supports voice and data simultaneously, with a data rate of 768 Kbps in both directions.

***Very High Speed DSL.*** VDSL provides 25 to 50 Mbps to the user (downstream), and 1.5 to 3 Mbps from the user to the service provider (upstream).

**Asymmetrical Digital Subscriber Line (ADSL)**

**Asymmetrical Digital Subscriber Line (ADSL)** is a new modem that uses an existing twisted-pair telephone line to access the Internet for transferring information such as multimedia. ADSL transfers data at a higher rate **downstream** (from the telephone company switch) to the subscriber than **upstream** (from subscriber to telephone company switch). The upstream and downstream data rate is a factor of the distance between the telephone company switch and the subscriber. The downstream data rate is between 1.5 to 8 Mbps and upstream data rate is between 16 Kbps to 640 Kbps, as shown in Figure 7.12. The advantage of ADSL is that it uses the present twisted-pair wire of telephone lines to transmit data in the range of 6 to 8 Mbps. It is important to note that ADSL does not affect the current telephone voice channel.

**FIGURE 7.12**
ADSL modem connection

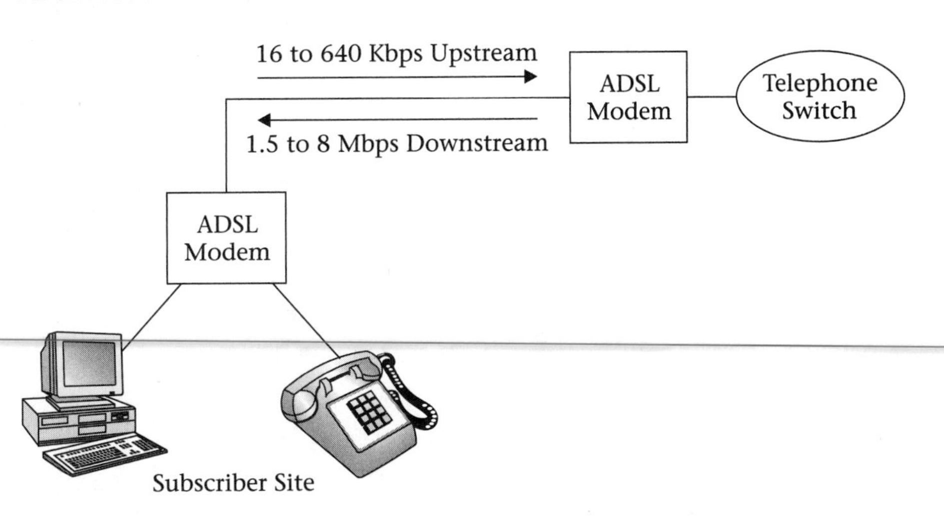

## ADSL Modem Technology

ADSL uses Discrete Multi-Tone (DMT) encoding methods, which use FDM to divide the bandwidth of the channel into multiple subchannels, with each channel transmitting information using QAM modulation. The twisted-pair cable used in telephone wire has a frequency spectrum of 1.1 MHz. Figure 7.13 shows the frequency spectrum of ADSL. DMT uses the frequency spectrum from 26 KHz and 1.1 MHz for broadband data. For POTS it uses the frequency spectrum from 0 to 4 KHz. The frequency spectrum from 26 KHz to 138 KHz is used for upstream transmission, and the frequency spectrum from 138 KHz to 1.1 MHz is used for downstream transmission, as shown in Figure 7.13. The frequency spectrum above 26 KHz is divided into 249 independent subchannels, each containing 4.3 KHz bandwidth. Each sub-frequency is an independent channel and has its own stream of signals. The lower 4 KHz channel is separated by an analog circuit and used in POTS; 25 channels are used for upstream transmissions, and 224 channels are used for downstream transmissions.

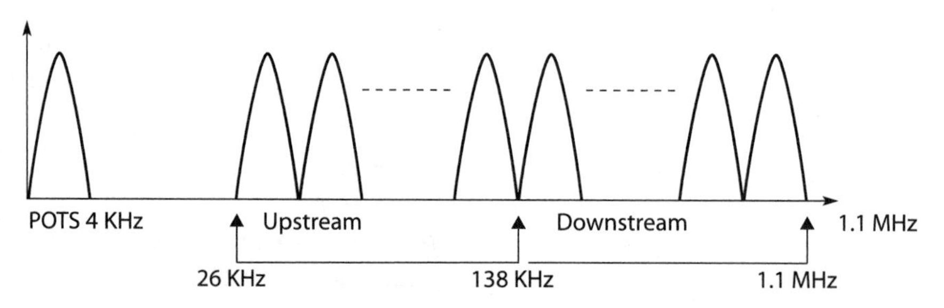


**FIGURE 7.13**
Frequency spectrum of ADSL

Figure 7.14 shows the ADSL modem architecture. The function of the POTS filter is to separate the voice channel from the data channel.

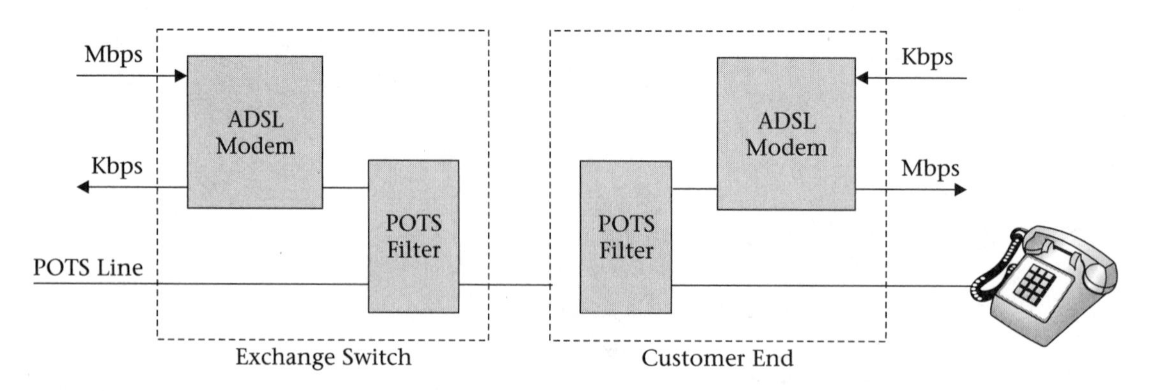


**FIGURE 7.14**
ADSL modem architecture

Each subchannel can modulate from 0 to 15 bits per signal. This allows up to 60 Kbps per channel (15*4 KHz ). Therefore, the data rate is calculated by:

Data Rate = Number of Channels * Number of Bits/Channel
* Frequency of Channel

Using this equation, the upstream and downstream data rates can be computed as follows:

Maximum Upstream Data Rate = 25*15*4.3 KHz = 1.6 Mbps
Maximum Downstream Data Rate = 224*15*4.3 KHz = 14.4 Mbps

The data rate of an ADSL modem is a factor of the distance between the subscriber and the telephone switch. Table 7.3 shows the data rate versus the distance for an ADSL modem.

**TABLE 7.3** Data Rate of ADSL Modem versus Distance

| Data Rate | Wire Gage | Distance |
|---|---|---|
| 1.5–2 Mbps | 24 AWG | 5.5 Km or 18000 ft |
| 1.5–2 Mbps | 26 AWG | 4.6 Km or 15000 ft |
| 6.1 Mbps | 24 AWG | 3.7 Km or 12000 ft |
| 6.1 Mbps | 26 AWG | 2.7 Km or 9000 ft |

**Rate Adaptive Asymmetric DSL (RADSL)**

**Rate Adaptive Asymmetric DSL (RADSL)** offers downstream transmission of 7.0 Mbps and upstream transmission of 1.0 Mbps. The rate of RADSL is dynamically adapted by the condition of the line.

Before transmitting information, RADSL makes an initial test to check the condition of the channels. Some of the channels may not be used due to the presence of strong noise. Table 7.4 shows xDSL and cable distance.

## 7.3 Cable Modem

The **cable modem** is another technology used for remote connection to the Internet. Residential access to the Internet is growing, and current modem technology can transfer data at only 56 Kbps. Local telephone companies also offer a service known as *Basic Rate ISDN*, which has a transmission rate of 128 Kbps. The cable modem offers high-speed access to the Internet using a medium other than phone lines.

**TABLE 7.4** xDSL and Cable Distance

| Technology | Cable Distance in Feet | Data Rate Downstream/Upstream |
|---|---|---|
| ADSL | 3000 | 9 Mbps/1 Mbps |
| ADSL | 5000 | 8.448 Mbps/1 Mbps |
| ADSL | 9000 | 7 Mbps/1 Mbps |
| ADSL | 12000 | 6.312 Mbps/640 Kbps |
| ADSL | 18000 | 1.544 Mbps/16–64 Kbps |
| HDSL | 5000 | 1.544 Mbps/1.544 Mbps |
| HDSL | 12000 | 1.544 Mbps/1.544 Mbps |
| RADSL | 3000 | 12 Mbps/1 Mbps |
| RADSL | 9000 | 7 Mbps/1 Mbps |
| RADSL | 12000 | 6 Mbps/1 Mbps |
| RADSL | 18000 | 1 Mbps/128 Kbps |
| UDSL | 0–15000 | 2 Mbps/2 Mbps |
| UDSL | 15000–18000 | 1 Mbps/1 Mbps |
| VDSL | 1000 | 51.84 Mbps/2.3 Mbps |
| VDSL | 3000 | 25.82 Mbps/2.3 Mbps |
| VDSL | 4000 | 12.98 Mbps/1.6 Mbps |

Source: *Data Communication Magazine*, April 1998.

## Cable TV System Architecture

Cable TV is designed to transmit broadband TV signals to homes using coaxial and fiber-optic cable. Figure 7.15 shows the full coaxial cable TV system architecture. As shown in this diagram, cable TV uses tree and branch bus topology. The tree and branch cable is constructed of 75-ohm coaxial cable connected to the trunk cable.

The head end transmits TV signals over a **trunk cable** to a group of subscribers. The medium can be either coaxial or fiber-optic cable. The function of a **coaxial amplifier** is to amplify the signal, and it works in either direction. Feeder and drop cables are both coaxial cables. The **drop cable** is the part of a cable system that connects the subscriber to the feeder cable. **Feeder cables** are connected to trunk cable to cover a large area. The maximum distance between head end and subscriber is 10 to 15 km. The maximum number of cascaded amplifiers is 35 and the maximum number of connections are 125,000.

**FIGURE 7.15**
Full coaxial cable TV system architecture

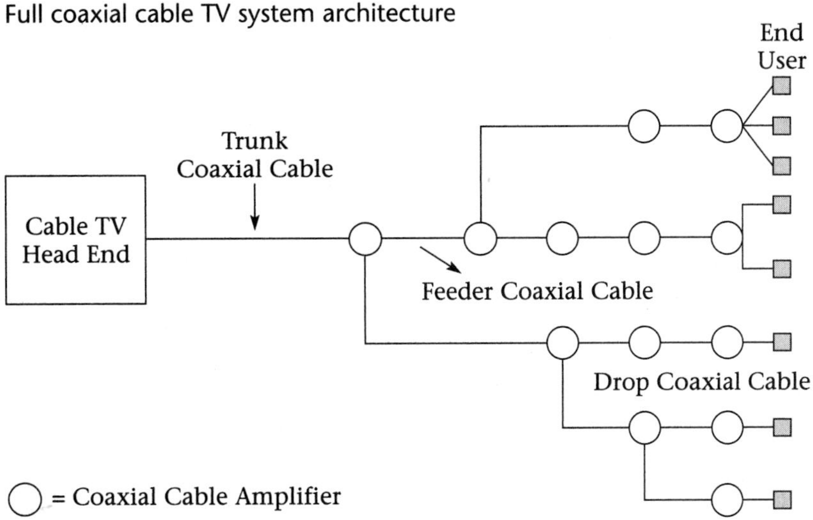


A TV signal transmits two kinds of frequency bands: VHF (Very High Frequency) and UHF (Ultra High Frequency). The VHF channels use lower frequencies to generate stronger signals needed to transmit longer distances. The VHF channels start at channel 2 with frequency of 54 MHz and end at channel 13 with frequency of 216 MHz. The VHF channels use the frequency range of 54 MHz to 216 MHz. UHF channels start at channel 14, with a frequency of 470 MHz and end at channel 83 with frequency of 890 MHz. Each TV channel occupies 6 MHz of the TV radio frequency (RF) spectrum.

The bandwidth of coaxial cable is 500 MHz, with each TV channel requiring 6 MHz of bandwidth. The number of TV channels that can be transmitted is $(500 - 54)/6 = 75$ channels. In order to increase the bandwidth of cable TV, cable TV corporations use **Hybrid Fiber Cable (HFC),** which is a combination of fiber-optic cable and coaxial cable, as shown in Figure 7.16. The bandwidth of a cable TV system using HFC cable is 750 MHz to 1 GHz. Therefore, the number of channels computed by $(750 - 54)/6 = 116$ channels. The TV signal is transmitted to a fiber node using optical cable. The fiber node converts the optical signal to an electrical signal, and also converts electrical to optical. The coaxial amplifiers are two-way devices used to amplify the incoming signal. The maximum distance from head end to end user is 80 km and the maximum number of end users per fiber node connection is between 500 to 3000 (depending on the vendor). A channel between 5 MHz and 42 MHz is used to carry upstream signals (from subscriber to the head end).

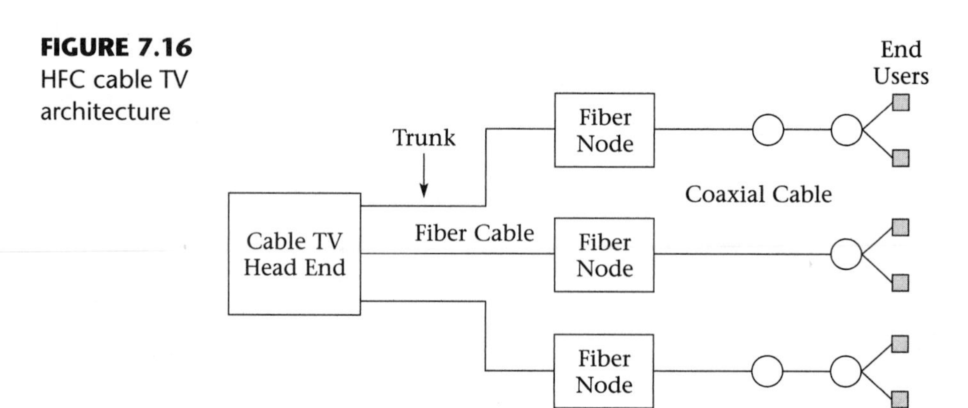


**FIGURE 7.16** HFC cable TV architecture

## Cable Modem Technology

Figure 7.17 shows the components of a cable network consisting of a coaxial cable, **head end**, and a cable modem. The connection between cable modem and user is 10Base-T. The user requires a **10Base-T** NIC card in order to use the cable modem. A cable modem can support more than one station, using a router, as shown in Figure 7.18.

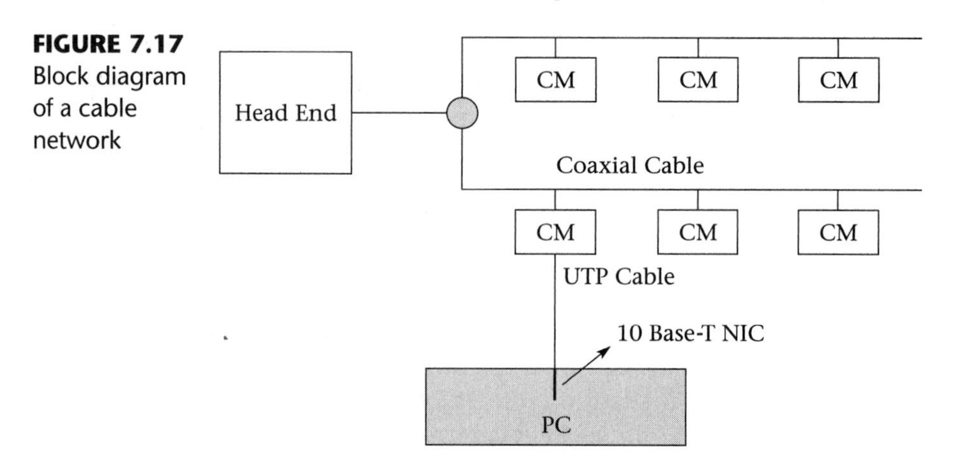


**FIGURE 7.17** Block diagram of a cable network

A cable modem uses 64-QAM or 256-QAM modulation techniques to transmit information from the head end to the cable modem (downstream transmission). If a cable modem uses 256-QAM, this means there are 8 bits per signal and each signal is transmitted at 6 MHz. Theoretically the data rate of a cable modem is:

$$8*6 \text{ Mhz} = 48 \text{ Mbps}$$

By using 64-QAM modulation, the data rate becomes:

$$6*6 \text{ Mhz} = 36 \text{ Mbps}$$

Upstream transmission (from cable modem to head end) uses a 600-KHz channel between 5 to 42 MHz. This low frequency is close to the CB radio frequency. The Quadrature Phase Shift Keying (QPSK) modulation method is used. The data rate of the cable modem for upstream transmission becomes:

$$2*600 \text{ Khz} = 1200 \text{ Kbps}.$$

Downstream and upstream bandwidths are shared by 500 to 5000 cable modem subscribers. If 100 subscribers are sharing a 36-Mbps connection, each user will receive a data at rate of 360 Kbps. A cable modem provides a constant connection (like a LAN); it does not require any dialing. The cable modem head end communicates with the cable modem, and when the cable modem is commanded by the cable modem head end, the modem will select an alternate channel for upstream transmission.

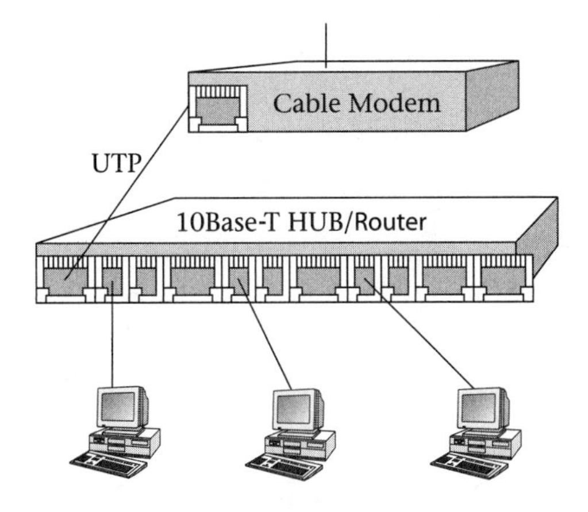


**FIGURE 7.18**
Connection of more than one station to a cable modem

**IEEE 802.14**

A cable modem operates at the Physical and Data Link layers of the OSI model. The **IEEE 802.14 standard** provides a network logical reference model for the Media Access Control (MAC) and Physical layer. The following are general requirements defined for cable modems by IEEE 802.14:

- Cable modems must support symmetrical and asymmetrical transmission in both directions
- Support of operation, administration, and maintenance (OAM) functions
- Support the maximum 80-km distance for transmission from the head end to the user
- Support a large number of users
- MAC layer should support multiple types of service, such as data, voice, and images
- MAC layer must support unicast, multicast, and broadcast service
- MAC layer should support fair arbitration for accessing the network

## 7.4 Integrated Services Digital Network (ISDN)

The **Integrated Services Digital Network (ISDN)** is a set of digital transmission standards that are used for end-to-end digital connectivity. ISDN supports voice and data. The integration of different services has become an ISDN hallmark. In the past, video, audio, voice, and data services required at least four separate networks. ISDN integrates voice, data, video, and audio over the same network. ISDN uses a digital signal, which is less vulnerable to noise than the analog signal used by a modem. ISDN brings the digital network to users. There are two types of ISDN: **Narrowband ISDN (N-ISDN)** and **Broadband ISDN (B-ISDN).**

The International Telecommunications Union (ITU), formerly known as CCITT, has defined standards for ISDN that provide end-to-end digital connection to support a wide range of services including voice and non-voice transmission. ISDN offers digital transmission over existing telephone wiring as provided by telephone companies. ISDN offers the Basic Rate Interface (BRI) and Primary Rate Interface (PRI).

### Basic Rate Interface (BRI)

The **Basic Rate Interface (BRI)** is made of two B-channels (bearer) and one D-channel. Therefore, the total rate is 2B + D. B-channels are 64 Kbps and can be used for voice and data communications. The D-channel is 16 Kbps and is used for call initialization and signaling connection. Figure 7.19 shows ISDN Basic Rate Interface (BRI).

**FIGURE 7.19**
ISDN BRI

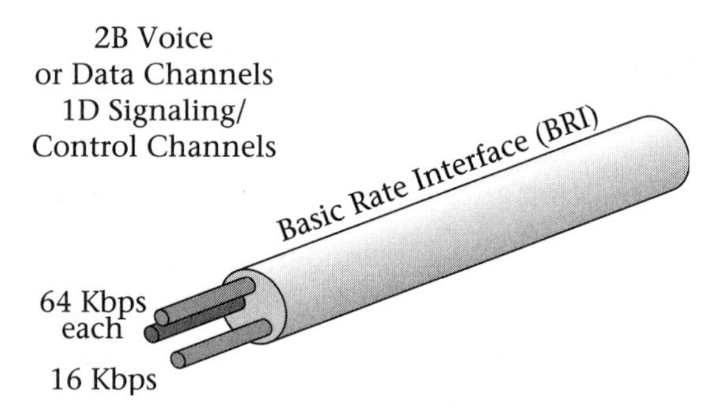


### Application of the BRI

ISDN can carry multiple services voice, video, and data—on a single telephone line over existing twisted-pair copper wire. ISDN's BRI uses two 64-Kbps B-channels and one 16-Kbps D-channel. By combining two B-channels, the total data transmission rate is 128 Kbps. Figure 7.20 shows an application of BRI in ISDN.

**FIGURE 7.20**
Application of BRI

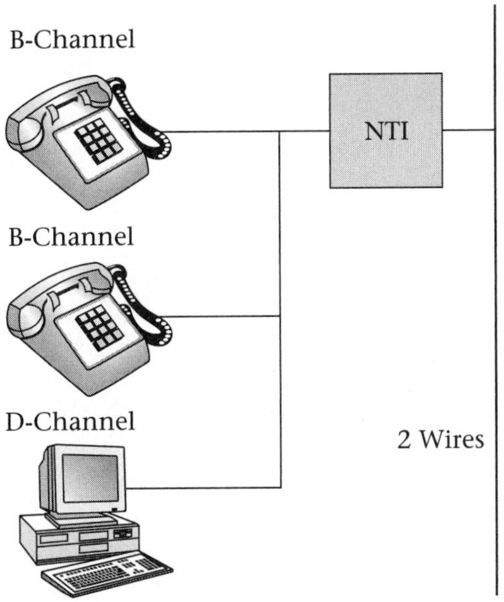

A **Network Terminator Type 1 (NT1)** and a power supply are required for every ISDN line. The NT1 is a device that is physically connected to the ISDN line. A special terminal adapter can combine the two B-channels to be used as a 128-Kbps channel and can then be connected to the computer.

The NT1 device works as a multiplexer and demultiplexer. Figure 7.21 shows the frame format of ISDN BRI. The size of a frame is 48 bits. Frame size = 4*8 + 4 + 12 bits overhead.

**FIGURE 7.21**
ISDN BRI frame format

| 8 bits | 1 bit | 8 bits | 1 bit | 8 bits | 1 bit | 8 bits | 1 bit |
|---|---|---|---|---|---|---|---|
| B1 | D | B2 | D | B1 | D | B2 | D |

**Primary Rate Interface (PRI)**

The **Primary Rate Interface (PRI)** in North America has 23 B-channels and one 64-K D-channel or 23 B + D. 23 B + D channels have a total bandwidth of 1.544 Mbps and are designed to replace T1 links. PRI in Europe uses 30 B-channels and one D-channel or 30 B + D with total rate of 2.048 Mbps as shown in Figure 7.22.

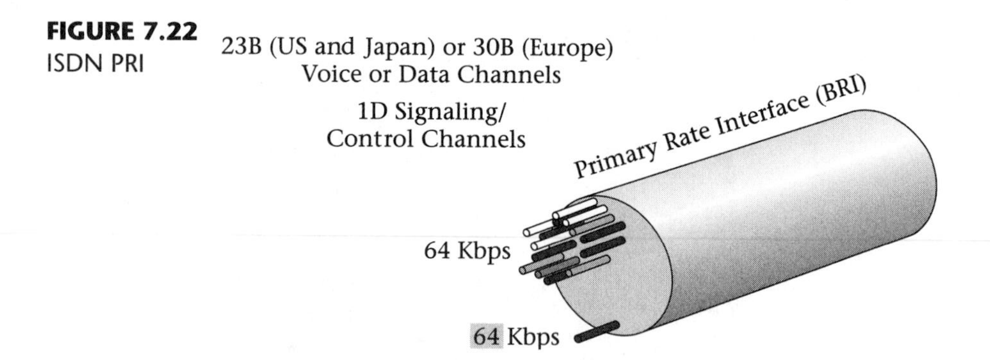


**FIGURE 7.22**
ISDN PRI

**Application of PRI**

An application of PRI is to connect two central switches together or to use it as a T1 link, as shown in Figure 7.23. The devices that handle switching and multiplexing (such as PBX) are called *Network Terminator Type 2 (NT2)*. ISDN PRI can connect the customer directly through an NT2 device whereas ISDN BRI requires an NT1 device.

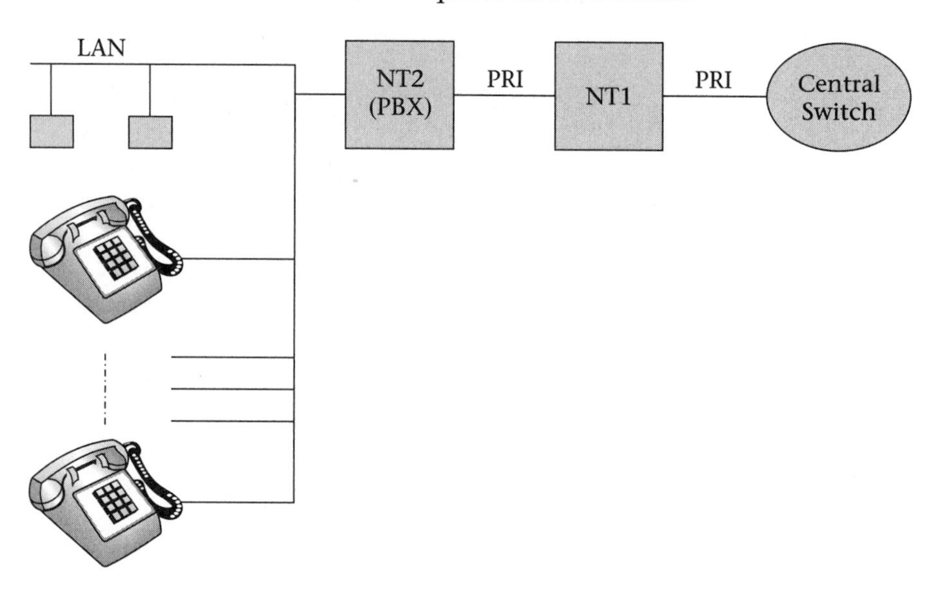


**FIGURE 7.23**
ISDN PRI application

Four types of modem technology have been presented: dial-up, DSL, cable, and ISDN. There are some features of each that you need to remember.

The maximum data rate for a dial-up modem is 56 Kbps. The maximum data rate for an ISDN modem is 128 Kbps, and the cable modem data rate is variable and depends on how many users are accessing the link at the same time. Also, a cable modem link is shared with other users, making security and privacy an issue to take into account.

The xDSL modem is the latest technology; this modem uses the telephone line and can transmit data at a higher rate than an ISDN modem, or dial-up modem. One of the biggest advantages of the xDSL modem over cable modem is security and privacy. The xDSL modem is more cost effective, since it can replace the expensive T1 link. Because it is cost effective, it can be used for connecting libraries and schools to the Internet. The CATV coaxial cable can carry a much higher bandwidth than POST and UTP.

## Summary

- The function of a modem is to convert an analog signal to a digital signal and a digital signal to an analog signal.
- Modulation is used to convert a digital signal to an analog signal.
- Modulation methods are Amplitude Shift Keying (ASK), Frequency Shift Keying (FSK), Phase Shift Keying (PSK), and Quadrature Amplitude Modulation (QAM).
- Amplitude Shift Keying (ASK) changes the amplitude of carrier signals in order to represent a digital signal.
- Frequency Shift Keying (FSK) changes the frequency of carrier signals in order to represent a digital signal.
- Phase Shift Keying (PSK) changes the phase of a carrier signal.
- QAM modulation is a combination of ASK and PSK, used in high-speed modems.
- Baud rate is the number of signals per second that a modem can transmit.
- Data rate is the number of bits per second that a modem can transmit.
- Digital Subscriber Line (DSL) is a type of modem that can transfer data at higher speeds than a modem.
- DSL technology divides the bandwidth of UTP cable into 250 channels. The bandwidth of each channel is 4 KHz. The first channel is used for telephone, and other channels are used for transmitting information.
- Types of DSL are ADSL, SDSL, HDSL, and VDSL.
- ADSL can transfer data downstream at rates of 1.5 to 8 Mbps.
- ADSL can transfer data upstream at rates of 16 to 640 Kbps.

- Cable modems use a cable TV network to connect residential computers to the Internet.
- The head end of a cable TV system uses TV channels to transmit information to a cable modem at the subscriber site.
- Each cable TV channel requires 6-MHz bandwidth.
- Connecting a computer to a cable modem requires a 10Base-T network card.
- More than one station can be connected to a cable modem by using a hub or repeater.
- Cable modem operates in layers 1–3 of the OSI model.
- IEEE 802.14 has developed the standard for cable modems.
- Integrated Digital Network (ISDN) provides end-to-end digital connection.
- Narrowband ISDN offers Basic Rate Interface (BRI) and Primary Rate Interface (PRI).
- The Basic Rate Interface (BRI) is made of two B-channels and one D-channel. B-channel data rate is 64 Kbps and D-channel is 16 Kbps.
- The Primary Rate Interface (PRI) is made of 23 B-channels and one 64-K D-channel in the US.
- The BRI offers to the telephone subscriber line two telephone lines and one data line. The two telephone lines can be combined and used as a 128-Kbps data line.
- The application of PRI is to connect two central switches together or it can be used as a T1 link.
- Network Termination Device Type 1 (NT1) is connected to a subscriber telephone line in order to provide service for two telephones and one computer.

## Key Terms

10Base-T
Amplitude Shift Keying (ASK)
Asymmetrical Digital Subscriber Line (ADSL)
Basic Rate Interface (BRI)
Baud Rate
Broadband ISDN (B-ISDN)
Cable Modem
Coaxial Amplifier
Constellation Diagram
Data Rate
Demodulation
Digital Subscriber Line (DSL)
Downstream

Drop Cable
External Modem
Feeder Cables
Frequency Shift Keying (FSK)
Hybrid Fiber Cable (HFC)
Head End
IEEE 802.14 Standard
Integrated Services Digital Network (ISDN)
Internal Modem
Modem
Modulation
Narrowband ISDN (N-ISDN)
Network Terminator Type 1 (NT1)
Phase Shift Keying (PSK)
Primary Rate Interface (PRI)
Quadrature Amplitude Modulation (QAM)
Rate Adaptive Asymmetric DSL (RADSL)
Trunk Cable
Upstream

## Review Questions

### • Multiple Choice Questions

1. A modem converts ______.
   a. analog signal to digital
   b. digital signal to analog
   c. a and b
   d. analog to analog
2. ______ is responsible for developing standards for modems.
   a. ITU
   b. IEEE
   c. EIA
   d. ISO
3. The maximum theoretical data rate for a modem is ______.
   a. 33.6 Kbps
   b. 56 Kbps
   c. 28 Kbps
   d. 24 Kbps
4. ______ is the latest technology in modems.
   a. DSL
   b. Cable
   c. Dial-up
   d. LAN
5. ADSL uses ______ encoding.
   a. QAM
   b. DMT
   c. PSK
   d. FDM
6. Cable TV is designed to transmit a ______ signal.
   a. baseband
   b. broadband
   c. digital
   d. optical
7. The data rate of a communication channel with bandwidth of 40 Khz and each signal represented by 4 bits is ______.
   a. 40 Kbps
   b. 80 Kbps
   c. 160 Kbps
   d. 10 Kbps

8. QAM modulation is a combination of ______.
   a. ASK and FSK
   b. ASK and PSK
   c. PSK and FSK
   d. none of the above
9. The ______ device uses twisted-pair cable.
   a. cable modem
   b. DSL modem
   c. 10Base-2
   d. 10Base-5
10. The type of modulation method used in cable modems for downstream transmission is ______.
    a. DMT
    b. QAM
    c. QPSK
    d. ASK
11. The type of modulation used in cable modems for upstream transmission is ______
    a. QAM
    b. DMT
    c. QPSK
    d. FSK
12. DSL operates with ______.
    a. an analog signal
    b. a digital signal
    c. an optical signal
    d. baseband
13. The bandwidth of each TV channel is ______.
    a. 4 MHz
    b. 2 MHz
    c. 6 MHz
    d. 1 MHz
14. The lowest frequency of TV Channel 2 is ______.
    a. 40 MHz
    b. 54 MHz
    c. 60 MHz
    d. 30 MHz

## • Short Answer Questions

1. What does the word modem stand for?
2. Explain the function of a modem.
3. Define data rate.
4. Define baud rate.
5. Explain ASK modulation.
6. Explain FSK modulation.
7. Explain PSK modulation.
8. The speed of a modem is represented by ______.
9. What is the speed of modems currently being produced?
10. Explain 56-K modem operation.
11. Distinguish between data rate and baud rate.

12. Draw a constellation diagram for 32QAM using two amplitudes.
13. What does DSL stand for?
14. What does ADSL stand for?
15. Explain ADSL operation.
16. What type of modulation is used for ADSL?
17. Explain xDSL.
18. Why can ADSL transfer information faster than a modem?
19. Is ADSL dependent on the length of the cable?
20. What type of cable is used for ADSL?
21. What are the components of a cable TV system?
22. What does HFC stand for?
23. What is the bandwidth of a TV channel?
24. What type of modulation is used in cable TV modems for upstream transmission?
25. What type of modulation is used in cable TV modems for downstream transmission?
26. What type of NIC is used in a computer connected to a cable TV modem?
27. List the devices that can be connected to a cable modem.
28. What is the baud rate of ASK with a data rate a 600 bits per second?
29. What is the data rate of a modem using frequency shift keying with the baud rate of 300 signals per second?
30. What is the data rate of a QAM signal with baud rate of 1200 and each signal represented by 4 bits?
31. Calculate the number of the bits represented by each signal for a PSK signal with a data rate of 2400 bps and baud rate of 600.
32. How many bits per signal can be represented by a 32-QAM signal?
33. Calculate the baud rate of a 32-QAM signal with a data rate of 25 Kbps.
34. What does ISDN stand for?
35. List the types of ISDN.
36. How many channels does BRI have?
37. List the data rates of the B-channel and D-channel for BRI.
38. How many devices can be connected to BRI ISDN?
39. What is the data rate of the D-channel for PRI?

CHAPTER 8

# Ethernet and IEEE 802.3 Networking Technology

**OUTLINE**

8.1 Ethernet Operation

8.2 IEEE 802.3 Frame Format

8.3 Ethernet Characteristics, Cabling, and Components

**OBJECTIVES**

After completing this chapter, you should be able to:

- Describe Ethernet access methods
- Discuss the function of each field in the Ethernet frame format
- Distinguish among unicast, multicast, and broadcast addresses
- Explain the different types of Ethernet media

**INTRODUCTION**

**Ethernet** was invented by the Xerox Corporation in 1972. It was further modified by Digital, Intel, and Xerox in 1980 and it was called *Ethernet version* I or DIX (Digital, Intel, and Xerox). The IEEE (Institute of Electrical and Electronics Engineers) was assigned to develop a standard for local area networks. The committee that standardized Ethernet, Token Ring, fiber optic, and other LAN technology called it "802." Ethernet can be the least expensive LAN to implement. The IEEE developed the standards for Ethernet in 1984. It is called **IEEE 802.3** and uses the bus topology and logically a star topology. Figure 8.1 shows Ethernet bus topology.

Figure 8.2 shows how Ethernet fits into the OSI model. The Data Link layer is divided into two sublayers, the **Logical Link Control (LLC)** and the **Media Access Control (MAC)** layers. The function of the LLC is to establish a logical connection between the source and the destination. The IEEE standard for the

LLC is IEEE 802.2. The function of the MAC is to access the network, which uses **CSMA/CD (Carrier Sense and Multiple Access with Collision Detection).**

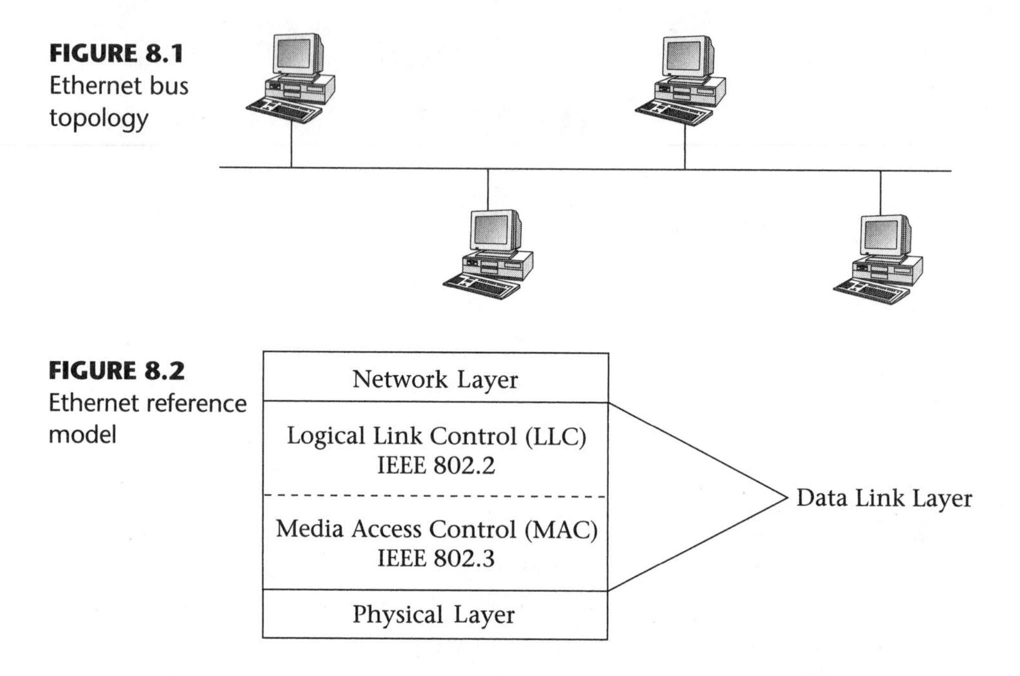


**FIGURE 8.1** Ethernet bus topology

**FIGURE 8.2** Ethernet reference model

## 8.1 Ethernet Operation

Each network card has a unique physical address. When a station transmits a frame on the bus, all stations connected to the network will copy the frame. Each station checks the address of the frame, and if it matches the station's NIC address, it accepts the frame. Otherwise the station discards the frame.

In an Ethernet network, each station uses the CSMA/CD protocol to access the network in order to transmit information. CSMA/CD works as follows:

1. If a station wants to transmit, the station senses (listens to) the channel. If there is no carrier, the station transmits and checks for a collision as described next. If the channel is in use, the station keeps listening until the channel becomes idle. When the channel becomes idle, the station starts transmitting again.
2. If two stations transmit frames at the same time on the bus, the frames will collide. The station that detected the collision first sends a jamming code on the bus (a jam signal is 32 bits of all

ones), in order to inform the other stations that there is a collision on the bus.

3. The two stations that were involved in the collision wait according to the back-off algorithm (a method used to generate random waiting times for stations that were involved in a collision), and then start retransmission. Figure 8.3 shows the flow chart of CSMA/CD.

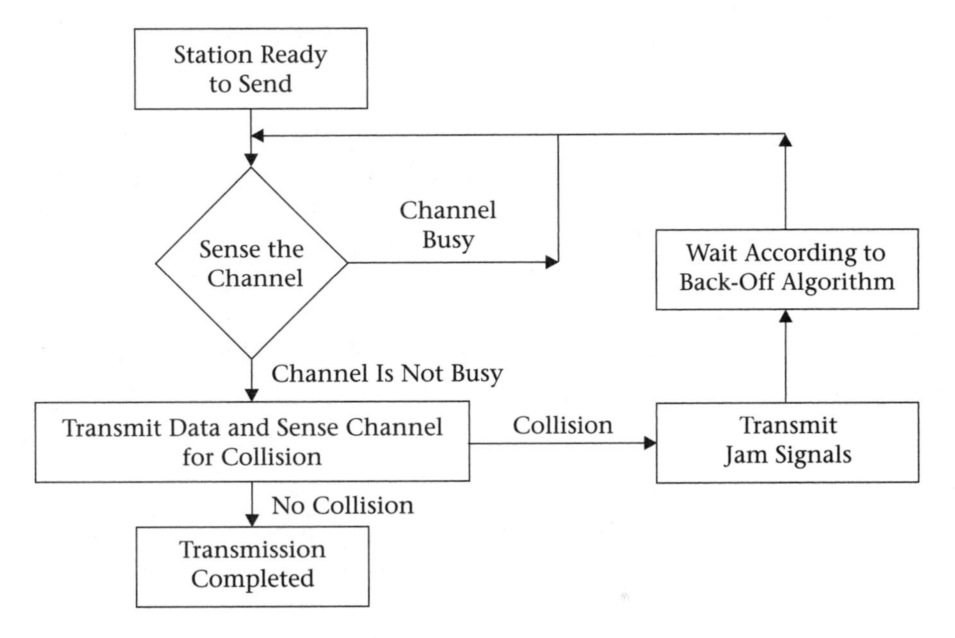


**FIGURE 8.3**
CSMA/CD flow chart

## 8.2 IEEE 802.3 Frame Format

A block of data transmitted on the network is called a *frame*. Figure 8.4 shows the IEEE 802.3 **Ethernet frame format**.

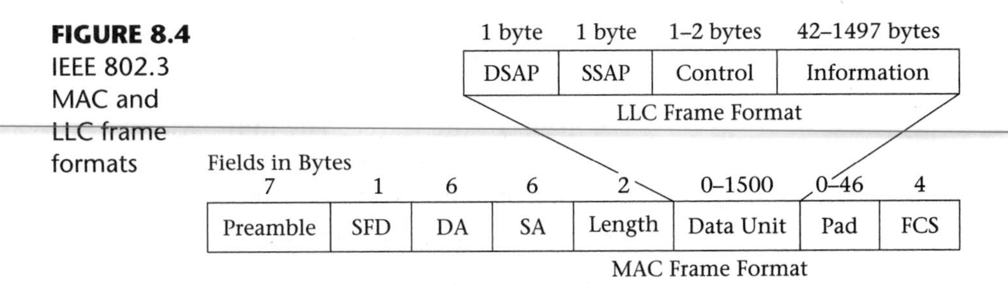


**FIGURE 8.4**
IEEE 802.3 MAC and LLC frame formats

Most network manufacturers conform to the IEEE standards. The items listed below describe each field of the IEEE 802.3 frame format:

**Preamble:** The preamble provides signal synchronization and consists of seven bytes of alternating 1 and 0 bits.

**Start of Fame Delimiter (SFD):** The SFD represents the start of the frame and is always set to 10101011.

**Destination Address (DA):** The destination address is the hardware address of a recipient station and it is six bytes (48 bits). This address is a unique address (in the entire world). The hardware address of the Network Interface Card (NIC) is also called a *MAC address* or *physical address*. The IEEE oversees the physical addresses of NICs worldwide by assigning 22 bits of physical address to the manufacturers of network interface cards. The 46-bit address is burned into the ROM of each NIC and is called the *universal administered address*. Figure 8.5 shows the format of the destination address. The destination address can have the following types of addresses:

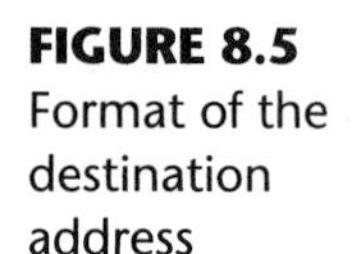

**FIGURE 8.5**
Format of the destination address

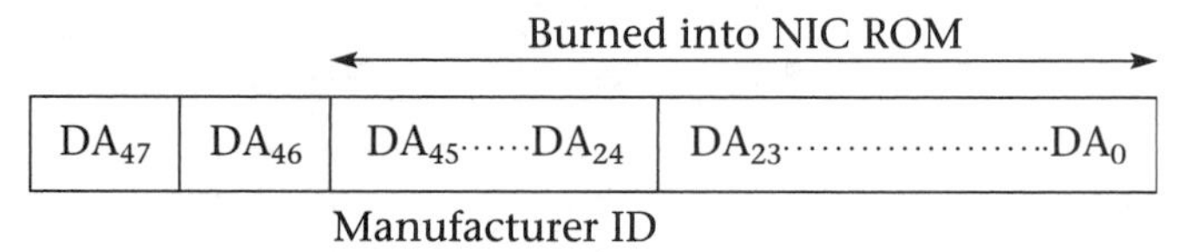


- Recipient is an individual station (**unicast**).
- Recipient is a group of stations (**multicast**).
- Recipients are all stations in the network (**broadcast**). The 48 bits of destination address are all set to ones, meaning that the DA address is FFFFFFFFFFFF Hex for a broadcast address.
- In Figure 8.5 the $DA_{47}$ is set to zero, which means that the address is used for an individual recipient of the frame. A multicast address is a special address that is assigned by the network operating system to a group of stations. When a frame is transmitted to the network with a multicast address, stations having multicast addresses will accept the frame. The multicast address is assigned by IEEE to the manufacturer of NICs. For example, the multicast address for Cabletron NICs is 010010FFFF20 Hex.
- $DA_{46}$ set to one means that the address is a universally administered address.

- $DA_{46}$ set to zero means that the address is locally administered, and a network administrator can assign an address to the NIC. This type of addressing is used only in a closed network.
- $DA_{45}$ to $DA_{24}$ represent the manufacturer's identifier.
- $DA_{45}$ to $DA_0$ are burned into the ROM of the NIC card and indicate the physical address of the card.

**Source Address (SA):** The SA shows the address of the source from which the frame originated.

**Length Field:** This two-byte field defines the number of bytes in the data field.

**Data Field:** According to Figure 8.4, the data field contains the actual information. The IEEE specifies that the minimum size of the data field must be 46 bytes, and the maximum size is 1500 bytes.

**Pad Field:** If the information in the data field is less than 46 bytes, extra information is added in the pad field to increase the size to 46 bytes.

**Frame Check Sequence (FCS):** The FCS is used for error detection to determine if any information was corrupted during transmission. IEEE uses CRC-32 for error detection.

**Destination Service Access Point (DSAP):** The MAC layer passes information to the LLC layer, which must then determine which protocol the incoming information belongs to, such as IP, NetWare, or DecNet.

**Source Service Access Point (SSAP):** The SSAP determines which protocol is sent to the destination protocol, such as IP or DecNet.

**Control Field:** The control field determines the type of information in the information field, such as the supervisory frame, the unnumbered frame, and the information frame.

## 8.3 Ethernet Characteristics, Cabling, and Components

Table 8.1 gives the Ethernet characteristics. The gap between each frame should not be less than 9.6 msec. A station can have a maximum of ten successive collisions. The size of the jam signal is 32 bits of all ones. The maximum size of the frame is 1512 bytes, including the header. Slot time is the propagation delay of the smallest frame. The smallest frame is 512 bits and each bit time is $10^{-7}$ second; therefore the propagation delay is 512-bit time.

**TABLE 8.1** Ethernet Characteristics

| | |
|---|---|
| Data rate | 10 Mbps |
| Encoding | Manchester encoding |
| Slot time or propagation delay | 512-bit time |
| Interframe gap | 9.6 msec |
| Backoff limit | 10 |
| Jam size | 32 bits |
| Maximum size | 1512 bytes |
| Minimum frame size | 64 bytes |

## Cabling and Components

The Ethernet network uses three different media, called *10BaseT*, *10Base2*, and *10Base5*. Figure 8.6 illustrates the ports of an NIC that are used to connect a computer to a network. There are three different types of connectors that come with an NIC: BNC, RJ-45, and DIX. RJ-45 is used for 10BaseT connections, DIX is used for 10Base5 connections, and BNC is used for 10Base2 connections, all of which are described in the next section.

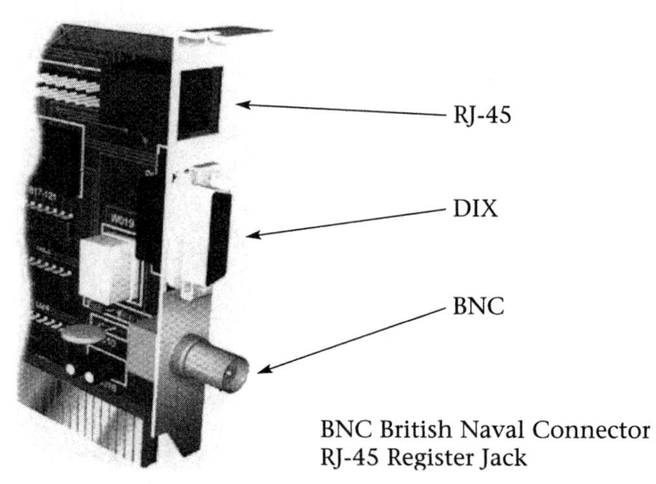


**FIGURE 8.6** Network interface card (NIC)

***ThinNet.*** The specifications of **10Base2 (ThinNet),** as shown in Figure 8.7, are as follows:

- 10Base2 uses thin coaxial cable with BNC connectors.
- The maximum length of one segment is 185 meters.

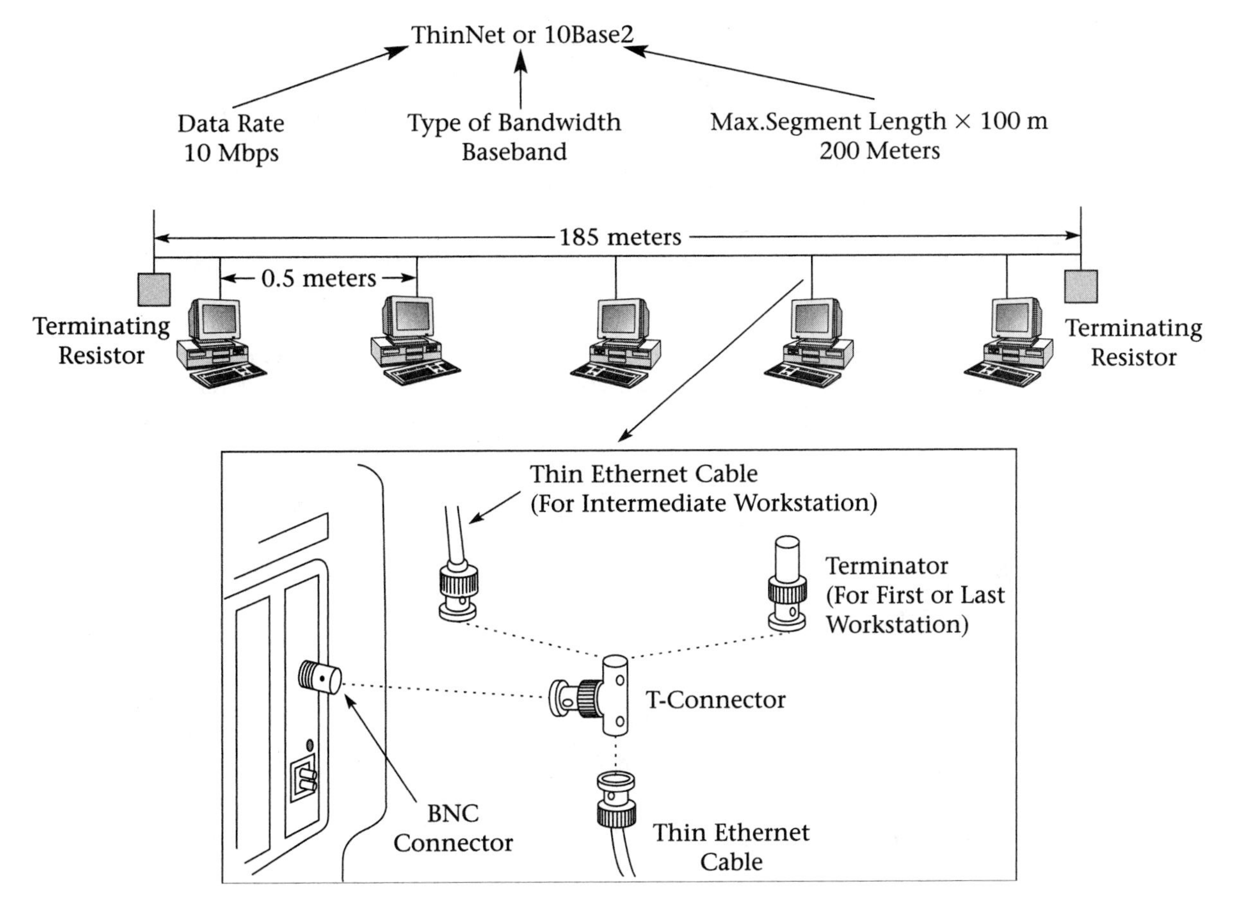


**FIGURE 8.7**
Components of ThinNet

- The maximum length of a network cable is 925 meters, by using four repeaters.
- The transceiver is built into the NIC.
- The minimum distance between T-connectors is 0.5 meters.
- No more than 30 connections are allowed per segment.
- The first and the last device on each segment must be terminated with a 50-Ω resistor called a *BNC terminator*. The function of a terminator is to prevent signal reflection on the cable.
- A T-connector must plug directly into the Ethernet device.

***ThickNet.*** 10Base5 is occasionally used for network backbones. The **transceiver** is a separate component attached to a coaxial cable, as shown in Figure 8.8. The function of a transceiver is to transmit information on the network, receive information from the network, and detect collisions on the network.

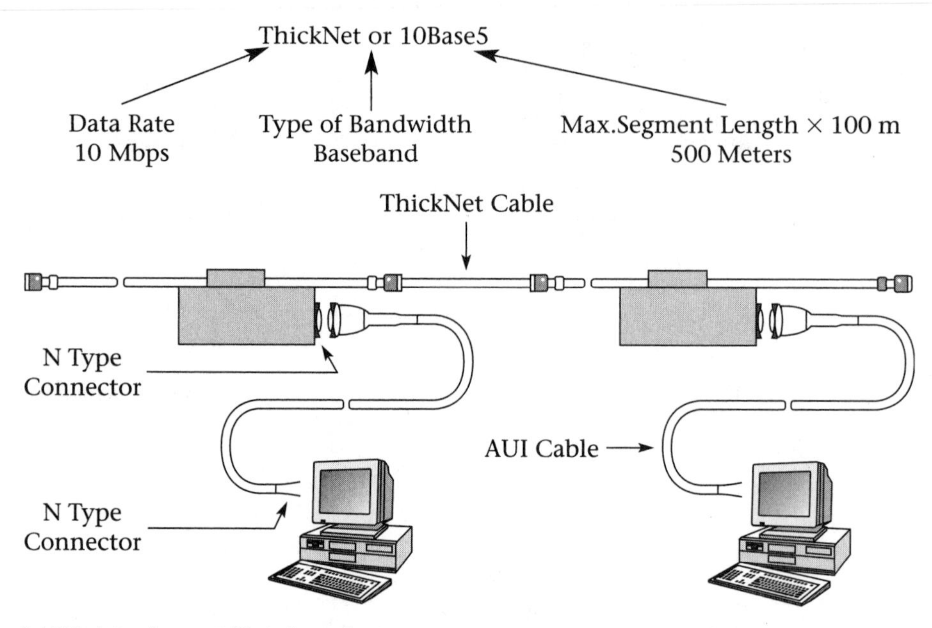


**FIGURE 8.8**
Location of a transceiver for 10Base5 (ThickNet)

The specifications of **10Base5 (ThickNet)** are as follows:

- The maximum length of one segment is 500 meters (without repeaters).
- Devices are attached to the backbone via a transceiver.
- The maximum length of the Attachment Unit Interface (AUI) cable is 50 meters.
- The minimum distance between transceivers is 2.5 meters.
- No more than 100 transceivers are allowed in one segment of the network.
- Both ends of the segments must be terminated by 50-Ω resistor.

***Thick Coax Transceiver with Signal Quality Error (with SQE).*** Figure 8.9 shows a thick coax transceiver. Some transceivers come with a single port and others have two or four ports for output. The function of a transceiver is to interface the Ethernet coax cable with an Ethernet NIC.

**FIGURE 8.9**
A transceiver with SQE

The objective of the **Signal Quality Error (SQE)** is to inform the station that the collision detection section of the transceiver is working properly. A user-selectable switch is provided, permitting the network manager to disable or enable SQE. Figure 8.10 shows the internal architecture of a transceiver. The function of the jabber control is to disconnect the transceiver from the cable in case of a short circuit or malfunction in the transceiver.

**FIGURE 8.10**
Internal architecture of a transceiver

Receive Data
SQE
Signal Quality Error
Collision Detection
Collision Detection Signal
Coaxial Cable
Jabber Control
Transmit Data

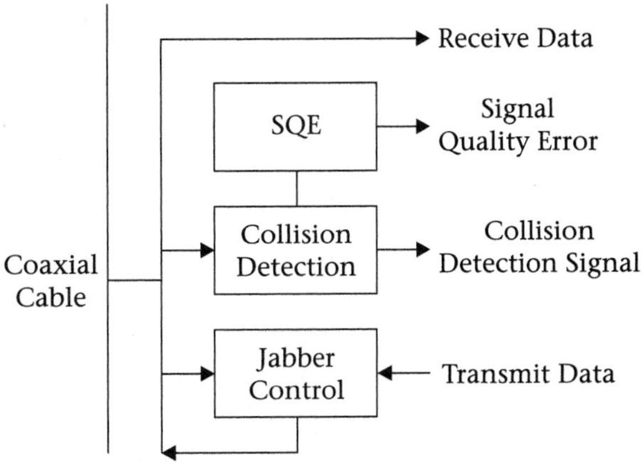

***10BaseT.*** 10BaseT uses UTP cable for transmission media, and all stations are connected to a repeater or hub, as shown in Figure 8.11. The function of the **repeater** (hub) is to accept frames from one port and retransmit

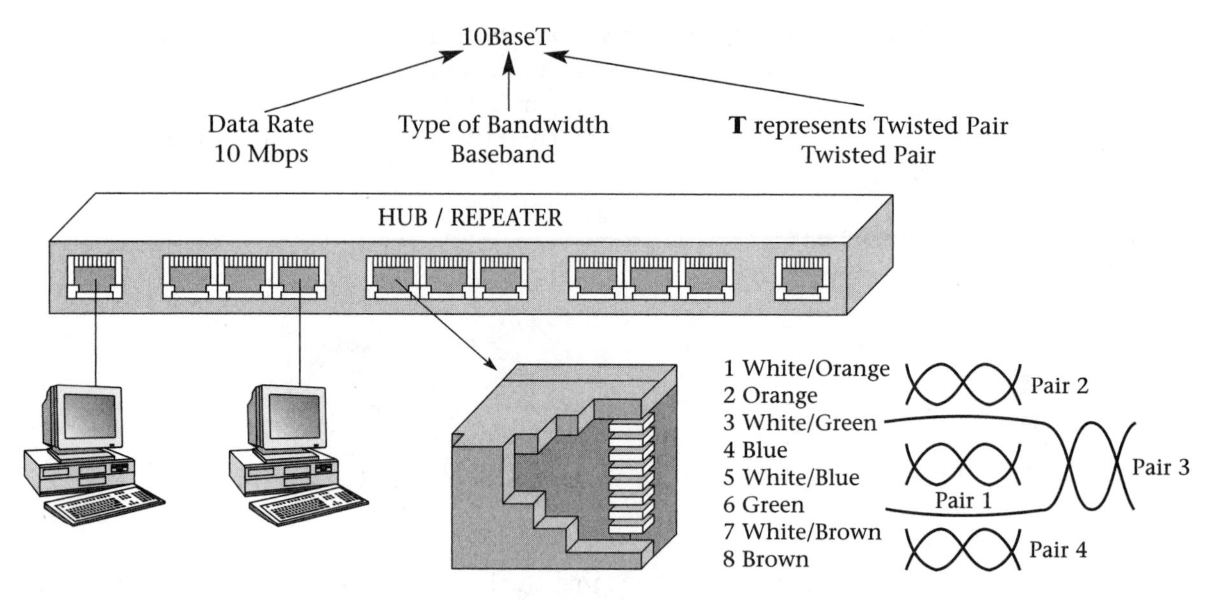


**FIGURE 8.11**
10BaseT connection

the frames to all the other ports. Table 8.2 shows the pin connection of a RJ-45 connector.

**TABLE 8.2** RJ-45 Connector Pins

| PIN | Signals |
|---|---|
| 1 | RD+ pair 2 |
| 2 | RD− pair 2 |
| 3 | TD+ pair 3 |
| 4 | NC pair 1 |
| 5 | NC pair 1 |
| 6 | TD− pair 3 |
| 7 | NC pair 4 |
| 8 | NC pair 4 |

The specifications of **10BaseT** are as follows:

- The maximum length of one segment is 100 meters.
- The transceiver for 10BaseT is built into the NIC.

- The cable used is 22 to 26 AWG unshielded twisted-pair cable, category 4 or 5.
- Devices are connected to a 10BaseT HUB in a physical star topology (electrically, they are in a bus topology).
- Devices with standard AUI connectors may be attached to the HUB by using a 10BaseT transceiver.
- 10BaseT topology allows a maximum of four repeaters connected together, and the maximum diameter is 500m.

A quick reference chart of IEEE 802.3 networks is provided in Table 8.3.

**TABLE 8.3** Comparison of Different Types of IEEE 802.3 Networks

| | 10Base5 (ThickNet) | 10Base2 (ThinNet) | 10BaseT Unshielded Twisted Pair |
|---|---|---|---|
| Medium | Coaxial Cable<br>50 Ohms-10 mm (dia) | Coaxial Cable<br>50 Ohms-5 mm (dia) | Twisted Pair<br>Category 3, 4, 5 |
| Signal | 10 Mbps<br>Baseband/Manchester | 10 Mbps<br>Baseband/Manchester | 10 Mbps<br>Baseband/Manchester |
| Maximum Segment | 500 m | 185 m | 100 m |
| Maximum Distance | 2.5 km | 0.925 km | 500 m |
| Node per Segment | 100 | 30 | n/a |
| Topology | Bus | Bus | Star |

## Summary

- Ethernet (IEEE 802.3) uses bus topology.
- In the bus topology, the medium is shared by all stations.
- Ethernet uses Carrier Sense Multiple Access/Collision Detection (CSMA/CD) to access the network.
- Ethernet data rate is 10 Mbps.
- The maximum size of an Ethernet frame is 1512 bytes.
- An Ethernet network card comes with three types of connectors: RJ-45, BNC, and DIX.
- ThinNet or 10Base2 has these features: 10 Mbps, baseband, and 200 meters per segment using thin coaxial cable with a BNC connector.

- 10BaseT is a medium with these features: 10 Mbps, baseband, and UTP cable with RJ-45 connectors.
- 10BaseT requires a repeater or hub.
- A unicast address tells you that the recipient of the frame is an individual station.
- A multicast address indicates that the recipient of the frame is a group of stations.
- A broadcast address means the recipient of the frame is every station in the network.
- ThickNet or 10Base5 has these features: 10 Mbps, baseband, and 500 meters per segment using thick coaxial cable with DIX connector.
- 10Base5 is used for the backbone, and an external transceiver makes the connection between the NIC and the cable.
- There is a maximum of four repeaters allowed for use in Ethernet.

## Key Terms

Broadcast

Carrier Sense and Multiple Access with Collision Detection (CSMA/CD)

Destination Address (DA)

Destination Service Access Point (DSAP)

Ethernet

Ethernet Frame Format

Frame Check Sequence (FCS)

IEEE 802.3

Logical Link Control (LLC)

Media Access Control (MAC)

Multicast

Repeater

Signal Quality Error (SQE)

Source Address (SA)

Source Service Access Point (SSAP)

10Base2

10Base5

10BaseT

ThickNet

ThinNet

Transceiver

Unicast

## Review Questions

### • Multiple Choice Questions

1. ______ is the least expensive LAN.
   a. Ethernet
   b. Token Ring
   c. a and b
   d. Gigabit Ethernet
2. The standard for Ethernet is ______.
   a. IEEE 802.3
   b. IEEE 802.4
   c. IEEE 802.5
   d. IEEE 802.2
3. Ethernet uses ______ to access channels.
   a. CSMA/CD
   b. token passing
   c. demand priority
   d. full-duplex
4. A destination address has ______ bytes.
   a. 2
   b. 3
   c. 6
   d. 8
5. Ethernet uses ______ encoding.
   a. Manchester
   b. Differential Manchester
   c. NRZ
   d. NRZ-I
6. Ethernet networks come with ______ different media.
   a. 3
   b. 6
   c. 8
   d. 2
7. An RJ-45 connector is used for ______.
   a. 10BaseT
   b. 10Base2
   c. 10Base5
   d. none of the above
8. 10Base2 uses ______ cable.
   a. UTP
   b. STP
   c. thin coaxial
   d. thick coaxial
9. The maximum length of one segment of 10Base2 is ______ meters.
   a. 10
   b. 185
   c. 100
   d. 300
10. Of the following networks, ______ requires a separate transceiver.
    a. 10BaseT
    b. 10Base5
    c. 10Base2
    d. none of the above
11. The type of Ethernet you will choose for connecting two computers is ______.
    a. 10BaseT
    b. 10Base2
    c. 10Base5
    d. none of the above

## • Short Answer Questions

1. Define the following terms:
   a. 10BaseT
   b. 10Base5
   c. 10Base2
2. What type of Ethernet uses coaxial cable?
3. What do UTP and STP stand for?
4. What is 10BaseT topology?
5. What is a network segment?
6. What does AUI stand for?
7. Explain the function of a repeater or a hub.
8. Show the IEEE 802.3 frame formats and functions of each field.
9. Describe the access method for Ethernet.
10. What does CSMA/CD stand for?
11. What is IEEE 802.2?
12. Determine the location of a transceiver for the following network interface cards:
    a. 10Base5
    b. 10Base2
    c. 10BaseT
13. What is a MAC address?
14. Explain the function of jabber control in a transceiver.
15. Explain collisions in Ethernet.
16. What is a jam signal?
17. Explain broadcast addresses.
18. Describe unicast addresses.
19. What is the size of an NIC address?
20. What is the application of CRC (Cyclic Redundancy Check)?
21. What is the function of a transceiver?
22. What is SQE used for in a transceiver?
23. Determine the maximum size of a network using three repeaters for the following media:
    a. 10BaseT
    b. 10Base5
    c. 10Base2

24. What is the maximum number of repeaters allowed for the following media?
    a. 10Base2
    b. 10Base5
    c. 10BaseT
25. What is the maximum size of a frame for IEEE 802.3?
26. How many bits of a network address represent the manufacturer's ID?
27. How do computers distinguish one another on an Ethernet network?
28. What happens when two or more computers simultaneously transmit frames on an Ethernet network?
29. What is the function of the FCS field in the Ethernet frame format?
30. What is the function of the back-off algorithm in an Ethernet network?
31. What is the function of the length field in an Ethernet frame?
32. What is the function of the terminating resistor in 10Base5 and 10Base2?
33. List the IEEE sublayers of the Data Link layer.
34. What is the function of the PAD field in IEEE 802.3 frame format?

CHAPTER 9

# Token Ring and Token Bus Networking Technology

**OUTLINE**

9.1 Token Ring Operation
9.2 Physical Connections
9.3 Ring Management
9.4 Token and IEEE 802.5 Frame Format
9.5 Token Ring NIC and Cable Specifications
9.6 Token Bus (IEEE 802.4)

**OBJECTIVES**

After completing this chapter, you should be able to:

- Discuss token ring technology
- Discuss the token ring access method
- Explain the frame format of the token and the function of each field
- List the components of a token ring
- Explain the token ring MAU and cabling
- Understand the function of the active monitor
- Discuss token bus operation

**INTRODUCTION**

The token ring was originally developed by IBM in the 1970s. **Token ring** was a powerful LAN topology that was designed to handle heavy loads. Token ring networking was introduced in 1985, at a data rate of 4 Mbps. The IEEE 802.5 specification was modeled after the IBM token ring. IBM introduced a second type of token ring in 1989 with a data rate of 16 Mbps.

Currently, the term *token ring* is generally used to refer to both IBM's token ring network and IEEE 802.5 networks, as there is little difference between the two types. **Token ring topology** consists of a ring station and transmission medium, as shown in Figure 9.1. A ring is a combination function (physically a star and logically a ring topology) that allows a device to connect to the ring. A token ring network uses a wiring concentrator device called a **Multistation Access Unit** (**MAU**). The token ring network can be configured with one or more rings and with up to three servers connected to each ring.

**FIGURE 9.1**
Token ring topology

## 9.1 Token Ring Operation

Figure 9.2 shows the flow of information within token ring networks. Traffic passes through each station on the ring, and each station repeats the information to the next station on the ring.

Physically, token ring is a star topology, and electrically it is a ring topology. The **token** itself is a three-byte frame circulating around the network. Any station that wants to transmit information must seize the token, and only then can it transmit information. When a station does not have any information to transmit, it passes the token to the next station. If a station possesses the token and has information to transmit, it inserts information into the token and transmits the frame on the ring. The next station checks the destination address of the frame, and if it matches with the station address, it performs following the functions:

The node copies the frame into its buffer.

The buffer sets the last two bits of the frame to inform the source that the frame was copied by the destination.

The frame is retransmitted on the ring.

The frame circulates on the ring until it reaches the source, which removes the frame from the ring.

The source releases the token by changing the **token bit** (T bit) to one.

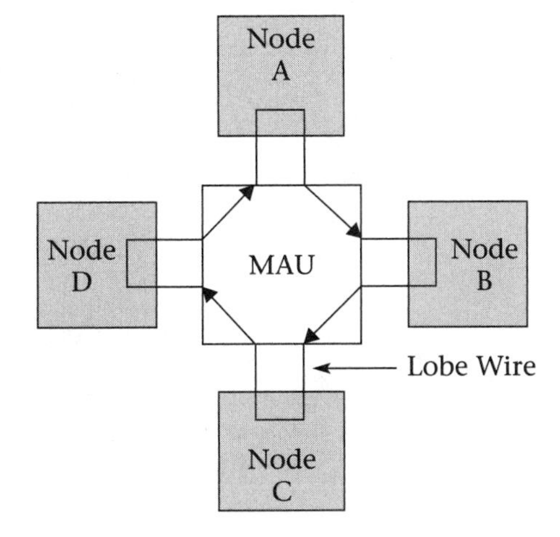


**FIGURE 9.2**
Flow of information on a token ring network

## 9.2 Physical Connections

A ring station is called a **Multistation Access Unit,** or **Multiple Access Unit (MAU),** as shown in Figures 9.3(a) and 9.3(b). Figure 9.3(a) shows an MAU using IBM type connectors, and Figure 9.3(b) shows an MAU using RJ-45 connectors. Each computer is connected to the MAU and each MAU can accommodate up to eight stations. **Shielded twisted-pair (STP) cabling** is used for a 16-Mbps token ring network, and **unshielded twisted-pair (UTP) cabling** is used for a 4-Mbps token ring network. When a station is attached to the ring, it performs the following functions:

1. The station receives a frame from the ring and passes it to next station.
2. The station that generates the frame will remove the frame from the ring and check for error.
3. If the station does not have any frame to transmit, then put the token on the ring.

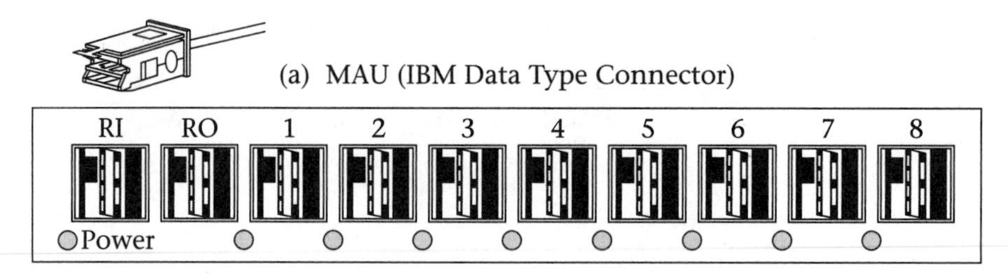


**FIGURE 9.3(a)**
MAU IBM data type connector

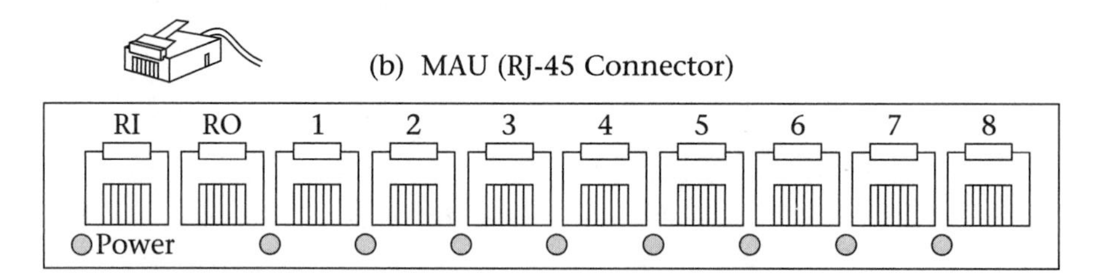


**FIGURE 9.3(b)**
MAU RJ-45 connector

**Expanding the Ring**

A maximum of eight stations can be connected to each MAU. In order to connect sixteen stations to a ring, two MAUs (MAU1 and MAU2) are required. Each MAU has a ring-out port and a ring-in port. Connecting the ring-out of MAU1 to the ring-in of MAU2 and then connecting the ring-in of MAU1 to the ring-out of MAU2 results in a larger ring, to which sixteen stations can be connected, as shown in Figure 9.4.

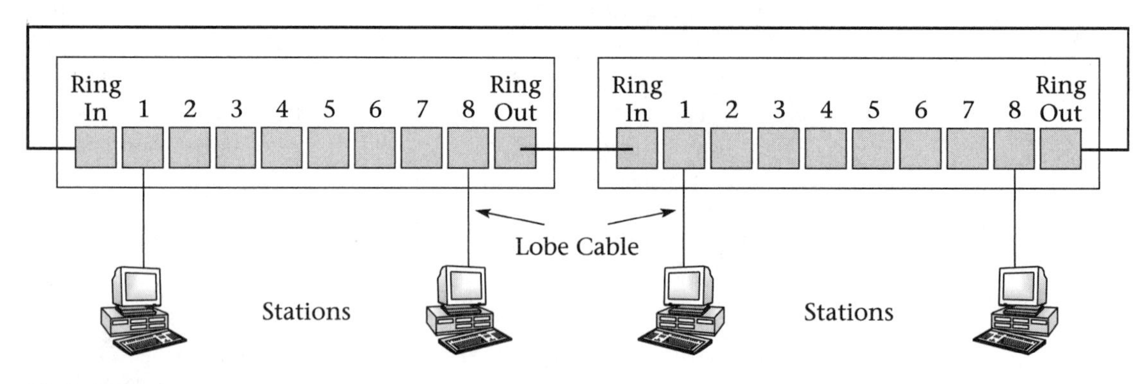


**FIGURE 9.4**
Connection of two MAUs

## 9.3 Ring Management

The token ring technology and protocol were designed to make a token ring network self-managing. Token ring network management is accomplished by using a token ring network card. All token NICs have a network management function, such as adding a new station to the ring without network disruption, generating a token, monitoring activity of the ring, and reporting the loss of a token. These functions are specified by IEEE 802.5.

### Error Detection and Correction

One station acts as the **active monitor** on a ring. All other stations on the ring are standby monitors. The functions of the active monitor are as follows:

1. Generate a 24-bit (3-bytes) token.
2. Broadcast the active monitor MAC frame every seven milliseconds to all the stations in the ring, which informs stations that there is an active monitor on the ring.
3. Determine the address of any active upstream neighbors.
4. Detect a lost token and frame.
5. Maintain the ring master and check for control timing.
6. Purge the ring: the active monitor sends a beacon frame on the ring to inform all stations when there is a problem in the network and token passing has stopped.

When an active monitor fails, it is the responsibility of the **standby monitor** to become the active monitor.

### Adding a Station to the Ring

To add a station to the ring, the following processes must take place:

1. The station must be physically inserted into the ring.
2. The station then sends multiple MAC frames to the MAU to test the wiring lobe.
3. The station must verify that it has a unique address by sending a MAC frame with the same source and destination address and check the frame status (FS) field to see if any station copied the frame.
4. The station determines its upstream neighbor and informs its downstream neighbor of its address.

***Physically Inserting a Station into the Ring.*** Figure 9.5(a) shows a ring before insertion of new stations, and Figure 9.5(b) shows stations after insertion to the ring.

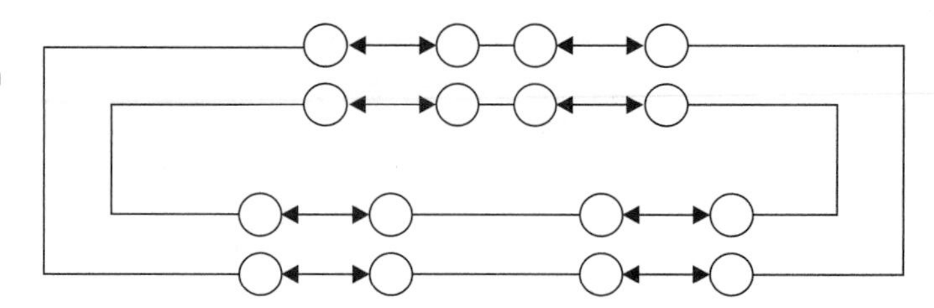

**FIGURE 9.5(a)**
Ring before insertion of new stations

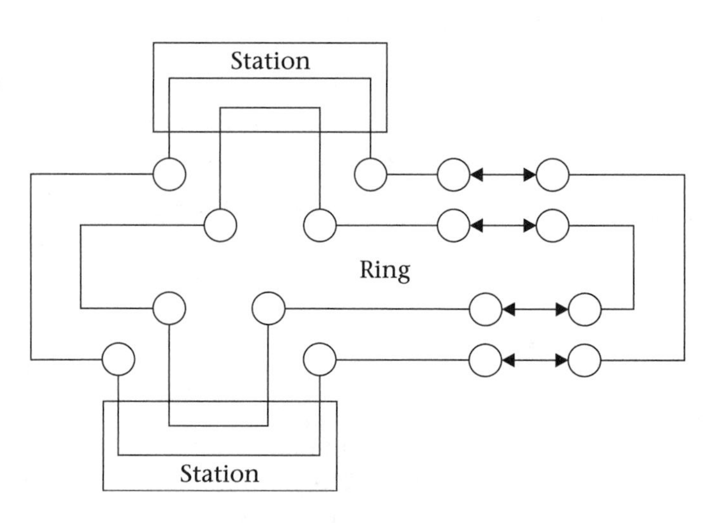


**FIGURE 9.5(b)**
Ring after insertion of new stations

## 9.4 Token and IEEE 802.5 Frame Format

Remember, the **token frame** is a three-byte frame, as shown in Figure 9.6, that is passed from station to station on the network. Only one token is allowed on the network at any given time.

**FIGURE 9.6**
Token frame format

| 8 bits | 8 bits | 8 bits |
|---|---|---|
| SD | AC | ED |

The following describes the functions of each field of the token frame format:

**SD:** Start delimiter of token or frame set to JK0JK000.

**ED:** End delimiter and set to JK1JK10E.

J and K are non-data bits. These bits are Differential Manchester code violations; they do not have mid-point transitions. The signal level for J is the same as for the previous bit and the signal level for K is the opposite of the J signal. This is shown in Figure 9.7. The purpose of this pattern is to keep **Start Delimiter (SD)** bytes or **End Delimiter (ED)** bytes from repeating in the information field of a token ring frame. The ED field is JK1JK10E. E is initially set to zero, and if any station detects an error, it will set this bit to one to alert the other stations.

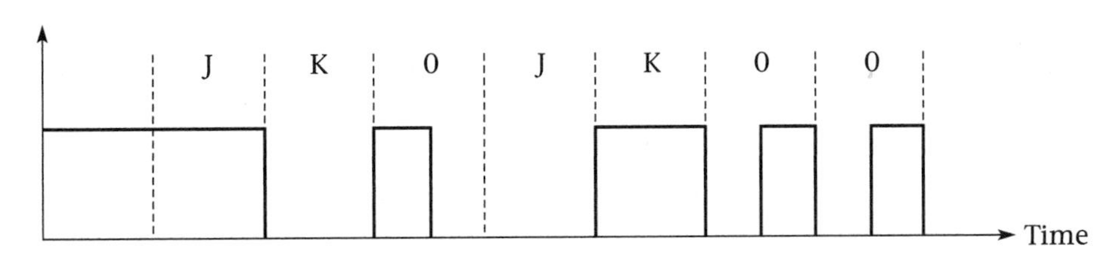


**FIGURE 9.7**
JK0JK00 timing diagram

## Access Control Byte (AC)

Figure 9.8 shows the **Access Control Byte (AC)**. The access control field contains three priority bits, and a token bit that informs stations if the data is in token or frame format. The R bits are reserved.

P = priority bits (000–111)

T = token bit (0 = frame; 1 = token)

R = reserve bits

**FIGURE 9.8**
Access control byte

| P | P | P | T | M | R | R | R |
|---|---|---|---|---|---|---|---|

The **monitor bit (M)** is used to prevent frames from circulating onto the ring. When a frame passes through the active monitor, the M bit is set to 1. If a frame is passed by an active monitor with the M bit set to 1, the active monitor assumes that the frame has already circulated the ring. The active monitor then removes the frame from ring.

## IEEE 802.5 Frame Format

A frame is a unit of information that is used by a token ring network for transferring information between stations. **IEEE 802.5** defined the frame format for the token ring network, which is shown in Figure 9.9.

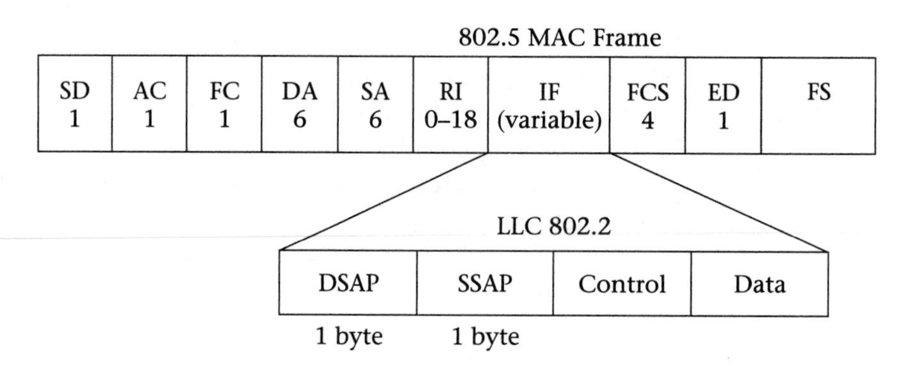


**FIGURE 9.9**
IEEE 802.5 MAC and LLC 802.2 frame

The following are the functions of each field of a token ring frame format:

**Frame Control (FC):** FC is 8 bits, as shown in Figure 9.10.

**FIGURE 9.10**
Frame control (FC)

| F F | R R | Z Z Z Z |
|---|---|---|

**FF Field:** Frame type bits:

00 = MAC frame

01 = LLC frame

10 = reserved

11 = reserved

**RR Reserved Bits:** RR is set to 00.

**ZZZZ Bits:** ZZZZ is 0000 (meaning a normal buffer) or ZZZZ is 0001 (meaning an express buffer).

**Destination Address Field (DA):** The destination address consists of six bytes and is the hardware address of the recipient station. DA uses the same address format as IEEE 802.3.

**Source Address (SA):** The source address is six bytes and represents the originator of the frame.

**Routing Information (RI):** Routing information is sometimes included between the source address and the data. This information is optional and not standard in the IEEE 802.5 standard.

**Information Field (IF):** If the information field contains a MAC frame, the frame is called a *MAC protocol data unit.* If the information field contains an LLC frame, the field is called an *LLC protocol* unit (LPDU).

**Frame Check Sequence (FCS):** The FCS is four bytes.

**Frame Status Field (FS):** The frame status is one byte, as shown in Figure 9.11.

**FIGURE 9.11**
Frame status field (FS)

| A | C | R | R | A | C | R | R |
|---|---|---|---|---|---|---|---|

**Access Control (AC) Value:**

00 = No station recognized the address and frame.

11 = Station recognized the DA and copied the frame.

10 = Station recognized the DA but did not copy the frame. This may also indicate that the buffer is full.

01 = invalid.

There are two types of framed data: one is generated by higher layers of protocol and the other is generated by application software. These frames usually contain application data or network commands. This kind of frame is called an **LLC frame**. **MAC frames** contain major vector commands such as active monitor setting, beacon frame, and ring purge frame and are used for controlling the network.

## 9.5 Token Ring NIC and Cable Specifications

Figure 9.12 shows a token ring NIC with two types of connectors: the RJ-45 used for a UTP cable, and DB-9 used for an STP cable. Table 9.1 shows token ring specifications and Table 9.2 shows the token ring cable specification.

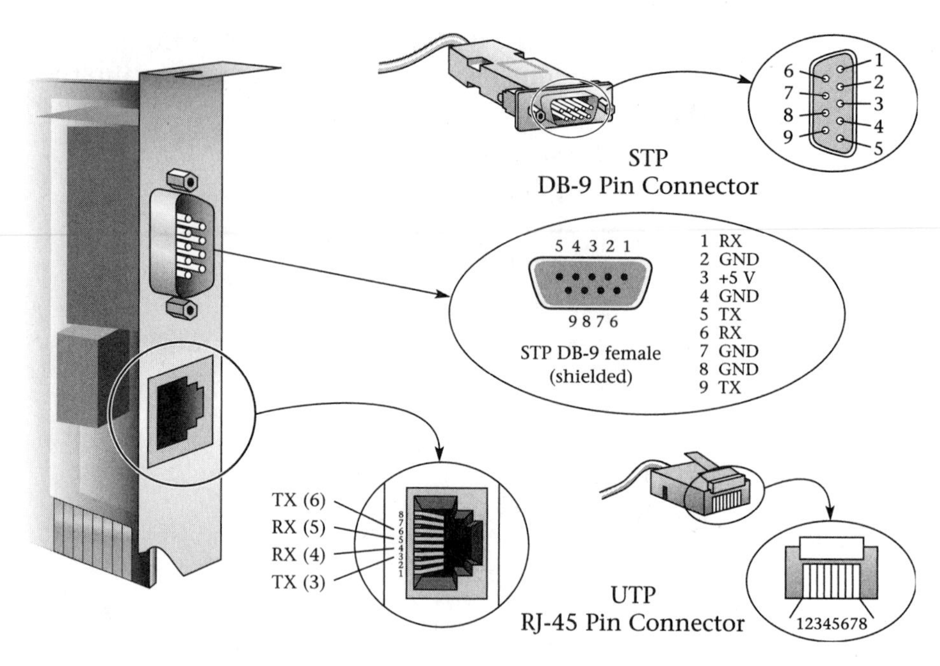


**FIGURE 9.12**
Token ring NIC

**TABLE 9.1** Token Ring Specifications

| | |
|---|---|
| Maximum number of stations | 260 stations on one ring using shielded twisted-pair (STP) cable, and 72 stations on one ring using unshielded twisted-pair (UTP) cable |
| Transmission media | UTP or STP |
| Data rate | 4 or 16 Mbps |

**TABLE 9.2** Token Ring Cable Specifications

| Speed/Cable Type | Max. Length of Cable (in meters) | Max. Length between MAUs (in meters) |
|---|---|---|
| 4 Mbps/UTP | 100 | 100 |
| 16 Mbps/UTP | 75 | 75 |
| 4 Mbps/STP | 300 | 200 |
| 16 Mbps/STP | 100 | 100 |

Table 9.3 compares the token ring with the Ethernet network.

**TABLE 9.3** Comparison of Token Ring and Ethernet

| | Token Ring | Ethernet |
|---|---|---|
| Priority | Yes | No |
| Routing information field | Yes | No |
| Frame type | IEEE 802.5 | IEEE 802.3 |
| Frame size | 1–18000 bytes | 64–1500 bytes |
| Performance | Deterministic | Variable |
| Cable | UTP/STP/Fiber/Coax | UTP/STP/Fiber/Coax |
| Speed | 4, 16 Mbps | 10, 100, 1000, 10000 Mbps |

## 9.6 Token Bus (IEEE 802.4)

The **token bus** is a type of network that uses bus topology and token passing for accessing the network. The **IEEE 802.4** committee has defined token bus standards.

Figure 9.13 shows four stations connected to a bus. All stations are numbered, which indicates a logical ring. The token is passed from station to station; therefore, each station can access the bus by holding the token as it is passed around the logical ring. The applications of token bus network are in manufacturing. Token bus was designed for use in factory automation because of its deterministic delay property, but this technology has not been widely used.

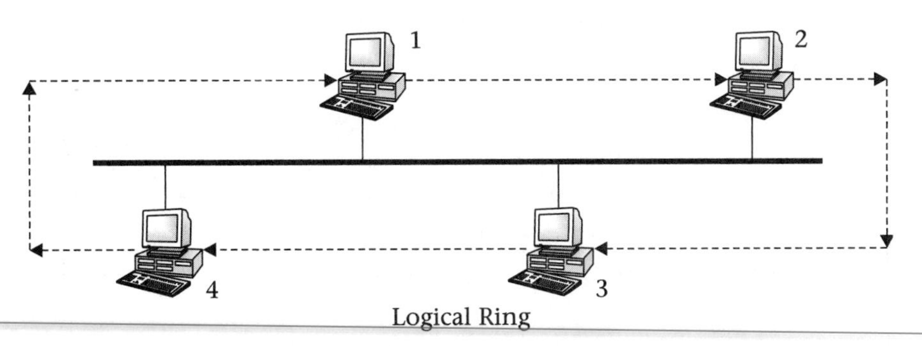


**FIGURE 9.13**
Token bus

## Summary

- The token ring was initially developed by IBM.
- A token ring uses Multiple Access Unit (MAU) as a ring.
- A token ring or IEEE 802.5 uses token passing as its access method.
- A token is a three-byte frame. The token circulates on the ring and any station that wants to use the ring seizes the token, inserts its frame into the token, and then transmits the frame onto the ring.
- The functions of the active monitor are to generate the token, detect the loss of the token, and control the ring operation, removing frames from the ring and purging the ring.
- The function of the standby monitor is to replace the active monitor should it fail.
- Token ring transmission media are shielded twisted-pair and unshielded twisted-pair cables.
- A token ring uses Differential Manchester signal encoding.

## Key Terms

Access Control (AC) Byte
Active Monitor
End Delimiter (ED)
IEEE 802.5
IEEE 802.4
Informative Field (IF)
LLC Frame
MAC Frames
Monitor Bit (M)
Multistation Access Unit (MAU)
Routing Information (RI)
Shielded Twisted-Pair (STP) Cabling
Standby Monitor
Start Delimiter (SD)
Token
Token Bit
Token Bus
Token Frame
Token Ring
Token Ring Topology
Unshielded Twisted-Pair (UTP) Cabling

## Review Questions

### • Multiple Choice Questions

1. A token ring is also known as ______.
   a. IEEE 802.2 c. IEEE 802.5
   b. IEEE 802.3 d. IEEE 802.4

2. A token bus is also known as ______.
   a. IEEE 802.5 c. IEEE 802.3
   b. IEEE 802.4 d. IEEE 802.2

3. A ______ is a combination function that allows a device to connect to the ring.
   a. bus c. token
   b. ring d. hub

4. A token is a ______ byte frame.
   a. 7- c. 2-
   b. 4- d. 3-

5. A ______ station is called a MAU.
   a. token bus c. bus
   b. token ring d. ring

6. There is/are ______ active monitor(s) in a ring.
   a. 1 c. 2
   b. 6 d. a and c

7. The maximum number of stations that can be connected to ring is ______.
   a. 7 c. 8
   b. 12 d. 10

8. The token ring uses ______ encoding.
   a. Manchester c. a and b
   b. Differential Manchester d. NRZ

9. Of the following access methods, ______ is used in token ring.
   a. token passing c. demand priority
   b. CSMA/CD d. full-duplex

10. A station distinguishes between an incoming frame and a token ______.
    a. from the M bit of the AC field c. from the P bits in the AC field
    b. from the T bit in the AC field d. none of the above

11. The station that generates a token is ______
    a. a station close to the MAU c. the station that turns on last
    b. the station that turns on first d. all stations in the ring

## • Short Answer Questions

1. What is the function of a MAU?
2. What is IEEE 802.5?
3. How many stations can be connected to each MAU?
4. What is the access method used by a token ring?
5. How many stations can be connected to a token ring network:
   a. When the stations are connected to a MAU by UTP?
   b. When the stations are connected to a MAU by STP?
6. Show the frame format of the token and explain the function of each field.
7. Explain the function of an active monitor.
8. Describe the function of standby monitor.
9. Explain the process that takes place when inserting a station into the ring.
10. How can a station distinguish between a token and a frame?
11. What type of signaling is used in a token ring network?
12. Show the IEEE 802.2 frame format and explain the function of each field.
13. How many bits are in a token?
14. What is the difference between a token bus and a token ring?
15. What is purpose of the J and K bits in the token ring frame format?

CHAPTER 10

# Fast Ethernet Networking Technology

**OUTLINE**



**OBJECTIVES**

After completing this chapter, you should be able to:

- Discuss Fast Ethernet technology
- Distinguish between the different types of Fast Ethernet media
- Explain the differences and similarities among 100BaseT4, 100BaseTX, and 100BaseFX
- Distinguish between different types of repeaters and know the maximum network diameter

**INTRODUCTION**

**Fast Ethernet** offers a data rate of 100 Mbps. A group of leading network corporations has formed a consortium to draft specifications for Fast Ethernet. This consortium proposed several specifications for Fast Ethernet to the IEEE and the **IEEE 802.3u** committee approved the standard for Fast Ethernet in 1995.

Fast Ethernet is an extension of the Ethernet standard, with a data rate of 100 Mbps, using the Ethernet protocol. The goal of Fast Ethernet is to increase the bandwidth of Ethernet networks while using the same CSMA/CD transmission protocol. Using the same protocol for Fast Ethernet allows users to connect an existing 10BaseT LAN to a 100BaseT LAN with switching devices. Figure 10.1 shows the Fast Ethernet protocol architecture.

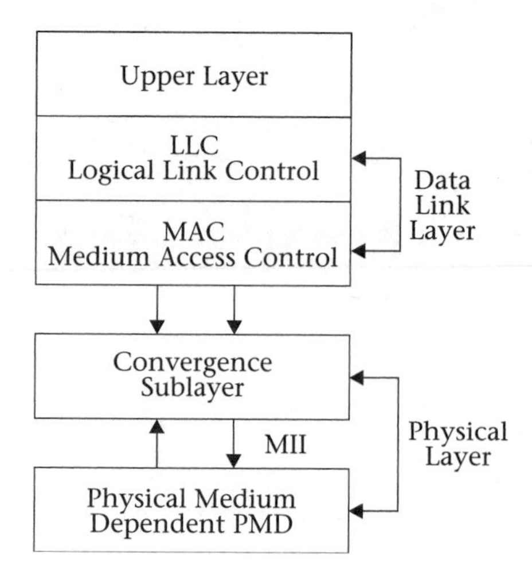


**FIGURE 10.1**
Fast Ethernet protocol architecture

The role of the **Convergence Sublayer (CS)** is to interface the MAC sublayer to the **Physical Media Dependent (PMD) sublayer** in order to transmit with a higher bit rate, using different media. The **Media Independent Interface (MII)** is defined as the interface between the CS layer and the PMD layer.

## 10.1 Fast Ethernet Media Types

One of the most popular media for a Fast Ethernet network is unshielded twisted-pair wire, because it is easy to work with and it is a less expensive medium. The IEEE has approved specifications for the following three types of media for Fast Ethernet:

- 100BaseT4
- 100BaseTX
- 100BaseFX

**100BaseT4**

**100BaseT4** cable is designed to be used with Category 3 unshielded twisted-pair (UTP) wire consisting of four pairs of wire. Categories 4 and 5 can also be used for 100BaseT4. An RJ-45 connector is used to connect a station to a repeater.

100BaseT4

Data Rate
100 Mbps

Type of Bandwidth
Baseband

**T** Represents Twisted Pair
4 Pair UTP

100BaseT4 requires four pairs of twisted-pair wires, as shown in Figure 10.2. Pairs 3 and 4 are used for the transmission of data. One wire from pair 1 and one from pair 2 are used for CSMA/CD. Only three pairs of wire are used for transmission of data and carrier sense detection. Each pair transmits data at a rate of 33.3 Mbps, which exceeds the 30-Mbps specification for UTP Cat-3. Ternary code (also known as *8B6T*) is used to convert eight bits to six bits. A partial list of 8B6T code is shown in Table 10.1. The ternary code is represented by three voltage levels, where + is represented by positive voltage, 0 is represented by zero volts, and − is represented by negative voltage. This method reduces the data rate of each pair from 33.3 Mbps to 25 Mbps.

**FIGURE 10.2**
Physical connection of 100BaseT4

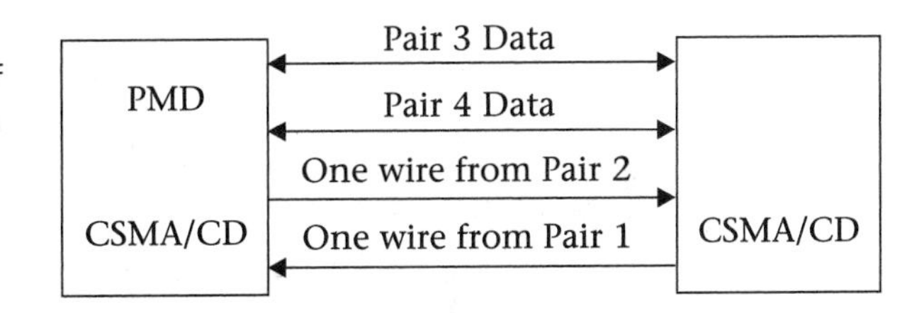


**TABLE 10.1** 8-Bit Binary to 6 Ternary (8B/6T) Code

| 8-bit (8B) Binary in Hex | Ternary Code (6T) |
|---|---|
| 00 | − + 0 0 − + |
| 01 | 0 − + − + 0 |
| 02 | 0 − + 0 − + |
| 03 | 0 − + + 0 − |
| 04 | − 0 0 + − + |
| 05 | 0 + − − 0 + |
| 06 | + − 0 − 0 + |

**100BaseTX** **100BaseTX** technology supports 100 Mbps over two pairs of Cat-5 UTP or Cat-1 STP. Cat-5 UTP cable is the most common media for transmission and is designed to handle frequencies up to 100 MHz. Manchester

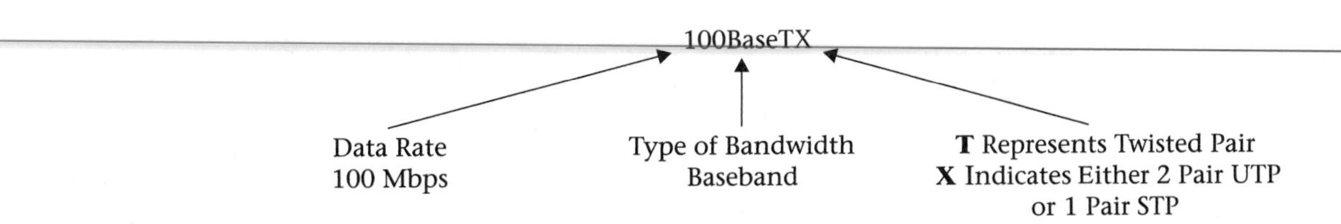

encoding, which is used for 10BaseT, is not suitable for 100BaseT, because it doubles the frequency of the original signal. 100BaseT uses 4B/5B encoding with **Multi-Level Transition-3 (MLT-3)** levels for signal encoding. Figure 10.3 shows $(0E)_{16}$ converted to 10 bits 1111011100 using Table 10.2 and then converted to MLT-3. MLT-3 reduces the frequency of the signal by a factor of four.

**FIGURE 10.3**
MLT-3 signal for binary value 11110 11100

0-03a

**TABLE 10.2** 4B/5B Encoding

| 4 Bits Binary | 5 Bits Symbol | Control Symbols | 5 Bits Symbol |
|---|---|---|---|
| 0000 | 11110 | Idle | 11111 |
| 0001 | 01001 | Halt | 00100 |
| 0010 | 10100 | J | 11000 |
| 0011 | 10101 | K | 10001 |
| 0100 | 01010 | T | 01101 |
| 0101 | 01011 | Set | 11001 |
| 0110 | 01110 | Reset | 00111 |
| 0111 | 01111 | Quiet | 00000 |
| 1000 | 10010 | | |
| 1001 | 10011 | | |
| 1010 | 10110 | | |
| 1011 | 10111 | | |
| 1100 | 11010 | | |
| 1101 | 11011 | | |
| 1110 | 11100 | | |
| 1111 | 11101 | | |

MLT-3 encoding uses three voltage levels: +V, −V, and zero. The MLT-3 encoding rules are as follows:

1. If the next bit of the original signal is zero, then the next output is the same as the preceding value.

2. If the next bit of the original signal is one, then the next output value has a transition (high to low or low to high).
3. If the preceding output was either +V or −V, then the next output value is zero.
4. If the preceding output was zero, then the next output is nonzero (it is the opposite sign of the last non-zero output).

**100BaseFX**

**100BaseFX** technology transfers data at a rate of 100 Mbps using fiber-optic media for transmission. The standard cable for 100BaseFX is one pair of multimode fiber-optic cable with a 62.5-micron core and 125-micron cladding. The EIA recommends an SC plug-style connector.

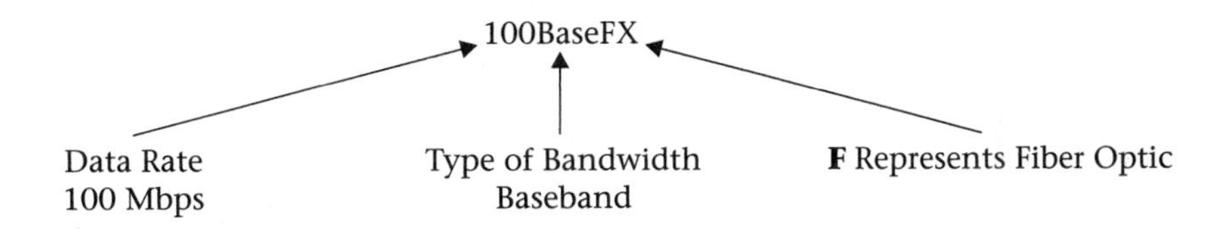


The SC connector uses push-on and push-off to connect and disconnect from the repeater. Figure 10.4 illustrates how a 100BaseFX connects to a repeater.

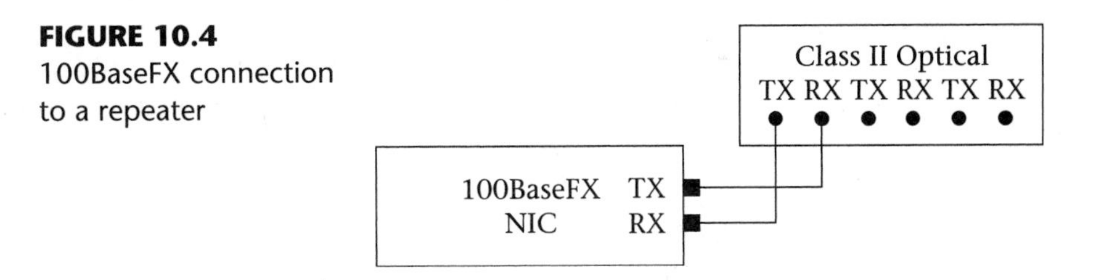


**FIGURE 10.4**
100BaseFX connection to a repeater

100BaseFX uses 4B/5B encoding with NRZ-I signal encoding. In this type of encoding, four bits of information are converted to five bits, as shown in Table 10.2; the five bits are converted to NRZ-I digital signals. These signals are then converted to an optical ray for transmission over the fiber-optic cable to the receiver.

At the receiver side, the optical signal is sampled every eight nanoseconds. If there is a change of light (from on to off or from off to on) in the sample, binary one is represented; if there is no change of light, it is represented by binary zero.

The conversion from four bits to five bits changes the data rate from 100 Mbps to 125 Mbps, respectively. NRZ-I digital encoding reduces the frequency of transmission by half.

## 10.2 Fast Ethernet Repeaters

Repeaters are used to expand network diameter. There are two types of repeaters used in Fast Ethernet: a Class I repeater and a Class II repeater.

**Class I Repeater**

A **Class I repeater** converts line signals from the incoming port to digital signals. This conversion allows different types of Fast Ethernet technology to be connected to LAN segments. For example, it is possible to connect a 100BaseTX station to a 100BaseFX station by using a Class I repeater, which has a larger internal delay. Only one Class I repeater can be used in a Fast Ethernet segment, as shown in Figure 10.5.

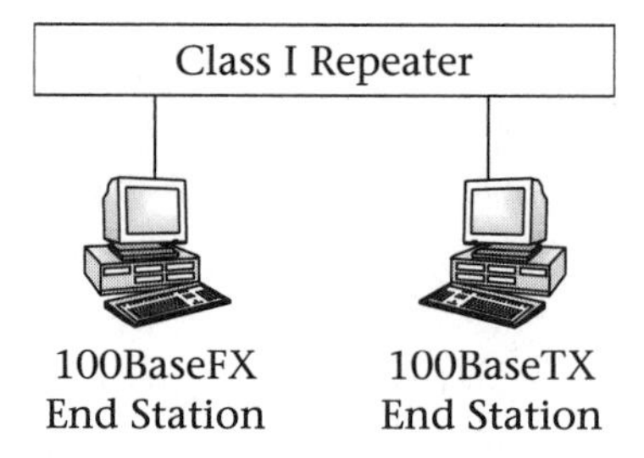


**FIGURE 10.5** Class I repeater

**Class II Repeater**

A **Class II repeater** repeats the incoming signal, sending to every other port on the repeater. Class II repeaters are used to connect the same media to the collision domain (all stations connected to the repeater are the same type, such as 100BaseTX). Only two Class II repeaters are permitted in one Fast Ethernet segment, as shown in Figure 10.6.

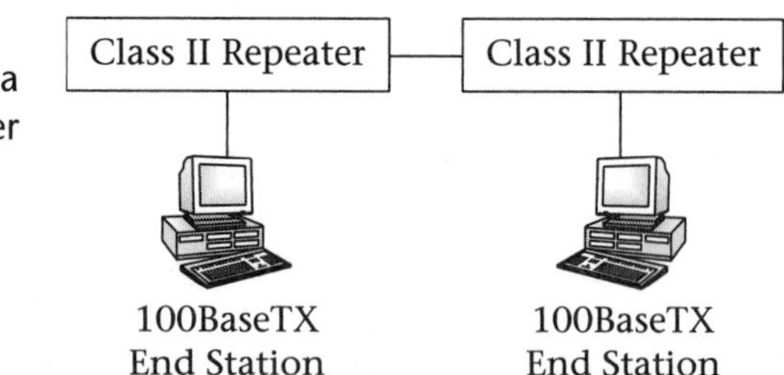


**FIGURE 10.6** Application of a Class II repeater

## 10.3 Fast Ethernet Network Diameter

**Network diameter** is a term used to describe the distance between two end stations connected together through a repeater, switch, or bridge.

The network diameter is a function of propagation delay. **Propagation delay** is defined as the difference between the transmission time and the receiving time of a signal. This difference is caused by network components such as cable, NIC, and repeater.

The propagation delay in a network is measured in a unit called a **bit time.** One bit time is the duration of one bit on the network. For example, Ethernet's bit time is $1/10^7$ second and Fast Ethernet's bit time is $1/10^8$ second. In Ethernet networks, the maximum propagation delay is defined as 512 bit times and the first bit should reach the destination before the last bit is transmitted by the source. Table 10.3 shows the propagation delay of Fast Ethernet components.

**TABLE 10.3** Propagation Delay of Fast Ethernet Components

| Component Type | Bit Times |
|---|---|
| Two TX NIC or Two FX NIC | 100 |
| Two T4 NIC | 138 |
| One T4 NIC with One TX NIC | 127 |
| One TX NIC and One FX NIC | 127 |
| 100 Cat-3 UTP Wires | 114 |
| 100 Meters Cat-4 UTP Wires | 114 |
| 100 Meters Cat-5 UTP | 111 |
| 100 Meters STP (IBM type 1) | 111 |
| 412 Meters Fiber-Optic cable | 1 per Meter or 412 |
| Class I Repeater | 140 |
| Class II Repeater for TX or FX | 92 |

To calculate the network diameter, the total propagation delay of the network components must be less than 512 bit times.

Due to the Electronic Industries Association (EIA), wiring is limited to rules that specify the diameter of Fast Ethernet using twisted-pair wire (100BaseTX and 10BaseT4) at 205 meters. Table 10.4 shows the maximum network diameter using different types of Fast Ethernet repeaters.

**TABLE 10.4** Maximum Network Diameter

| Repeater Type | 100BaseTX or 100BaseT4 | 100BaseFX |
|---|---|---|
| Host to Host Connection (without Repeater) | 100 m | 412 m |
| One Class I Repeater | 200 m | 272 m |
| One Class II Repeater | 200 m | 320 m |
| Two Class II repeaters | 205 m | N/A |

The following information can be used to determine the diameter of a Fast Ethernet network:

- The maximum distance between two repeaters using UTP is five meters.
- The length of all UTP cables should not exceed 100 meters.
- The distance between a repeater connected to a switch with UTP cable must not exceed 100 meters.
- The maximum distance between two switches connected by fiber-optic cable is 2000 meters (full-duplex operation).

Figure 10.7 shows the connection of 100BaseTX using one Class I repeater.

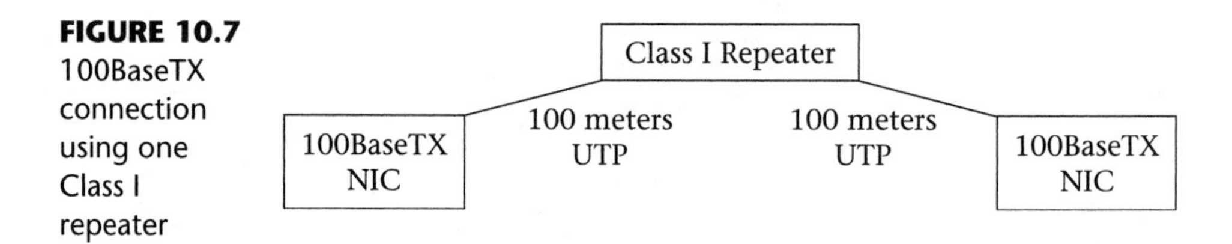


**FIGURE 10.7** 100BaseTX connection using one Class I repeater

Figure 10.8 shows the connection of 100BaseTX using two Class II repeaters.

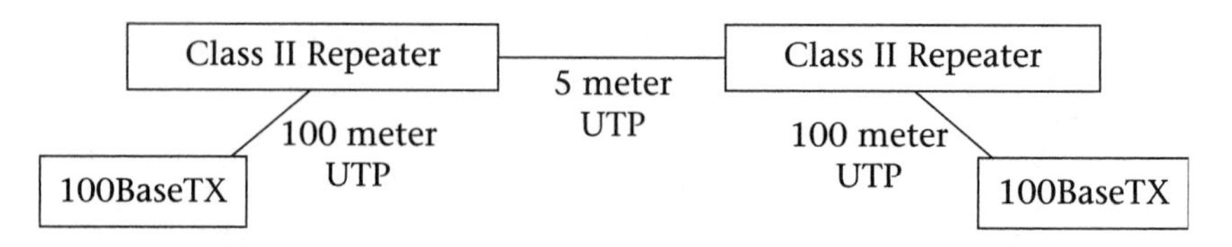


**FIGURE 10.8** 100BaseTX connection using two Class II repeaters

## 10.4 Expanding Fast Ethernet

Fast Ethernet allows the use of only two repeaters. A switch can be used to expand the Fast Ethernet network diameter. Figure 10.9 shows several segments of Fast Ethernet that are connected together using Fast Ethernet switches.

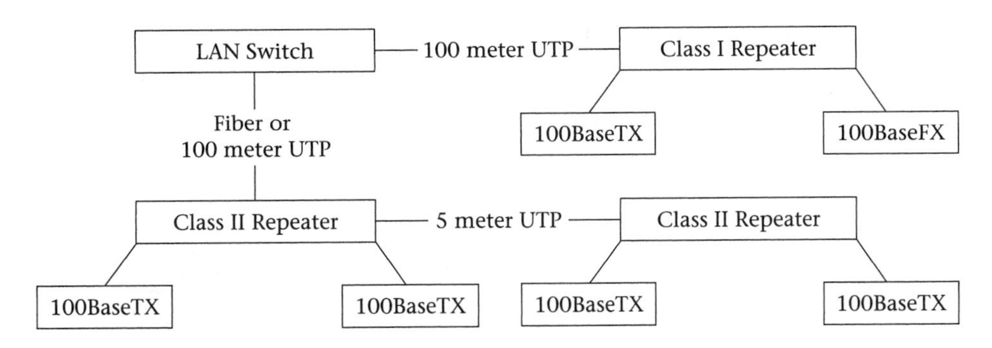


**FIGURE 10.9**
Expansion of Fast Ethernet

## Summary

- The data rate of Fast Ethernet is 100 Mbps.
- Fast Ethernet uses three types of media: 100BaseT4, 100BaseTX, and 100BaseFx.
- Fast Ethernet uses the same frame format as Ethernet.
- 100BaseT4 uses four pairs of Cat-3 UTP wires; 100BaseTX uses two pairs of Cat-5 UTP wires; and 100BaseFX uses fiber-optic cable.
- Fast Ethernet uses Class I repeaters to connect NIC cards with different types of media.
- Fast Ethernet uses Class II repeaters to connect stations having the same type of NIC.
- In Fast Ethernet, only one Class I repeater is allowed.
- In Fast Ethernet, only two Class II repeaters are allowed.

## Key Terms

Bit Time
Class I Repeater
Class II Repeater
Convergence Sublayer (CS)
Fast Ethernet
100BaseFX
100BaseT4
100BaseTX
IEEE 802.3u
Media Independent Interface (MII)
Multi-Level Transition-3 (MLT-3)
Network Diameter
Physical Media Dependent (PMD) Sublayer
Propagation Delay

## Review Questions

### • Multiple Choice Questions

1. Fast Ethernet uses ______.
   a. IEEE 802.2
   b. IEEE 802.5
   c. IEEE 802.3u
   d. IEEE 802.4
2. The goal of Fast Ethernet is to increase ______.
   a. the number of stations in a network
   b. the frequency of signals
   c. bandwidth in a network
   d. network diameter
3. The role of ______ is to interface the MAC sublayer to the PMD sublayer.
   a. 100BaseT4
   b. 100BaseX
   c. convergence sublayer
   d. LLC
4. ______ is the most popular media for Fast Ethernet.
   a. UTP
   b. STP
   c. Fiber-optics
   d. Coaxial cable
5. The data rate of 100BaseTX is ______ Mbps.
   a. 100
   b. 10
   c. 200
   d. 1000
6. 100BaseFX uses ______ cable.
   a. UTP
   b. STP
   c. coaxial
   d. fiber-optic

7. There are ______ types of repeaters.
   a. five
   b. two
   c. three
   d. four

8. The maximum distance between two repeaters using UTP cable is ______ meters.
   a. 10
   b. 5
   c. 100
   d. 200

9. Fast Ethernet's data rate is ______ Mbps.
   a. 100
   b. 10
   c. 400
   d. 200

10. The type of access method used in Fast Ethernet is ______.
   a. token
   b. CSMA/CD
   c. demand priority
   d. full-duplex

11. The Class I repeaters that can be used for Fast Ethernet is ______.
   a. 1
   b. 2
   c. 3
   d. 4

12. The Class II repeaters that can be used for Fast Ethernet is ______.
   a. 1
   b. 2
   c. 3
   d. 4

## • Short Answer Questions

1. Explain the following terms:
   a. 100Base4T
   b. 100BaseTX
   c. 100BaseFX
2. What is the cable type of 100BaseTX?
3. What is the difference between 100BaseTX and 100BaseT4?
4. What is the application of a Class I repeater?
5. What is the application of a Class II repeater?
6. What is the maximum network diameter using two Class II repeaters in a 100BaseT network?
7. Name the IEEE committee that developed the standard for Fast Ethernet.
8. Identify and explain the access method for Fast Ethernet.
9. What is the function of the convergence sublayer?

10. What are the types of media used for Fast Ethernet?
11. What type of signal encoding is used for 100BaseT4?
12. What type of signal encoding is used for 100BaseFX?
13. Convert 84 Hex to 5 bit symbols and then show the corresponding MLT digital signals.
14. Show the binary value for ternary code 0−+−+0 ternary code.

CHAPTER

# LAN Interconnection Devices

OUTLINE



OBJECTIVES

After completing this chapter, you should be able to:

- List LAN interconnection devices
- Describe the function and operation of a repeater
- Describe the function and application of a bridge
- Understand the function of a router and the layers of the OSI model corresponding to a router
- Describe the function and application of a gateway
- Explain switch operation
- Discuss the applications of LAN switching
- Distinguish between symmetric and asymmetric switches
- Understand the technology of cut-through switches and store-and-forward switches

- Identify the application of an L2 switch, L3 switch, and L4 switch
- Discuss the application of virtual LANs

## INTRODUCTION

A **LAN interconnection** is the linking of Local Area Networks (LANs) to form a single network. LANs on different floors of a building or LANs in separate buildings can be connected so that all computers on the site are linked.

There are two reasons for linking LANs together; one is to expand the geographic coverage of the network, and the other reason is to divide the traffic load by creating internetworking.

The devices discussed in this chapter are used for linking LANs together and can be distinguished by the OSI layer at which they are operating.

## 11.1 Repeaters

A **repeater** is a device that is used to connect several segments of LAN in order to extend the allowable length of the network. A repeater accepts traffic from its input port and then retransmits the traffic at its output port. A hub is a multiple output repeater. A repeater works in the Physical layer of the OSI model. Figure 11.1 shows a repeater connecting two segments of a LAN together.

**FIGURE 11.1**
Two segments of a LAN connected by a repeater

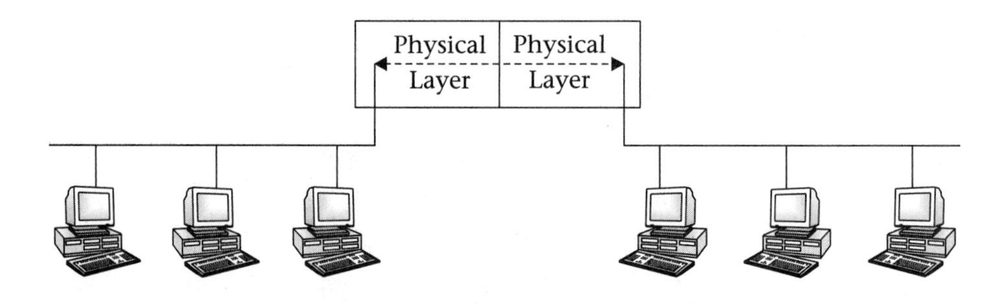


## 11.2 Bridges

A **bridge** is used to connect some segments of a network together (homogeneous network). A bridge operates in the Data Link layer, as shown in

Figure 11.2. Bridges forward frames based on the destination addresses of the frames, and can control data flow and detect transmission errors.

**FIGURE 11.2**
OSI reference model of a bridge

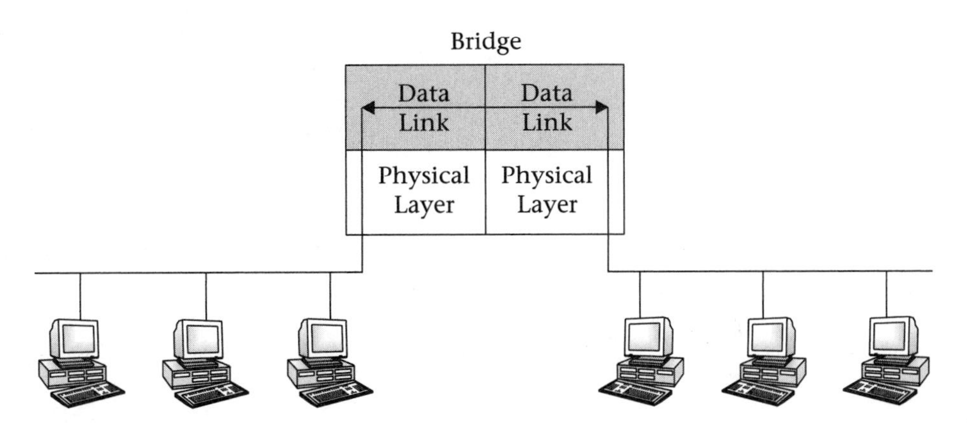


## Functions of a Bridge

The purpose of a bridge is to analyze the incoming destination address of a frame and make a forwarding decision based on the location of the station. Figure 11.3 shows a bridge that is used to connect two Ethernet LANs together. For example, if station A sends a frame to station B, the bridge gets the frame and sees station B in the same segment of A and discards the frame. However, if station A forwards a frame to station C, the bridge would realize that station C is in different LAN segments, then the bridge forwards the frame to station C. The bridge forwards the data from one LAN to another without alteration of the frame. Bridges allow network administrators to segment their networks transparently. That means the individual station does not need to know that there is a bridge in the network.

Bridges are capable of filtering. **Filtering** is useful for eliminating unnecessary broadcast frames. Bridges can be programmed not to forward frames from specific sources.

By dividing a large network into segments and using a bridge to link the segments together, the throughput of the network will increase. If one segment of the network has failed, the other segments connected to the bridge can keep the network alive.

Bridges extend the length of the LAN. While stations A and B are communicating with each others, stations C and D can communicate with each other.

**FIGURE 11.3**
A bridge connecting two Ethernet LANs

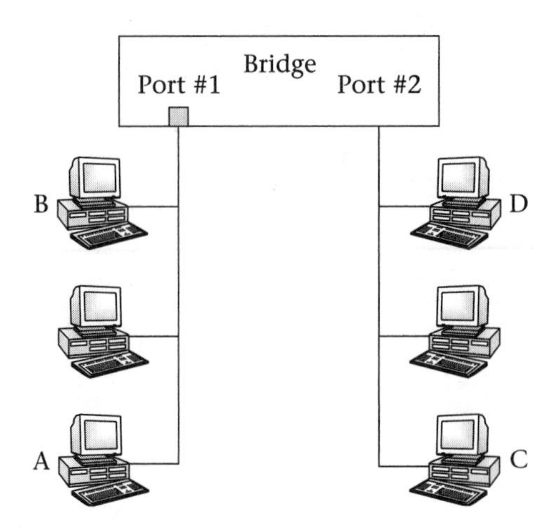


There are two types of bridges: the **learning bridge** or **transparent bridge** and the **source routing bridge**.

***Learning Bridge or Transparent Bridge.*** The learning bridge requires no initial programming. It can learn the location of each device by accepting a frame from the network segment and recording the MAC address and the port number. The frame comes to the bridge, which then retransmits the frame to all the segments of the network except the segment that sent the frame. By using this method the learning bridge learns which station is connected to which segment of the network.

***Source Routing Bridge.*** The frame contains the entire route to the destination. A source routing bridge is used for a token ring network because a token ring frame has a field that specifies the routing of the frame.

## 11.3 Routers

Routers are more complex internetworking devices than bridges. The **router** works in the Network layer of the OSI model to route a frame from one LAN to another, as shown in Figure 11.4. To do this, a router must recognize each Network layer of the LAN segments connected to the router. Therefore, a router that recognizes multiple Network layers is called a *multi-port protocol router.*

The main function of a router is to determine the optimal data path and transfer the information using that path. Figure 11.5 shows how routers can be used to connect several LANs together at different locations. Another function of a router is to convert one type of frame to another type. Station B is connected to a token ring network and has a

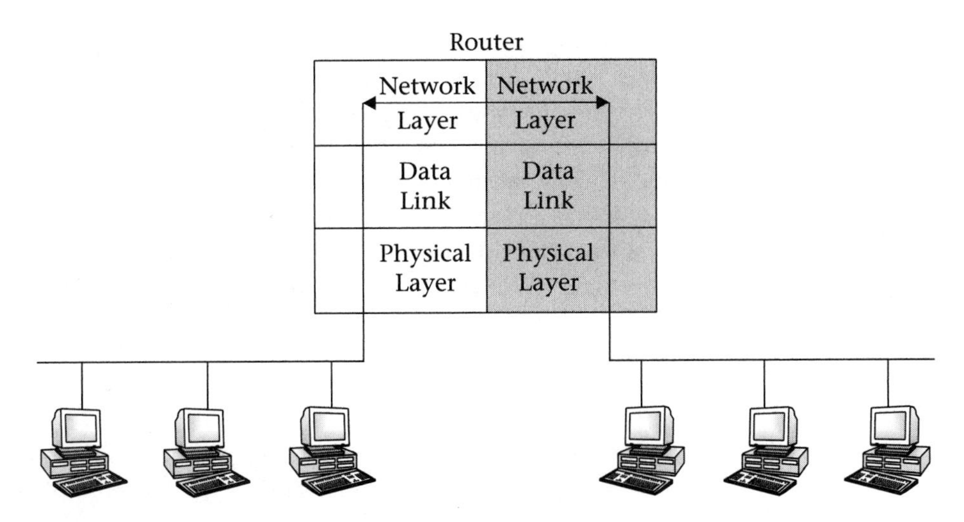


**FIGURE 11.4**
OSI reference model for router

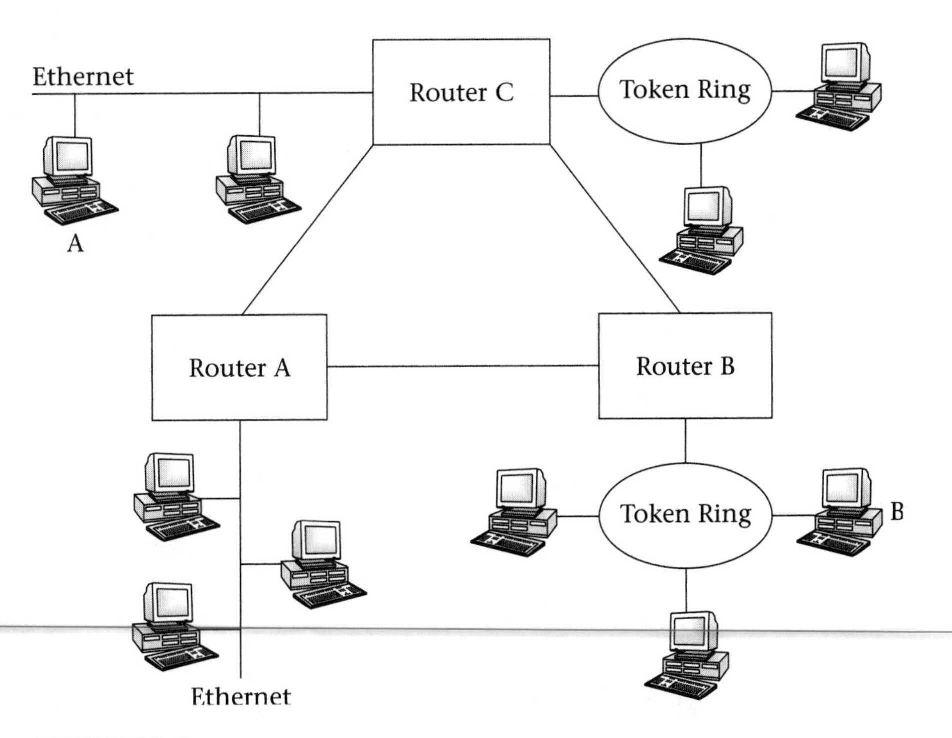


**FIGURE 11.5**
Several LANs connected together using routers

frame for station A. Router C is capable of converting the token ring frame format to an Ethernet frame format.

A router that can be configured manually by a network administrator is called a **static router** and a router that is configured by itself is called a **dynamic router.**

With a static router, the routing table is administered manually by the network administrator who determines the route. A dynamic router sets up its own routing table and updates the routing table automatically. The dynamic router also exchanges information with the next router on the network.

## 11.4 Gateways

Gateways operate up to the Application layer, as shown in Figure 11.6. The purpose of a **gateway** is to convert from one communication protocol to another communication protocol. In Figure 11.6, a network with IBM SNA architecture is connected, through a gateway, with a LAN running the TCP/IP protocol.

**FIGURE 11.6**
OSI reference model for a gateway

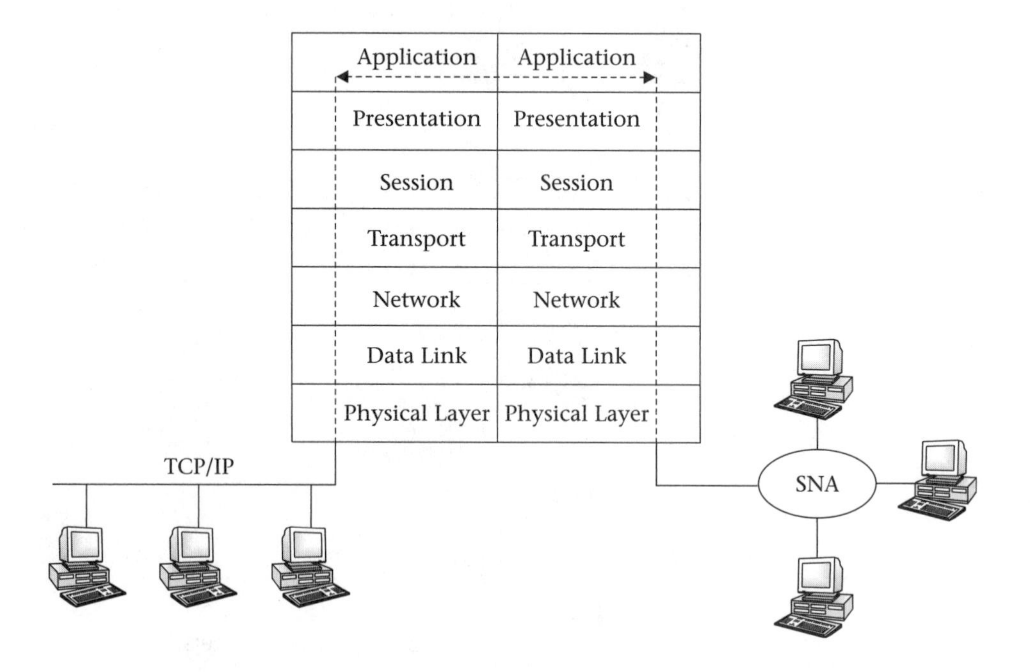

## 11.5 CSU/DSU

A **Channel Service Unit/Data Service Unit (CSU/DSU)** is used to connect a LAN to a WAN link. Figure 11.7 show the application of a CSU/DSU. In Figure 11.7, office LANs are connected together by a router, and the router is connected through a T1 link to the frame relay network. The format of the frame and signal types is different from LAN to WAN; therefore, a CSU/DSU is used to make this conversion.

**FIGURE 11.7**
Application of a CSU/DSU

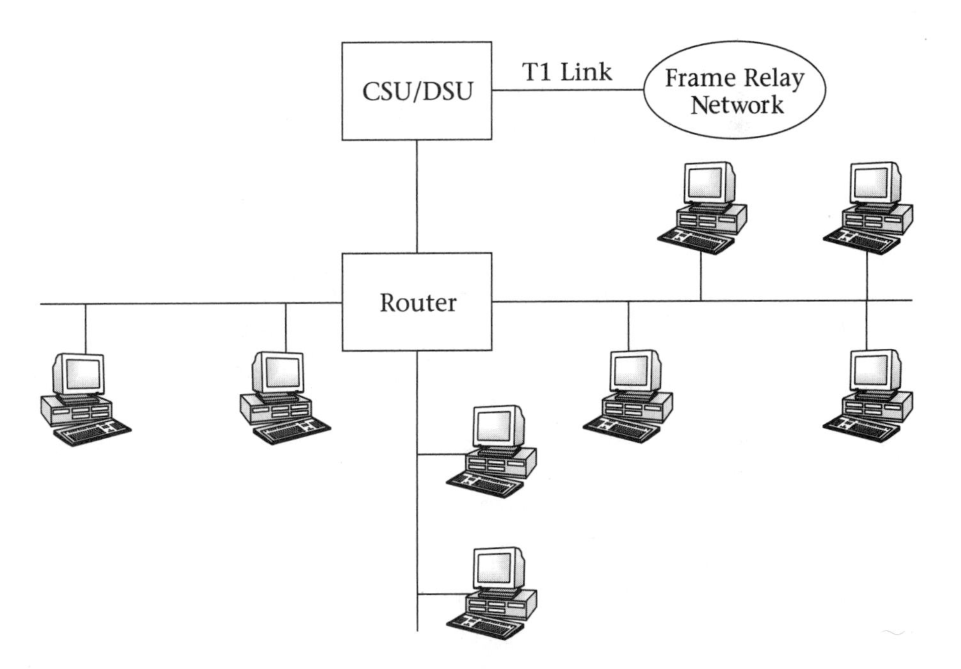


## 11.6 Switches

LAN switching is the fastest growing technology in the networking industry. The switch is used to connect LAN segments together in order to increase the network throughput. A **switch** is a device with multiple ports, which accepts packets from one port, examines the destination address, and then transmits the packets to the intended port having a host with the same destination address, as shown in Figure 11.8. Most LAN switches operate at the Data Link layer of the OSI model. Figure 11.9 shows a symbolic representation of a switch. When a switch operates in the Data Link layer, it performs the function of a bridge.

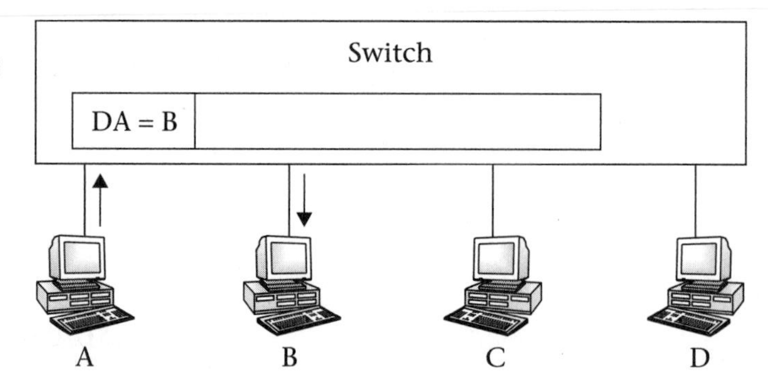


**FIGURE11.8**
Switch operation

**FIGURE 11.9**
Symbolic representation of a switch

## Ethernet LAN Switching

Ethernet is one of the most popular LAN technologies because it uses unshielded twisted-pair cable. However, when the number of stations increases in an Ethernet LAN, the number of collisions also increases and performance decreases accordingly. In order to increase the performance of an Ethernet LAN, it can be segmented, with the segments connected to switch ports. In Figure 11.10, each segment acts as an independent LAN, and each segment has its own collision domain.

A **LAN switch** is similar to multi-port bridge. As each LAN frame enters the switch, the switch compares the frame's destination with a table of previously learned addresses, and the frame is sent to the proper port.

If an Ethernet LAN is comprised of 20 stations, the bandwidth of each station is equal to the network bandwidth divided by 20. If this LAN were divided into five segments, with each segment connected to a switch, the bandwidth of each station becomes the network bandwidth divided by five.

**FIGURE 11.10**
Connection of LAN segments to a switch

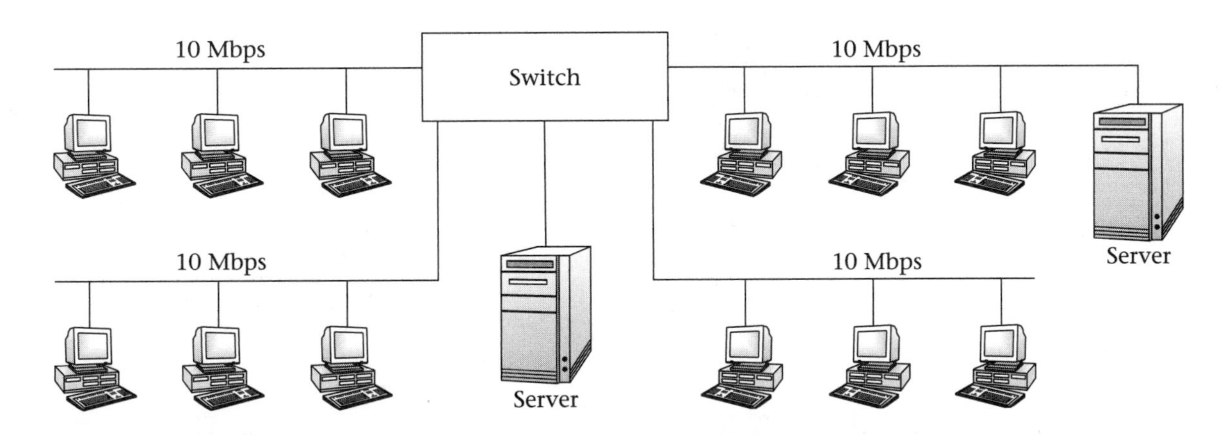


**Switch Classifications** The manufacturers of switches classify them based on their applications: symmetrical or asymmetrical:

A **symmetric switch** provides switching between segments having the same bandwidth; for example, 10 Mbps to 10 Mbps or 100 Mbps to 100 Mbps, as shown in Figure 11.11.

**FIGURE 11.11**
Symmetric switch

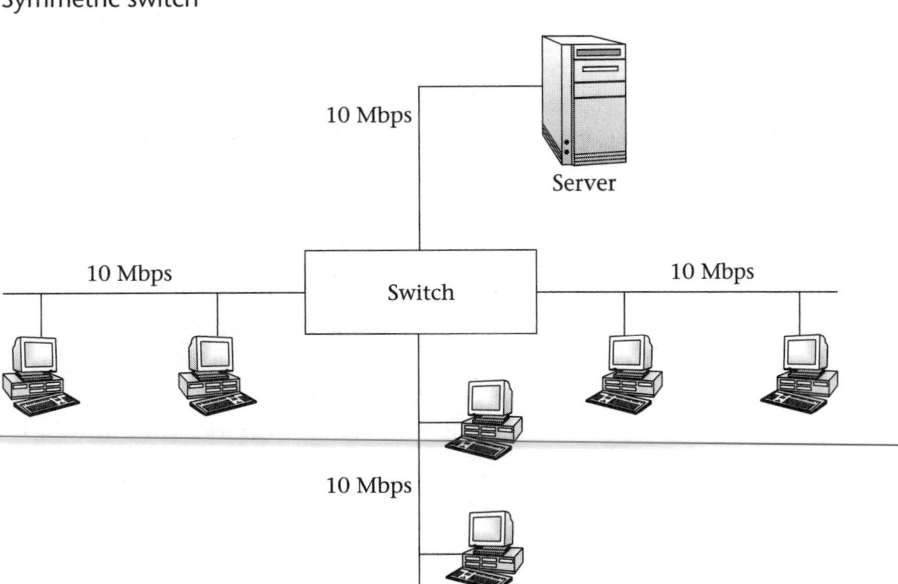

An **asymmetric switch** provides switching between segments of different bandwidth; for example, 10 Mbps to 100 Mbps or 100 Mbps to 10 Mbps, as shown in Figure 11.12.

**FIGURE 11.12**
Asymmetric switch

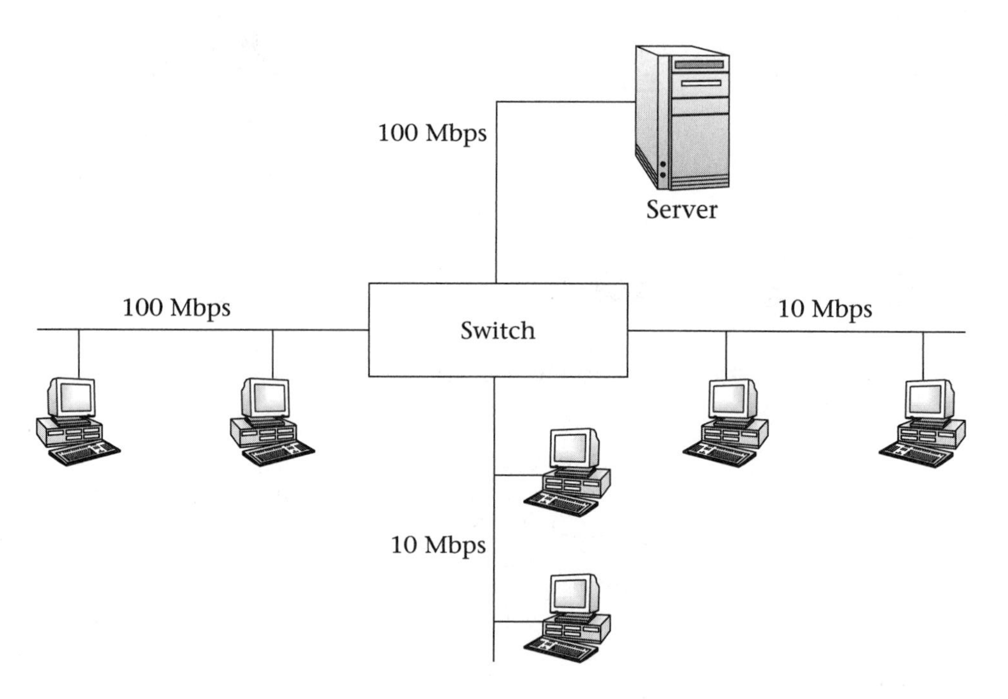


### Switch Operations

A LAN switch uses RISC (Reduced Instruction Set Computer) processors and ASIC (Application Specific Integrated Circuit) processors to increase performance. RISC processors are not as fast as ASIC processors, but they are less expensive. ASIC switches are custom designed for specific operations, and all of their operations are accomplished through hardware. There are two types of switches.

***Cut-Through Switch.*** A **cut-through switch** reads the first few bytes of the packet to obtain the source and destination addresses. The packets are sent to the destination segment without checking the rest of the packet for errors. The cut-through switch uses an ASIC processor for processing the packet.

***Store-and-Forward Switch.*** The **store-and-forward switch** stores the entire packet, and then checks for errors in the packet. If a packet contains errors, it is discarded; otherwise the switch forwards the packet to the specified destination. The store-and-forward switch is more suitable for an Ethernet LAN because it will filter out any corrupted packets to the other segments and therefore reduce collision.

## Switch Architecture

Switch architecture is based on the OSI model. The different types of switch architectures are described next.

***Layer 2 (L2) Switch.*** A Layer 2 switch operates in the Data Link layer of the OSI model. It is used for network segmentation and for creating workgroups. An L2 switch is similar to a multi-port bridge. The frame enters from one port of the switch and is forwarded, based on the MAC address of the frame, to the proper port. A frame with a broadcast address will be repeated to all ports of the switch. An L2 switch learns the MAC addresses of the hosts connected to every port and creates a switching table that contains MAC addresses belonging to each port. The switch uses this table to forward the frame to the proper port.

***Layer 3 (L3) Switch.*** A Layer 3 switch is a type of router that uses hardware rather than software. An L3 switch, sometimes called a *routing switch,* uses ASIC switching technology such as a crossbar switch. This switch operates on the Network layer of the OSI model.

The function of an L3 switch is to route the packet based on the logical address (or layer 3) information. An L3 switch accepts the packet from the incoming port and forwards the packet to the proper port based on a logical address such as an IP address. In order to increase performance, the switch finds the route for the first packet and establishes a connection between the incoming and outgoing ports for transferring the rest of the packets. This is called "route once and switch many." Figure 11.13 shows an application of L3 switches in which they are used to connect the networks of two buildings.

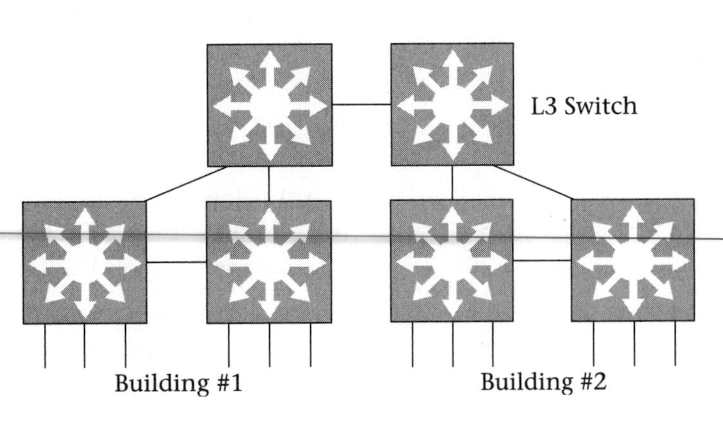


**FIGURE 11.13** Connecting the networks of two building using L3 switches

***Layer 4 (L4) Switch.*** A Layer 4 switch operates on the Transport layer of the OSI model. The Internet uses the Transport layer with TCP and UDP. TCP is used for reliable communication and UDP is used for unreliable communication. Application protocols running on top of TCP are Telnet, FTP, HTTP, and SMTP. The TCP header contains fields called *source port number* and *destination port number*. The source port number identifies the source protocol of an incoming packet and the destination port number identifies the destination protocol for an incoming packet. An L4 switch operates on the port number to forward a packet to the destination. An L4 switch is used for network security and for filtering packets based on application protocol.

## 11.7 Virtual LAN (VLAN)

A **Virtual LAN (VLAN)** (or IEEE 802.1q) is a configuration option on a LAN switch that allows network managers the flexibility to group or segment ports on an individual switch into logically defined LANs. There are two immediate benefits from VLAN. First, it provides a way for network administrators to decrease the size of a broadcast domain and second, VLANs can provide security options for administrators. A VLAN is one way to prevent hosts on virtual segments from reaching one another. Another application of VLAN is for logical segmentation of workgroups within an organization.

There are three methods for assigning a packet to a VLAN membership:

**Port-based VLAN:** In this method VLAN membership is based on a switch port. The network administrator assigns each port of the switch to a spec ific VLAN. For example, in an eight-port switch, ports 1, 2, and 3 are assigned to VLAN1; ports 4 and 5 are assigned to VLAN2; and ports 6, 7, and 8 are assigned to VLAN3.

**MAC address-based VLAN:** In this method VLAN membership is based on the source or destination MAC address of the frame. Each switch contains a table that consists of the MAC address of the stations connected to the switch with the corresponding VLAN.

**Layer 3-based VLAN:** In this method VLAN membership is based upon protocols such as IP, IPx, and NetBios or IP address. The switch contains a table that consists of IP addresses of the sources and the corresponding VLAN.

IEEE 802.11q developed the standard for tagging frames for use in VLAN. The IEEE 802.1q defines a method in which the switch adds a tag to the frame and the switch can process an untagged frame or a tagged frame. Figure 11.14 shows an IEEE801q frame format. The tag is 4 bytes and is inserted between the source address (SA) and the Type/Length field on an Ethernet frame format.

**FIGURE 11.14**
IEEE 801q frame format

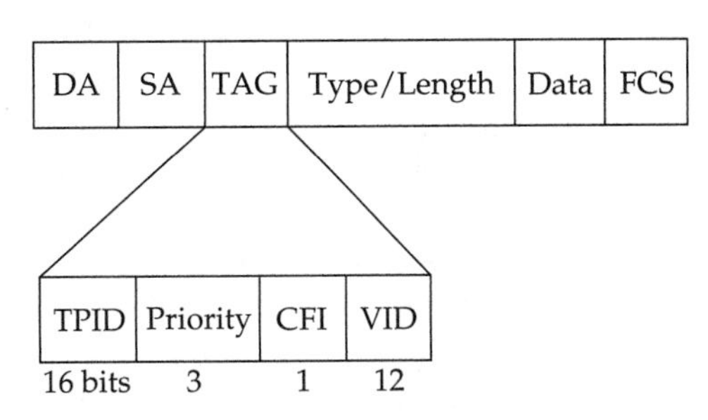


The following describes the function of each field in the tag.

**TPID (Tag Protocol Identifier):** This field is 16 bits and is set to 8100 (Hex) to identify the frame in IEEE 802.1q.

**Priority:** This field is 3 bits and identifies the priority of the frame.

**CFI (Canonical Format Indicator):** The CFI bit is mainly used for compatibility between Ethernet and token ring network and is set to 0 for an Ethernet switch.

**VID (VLAN ID):** This field is 12 bits and represents the VLAN number that the frame belongs to.

Table 11.1 shows the general default connectivity matrix for an eight-port Ethernet switch. A + represents connectivity between ports and a – represents no connectivity between ports. This matrix demonstrates that each port has the ability to see and pass packets to every other port on the switch.

**TABLE 11.1** General Matrix Connectivity of an Eight-Port Switch

| Port # | 1 | 2 | 3 | 4 | 5 | 6 | 7 | 8 |
|---|---|---|---|---|---|---|---|---|
| 1 | – | + | + | + | + | + | + | + |
| 2 | + | – | + | + | + | + | + | + |
| 3 | + | + | – | + | + | + | + | + |
| 4 | + | + | + | – | + | + | + | + |
| 5 | + | + | + | + | – | + | + | + |
| 6 | + | + | + | + | + | – | + | + |
| 7 | + | + | + | + | + | + | – | + |
| 8 | + | + | + | + | + | + | + | – |

One advantage of Ethernet switches is that they may be configured to isolate ports from one another, thereby creating virtual LANs. These VLANs can provide isolation from errant broadcasts as well as introduce additional security on the switch.

A network manager can create several VLANs in an Ethernet switch. Table 11.2 shows the matrix connectivity of a switch containing several VLANs, and Table 11.3 shows the VLANs generated by Table 11.2.

**TABLE 11.2** Matrix Connectivity of an Eight-Port Switch

| Port #<br>VLAN # | 1 | 2 | 3 | 4 | 5 | 6 | 7 | 8 |
|---|---|---|---|---|---|---|---|---|
| 1 | + | + | − | − | + | − | − | − |
| 2 | + | − | + | − | − | − | − | − |
| 3 | + | − | − | + | − | − | − | − |
| 4 | + | − | + | − | + |  | + | − |
| 5 | − | − | − | + | − | + | − | − |
| 6 | + | − | − | − | − | − | + | + |
| 7 | + | − | + | − | − | − | − | + |
| 8 | + | − | − | − | − | + | + | − |

**TABLE 11.3** VLANs Generated for Table 11.2

| VLAN # | Port Logically Connected Together |
|---|---|
| 1 | 1,2,5 |
| 2 | 1,3 |
| 3 | 1,4 |
| 4 | 1,3,5,7 |
| 5 | 4,6 |
| 6 | 1,7,8 |
| 7 | 1,3,8 |
| 8 | 1,6,7 |

## Using Switches as Firewalls

Most of an organization's networks are connected together through the Internet, frame relay, or leased lines. Security is important and necessary to organizations to prevent hackers from accessing the private network.

One of the most popular systems used to implement greater security is the **firewall**. A firewall is a system (a combination of hardware and software) used to prevent unauthorized access to a private network. All incoming and outgoing packets from the network must pass through the firewall. The firewall examines each packet that leaves or enters the network, and then accepts or rejects each packet based on information in the packet header, such as the IP address.

***Example.*** Suppose a corporation must have data connectivity to many business partners over dedicated lines. These circuits could all terminate on a switch. A single firewall is used to protect the organization's network, as shown in Figure 11.15.

**FIGURE 11.15**
Application of a firewall

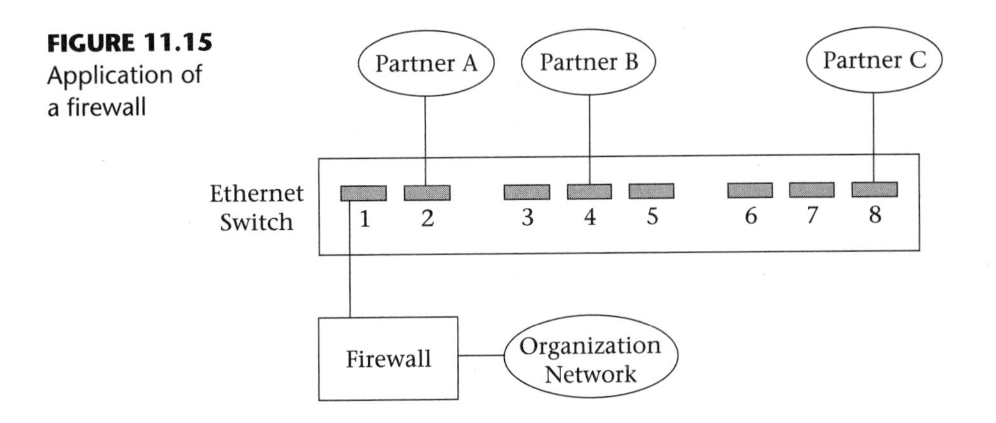


While this configuration would provide protection for the corporation's internal network, it would do nothing to prevent one business partner from trying to connect or hack into another business partner's network. While it is the responsibility of each business partner to protect its own network, there may be some sort of legal liability to the corporation should some attack or disruption occur because of the inter-partner connectivity provided by this switch.

A solution to this problem is neatly provided by the inherent configuriability of switches. Network administrators need only configure multiple virtual LANs to prevent connectivity among the business partner networks. Table 11.4 shows the connectivity matrix of such a virtual LAN. This configuration was adopted to allow a firewall connected to port 1 to see and communicate with the other networks connected to ports 2 through 8. However, ports 2 through 8 cannot see or communicate with one another. The only port they can communicate with is port 1. A direct connection to the corporation's internal network is provided by a second Ethernet interface on the firewall, thereby providing connectivity between the corporate network and its business partners.

**TABLE 11.4** Connectivity of VLAN

| Port # | 1 | 2 | 3 | 4 | 5 | 6 | 7 | 8 |
|---|---|---|---|---|---|---|---|---|
| 1 | − | + | + | + | + | + | + | + |
| 2 | + | − | − | − | − | − | − | − |
| 3 | + | − | − | − | − | − | − | − |
| 4 | + | − | − | − | − | | − | − |
| 5 | + | − | − | − | − | − | − | − |
| 6 | + | − | − | − | − | − | − | − |
| 7 | + | − | − | − | − | − | − | − |
| 8 | + | − | − | − | − | − | − | − |

## Summary

- LAN interconnection devices are repeaters, bridges, routers, switches, and gateways.
- A repeater is used to extend the length of the network and operates at the Physical layer of an OSI model. A repeater accepts traffic from its input and repeats it at its output.
- A bridge is used to connect segments of same-type networks; the function of the bridge is to analyze the incoming frame's destination address and forward the frame to the proper segment. Bridges operate on the Data Link layer of the OSI model.
- A learning bridge or transparent bridge learns the location of each station by recording the NIC address and the port number of which frame enters the bridge.
- A source routing bridge routes the frame based on information in the routing field of the frame.
- A router is used to route a frame from one LAN to another LAN according to its routing table. Routers operate in the Network layer of the OSI model.

- A gateway is used to convert one protocol to another protocol and operates in all seven layers of the OSI model.
- A switch accepts a packet from one port and examines the destination address; it then retransmits the packet to the port having a host with the same destination address.
- When the number of users is increased in an Ethernet LAN, the number of collisions will increase. To overcome this problem, Ethernet LAN can be segmented, with each segment connected to a port on a switch.
- A symmetric switch provides switching between LAN segments with the same data rate.
- An asymmetric switch provides switching between LAN segments with different data rates.
- A cut-through switch reads the first few bytes of the frame to determine by which output port the frame must leave.
- A store-and-forward switch stores the entire frame and checks for errors. If the frame is corrupted, then it is discarded; otherwise the frame is sent to the proper port for its destination.
- The IEEE 802.10 committee approved the standard for Virtual LAN (VLAN). In VLAN the switch port can be enabled and disabled by a network administrator. The administrator can also connect several ports to make a VLAN.
- A Layer 2 switch is a multi-port device that operates on layer 2 of the OSI model.
- A Layer 3 switch is a type of router that uses integrated switching technology.
- A Layer 4 Switch is a type of switch that operates on layer 4 (Transport layer) of the OSI model.

## Key Terms

Asymmetric Switch
Bridge
Channel Service Unit/Data Service Unit (CSU/DSU)
Cut-Through Switch
Dynamic Router
Filtering
Firewall
Gateway
LAN Interconnection
LAN Switch
Layer 2 (L2) Switch

Layer 3 (L3) Switch
Layer 4 (L4) Switch
Learning Bridge
Repeater
Router
Source Routing Bridge
Static Router
Store-and-Forward Switch
Switch
Symmetric Switch
Transparent Bridge
Virtual LAN (VLAN)

## Review Questions

### • Multiple Choice Questions

1. A hub is a multiple port ______.
   a. server
   b. client
   c. modem
   d. repeater
2. ______ operate in the Data Link layer.
   a. Bridges
   b. Repeaters
   c. Switches
   d. Gateways
3. ______ are capable of filtering.
   a. Bridges
   b. Repeaters
   c. Switches
   d. Hubs
4. In a ______, the frame contains the entire route to the destination.
   a. source routing bridge
   b. learning bridge
   c. repeater
   d. gateway
5. ______ are more complex Internet working devices than bridges.
   a. Switches
   b. Routers
   c. Gateways
   d. Hubs
6. A ______ operates up to the Application layer.
   a. router
   b. switch
   c. gateway
   d. repeater
7. A ______ bridge learns the location of each station by recording the NIC address and the port number.
   a. source routing
   b. transparent
   c. a and b
   d. none of the above
8. A ______ is used to convert one protocol to another.
   a. router
   b. switch
   c. gateway
   d. hub

9. The _____ is used to connect segments of a LAN.
   a. router
   b. hub
   c. switch
   d. gateway

10. A switch is a device with _____ port(s).
    a. a single
    b. two
    c. multiple
    d. none of the above

11. _____ provides switching between different bandwidth segments.
    a. A symmetric switch
    b. An asymmetric switch
    c. A store-and-forward switch
    d. A cut-through switch

12. A(n) _____ switch reads only the first few bytes of the packet.
    a. cut-through
    b. store-and-forward
    c. symmetric
    d. asymmetric

13. Layer 3 switches or routing switches work on the OSI Physical layer, Data Link layer, and _____ layer.
    a. Application
    b. Session
    c. Presentation
    d. Network

14. A _____ is a configuration option on a LAN switch.
    a. VLAN
    b. firewall
    c. repeater
    d. router

15. A(n) _____ server is one of the firewall techniques.
    a. application
    b. communication
    c. file
    d. proxy

16. A _____ is a system that is used to prevent unauthorized users to access an organization's network.
    a. VLAN
    b. firewall
    c. a and b
    d. router

17. The type of switch used to connect the segments of a LAN is an _____.
    a. L2 switch
    b. L3 switch
    c. L4 switch
    d. all of the above

18. A Layer 2 switch operates at the _____ layer.
    a. Physical
    b. Data Link
    c. Network
    d. Application

19. The type of switch used to connect a token ring LAN and an Ethernet LAN is an ______.
    a. L2 switch
    b. L3 switch
    c. L4 switch
    d. none of the above

20. Of the following switches, a/an ______ is fastest.
    a. store-and-forward
    b. cut-through
    c. L3
    d. L4

21. Of the following switches, a/an ______ can check for errors in an incoming frame.
    a. store-and-foward
    b. cut-through
    c. L3
    d. L2

22. The application of an L4 switch is for ______.
    a. connecting LAN segments
    b. connecting two different LAN technologies
    c. security
    d. routing

## • Short Answer Questions

1. List the LAN interconnection devices.
2. What is the function of a repeater?
3. Describe the function of a bridge.
4. In what layer of the OSI model does a bridge operate?
5. Explain the operation of a transparent bridge.
6. Explain the operation of source routing bridge.
7. Explain the function of a router.
8. Explain a static router.
9. What is the difference between a router and an L2 switch?
10. A router works in which layer of the OSI model?
11. Explain the dynamic router.
12. What is the application of a gateway?
13. A gateway operates in which layers of the OSI model?
14. What is the difference between a gateway and a router?
15. Explain switch operation.
16. What is the application of a symmetric switch?
17. What is the application of an asymmetric switch?

18. Explain the operation of a cut-through switch.
19. Explain the operation of a store-and-forward switch.
20. What does VLAN stand for?
21. What is the difference between a router and an L3 switch?
22. What is the application of an L4 switch?
23. Suppose a company has two working groups, A and B. Group A has 4 computers and group B has 3 computers, all connected to an eight-port Ethernet switch. Both groups need to access a common file server FS1. There is an in-house requirement that group A computers should not be able to see Group B computers in the network.
    a. Draw a diagram showing an Ethernet switch with seven computers and a file server.
    b. Show the VLAN connectivity matrix for these requirements.

CHAPTER 12

# Gigabit and 10 Gigabit Ethernet Technology

**OUTLINE**



**OBJECTIVES**

After completing this chapter, you should be able to:

- Recognize Gigabit standards and the Gigabit Ethernet architecture
- Identify the components of Gigabit Ethernet
- Discuss the different types of Gigabit Physical layers
- List some of the applications for Gigabit Ethernet
- List 10 GbE Physical layers
- Identify applications for 10 GbE

**INTRODUCTION**

With recent advances in the PCI bus and CPU, workstations are getting faster. Today's PCI bus can transfer data at gigabit speed. A 64-bit PCI bus runs at 533 MHz and can transfer data at up to 6.4 gigabits per second. **Gigabit Ethernet** transfers data at one gigabit per second, or 10 times faster than Fast Ethernet. Gigabit Ethernet is technology compatible with Ethernet and Fast Ethernet, and it is used for backbones with gigabit switches. By adding **Quality of Service (QoS)**, the Gigabit Ethernet has become the next generation of LAN and campus backbone.

Gigabit Ethernet is used for the campus **backbone** by connecting gigabit switches together. The switches operate in store-and-forward or cut-through technology. The IEEE 802 committee has developed a standard called *Quality of Service* (IEEE 802.1p). This protocol corresponds to the Network layer of the OSI model. The IEEE 802.1p standards provide tagging for each frame, indicating the priority or class of the service desired for the frame to be transmitted. By adding QoS to Gigabit Ethernet, the Gigabit Ethernet is able to handle all types of data transmission, even voice and video information.

## 12.1 Gigabit Ethernet Standards

In 1995, the **IEEE 802.3** committee formed a study group (the IEEE 802.3z task force) to research Gigabit Ethernet. The IEEE 802.3z task force developed standards for Gigabit Ethernet. In 1996, the Gigabit Ethernet Alliance was formed by more than 60 companies to support the development of Gigabit Ethernet.

## 12.2 Characteristics of Gigabit Ethernet

Gigabit Ethernet is used for linking Ethernet switches and Fast Ethernet switches and for interconnecting very high-speed servers. Gigabit Ethernet enables organizations to upgrade their networks to 1000 Mbps while using the same operating systems and the same application software. The following are the characteristics of Gigabit Ethernet:

- Operates at 1000 Mbps (1 Gbps)
- Uses the IEEE 802.3 frame format
- Uses the IEEE 802.3 maximum frame size
- Supports full-duplex and half-duplex operation
- Uses CSMA/CD access method for half-duplex operation and supports one repeater per collision domain
- Uses optical-fiber and copper wire for transmission media
- Supports 200-meter collision domain diameters

**Gigabit Ethernet Components**

There are four hardware components required in order to achieve Gigabit Ethernet:

1. Gigabit Ethernet NICs
2. Switches that can handle 1000-Mbps Ethernet

3. Gigabit Ethernet repeater
4. Transmission media able to handle 1000 Mbps

## 12.3 Gigabit Ethernet Protocol Architecture

Figure 12.1 illustrates the **IEEE 802.3z** Gigabit protocol architecture. The MAC layer offers two types of connections: full-duplex and half-duplex. Half-duplex uses the CSMA/CD access method.

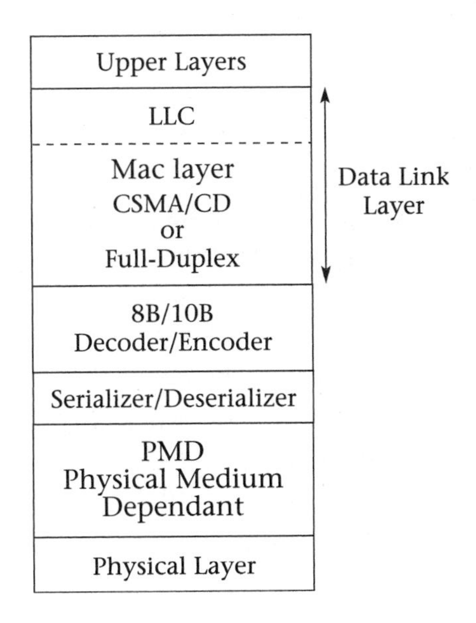


**FIGURE 12.1** The IEEE 802.3z Gigabit Ethernet protocol architecture

**Physical Interface Layer** The Physical Interface layer defines the physical characteristics of the interface media, including the connector type, cable type, transmitter, and receiver. The Gigabit Ethernet Physical layer is designed for the following types of transmission media:

1. **1000BaseLX:** 1000BaseLX is designed for long-wavelength laser (LW) over Single-Mode Fiber (SMF) and Multi-Mode Fiber (MMF), and having a wavelength of 1300 nanometers.
2. **1000BaseSX:** 1000BaseSX uses Short-Wavelength laser (SW) with a wavelength of 850 nanometers over Multi-Mode Fiber.
3. **1000BaseCX:** 1000BaseCX uses twinax cable (150-ohm balanced shielded cable) for transmission media. 1000BaseCX supports both 9 pin D-connector and 8 pin fiber channel connectors.

4. **1000BaseT:** 1000BaseT uses four pairs of Cat-5 UTP cable with RJ-45 connectors for transmission media.

### Long-Wavelength (LW) and Short-Wavelength Laser (SW) over Fiber-Optic Cable

Short-wavelength lasers are used for short distance transmission, while long-wavelength lasers are used for long distance transmission. The wavelengths of lasers are classified as follows:

The wavelength of 850 nanometers is called *Short-Wavelength (SW).*

The wavelength of 1310 nanometers is called *Long-Wavelength (LW).*

The wavelength of 1550 nanometers is called *Extended-Wavelength (EW).*

Short-wavelength and long-wavelength lasers can use multi-mode fiber. There are two diameters of multi-mode fiber cable cores: 50 micro meters, and 62.5 micron meters with 125 micrometer cladding.

***Single-Mode Fiber.*** Single-mode fiber uses a 9-micrometer diameter (core) with 1300-nm wavelength lasers, and is used for long distance transmission. 1000BaseLX uses 1300-nm long-wavelength lasers with SMF and can transmit data to a maximum distance of 5 Km.

***Multi-Mode Fiber.*** 1000BaseSX uses 850-nm short-wavelength lasers with MMF and can transmit information to a maximum distance of 500 meters. 1000BaseLX uses 1300-nm long-wavelength with MMF and can transmit information up to 500 meters.

***Copper Wire.*** 1000BaseT uses 4 pairs of unshielded twisted-pair wire with a maximum distance of 100 meters. 1000BaseCX uses shielded balanced twinax pair cable for a maximum distance of 25 meters.

### Serializer and Deserializer

The serializer and deserializer sublayer accepts information in parallel form from the upper layer (8B/10B encoder/decoder sublayer) and converts it into serial form. It then passes the information to the Physical layer. The Physical layer transfers information to the serializer and deserializer in a serial form where it is converted into parallel form and passed to the 8B/10B decoder/encoder sublayer.

***8B/10B Encoding.*** 8B/10B encoding is used for fiber-optic transmission media. It converts 8-bit information to 10-bit code and then transmits it.

### MAC Layer

Gigabit Ethernet supports both half-duplex and full-duplex transmission. Gigabit Ethernet half-duplex uses the CSMA/CD access method and full-duplex uses a point-to-point connection. CSMA/CD defines the smallest frame for Ethernet as 64 bytes, because the receiver should receive the first bit of the frame before the transmitter completes the transmission. By increasing the speed of the transmission from 100 Mbps to 1000 Mbps, the transmitter can complete the transmission before the receiver receives the first bit of the frame. Fast Ethernet overcomes this problem by reducing the size of the cable, and Gigabit Ethernet increases the minimum size of the frame from 64 bytes to 512 bytes by adding carrier extensions to the Ethernet frame. Figure 12.2 shows the minimum frame size for Gigabit Ethernet.

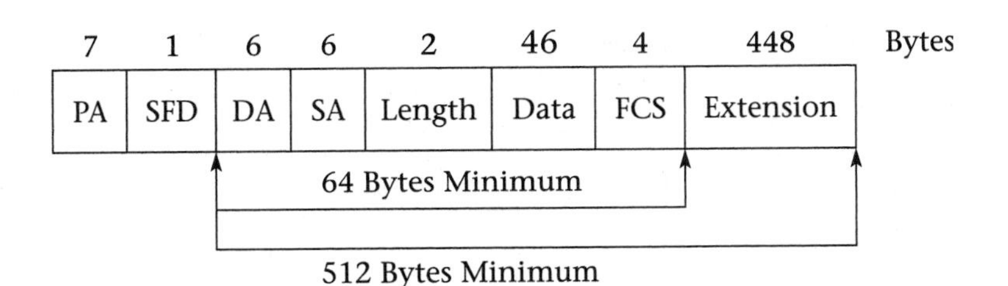


**FIGURE 12.2**
Minimum frame size format for Gigabit Ethernet

***Full-Duplex Transmission.*** In full-duplex transmission, signals travel in both directions simultaneously, and the data rate of the Ethernet is therefore doubled in a full-duplex connection. Full-duplex transmission is used only in point-to-point connections and the CSMA/CD access method is not needed. Figure 12.3 shows a point-to-point connection using full-duplex.

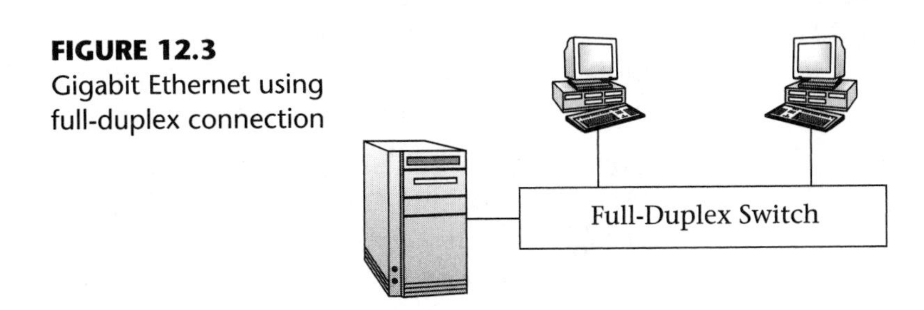


**FIGURE 12.3**
Gigabit Ethernet using full-duplex connection

Figure 12.4 shows detailed architecture of the Gigabit Ethernet MAC and Physical layers. Information is transferred between the MAC layer, the decoder/encoder, and the serializer/deserializer in the form of bytes.

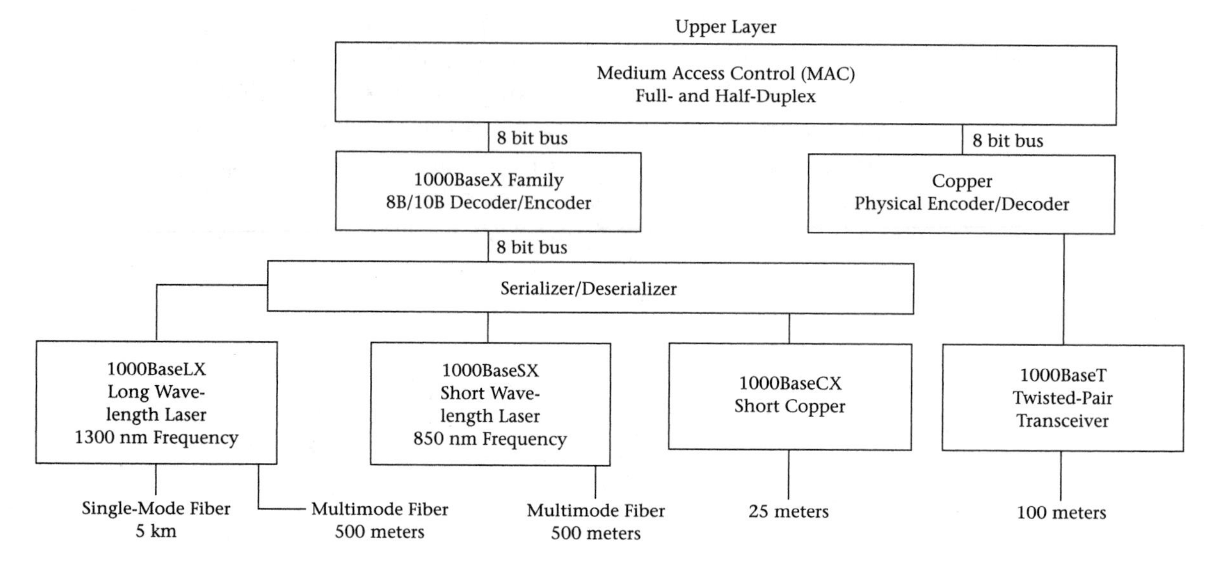


**FIGURE 12.4**
Components of Ethernet Physical layer

## Gigabit Ethernet Network Diameters

Unlike Fast Ethernet and Ethernet, Gigabit Ethernet does not support the use of repeaters. Gigabit Ethernet supports point-to-point connections using fiber-optic cable, twinax cable, and UTP. In Gigabit Ethernet, network diameters can be expanded only by using Gigabit Ethernet switches. Table 12.1 shows Gigabit Ethernet cable types and distances.

## Gigabit Ethernet Applications

Most organizations are currently using Fast Ethernet, which can be upgraded to network with Gigabit Ethernet. Gigabit Ethernet is used for campuses or buildings that require greater data rate. Gigabit Ethernet applications are switch-to-switch, switch-to-server, and repeater-to-switch connections. The following is a list of Gigabit Ethernet applications:

- Upgrading Fast Ethernet switches to Gigabit Ethernet switches and installing Gigabit Ethernet cards in high-speed servers, as shown in Figure 12.5(a) and 12.5(b). Figure 12.5(a) shows a 100 Mbps network and Figure 12.5(b) illustrates upgrading to 1000 Mbps.
- Upgrading a switch to a server link at 1000 Mbps, as shown in Figure 12.6.
- Upgrading a Fast Ethernet backbone switch from 100 Mbps to 1000 Mbps, as shown in Figure 12.7.
- Upgrading a switch to a switch link, as shown in Figure 12.8.

**TABLE 12.1** Gigabit Ethernet Cable Types

| Standard | Cable Type | Core Diameter in Microns | Modal Bandwidth MHz × km | Maximum Distance in Meters |
|---|---|---|---|---|
| 1000BaseSX | MMF | 62.5 | 160 | 220 |
| | MMF | 62.5 | 200 | 270 |
| | MMF | 50 | 400 | 500 |
| | MMF | 50 | 500 | 550 |
| 1000BaseLX | MMF | 62.5 | 500 | 550 |
| | MMF | 50 | 400 | 550 |
| | MMF | 50 | 500 | 550 |
| | SMF | 9 | N/A | 5000 |
| 1000BaseCX | Twinax | – | – | 25 |
| 1000BaseT | UTP | – | – | 100 |

MMF means multi-mode fiber
SMF means single-mode fiber
SX means short-wavelength of 850 nm
LX means long-wavelength of 1300 nm
62.5 micron and 50 micron are diameters of the fiber cores

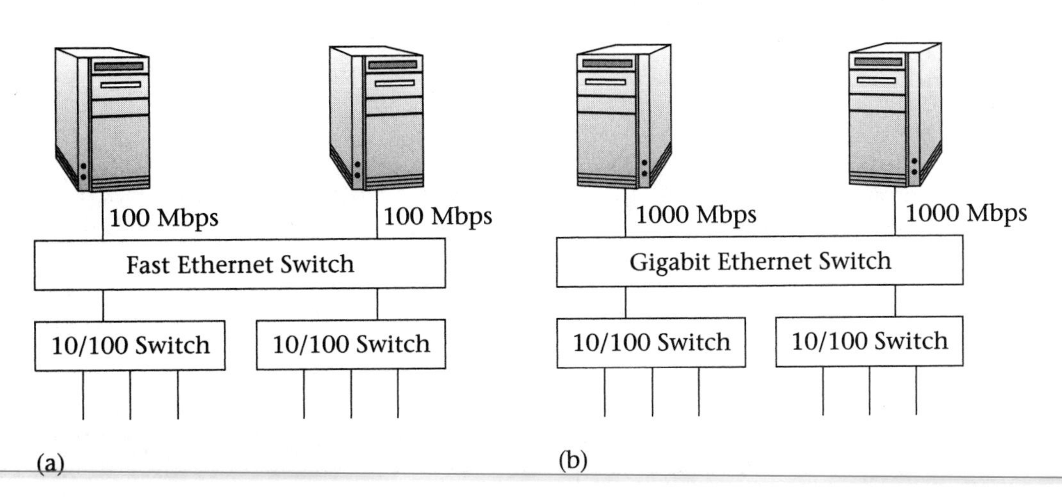


**FIGURE 12.5(a) and (b)**
Upgrading Fast Ethernet to Gigabit Ethernet

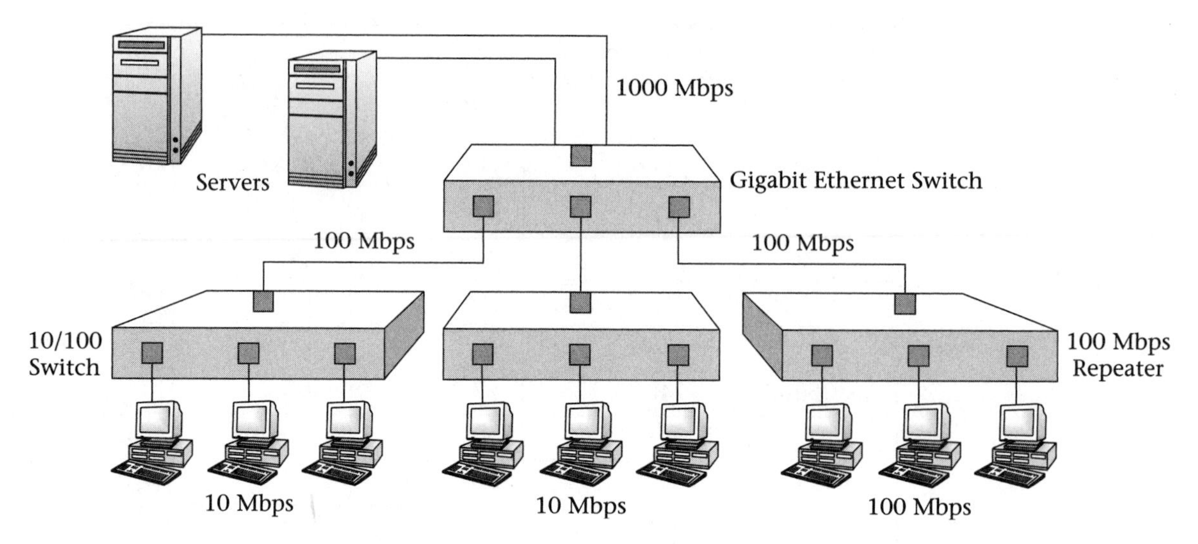


**FIGURE 12.6**
Upgrading a switch to a server link to 1000 Mbps

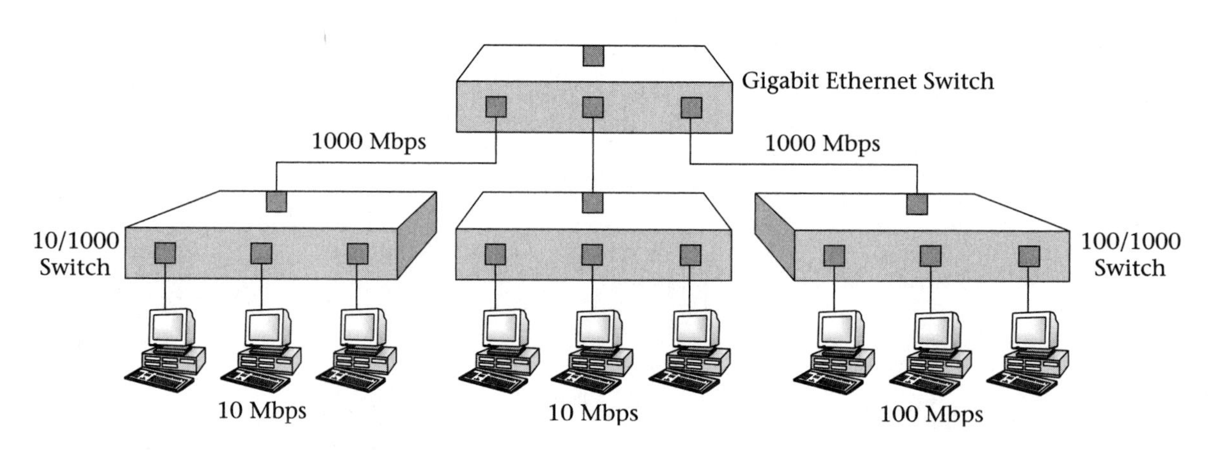


**FIGURE 12.7**
Upgrading Fast Ethernet backbone switch to 1000 Mbps

## 12.4 10 Gigabit Ethernet

The IEEE 802.3ae task force completed the standard for **10 Gigabit Ethernet (10 GbE)** in March 2002. The 10 Gigabit Ethernet standard defines two types of Physical layers: the **LAN Physical layer (LAN PHY)** and the **WAN Physical layer (WAN PHY)**.

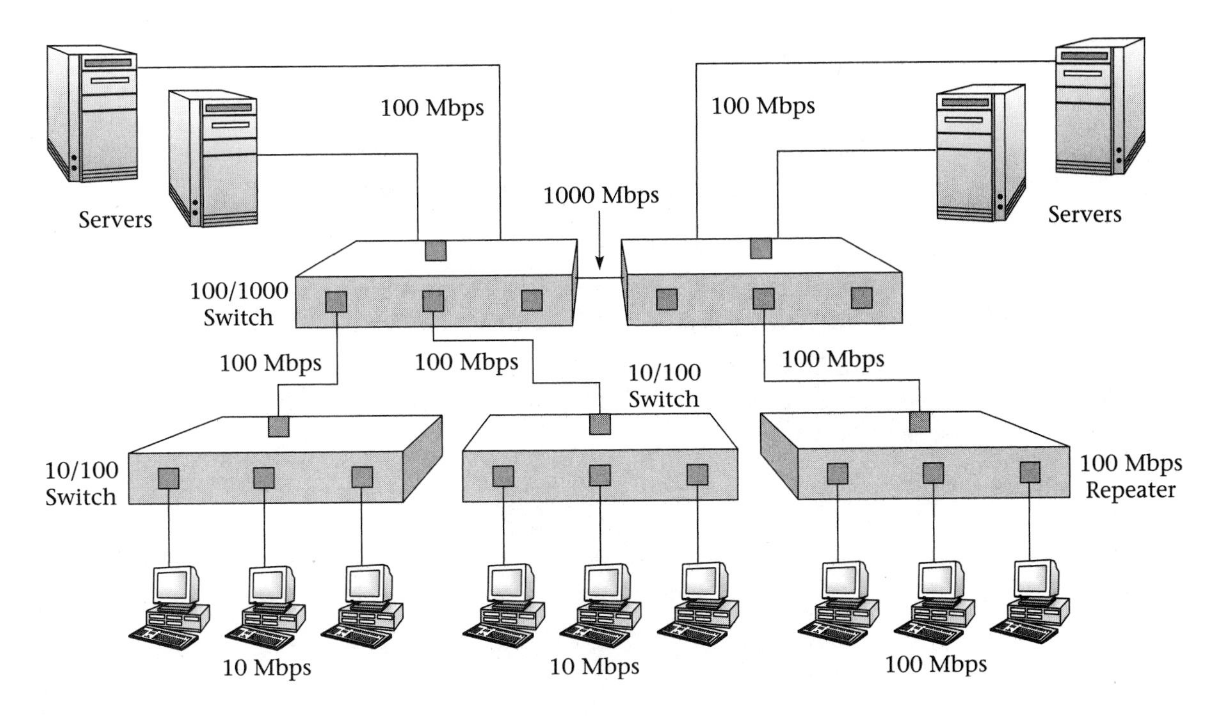


**FIGURE 12.8**
Upgrading a switch to a switch link

The WAN Physical layer operates at the rate that is compatible with OC-192C and it uses Wavelength Division Multiplexing (WDM).

Applications of 10 Gigabit Ethernet for LANs include the following:

- Connecting a server to a switch with 10 GbE
- Connections between switches

Applications of 10 Gigabit Ethernet for WAN include the following:

- Connecting two campus networks
- Storage Network Architecture (SNA)
- Connecting multiple networks in one metropolitan area with 10 GbE to offer services such as distance learning and video conferencing

## Characteristics of 10 Gigabit Ethernet

The following are characteristics of 10 Gigabit Ethernet:

- Uses the IEEE 802.3 frame format
- Uses minimum and maximum IEEE 802.3 frame sizes

- Supports only full-duplex connection
- Uses optical cable for transmission medium
- Supports LAN and WAN Physical layers
- Provides direct connection to OC-192C SONET

## 12.5 10 Gigabit Ethernet Physical Layer

Figure 12.9 shows the Physical layer for 10 Gigabit Ethernet. It consists of serial transmission and Wavelength Division Multiplexing (WDM).

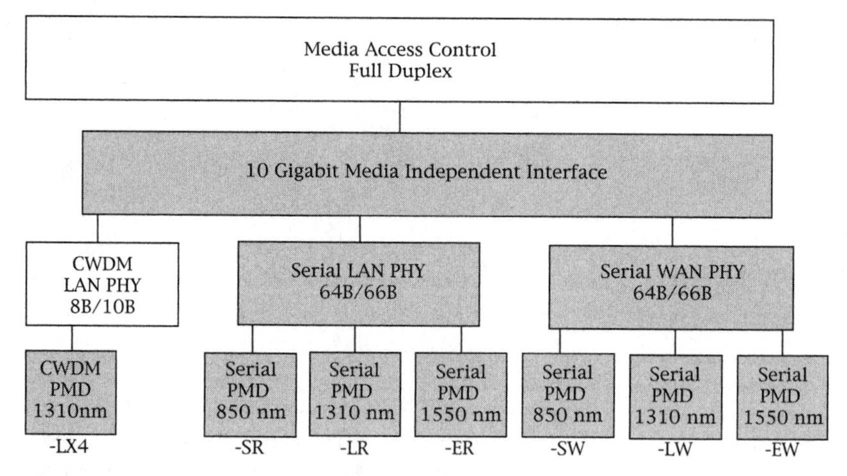


**FIGURE 12.9**
Physical layer of 10 Gigabit Ethernet

Serial transmission uses different types of laser wavelengths. The following are different physical medium definitions for 10 Gigabit Ethernet:

- 10GBaseSR
- 10GBaseSW
- 10GBaseLR
- 10GBaseLW
- 10GBaseER
- 10GBaseEW
- 10GBaseLX4

The suffix for Gigabit Ethernet consists of three characters. The first character indicates the wavelength of the laser, S, L, or E, where:

S represent short-wavelength laser with wavelength of 850 nm.

L represents long-wavelength laser with the wavelength of 1310 nm.

E represents extended long-wavelength laser with the wavelength of 1550 nm.

The second letter of the suffix represents the type of Physical layer:

R means Physical layer for LAN.

W means Physical layer for WAN.

***10GBaseLX4.*** L indicates long-wavelength laser, X refers to coding 8B/10B, and 4 indicates using wavelength division multiplexing transmitting and receiving over four separate wavelengths, which is also called *Wide Wavelength Division Multiplexing (WWDM).*

### 10 Gigabit Ethernet Fiber Cable and Distance

The 10 Gigabit Ethernet supports point-to-point connections using only fiber-optic cable as the transmission medium. Table 12.2 shows 10 Gigabit cable types and distances.

**TABLE 12.2** 10Gbe Types and Transmission Distance

| Standard | Cable Types | Core Diameter in Microns | Model Bandwidth | Distance |
|---|---|---|---|---|
| 10GBaseS | MM | 50 | 500 | 66 m |
| 10GBaseLX4 | MM | 62.5 | 160 | 300 m |
| 10GBaseL | SM | 9 | – | 10 Km |
| 10GBaseE | SM | 9 | – | 40 Km |
| 10GBaseLX4 | SM | 9 | – | 10 Km |

## Summary

- The IEEE 802.3z committee developed the standard for Gigabit Ethernet.
- Characteristics of Gigabit Ethernet are data rate of 1000 Mbps, uses IEEE 802.3 format, operates in full-duplex and half-duplex, uses optical fiber and copper for transmission media, and uses CSMA/CD for half-duplex operation.
- Gigabit Ethernet Physical layer can be 1000BaseLX, 1000BaseSX, 1000BaseCX, or 1000BaseT.
- 1000BaseLX means 1000 Mbps, baseband, L for long-wavelength laser with a wavelength of 1300 nm, and X means multi-mode or single-mode fiber.
- 1000BaseSX means 1000 Mbps, baseband, S for short-wavelength laser with a wavelength of 850 nm.
- 1000BaseCX uses shielded twisted-pair cable for transmission medium, and 1000BaseT uses unshielded twisted-pair cable Cat-5.
- The 10GbE Ethernet data rate is 10000 Mbps (10 Gbps).
- 10GbE operates only in the full-duplex mode.
- The 10GbE standard defines Physical layers for both WAN and LAN.
- 10GBaseSR means 10000 Mbps, baseband, using short-wavelength laser for LAN.
- 10GBaseSW means 10000 Mbps, baseband, using short-wavelength laser for WAN.
- 10GBaseLR means 10000 Mbps, baseband, using long-wavelength laser for LAN.
- 10GBaseLX4 means 10000 Mbps, baseband, using long-wavelength laser with WWDM.

## Key Terms

Gigabit Ethernet
Backbone
Extended-Wavelength (EW)
IEEE 802.1p
IEEE 802.3ae
IEEE 802.3z
LAN Physical Layer
Long-Wavelength (LW)
1000BaseCX
1000BaseLX

1000BaseSX
1000BaseT
Quality of Service (QoS)
Short-Wavelength (SW)
10 Gigabit Ethernet (10 GbE)
10GBaseER
10GBaseEW
10GBaseLR
10GBaseLW
10GBaseLX4
10GBaseSR
10GBaseSW
WAN Physical Layer

## Review Questions

### • Multiple Choice Questions

1. The data rate of Gigabit Ethernet is ________ Mbps.
   a. 100
   b. 1000
   c. 200
   d. 10,000
2. The standard for Gigabit Ethernet is ________.
   a. IEEE 802.2
   b. IEEE 802.3
   c. IEEE 802.3z (task force)
   d. IEE 802.3u
3. Gigabit Ethernet uses the ________ access method for half-duplex operation.
   a. CSMA/CD
   b. token passing
   c. demand priority
   d. none of the above
4. Gigabit Ethernet uses ________ encoding.
   a. Manchester
   b. Differential Manchester
   c. 8B/10B
   d. 4B/5B
5. 1000BaseLX uses ________ cable for transmission of data.
   a. UTP
   b. fiber-optic cable
   c. coaxial
   d. STP
6. The type of protocol that should be added to Gigabit Ethernet in order to carry voice and video information is ________.
   a. TCP
   b. IP
   c. 802.1p
   d. RSVP
7. Gigabit Ethernet can operate in ________.
   a. full-duplex
   b. half-duplex
   c. a and b
   d. none of the above
8. Gigabit Ethernet uses the CSMA/CD access method for ________.
   a. full-duplex
   b. half-duplex
   c. a and b
   d. none of the above

9. The transmission medium for 1000BaseT is ________.
   a. Cat-5 UTP
   b. Cat 4 UTP
   c. coaxial cable
   d. fiber cable
10. The maximum length of cable used for 1000BaseT is ________.
    a. 50 meters
    b. 100 meters
    c. 200 meters
    d. 1000 meters
11. The type of fiber cable used for Gigabit Ethernet is ________.
    a. multi-mode
    b. single-mode
    c. a and b
    d. none of the above
12. Gigabit Ethernet is used for ________.
    a. WAN
    b. campus backbone
    c. MAN
    d. WAN
13. The data rate of 10 GbE is ________ Mbps.
    a. 100
    b. 1000
    c. 10000
    d. 100000
14. Gigabit Ethernet operates in ________ mode(s).
    a. half-duplex
    b. full-duplex
    c. a and b
    d. CSMA/CD
15. 10 GbE is used for ________.
    a. LAN
    b. WAN
    c. Internet
    d. a and b
16. 10000BaseSR is used for ________.
    a. LAN
    b. WAN
    c. a and b
    d. none of the above
17. 10000BaseLW is used for ________.
    a. LAN
    b. WAN
    c. a and b
    d. none of the above
18. 10GBaseLX4 uses ________.
    a. WWDM
    b. CWDM
    c. DWDM
    d. none of the above

## • Short Answer Questions

1. What is the IEEE number for Gigabit Ethernet?
2. What is the data rate for Gigabit Ethernet?
3. What type of frame is used by Gigabit Ethernet?
4. What are the access methods for Gigabit Ethernet?
5. List the transmission media for Gigabit bit Ethernet.

6. Explain the following terms:
   a. 1000BaseCX
   b. 1000BaseLx
   c. 1000BaseSX
7. What are the hardware components of Gigabit Ethernet?
8. What are some applications of 10 GbE?
9. Explain the following terms:
   a. 10GBaseSR
   b. 10GBaseSW
   c. 10GBaseLR
   d. 10GBaseLW
   e. 10GBaseER
   f. 10GBaseEW
   g. 10GBaseLX4

CHAPTER 13

# Fiber Distributed Data Interface

**OUTLINE**



**OBJECTIVES**

After completing this chapter, you should be able to:

- Describe the Fiber Distributed Data Interface
- List FDDI rings and their function
- List the types of stations connected to FDDI
- Describe FDDI access methods and operation
- Describe the function of the dual and single attachment stations
- Show an FDDI frame format and explain the function of each field
- Understand FDDI bit transmission

**INTRODUCTION**

The Fiber Distributed Data Interface (FDDI) standard was developed by the ANSI X3T9.5 Standards Committee in 1980 and then submitted to the ISO. The ISO developed a new version of FDDI, which is compatible with the current ANSI standard.

## 13.1 FDDI Technology

**Fiber Distributed Data Interface (FDDI)** is a high-speed LAN using ring topology, with a data rate of 100 Mbps. FDDI uses two rings, and thus is termed a *dual-ring topology.* In this way FDDI is similar to IEEE 802.5 token ring. One of the most important features of FDDI is its use of optical fiber for transmission media. The main advantage of using fiber optics over copper wiring is security, because there is no electrical signal to tap on the media.

FDDI uses two rings, one primary and one secondary ring. Traffic travels through the rings in opposite directions, as shown in Figure 13.1. The **primary ring** is used for data transmission, and the **secondary ring** is used for backup in the case of primary breakdown. FDDI allows up to 1000 stations to be connected to the ring, with a maximum ring circumference of 200 Km.

**FIGURE 13.1**
FDDI data traffic flow

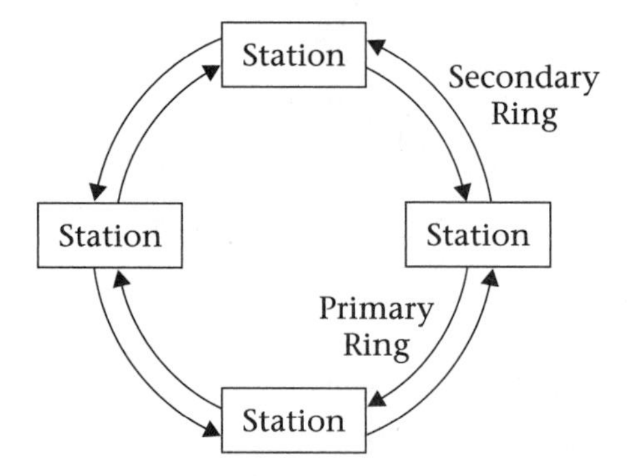


## 13.2 FDDI Layered Architecture

FDDI uses **token passing** as an access method. Any station that wants to transfer information holds the token and then transmits the information. The length of time a station holds the token is called *synchronous allocation time (SAT)* and this time is variable for each station. The allocation of this time for each station is achieved by station management. FDDI has added a new layer to the OSI model; this new layer is called *Station Management.* Figure 13.2 shows the FDDI layered architecture.

**Media Access Control (MAC):** Defines the medium access, frame format, addressing, token handling, and FCS calculation. FDDI also communicates with the higher-layer protocols, such as TCP/IP, SNA, and AppleTalk. The FDDI MAC layer accepts symbols from the upper layer, and then adds a MAC header and passes the data to the Physical layer. The maximum frame size is 4500 bytes.

**Physical Layer Protocol (PHY):** Handles data encoding/decoding, framing, and clock synchronization.

**Physical Medium Dependent (PMD):** Defines the transmission medium, such as fiber-optic link, optical component, and connectors.

**Station Management (SMT):** The function of SMT is ring control, ring initialization, station insertion, and station removal.

**FIGURE 13.2**
FDDI layered architecture

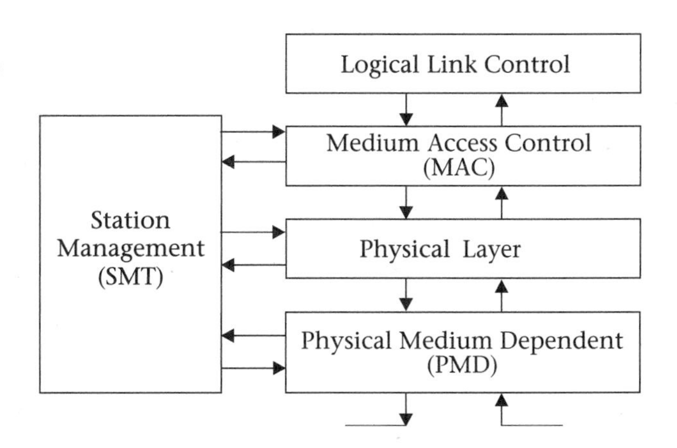


**Components of FDDI** The components of FDDI are fiber-optic cable, concentrator (rings), and the stations connected to the **concentrator,** as shown in Figure 13.3. There are two types of stations used in FDDI:

1. **Dual Attachment Station (DAS) or Class A:** DAS is attached to both rings, and has two ports to connect to the ring, one connected to the primary ring and the other to the secondary ring. Figure 13.4 shows the DAS ports.
2. **Single Attachment Station (SAS) or Class B:** SAS attaches to the primary ring.

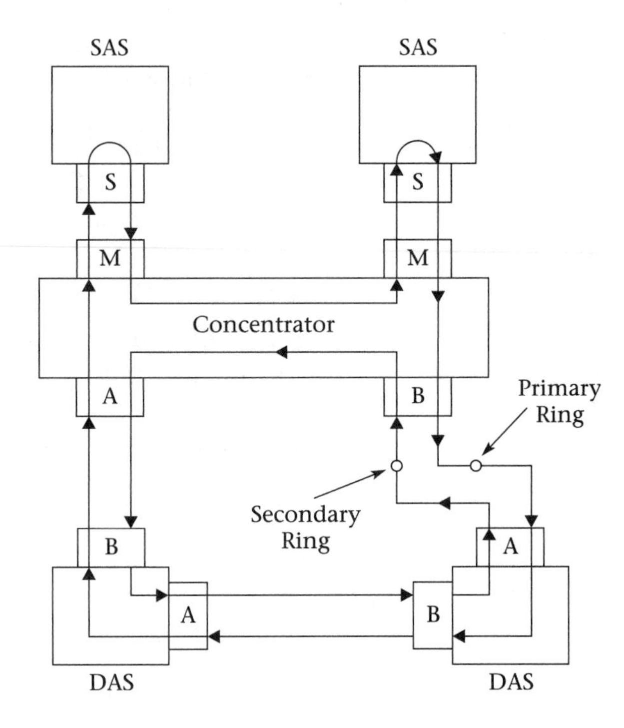


**FIGURE 13.3**
Components of an FDDI network

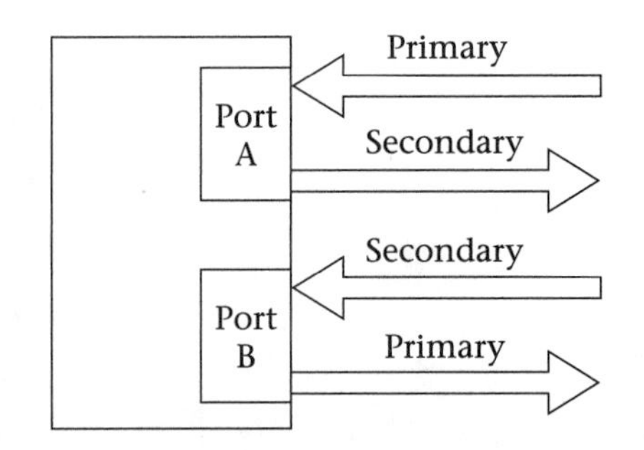


**FIGURE 13.4**
Dual attachment station (DAS)

## 13.3 FDDI Ports

FDDI defines four types of ports in order that the SAS and DAS can be connected to the FDDI concentrator, thus making two rings.

The four port types of ports used in FDDI topology are shown in Figure 13.3, and listed below:

**Type A:** Type A connects to the incoming primary ring and outgoing secondary ring of the FDDI dual rings. This port is used with a DAS.

**Type B:** Type B connects to the outgoing primary ring and incoming secondary ring of the FDDI dual rings.

**Type M:** Type M connects a concentrator to a SAS. This port is used in the concentrator.

**Type S:** Type S connects a SAS to a concentrator.

FDDI uses two rings and the rings are transmitting information opposite each other. If one ring breaks, the outer ring will handle the traffic. If both rings break at the same point, both rings can connect together to form a single ring, as shown in Figure 13.5.

**FIGURE 13.5**
FDDI wrapping

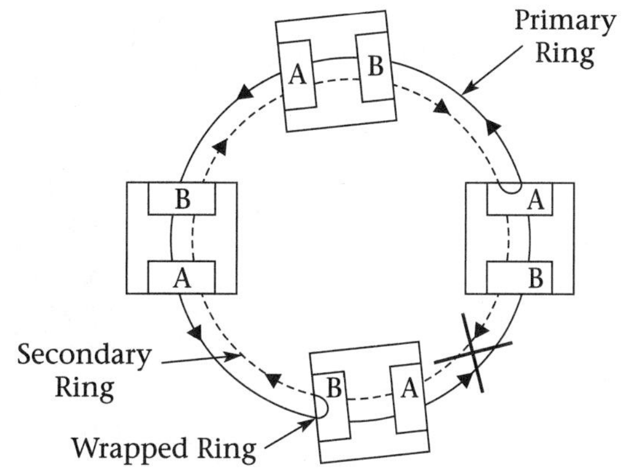


FDDI uses the principle of token ring topology. At any one time there is an active monitor in the ring to control the ring operations. FDDI uses the **4B/5B encoding** method that converts 4 bits of binary into a 5-bit symbol, as shown in Table 10.2, Chapter 10.

## 13.4 FDDI Access Method

A station wishing to transmit holds the token and starts a token timer, which determines how long a station can transmit. By combining a large frame size (4500 bytes) with a high transmission speed, FDDI can transfer data very efficiently from one point to another. It is suitable for backbone networks and file servers.

## 13.5 FDDI Fault Tolerance

A system able to respond to unexpected hardware and software failures and continue to provide service is said to be *fault tolerant*.

FDDI uses two fiber-optic rings, a primary ring, and a secondary ring. Traffic on these rings travels in opposite directions. The dual rings in an FDDI network provide fault tolerance for the FDDI.

If the primary ring fails, FDDI uses the secondary ring as backup. If both rings are damaged, the dual ring automatically "wraps" (the primary ring connects to the secondary ring) and changes the ring to a single ring, as shown in Figure 13.5. Figure 13.6 shows the cable broken down in two places, with FDDI wrapping the ring.

**FIGURE 13.6**
FDDI wrapped ring

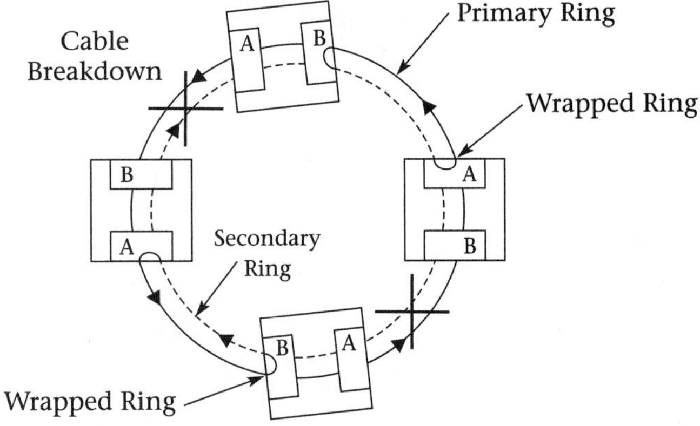


If a station on a dual ring fails or is powered down, the dual ring will wrap into a single ring. If two stations on the ring fail, the failure will cause ring segmentation. An optical bypass switch is used to avoid segmentations of the ring by eliminating the failed station(s) from the ring. Figure 13.7 shows DAS stations connected to the ring by an optical bypass switch.

## 13.6 FDDI Bit Transmission

FDDI uses 4B/5B for transmission as shown in Table 10.2, Chapter 10. It converts a 5-bit symbol to light pulses for transmission of information from source to destination. A bit can have two values, one or zero. FDDI determines bit changes by the state of the light on the receiver side. The receiver takes a sample of light every eight nanoseconds. The light is either on or off. If the light has changed since the last sample, then a one is received. If there is no change in the light, then a zero is received. Therefore, each time there is a transition of light, it is translated as one (from off to on and vice versa), as depicted in Figure 13.8.

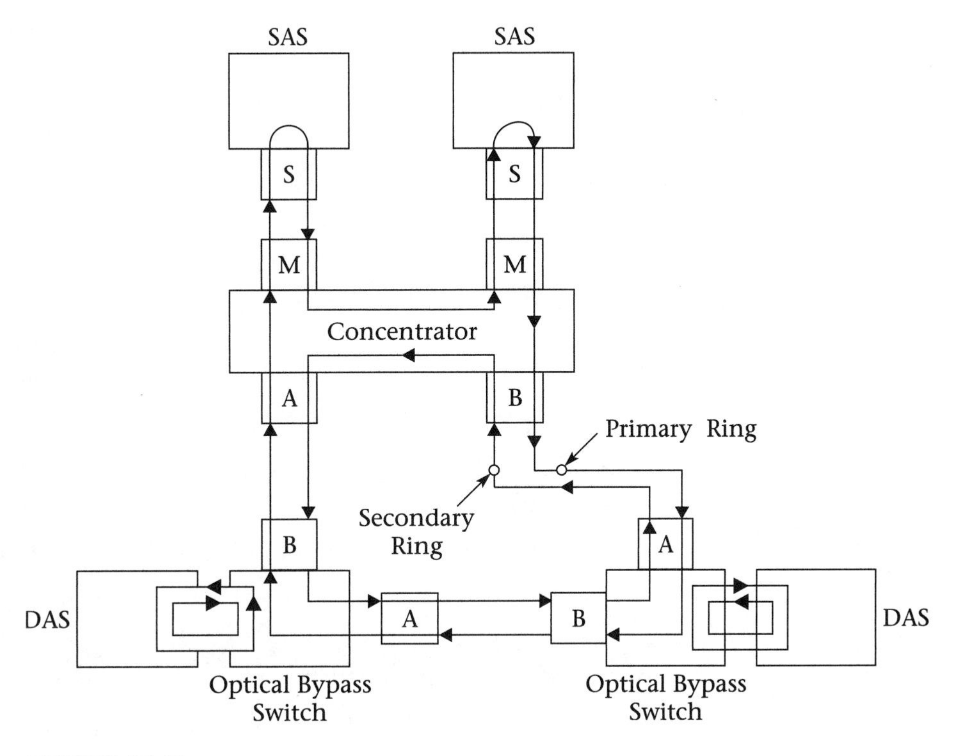


**FIGURE 13.7**
FDDI ring and optical bypass switch

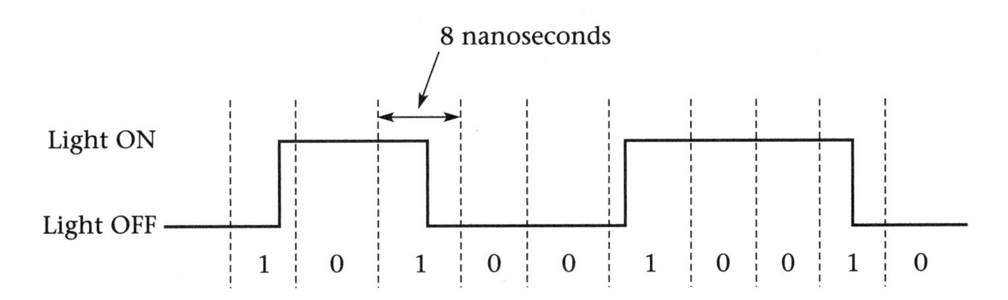


**FIGURE 13.8**
FDDI bit transmission

## 13.7 FDDI Frame and Token Formats

The FDDI frame and token formats are similar to those specified by **IEEE 802.5.** Figure 13.9 shows the FDDI frame format.

| Bytes >8 | 1 | 1 | 6 | 6 | | 4 | 1 | 1 |
|---|---|---|---|---|---|---|---|---|
| Preamble | Start Delimiter | Frame Control | DA | SA | Data | Checksum | End Delimiter | Frame Status |

**FIGURE 13.9**
FDDI frame format

The following describe the function of each field in an FDDI frame format:

- Preamble (PA) consists of 16 or more IDLE symbols
- Start Delimiter (SD) contains symbols J and K
- Frame Control (FC) contains two symbols used to define the type of token
- End Delimiter (ED) contains one or two control symbols (T)

## 13.8 FDDI Backbone

FDDI is used for LAN backbones in campuses and corporations having several buildings in one location. Figure 13.10 shows the application of FDDI as a backbone, using an asymmetric switch to convert from 1000 Mbps to 100 Mbps. An FDDI concentrator is used to connect two FDDI backbones together.

FDDI is one of the most expensive network backbones. With cheaper Ethernet switches and Fast Ethernet network cards, most network designers rely on a 100-Mbps Ethernet backbone rather than FDDI. Figure 13.11 shows a Fast Ethernet switch as a replacement for an FDDI backbone. The advantages of 100-Mbps Ethernet over FDDI are as follows:

- A Fast Ethernet NIC is less expensive than an FDDI NIC.
- Fast Ethernet can use UTP as a transmission media, which is less expensive than fiber-optic cable.

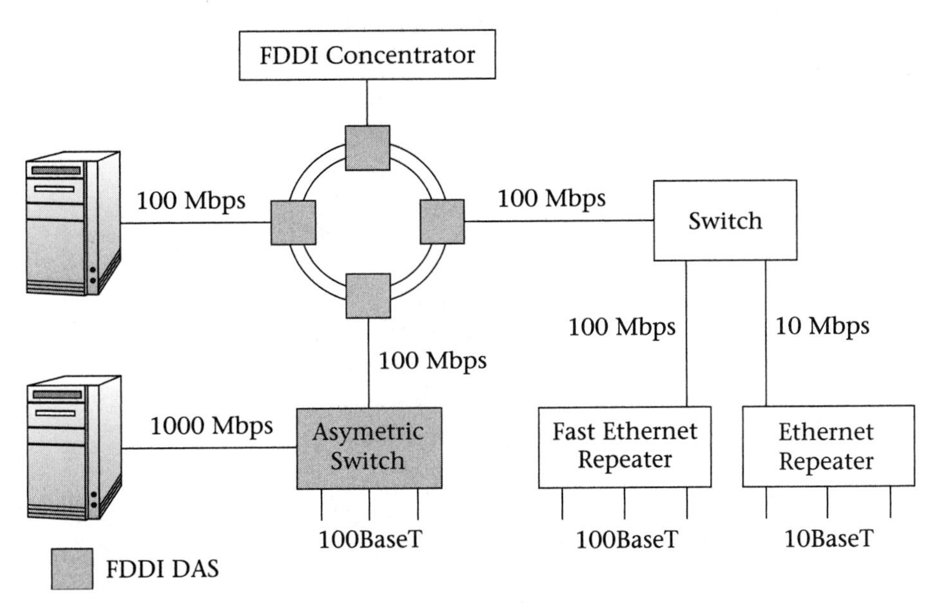


**FIGURE 13.10**
FDDI ring used for LAN backbone

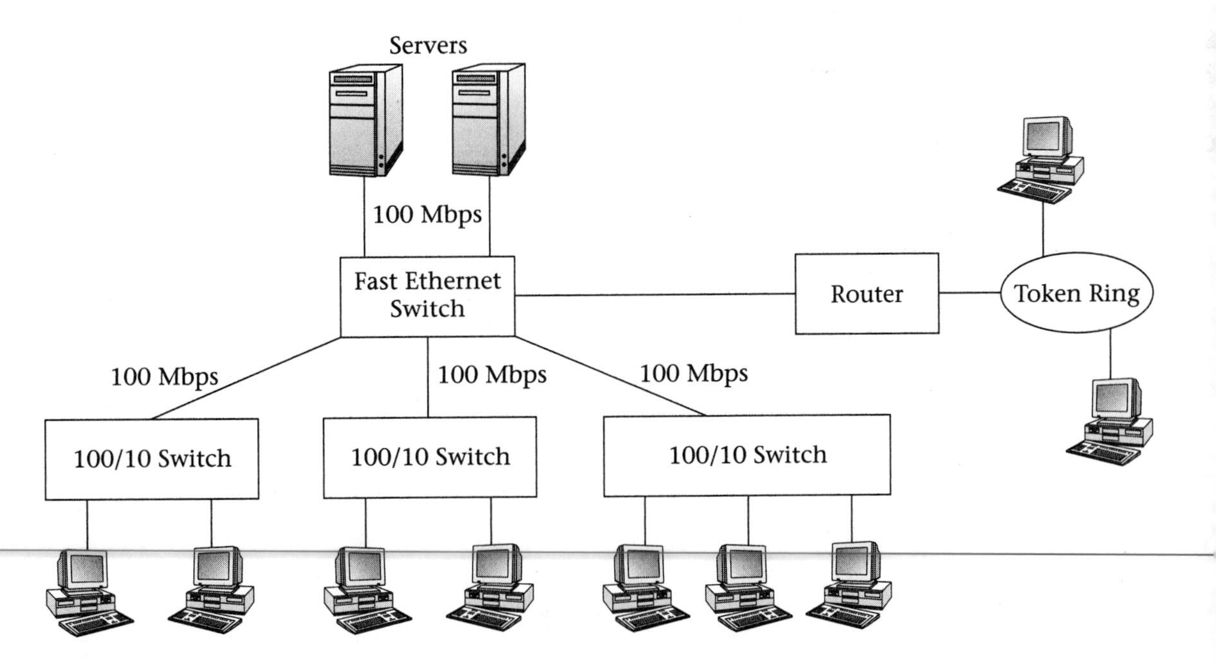


**FIGURE 13.11**
Fast Ethernet switch replacing FDDI ring

## Summary

- Fiber Distributed Data Interface (FDDI) is a high-speed LAN interface with a data rate of 100 Mbps using dual-ring topology and optical fiber as transmission media. FDDI is applied in network backbones.
- FDDI uses two fiber-optical rings, one primary ring for data transmission and a secondary ring for backup in the case of failure of the primary ring.
- FDDI uses token passing as its access method. The FDDI frame format is similar to the IEEE 802.5 frame format.
- Dual Attachment Stations (DAS) are connected to the primary and secondary ring.
- Single Attachment Stations (SAS) are connected only to the primary ring.
- FDDI uses 4B/5B encodings to convert four bits to a 5-bit symbol for transmission.

## Key Terms

Concentrator
Dual Attachment Station (DAS)
Fiber Distributed Data Interface (FDDI)
4B/5B Encoding
IEEE 802.5
Primary Ring
Secondary Ring
Single Attachment Station (SAS)
Token Passing

## Review Questions

- **Multiple Choice Questions**

1. FDDI is a high-speed ______ interface.
   a. LAN
   b. WAN
   c. a and b
   d. MAN
2. FDDI uses ______ topology.
   a. bus
   b. star
   c. ring
   d. hybrid
3. FDDI uses ______ rings.
   a. 2
   b. 3
   c. 4
   d. 1

4. FDDI's transmission medium is ______.
   a. UTP
   b. STP
   c. coaxial cable
   d. fiber-optic cable
5. FDDI allows up to ______ stations to be connected to a ring.
   a. 17
   b. 12
   c. 1000
   d. 500
6. The maximum frame size for FDDI is ______.
   a. 6 bytes
   b. 64 bytes
   c. 4500 bytes
   d. 1800 bytes
7. There are ______ types of station used in FDDI.
   a. 2
   b. 3
   c. 4
   d. 5
8. DAS is connected to the ______.
   a. primary ring
   b. secondary ring
   c. a and b
   d. none of the above
9. An application of FDDI is ______.
   a. office LAN
   b. campus backbone
   c. WAN
   d. Internet

## • Short Answer Questions

1. What does FDDI stand for?
2. Why does FDDI use two rings? What are the rings?
3. What is the data rate for FDDI?
4. Explain the FDDI access method.
5. Explain the function of a concentrator.
6. Explain the function of DAS.
7. Explain the function of SAS.
8. What is a wrapped ring?
9. Explain the function of a bypass relay switch.
10. Show the FDDI frame format, and explain the function of each field.
11. What type of encoding is used in FDDI?
12. What is the maximum number of stations that can be connected to an FDDI network?
13. List the advantages of a Fast Ethernet over FDDI.

CHAPTER 14

# Frame Relay

OUTLINE



OBJECTIVES

After completing this chapter, you should be able to:

- Discuss the application of a frame relay network
- List the components of a frame relay network
- Describe the function of a frame relay switch
- Describe the function of a frame assembler/disassembler
- Show the frame relay format and explain the function of each field

INTRODUCTION

Frame relay was originally designed for use over ISDNs. The initial proposal was submitted to ITU-T for standardization in 1984. The ITU-T ratified the proposal and this standard specification adds relay and routing functions to the Data Link layer of the OSI model.

Large organizations and corporations have multiple sites in different locations and they would like to connect LANs together. One solution is to lease data communication lines and connect their LANs together. A corporation with 100 offices in different locations must lease 100 lines to connect their LANs, and this method is not cost effective. The **frame relay** is a network that offers frame relay services to corporations, thus enabling their LANs to communicate with each other regardless of varying protocols.

The public carriers offer frame relay networking, which is less expensive than leased lines.

Frame relay is a packet-switching protocol service offered by telephone corporations to replace the X.25 protocol. It is a Wide Area Network (WAN). Figure 14.1 shows the architecture of a frame relay network consisting of a **frame relay network,** a **frame relay assembler/disassembler (FRAD),** and user LANs. The function of the FRAD (also called a *router*) is to convert the frame format from a site network to the frame format of the frame relay network and vice versa.

Figure 14.2 shows a reference model for a frame relay. The frame relay network operates in the Data Link and Physical layers of the OSI model.

**FIGURE 14.1**
Frame relay architecture

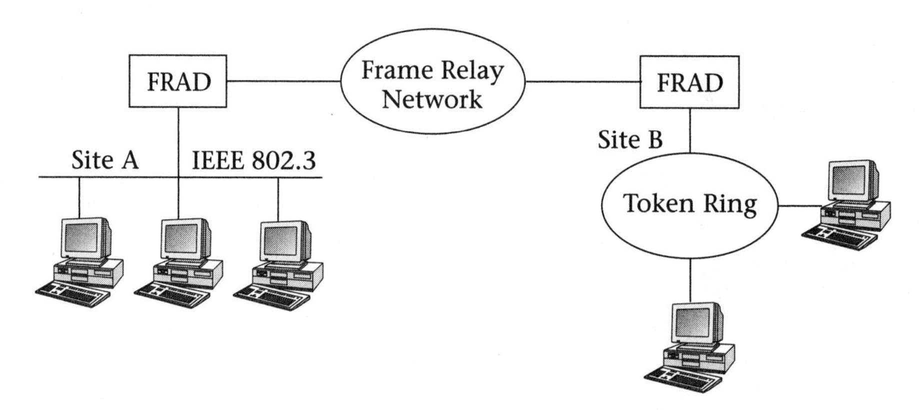


**FIGURE 14.2**
Frame relay network reference model

| Data Link Layer |
| --- |
| Physical Layer |

## 14.1 Frame Relay Network

Figure 14.3 shows a frame relay network, which consists of frame relay switches. The switches are connected to each other by T-1 or T-3 links, or by optical cables. These switches are located at the central office of each public carrier and the frame relay networks are controlled by telephone corporations.

**FIGURE 14.3**
Frame relay network

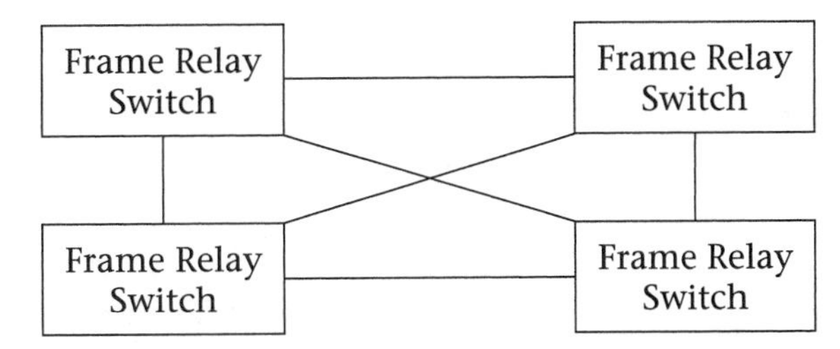


## 14.2 Components of Frame Relay

Frame relay is a service provided by public carriers to organizations so that the organizations can connect their networks, which are located in various locations. For a network to be connected to a frame relay network, the following components are required:

- Frame relay access equipment: (1) customer premises equipment such as a LAN, and (2) access equipment, which is a FRAD.
- Frame relay switches.
- Frame relay service provided by public carriers.

**Functions of the Frame Relay Switch**

The **frame relay switch** is located in the central office of a telephone company and performs the following functions:

- The switch checks for errors in a packet received from the user or the switch. If the packet contains an error, then the switch discards the packet; otherwise it transfers the packet to the next switch or destination.
- Frame relay switches use statistical packet multiplexing for multiplexing of information.
- When a packet is transmitted between switches, there is no acknowledgment between switches.
- The end user handles data integrity such as sequencing, error detection, and retransmission.

## 14.3 Frame Relay Frame Format

Figure 14.4 shows the frame format of a frame relay that consists of a starting flag (1 byte), an address (2 to 4 bytes), an information field (up to 4096 bytes), a frame check sequence (FCS) generated by a 16-bit CRC (2 bytes), and an ending flag (1 byte).

The following are descriptions and functions of each field in the frame relay frame format.

**The Starting Flag (SF) and Ending Flag (EF):** These are set to 0111110.

**Data Link Connection Identifier (DLCI):** The DLCI is 10 bits. This field determines the PVC (Permanent Virtual Circuit) value, which is used by the frame relay switch to find a path for the frame to reach to its destination.

**Command/Response (C/R):** This field determines whether the frame is a command frame or whether it is a response to a command.

**Extended Address (EA):** This bit identifies whether there is an extension address in the frame format.

EA = 0 means another address byte follows the current address byte.

EA = 1 means this is the last byte of the address field.

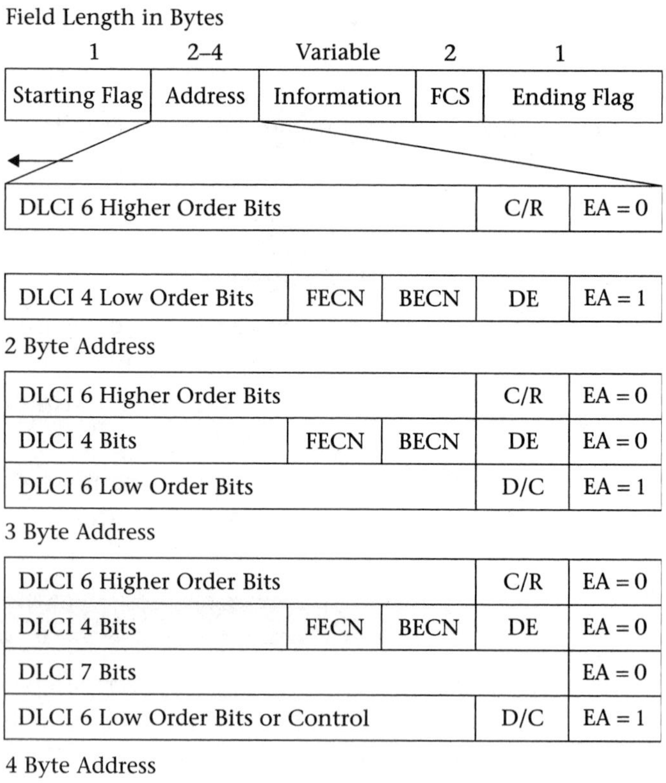

**FIGURE 14.4** Frame format of frame relay

Field Length in Bytes

| 1 | 2–4 | Variable | 2 | 1 |
|---|---|---|---|---|
| Starting Flag | Address | Information | FCS | Ending Flag |

| DLCI 6 Higher Order Bits | | | C/R | EA = 0 |
|---|---|---|---|---|
| DLCI 4 Low Order Bits | FECN | BECN | DE | EA = 1 |

2 Byte Address

| DLCI 6 Higher Order Bits | | | C/R | EA = 0 |
|---|---|---|---|---|
| DLCI 4 Bits | FECN | BECN | DE | EA = 0 |
| DLCI 6 Low Order Bits | | | D/C | EA = 1 |

3 Byte Address

| DLCI 6 Higher Order Bits | | | C/R | EA = 0 |
|---|---|---|---|---|
| DLCI 4 Bits | FECN | BECN | DE | EA = 0 |
| DLCI 7 Bits | | | | EA = 0 |
| DLCI 6 Low Order Bits or Control | | | D/C | EA = 1 |

4 Byte Address

**D/C:** This bit determines whether the low-order bits of the DLCI are control bits or DLCI.

**Congestion Control:** The following fields are used for congestion control:

- **Forward Explicit Congestion Notification (FECN):** This bit is used to notify the end user of congestion. It can be set by switches to show that there is congestion in the direction that the frame is traveling.
- **Backward Explicit Congestion Notification (BECN):** This bit can be set by switches to show that there is congestion in the opposite direction that the frame is traveling.

**Discard Eligibility (DE):** Because the congestion switch might discard some frames, this bit is set to one, which instructs the switch not to discard the frame.

## 14.4 Frame Relay Operation

Figure 14.5 shows a frame relay network. Network A is connected to switch A through a FRAD and Network B is connected to switch C through a FRAD. Each connection to the switch has a unique DLCI assigned by the service provider to each PVC. The path defined between two sites is called a **virtual circuit.** Frame relay supports **permanent virtual circuits (PVC)** and **switched virtual circuits (SVC).** Frame relay supports multiple PVCs simultaneously; therefore frame relay can connect multiple sites by a single connection.

For Network A to communicate with Network B, a permanent virtual connection is set up by the frame relay network administrator. Each switch has a table, which indicates the incoming frames DLCI, and an output port, as shown in Table 14.1.

**TABLE 14.1** Routing Table for Switches A, B, and C of Figure 14.5

| Switch A | | Switch B | | Switch C | |
|---|---|---|---|---|---|
| DLCI | Port # | DLCI | Port # | DLCI | Port # |
| 45 | 2 | 45 | 1 | 45 | 4 |

The following example tracks how a packet is transmitted from Network A to Network B, using Figure 14.5 and routing Table 14.1. When Network A transmits a packet to Network B, the following actions take place to reach Network B:

1. Network A transmits the packet to the FRAD and the FRAD converts the Ethernet frame format to the frame relay format; then the FRAD passes the frame to switch A.
2. Switch A checks for errors in the packet; if there is an error in the packet, the switch discards the packet. If the packet is correct, switch A uses its routing table (Table 14.1) to find out which port (based on DLCI 45) the packet has to exit.
3. The packet exits from port #2.
4. The packet reaches switch B.
5. Switch B checks for errors and transmits the packet, based on DLCI 45, to port #1.
6. The packet reaches switch C and switch C checks for errors. Then, using its routing table based on DLCI 45, switch C transmits the packet to port #4.
7. The packet reaches the FRAD. The FRAD changes the frame format of the packet from frame relay format to token ring format and transmits the frame to the token ring network.

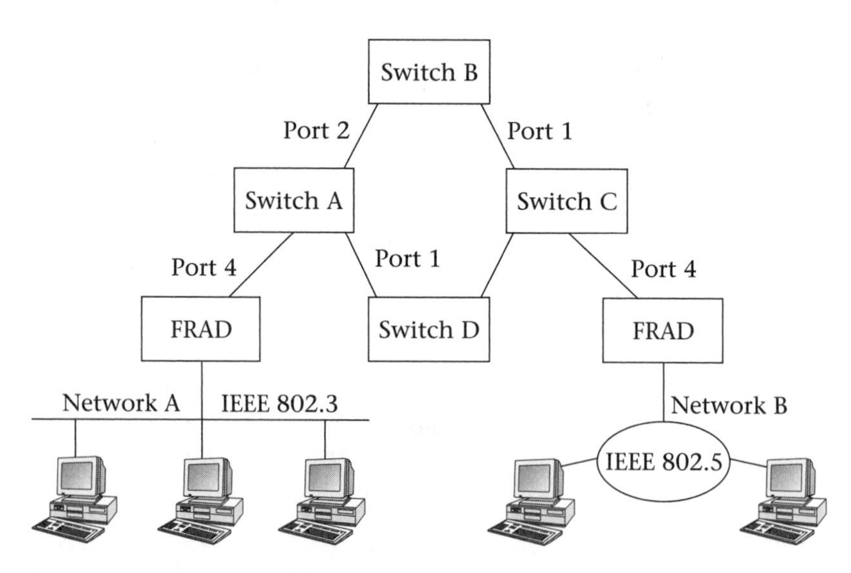


**FIGURE 14.5**
Frame relay network with three switches

## Summary

- The application of a frame relay network is to connect the networks of a corporation having several offices at different locations.
- Frame relay operates at the Data Link layer of the OSI Model.
- A frame relay network consists of frame relay switches, which are connected together by a T-1 or T-3 link.
- A frame relay network is a service offered by a telephone company.
- The components of a frame relay network are a customer LAN, frame relay assembler/disassembler (FRAD), and frame relay switch.
- The function of the FRAD is to convert a customer's LAN frame format to the frame format of a frame relay network and vice versa.
- A frame relay switch is located in the central switch of a telephone company. The function of the switch is to accept frames from FRADs and transmit them based on DLCI to the next switch.

## Key Terms

Frame Relay

Frame Relay Assembler/Disassembler (FRAD)

Frame Relay Network

Frame Relay Switch

Permanent Virtual Circuit (PVC)

Switched Virtual Circuit (SVC)

Virtual Circuit

## Review Questions

- **Multiple Choice Questions**

1. Frame relay was originally designed for use over a/an ______ network.
   a. Internet
   b. SONET
   c. ISDN
   d. WAN
2. Frame relay networks work in the ______ layer of the OSI model.
   a. Data Link
   b. Transport
   c. Session
   d. Physical

3. The frame relay switches are located ______.
   a. in a user's home
   b. at the central office of a telephone company
   c. at a government office
   d. at a corporation
4. EA = ______ means this is the last byte of the address field.
   a. 0
   b. 1
   c. 2
   d. 3
5. Frame relay is used as a/an ______.
   a. LAN
   b. WAN
   c. MAN
   d. Internet
6. Frame relay service is provided by ______.
   a. Internet service providers
   b. telephone corporations
   c. the U.S. government
   d. none of the above
7. A LAN can be connected to a frame relay network by a/an ______.
   a. repeater
   b. switch
   c. FRAD
   d. gateway
8. Frame relay uses ______ type of connection.
   a. permanent virtual circuit
   b. virtual circuit
   c. circuit switching
   d. message switching
9. Frame relay is used to connect ______.
   a. LANs that are located in different cities
   b. LANs that are located in a campus
   c. LANs that are located in the same city
   d. LANs that are located in one room

## • Short Answer Questions

1. What is the application of frame relay?
2. Explain the function of a FRAD.

3. Explain the function of the frame relay switch.
4. What are the components of frame relay networks?
5. Explain the function of the following fields in the frame relay frame format:
   a. DLCI
   b. EA
   c. C/R
   d. DE
   e. FECN and BECN

CHAPTER 15

# Synchronous Optical Network (SONET)

**OUTLINE**

15.1 SONET Characteristics
15.2 SONET Components
15.3 SONET Signal Rates
15.4 SONET Frame Format
15.5 SONET Multiplexing
15.6 Virtual Tributaries

**OBJECTIVES**

After completing this chapter, you should be able to:

- Describe the characteristics of a synchronous optical network (SONET)
- List the components of SONET and define the function of each component
- List SONET's optical signal rates
- Show the SONET frame format and explain the function of each overhead field

**INTRODUCTION**

**Synchronous Optical Network (SONET)** is a high-speed optical carrier using fiber optic-cable for transmission media. The term SONET is used in North America and is a standard established by the American National Standards Institute (ANSI). The ITU (International Telecommunications Union) has set a standard for SONET called *Synchronous Digital Hierarchy (SDH),* which is also used in Europe.

SONET optical architecture is based on a four-fiber bidirectional ring to provide the highest possible level of service assurance. New application software (such as Medical Images and CAD/CAM applications) require more bandwidth than other applications, and SONET provides high-speed transmission with a large bandwidth.

## 15.1 SONET Characteristics

The most significant characteristics of SONET are as follows:

- SONET uses byte multiplexing at all levels.
- SONET is a high-speed transport (carrier) technology with a self-correcting path.
- SONET uses multiplexing and demultiplexing.
- SONET provides **Operation Administration and Maintenance (OAM)** functions for network managers.
- The basic electrical signal for SONET is **Synchronous Transport Signal Level One (STS-1).**
- SONET transmits the STS-1 at the rate of 8000 frames per second.
- Slower signals can be multiplexed directly onto higher speeds.

## 15.2 SONET Components

Figure 15.1 shows SONET's components, which consist of a **Synchronous Transport Signal (STS)** multiplexer (MUX), a regenerator, an add/drop multiplexer, an electrical-to-optical converter, and an STS demultiplexer. The functions of these components are as follows:

**STS Multiplexer:** The function of the STS MUX is to multiplex electrical input signals to a higher data rate and then convert the results to an optical signal, as shown in Figure 15.1.

**Regenerator:** The regenerator performs the functions of a repeater. If the optical cable is longer than standard, the regenerator will be used to receive the optical signal and then to regenerate the optical signal.

**Add/Drop Multiplexer:** Add/drop multiplexers are used for extracting or inserting lower-rate signals from or into higher-rate multiplexed signals without completely demultiplexing the SONET signals.

**STS Demultiplexer:** STS demultiplexers convert and demultiplex optical signals to electrical signals for end users.

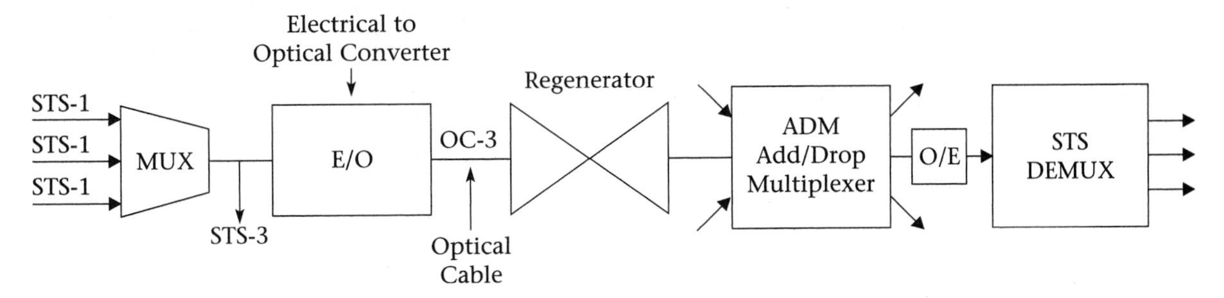


**FIGURE 15.1**
SONET components

## 15.3 SONET Signal Rates

The lowest-level signal in SONET is the Synchronous Transport Signal Level-one (STS-1), which has a signal rate of 51.84 Mbps. The STS-1 is an electrical signal, which is converted to an **optical carrier signal (OC)** called *OC-1*. The higher SONET data rate is represented by STS-*n* where *n* is 1, 3, 9, 12, 18, 24, 34, 48, 96, and 192. Table 15. 1 shows SONET and SDH signal rates.

**TABLE 15.1** Data Rate for OC, STS, and STM Signals

| Fiber Optical (OC) Signal OC-n Level | Synchronous Transport Signal (STS) for SONET | Synchronous Transport Module (STM) for SDH | Data Rate in Mbps |
|---|---|---|---|
| OC-1 | STS-1 | | 51.84 |
| OC-3 | STS-3 | STM-1 | 155.52 |
| OC-9 | STS-9 | STM-3 | 446.56 |
| OC-12 | STS-12 | STM-4 | 622.08 |
| OC-18 | STS-18 | STM-6 | 933.12 |
| OC-24 | STS-24 | STM-8 | 1244.16 |
| OC-36 | STS-36 | STM-12 | 1866.24 |
| OC-48 | STS-48 | STM-16 | 2488.32 |
| OC-96 | STS-96 | STM-32 | 4976.64 |
| OC-192 | STS-192 | STM-64 | 9953.28 |

OC = Optical Carrier
STS = Synchronous Transport Signal (Electrical Signal for SONET)
STM = Synchronous Transport Module (Electrical Signal for SDH)

## 15.4 SONET Frame Format

The basic transmission signal for the **SONET frame format** is the STS-1. The STS-1 format is shown in Figure 15.2. It is made up of 9 rows and 90 columns of bytes. The frame size is $90 * 9 = 810$ bytes or $810 * 8 = 6480$ bits. SONET transmits 8000 frames per second. The data rate for STS-1 is $6480 * 8000 = 51.84$ Mbps.

The first three columns are called *transport overhead,* which is $3 * 9 = 27$ bytes; 9 of these 27 bytes are used for section overhead, 18 bytes are used for line overhead, and 9 bytes are used for path overhead. The actual data rate is 86 columns * 9 rows * 8 bits * 8000 frames/sec = 49.536 Mbps.

The STS-1 frame is transmitted by the byte from row 1, column 1 to row 9, column 90 (scanning from left to right).

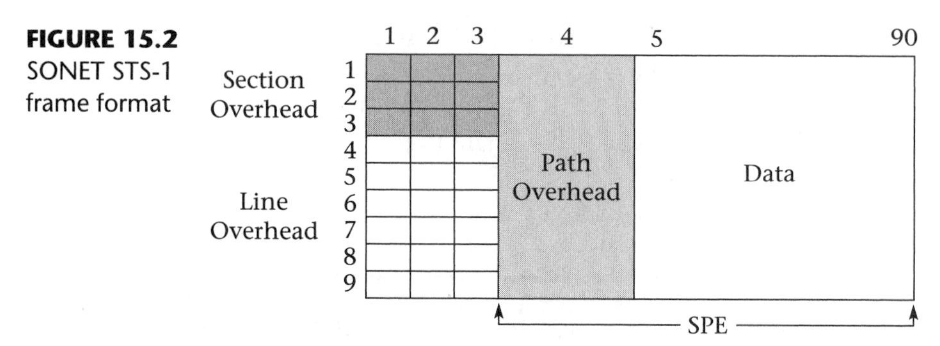


**FIGURE 15.2** SONET STS-1 frame format

**Path Overhead** is part of the **Synchronous Payload Signal (SPE)** and contains the following information: performance monitor of synchronous transport signal, path trace, parity check, and path status.

**Section Overhead** contains information about frame synchronization (informing destination of incoming frame) and frame identification, carries information about OAM, handles frame alignment, and separates data from the voice.

**Line Overhead** carries the payload pointers to specify the location of SPE in the frame and provides automatic switching (for standby equipment). It separates voice channels and provides multiplexing, line maintenance, and performance monitoring.

## 15.5 SONET Multiplexing

Higher levels of synchronous transport signals can be generated by using byte interleave multiplexing. The STS-3 is generated by multiplexing three

STS-1 signals, as shown in Figure 15.3. The output of STS-3 is converted to an optical signal called *OC-3*. Therefore, the STS-3 frame is made up of 3 * 90 or 270 columns and 9 rows with a total of 2430 bytes. The STS-3 is transmitted at 8000 frames per second; therefore, the bit rate of STS-3 is:

$$2430 \text{ bytes} * 8 \text{ bits} * 8000 \text{ frames/sec} = 155.52 \text{ Mbps}$$

Figure 15.4 shows the STS-3 frame format. The transport overhead is made up of 9 columns and 9 rows; the SONET payload envelop is 260 * 9 bytes. The STS-9 is generated by multiplexing three STS-3s, as shown in Figure 15.5.

**FIGURE 15.3**
Multiplexing three STS-1 to generate STS-3

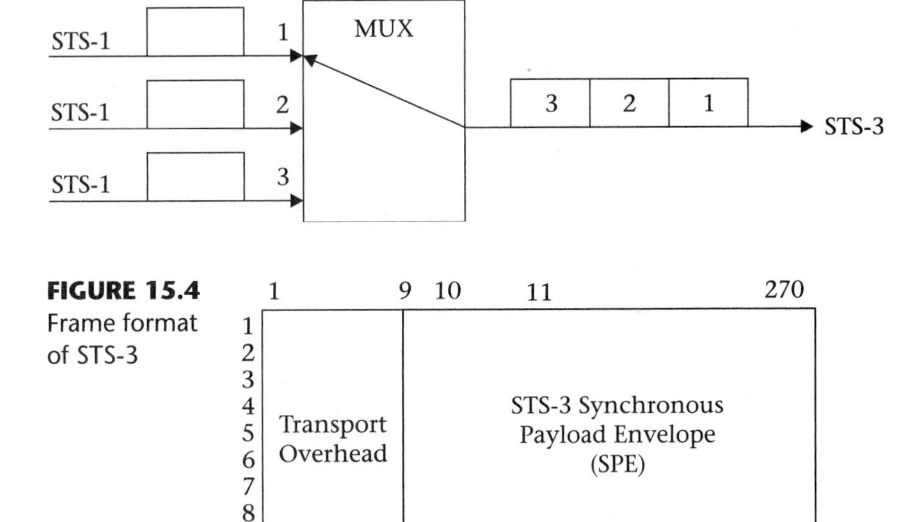


**FIGURE 15.4**
Frame format of STS-3

**FIGURE 15.5**
Generating an STS-9 frame

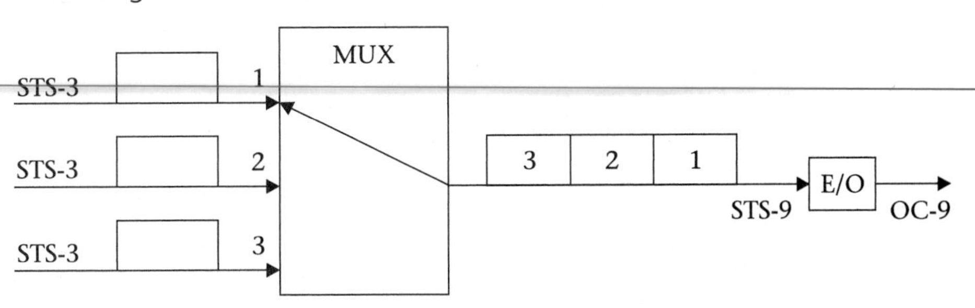

## 15.6 Virtual Tributaries

The basic frame for the SONET is STS-1 with a data rate of 51.84 Mbps. The payload of STS-1 is made up of 86 columns and 9 rows. In order for the SONET to carry lower data rate frames such as DS-1 and DS-2, the lower data rate frame is mapped to the STS-1 payload and is called a **virtual tributary (VT).** Figure 15.6 shows the VTs in STS-1 payload.

There are four types of VTs that map into the STS-1 payload:

VT1.5 is a frame of 27 bytes that is made up of 3 columns and 9 rows, as shown in Figure 15.7. The data rate of the VT1.5 is calculated as follows:

$$\text{Data Rate VT1.5} = 27 \text{ bytes} * 8 \text{ bits} * 8000 \text{ frames/sec}$$

$$= 1.728 \text{ Mbps}$$

The VT1.5 is used for transmission of DS-1 with a date rate of 1.54 Mbps. The STS-1 payload can transmit 28 VT1.5s.

VT2 is a frame of 36 bytes (made up of 4 columns and 9 rows) and is used for transmission of a European E-1 line (an E-1 line can carry 30 voice channels) with a data rate of 2.3048 Mbps.

VT3 is 54 bytes (made of 6 columns and 9 rows) that is used for transmission of DS-1C frame with a data rate of 3.152 Mbps.

VT6 is 108 bytes (made of 12 columns and 9 rows) that is used for transmission of a DS-2 frame with a data rate of 6.312 Mbps.

**FIGURE 15.6**
Virtual tributaries

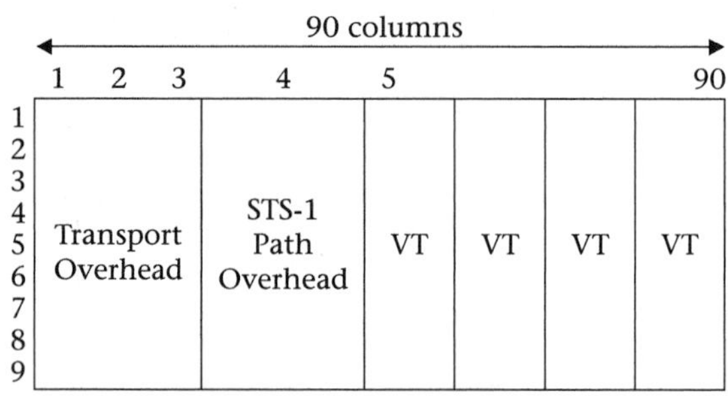


**FIGURE 15.7**
VT1.5 mapped into STS-1 frame

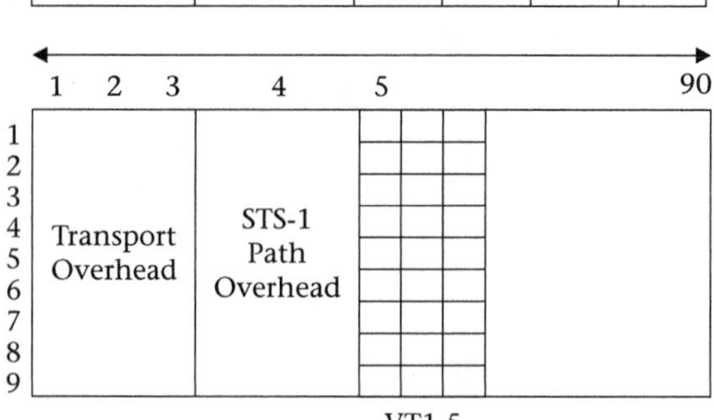

## Summary

- The synchronous optical network (SONET) is a high-speed data carrier. The term SONET is used in North America, and Synchronous Digital Hierarchy (SDH) is used in Europe.
- SONET uses fiber-optical cable for transmission media, since optical transmission is immune to interference and can transmit data over a long distance.
- SONET converts the synchronous transport signal (STS-1) to an optical signal. It is called *OC-1 (Optical Carrier)* and is transmitted at the rate of 8000 frames per second.
- SONET components are the STS multiplexer, the regenerator, the add/drop multiplexer (ADM), and the STS demultiplexer.
- The basic transmission signal for SONET is the STS-1. The frame format of SONET consists of 9 rows and 90 columns of bytes. The frame size is 810 bytes and this frame is transmitted at a rate of 8000 frames per second, which gives a bit rate of 51.84 Mbps for STS-1.
- Three STS-1 signals can be multiplexed and converted to optical to generate an OC-3 with a data rate of 155.52 Mbps.

## Key Terms

Add/Drop Multiplexer
Line Overhead
Operation Administration and Maintenance (OAM)
Optical Carrier Signal (OC)
Path Overhead
Regenerator
Section Overhead
SONET Frame Format
STS Demultiplexer
STS Multiplexer
Synchronous Optical Network (SONET)
Synchronous Payload Signal (SPE)
Synchronous Transport Signal (STS)
Synchronous Transport Signal Level One (STS-1)
Virtual Tributary

## Review Questions

- **Multiple Choice Questions**

1. ______ set the standard for SONET.
   a. IEEE
   b. ANSI
   c. ITU
   d. ISO
2. SONET uses byte multiplexing in ______ levels.
   a. upper
   b. mid
   c. all
   d. none of the above
3. The basic electrical signal for SONET is ______
   a. STS-1
   b. STS-2
   c. STS-3
   d. STS-n
4. SONET transmits the STS-1 at the rate of ______ frames/second.
   a. 6000
   b. 7000
   c. 8000
   d. 1000
5. The ______ performs the function of a repeater.
   a. regenerator
   b. STS multiplexer
   c. STS demultiplexer
   d. SONET
6. STS-1 has data rate of ______ Mbps.
   a. 810
   b. 8000
   c. 51.84
   d. 1.54
7. SONET is a/an ______
   a. LAN
   b. WAN
   c. optical carrier
   d. Internet
8. An STS-1 frame consists of ______.
   a. 9 columns and 90 rows
   b. 9 rows and 90 columns
   c. 10 rows and 100 columns
   d. none of the above
9. The optical signal for STS-1 is ______.
   a. OC-3
   b. OC-1
   c. OC-2
   d. OC-n
10. An STS-3 is generated by multiplexing ______.
    a. three STS-1s
    b. six STS-1s
    c. five STS-1s
    d. two STS-1s
11. An STS-9 is generated by multiplexing ______.
    a. six STS-1s
    b. three STS-3s
    c. three STS-1s
    d. two STS-3s

12. An STS-3 frame format consists of ____.
    a. 270 rows and 9 columns
    b. 9 rows and 270 columns
    c. 10 rows and 300 columns
    d. none of the above
13. SONET uses virtual tributaries to transmit ____.
    a. a frame with data rate higher than STS-1
    b. a frame with data rate less than STS-1
    c. a frame with STS-1data rate
    d. none of the above

## • Short Answer Questions

1. What does SONET stand for?
2. What does SDH stand for?
3. What is an application of SONET?
4. What is the basic electrical signal for SONET?
5. What is the transmission media for SONET?
6. List some of the advantages of SONET.
7. List the SONET components.
8. What does STS-1 stand for?
9. What is OC-1?
10. What is the data rate for STS-1?
11. How many bytes is STS-1?
13. How many STS-1s must be multiplexed to generate an STS-3?
14. SONET transmits how many frames per second?
15. Show the SONET frame format.
16. Calculate the VT2 data rate.
17. Calculate the VT3 and VT6 data rates.
18. Explain the function of add/drop multiplexing.
19. What is STS-n?
20. Why is the STS-1 bit rate 51.84 Mbps?

CHAPTER 16

# Internet Protocols (Part I)

OUTLINE



OBJECTIVES

After completing this chapter, you should be able to:

- Discuss the history of the Internet
- List the applications of the Internet and explain the function of each application protocol
- Explain the function of the Internet Architecture Board (IAB)
- List some TCP/IP protocols and describe the service of each protocol
- Explain the IP address classes and describe how IP addresses are assigned to a network of an organization
- Describe the Domain Name System (DNS)

- Show the TCP/IP reference model
- Show the User Datagram Protocol (UDP) packet format and define the function of each field
- List the applications protocol for TCP
- Describe the function of TCP, show the TCP packet format, and describe the function of each field
- Explain the function of IP and identify IP packet format
- Explain TCP connection and disconnection
- Show the IPv6 format and explain the function of each field
- Describe the advantages of IPv6
- Describe Internet II

## INTRODUCTION

The term *Internet*, short for *Internetwork*, describes a collection of networks that use the TCP/IP protocol to communicate among nodes. These networks are connected together through routers and gateways. Figure 16.1 shows an organization whose networks are connected together by an internal gateway that is, in turn, connected to the external gateway of the Internet. Today the term *gateway* defines a device that connects two different networks that have different protocols.

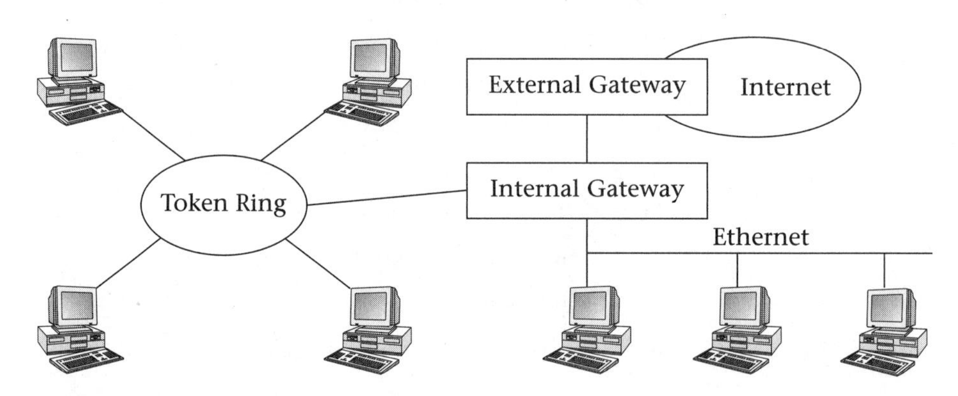


**FIGURE 16.1**
Connection of a network to the Internet

In 1968, the United States Department of Defense (DOD) created the Defense Advanced Research Project Agency (DARPA) for research on packet-switching networks. In 1969, DARPA created the **Advanced Research Project Agency (ARPA).** In the same year, ARPA selected Bolt Beranek and Newman (BBN), a research firm in Cambridge, Massachusetts, to build an experimental network (ARPANET) to provide a test bed for emerging network technology. **ARPANET** originally connected four nodes: the University of California at Los Angeles (UCLA), the University of California at Berkeley (UCB), Stanford Research Institute (SRI), and the University of Utah to share information and resources across long distances. ARPANET experienced rapid growth with the addition of universities. At that time, the protocol used in ARPANET was called the *network control protocol (NCP).* NCP did not scale well to the growing ARPANET, and in 1974 TCP/IP was introduced.

In 1980, the TCP/IP protocol became the only protocol that was in use on ARPANET. At the same time, most universities were using the UNIX operating system, which was created by Bell Labs in 1969. The University of California at Berkeley integrated the TCP/IP protocol into version 4.1 of its software distribution, later known as the *Berkeley Software Distribution (Berkeley UNIX or BSD UNIX).* The DOD separated the military network (MILNET) from the nonmilitary network (ARPANET).

In 1985 the National Science Foundation (NSF) connected the six super-computer centers together and named this network **NSFNET.** The NSFNET was then connected to ARPANET. Naturally, NSF encouraged universities to connect to NSFNET. Due to the growth of NSFNET, in 1987, NSF accepted a joint proposal from IBM, MCI Corporation, and MERT Corporation to expand the NSF backbone. Figure 16.2 shows the NSF backbone in 1993. By 1995, numerous companies were running commercial networks.

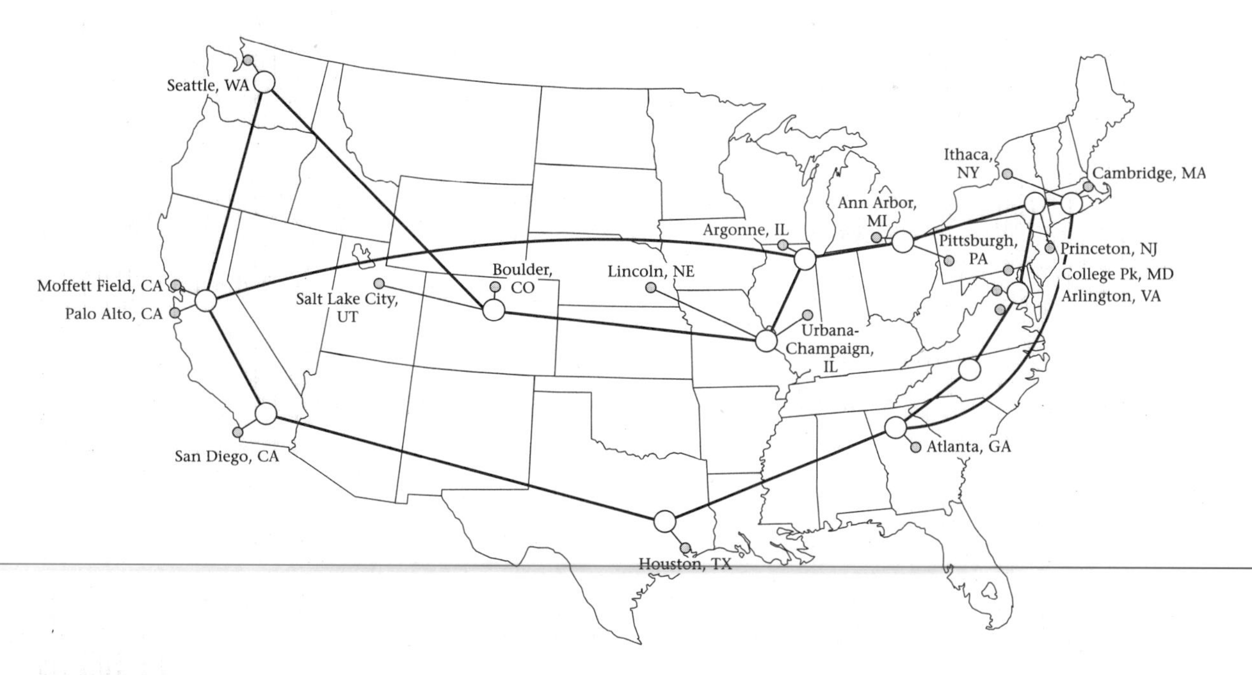


**FIGURE 16.2**
NSFNET backbone

The current Internet backbone is a connection of several backbones that belong to Internet network service providers such as MCI, AT&T, IBM, Sprint, and GTE. These backbones are connected through gateways. Figure 16.3 shows the GTE Internet backbone.

**Internet Address Assignment** Any organization wishing to connect its network to the Internet must contact the Internet Network Information Center (InterNIC) to obtain an Internet address (IP address). For example, the following is an InterNIC address.

www.internic.net
**Email:** Hostmaster@internic.net
**Mailing address:** Network Information Center
333 International
Menlo Park, CA 94025

Any organization that obtains a network IP address will submit its server's name to InterNIC. InterNIC will ensure that no two servers have the same name.

For example, elahia1@southernct.edu is the author's Internet address. Reading this domain name from right to left:

**edu.** Means that Elahi is at some U.S. educational site.

**Southernct.** Represents the machine that has information about Elahi's IP address.

## 16.1 The Internet Architecture Board (IAB)

The IAB is comprised of thirteen members; six are selected by the Internet Engineering Task Force (IETF). The functions of the IAB include the following:

- Determine the future of Internet addressing
- Provide leadership on the architecture of the Internet
- Decide the direction of IETF; management of a top-level DNS

The following is a list of IAB subcommittees and their functions:

**IESG:** The Internet Engineering Steering Group (IESG) works on Internet standards and oversees the work of all the other groups.

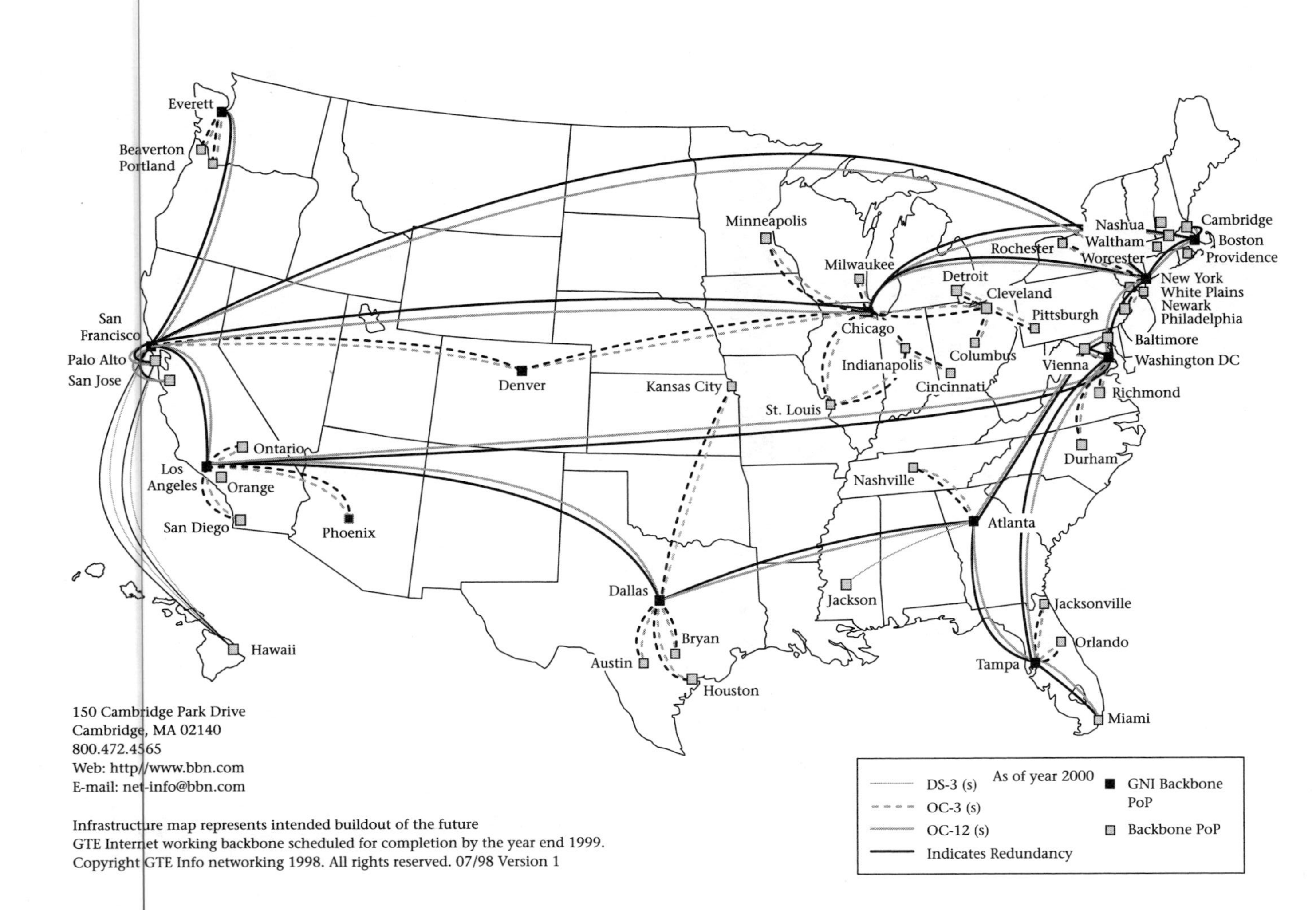


**FIGURE 16.3**
GTE Internet backbone (Courtesy of BBN technologies)

**IETF:** The Internet Engineering Task Force (IETF) is an open international committee of network designers, vendors, and researchers. The IETF is divided into subgroups, which are organized by their area of expertise, such as routing or transport.

**IRTF:** The function of the Internet Research Task Force (IRTF) is to promote long- and short-term research related to Internet protocols such as TCP/IP, Internet architecture, and IPv6.

**IANA:** The Internet Assigned Numbers Authority (IANA) works under the Internet Network Information Center (INIC). InterNIC consists of Network Solutions Inc. and AT&T Corp. The function of the INIC is DNS registration and education services. InterNIC manages registration of the second-level domain names under the following top-level domains: **gov, com, net, mil, biz, info, name, coop, aero, net,** and **org.**

## 16.2 TCP/IP Reference Model

Transmission Control Protocol and Internet Protocol (TCP/IP) essentially consists of four levels: Application level, Transport level, Internet level, and Network level as shown in Figure 16.4. Table 16.1 shows TCP/IP protocols and their functions.

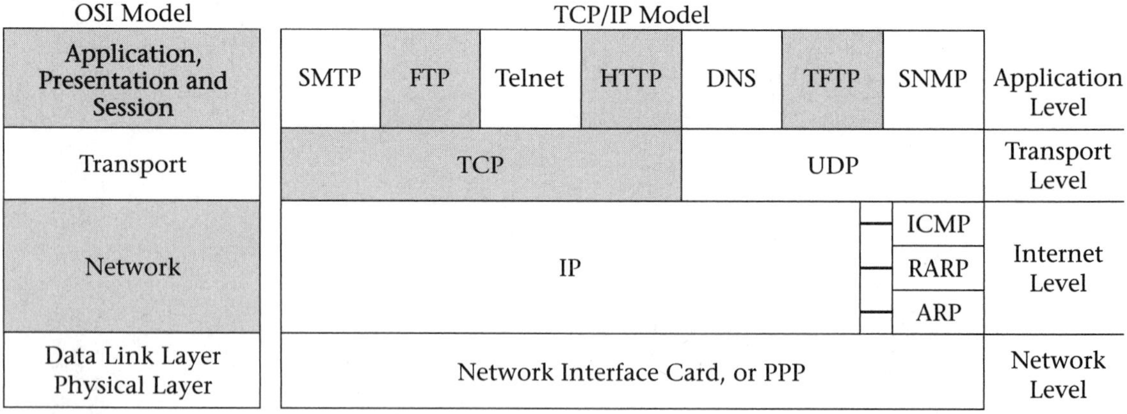

| OSI Model | TCP/IP Model | |
|---|---|---|
| Application, Presentation and Session | SMTP, FTP, Telnet, HTTP, DNS, TFTP, SNMP | Application Level |
| Transport | TCP, UDP | Transport Level |
| Network | IP (ICMP, RARP, ARP) | Internet Level |
| Data Link Layer<br>Physical Layer | Network Interface Card, or PPP | Network Level |

**FIGURE 16.4**
TCP/IP reference model

**TABLE 16.1** TCP/IP Protocols and Their Functions

| Protocol | Service |
|---|---|
| Internet Protocol (IP) | Provides packet delivery between networks |
| Internet Control Message Protocol (ICMP) | Controls transmission errors and controls messages between hosts and gateways |
| Address Resolution Protocol (ARP) | Requests physical addresses from source |
| Reverse Address Resolution Protocol (RARP) | Responds to the ARP |
| User Datagram Protocol (UDP) | Provides unreliable service between hosts (transfer of data without acknowledgment) |
| Transmission Control Protocol (TCP) | Provides reliable service between hosts |
| Simple Network Management Protocol (SNMP) | Used for diagnostics purposes between hosts |

## 16.3 TCP/IP Application Level

The Application level enables the user to access the Internet. The following are some of the Internet applications:

- Simple Mail Transfer protocol (SMTP)—email
- Telnet
- File Transfer Protocol (FTP)
- Hypertext Transfer Protocol (HTTP)
- Simple Network Management Protocol (SNMP)
- Domain Name System (DNS)

### Simple Mail Transfer Protocol (SMTP)

SMTP is used for email (electronic mail), transferring messages between two hosts. To send a mail message, the sender types in the address of the recipient and a message. The electronic mail application accepts the message/mail (if the address is right) and deposits it in the storage area/mailbox of the recipient. The recipient then retrieves the message from his/her mailbox.

***What Is an Email Address?*** An email address is made up of username @ mail server address. For example Elahia1@southernct.edu: "Elahia1" is the username and "southernct.edu" is the domain name of the mail server. The string "southernct" stands for Southern Connecticut State University, and "edu" stands for education.

Some email addresses are more complicated, for example Elahi@scsu1.southernct.edu where "Elahi" is the username, "scsu1" is a name of a workstation that is a part of "southernct," and "edu" tells you that it's an education center.

## Telnet or Remote Login

**Telnet** is one of the most important Internet applications. It enables one computer to establish a connection to another computer. Users can log into a local computer and then do a remote login across the network to any other host. The computer establishing the connection is referred to as the *local computer*; the computer accepting the connection is referred to as the *remote* or *host computer*. The remote computer could be a hardwired terminal or a computer in another country. Once connected, the commands typed in by the user are executed on the remote computer. What the user sees on his/her monitor is what is taking place on the remote computer.

**Remote login** was developed for Berkeley UNIX to work with the UNIX operating system only, but it has been ported to other operating systems. Telnet uses the client/server model. That is, a local computer uses a Telnet client program to establish the connection. The remote or host computer runs the Telnet server version to accept the connection and sends responses to the requests.

## File Transfer Protocol (FTP)

FTP is an Internet standard for file transfer. It allows Internet users to transfer files from remote computers without having to log into them. FTP establishes a connection to a specified remote computer using an FTP remote-host-address. Once connected, the remote host will ask the user for identification and a password. Upon compliance, the user can download or upload files.

Some sites make files available to the public. To access these files, users can enter *anonymous* or *guest* for identification and use their Internet address as a password. This application is called **anonymous FTP.**

## Hypertext Transfer Protocol (HTTP)

HTTP is an advanced file retrieving program that can access distributed and linked documents on the Web. Messages in HTTP are divided into request and response categories and work on the client/server principle. The request command is sent from the client to the server; and the response command is sent from the server to the client.

HTTP is a stateless protocol that treats each transaction independently. A connection is established between a client and a server for each transaction and is terminated as soon as the transaction is complete.

**Simple Network Management Protocol (SNMP)**

SNMP is used by network administrators to detect problems in the networks, such as routers and gateways. SNMP provides information for monitoring and controlling the network. SNMP is divided into two parts: SNMP management system and SNMP agent. The SNMP management system issues commands to the SNMP agent and the SNMP agent responds to the command. The SNMP management system can mange network devices remotely.

**Domain Name System (DNS)**

A **domain name** is a unique name used to identify and locate computers connected to the Internet (for example, "southernct.edu"). DNS is a collection of databases containing information about domain names and corresponding IP addresses. Top-level domain names can be categorized, as shown in Figure 16.5. A few are also listed here:

| | | | |
|---|---|---|---|
| **aero** | air transport industry | **mil** | military |
| **edu** | educational sites | **org** | organizations (non-profit organizations) |
| **gov** | government sites | **int** | international organizations |
| **com** | commercial sites | **biz** | businesses |
| **net** | major network providers | **name** | individuals |
| **coop** | cooperatives | | |
| **museum** | museums | | |

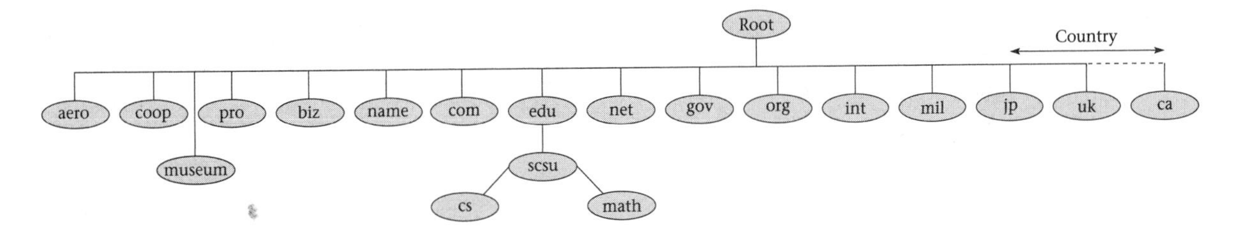


**FIGURE 16.5**
DNS top-level domain names

Country codes are used to identify international sites. A two-letter abbreviation can be used for a particular country, such as "uk" for United Kingdom, and "fr" for France.

## 16.4 Transport Level Protocols: UDP and TCP

The Transport level of the TCP/IP protocol consists of **User Datagram Protocol (UDP)** and **Transmission Control Protocol (TCP).** UDP performs an unreliable connection service for receiving and transmitting data. TCP performs reliable delivery of data. TCP adds a sequence number to each packet. When a packet reaches its destination, the destination acknowledges the sequence number of the next packet that it expects to receive.

## User Datagram Protocol (UDP)

UDP applications are Trivial File Transfer Protocol (TFTP) and Remote Call Procedure (RCP). UDP accepts information from the Application level and adds source port, destination port, UDP length, and UDP checksum. The resulting packet is called a *UDP datagram packet*. The total header of a UDP datagram is eight bytes. UDP passes the UDP packet to the IP protocol. The IP protocol adds its header to the packet and passes the packet to the Logical Link Control (LLC). The LLC generates an 802.2 frame (LLC frame) and passes the LLC frame to the Medium Access Control (MAC) layer, which adds its own header and transfers the frame to the Physical layer for transmission, as shown in Figure 16.6.

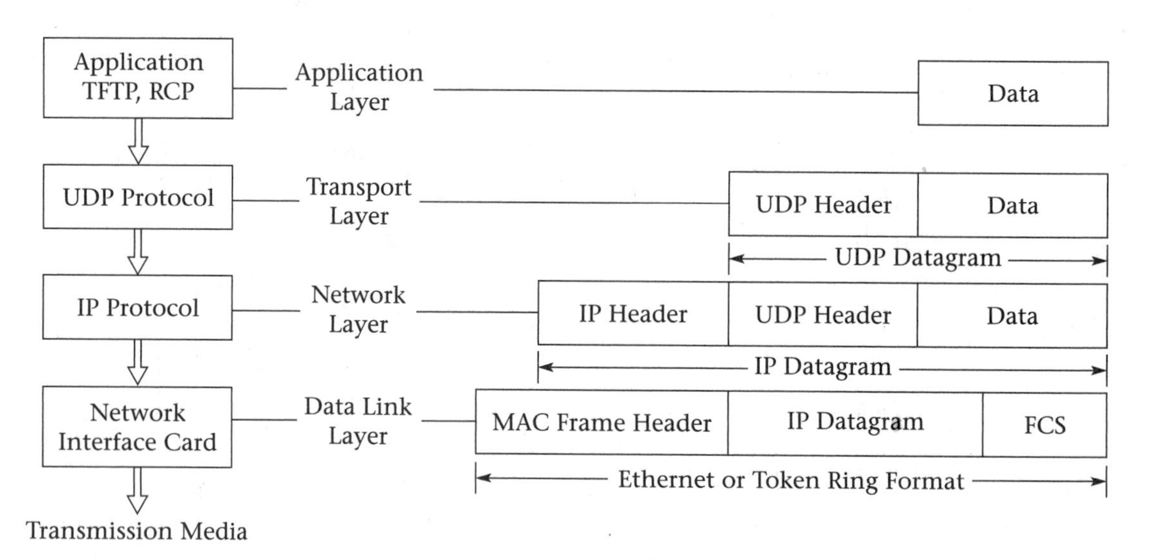


**FIGURE 16.6**
UDP operation

UDP allows applications to exchange individual packets over a network as datagrams. A UDP packet sends information to the IP protocol for delivery. There is no guaranteed reliability. Figure 16.7 shows the UDP packet format.

| 1 | 32 |
|---|---|
| Source Port 16 bits<br>Define application;<br>TFTP is port 69 | Destination Port 16 bits<br>Specifies destination port<br>on server |
| UDP Length 16 bits<br>Define number of bytes<br>in UDP header and Data | Checksum 16 bits<br>Checksum use for error detection<br>of UDP header and data |
| Data | |

**FIGURE 16.7**
UDP packet format

## Transmission Control Protocol (TCP)

Most applications prefer to use reliable delivery of information. TCP offers reliable delivery of data through the Internet. In TCP, before data is transmitted to the destination, a connection (not a physical connection) must be established before the information is transmitted. TCP assigns a sequence number to each packet. The receiving end checks the sequence number of all packets to ensure that they are received. When the receiving end gets a packet, it responds to the destination by acknowledging the next sequence number. If the sending node does not receive an acknowledgment within a given time, it retransmits the previous packet.

Figure 16.8 shows an application's data passing through TCP. TCP adds a 20-byte header and passes it to IP. IP adds its header and passes it to a

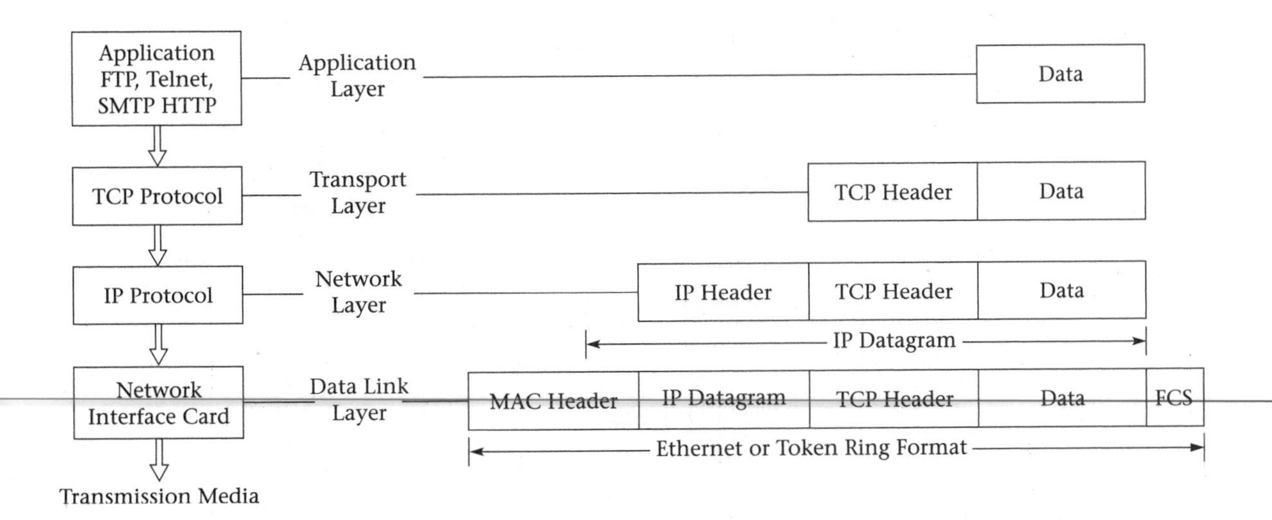


**FIGURE 16.8**
TCP operation

Network Interface Card (NIC). The NIC adds a MAC header to the information and transmits the packet. Figure16.9 shows the TCP packet format.

<table>
<tr><td colspan="4">Source Port 16 bits<br>Identifies source application program such as Telnet = 23, FTP = 21, and SMTP = 25</td><td colspan="3">Destination Port 16 bits<br>Identifies which application program on the receiving side receives the data</td></tr>
<tr><td colspan="7">Sequence Number 32 bits<br>A number assigned to the packet by the source</td></tr>
<tr><td colspan="7">Acknowledgment Number 32 bits<br>Acknowledges the next sequence number of the packet received from the source</td></tr>
<tr><td>Header Length 4 bits<br>Identifies number of 32-bit words in TCP header</td><td colspan="2">Reserved<br>6 bits</td><td colspan="2">Flag Bit<br>6 bits</td><td colspan="2">Window Size 16 bits<br>Size of the buffer source</td></tr>
<tr><td colspan="3">TCP Checksum 16 bits<br>Used for error detection in TCP header and data fields</td><td colspan="4">Urgent Pointer 16 bits<br>Field is valid if URG bit in flag is set</td></tr>
<tr><td colspan="7">Data (if any)</td></tr>
</table>

**FIGURE 16.9**
TCP packet format

The following describes the function of each field in a TCP packet:

**Sequence number:** The number label for each packet sent by the source.

**ACK sequence number:** Acknowledges the next packet expected to be received from the source.

**Header length:** Identifies the length of the header in 32-bit word.

**Flag bits:** Six-bits long and used for establishing a connection and disconnecting a connection.

| URG | ACK | PSH | RST | SYN | FIN |
|---|---|---|---|---|---|

**URG:** Urgent pointer is set to '1' when that field contains urgent data.

**ACK:** ACK bit is set to '1' to represent that the acknowledge number is valid.

**PSH:** Set to '1' means the receiver should pass the data to an application as soon as possible.

**RST:** Resets connection.

**SYN:** Set to '1' when a node wants to establish a connection.

**FIN:** Set to '1' means this is the last packet.

**Port Numbers** A **port number** is a logical channel in a communications system. TCP and UDP use port numbers to demultiplex messages to an application. Each application program has a unique port number associated with it. TCP/IP port numbers are between 1 and 65535; the well-known ports are from 0 to 1023, which are assigned by IANA; the registered ports are from 1024 through 49151; and the dynamic or private ports are from 49152 through 65535. Table 16.2 shows a few commonly used port num-

**TABLE 16.2** Commonly Used Port Numbers

| Network Services | Port Number | Network Services | Port Number |
|---|---|---|---|
| tcpmux | 1/tcp | netstat | 15/tcp |
| echo | 7/tcp | ftp-data | 20/tcp |
| echo | 7/udp | ftp | 21/tcp |
| discard | 9/tcp | telnet | 23/tcp |
| discard | 9/udp | smtp | 25/tcp |
| daytime | 13/tcp | http | 80/tcp |
| daytime | 13/udp | | |

## 16.5 Internet Level Protocols: IP and ARP

Internet level protocols consist of Internet Protocol (IP), Address Resolution Protocol (ARP), Reverse ARP (RARP), and Internet Control Message Protocol (ICMP).

**Internet Protocol (IP)** The function of the **Internet Protocol (IP)** is packet delivery with unreliable and connectionless service. These Internet datagrams also called *IP datagrams*. All TCP, UDP, ICMP, and ARP data are transmitted as IP datagrams. Figure 16.10 shows an IP datagram packet format.

The following describes the function of each field in an IP packet:

**Version:** Contains IP version number, which is 4 bits, and the current number is 4 (IPv4).

**Header length:** Represents the number of 32-bit words in the header. If there are no IP options and padding, header length is 20 bytes (5 words). IP is an unreliable service; there is no acknowledgment from the destination to the source. There is no physical connection between the source and destination. IP datagrams can arrive at the destination out of order.

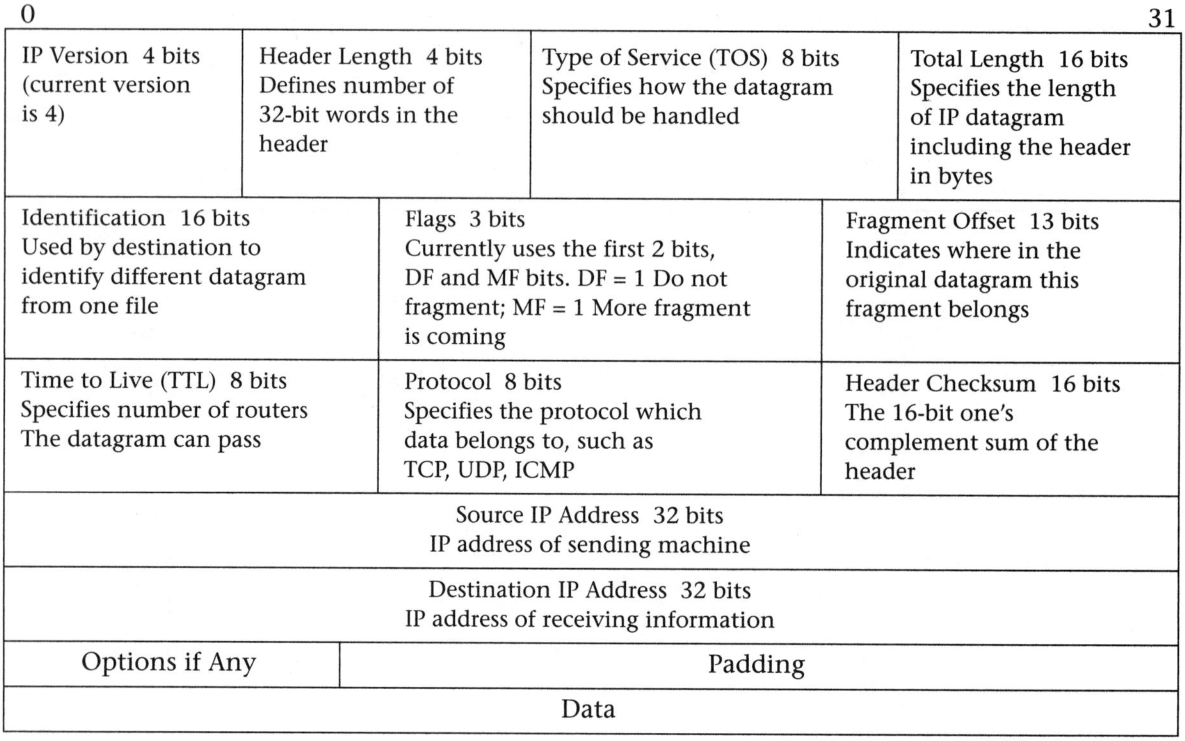

<table>
<tr><td colspan="6">0</td><td colspan="6" align="right">31</td></tr>
<tr><td colspan="3">IP Version 4 bits<br>(current version is 4)</td><td colspan="3">Header Length 4 bits<br>Defines number of 32-bit words in the header</td><td colspan="3">Type of Service (TOS) 8 bits<br>Specifies how the datagram should be handled</td><td colspan="3">Total Length 16 bits<br>Specifies the length of IP datagram including the header in bytes</td></tr>
<tr><td colspan="4">Identification 16 bits<br>Used by destination to identify different datagram from one file</td><td colspan="4">Flags 3 bits<br>Currently uses the first 2 bits, DF and MF bits. DF = 1 Do not fragment; MF = 1 More fragment is coming</td><td colspan="4">Fragment Offset 13 bits<br>Indicates where in the original datagram this fragment belongs</td></tr>
<tr><td colspan="4">Time to Live (TTL) 8 bits<br>Specifies number of routers The datagram can pass</td><td colspan="4">Protocol 8 bits<br>Specifies the protocol which data belongs to, such as TCP, UDP, ICMP</td><td colspan="4">Header Checksum 16 bits<br>The 16-bit one's complement sum of the header</td></tr>
<tr><td colspan="12" align="center">Source IP Address 32 bits<br>IP address of sending machine</td></tr>
<tr><td colspan="12" align="center">Destination IP Address 32 bits<br>IP address of receiving information</td></tr>
<tr><td colspan="4" align="center">Options if Any</td><td colspan="8" align="center">Padding</td></tr>
<tr><td colspan="12" align="center">Data</td></tr>
</table>

**FIGURE 16.10**
IP datagram packet format

**Type of Services (TOS):** TOS is 8 bits. For most purposes, the values of all bits in TOS are set to zero, meaning that in normal service the unused bits are always zeros.

| 0 1 2 | 3 | 4 | 5 | 6 7 |
|---|---|---|---|---|
| Precedence | D | T | R | Unused |

Precedence indicates the importance of a datagram (0 normal, 1 next important); TCP/IP ignores this field.

D, T, and R identify the type of transport the datagram requests. D is for Delay, T is for throughput, and R is for reliability.

D = 0 Normal delay
D = 1 Low delay
T = 0 Normal throughput
T = 1 High throughput
R = 1 High reliability
R = 0 Normal reliability

**Total length:** This field identifies the total length of the datagram (including the header) in bytes.

**Identification:** This is a number created by the sending node. It is required when reassembling fragmented messages. The identification field is used by the destination to put together related datagrams.

**Flag 3 bits:** The first bit is unused. The other two bits are: DF = Do Not Fragment, and MF = More Fragment. If DF is set to 1, the datagram cannot be fragmented. If the size of the data is more than the size of the MTU, the node cannot transfer data and an error message is generated.

| 0 | 1 | 2 |
|---|---|---|
| Unused | DF | MF |

**Fragmented offset field:** The offset field represents the offset of data in multiples of eight; therefore, the fragment size should be multiples of eight.

*Example*: 1,000 bytes are to be transferred over a network with an MTU of 256 bytes. Assume that the header of each datagram is 20 bytes. Find the number of datagrams if the following information is given:

- Identification: can be any number
- Total length
- Frame offset
- More fragment

$$256 - 20 = 236 \text{ bytes}$$

$$8 * 30 = 240$$

$8 * 29 = 232$ Each fragmented datum contains 232 bytes.

| Identification | 20 | 20 | 20 | 20 | 20 |
|---|---|---|---|---|---|
| Total length of each packet | 232 + 20 | 232 + 20 | 232 + 20 | 232 + 20 | 72 + 20 |
| Fragmented offset | 0 | 29 | 58 | 87 | 116 |
| MF | 1 | 1 | 1 | 1 | 0 |

**Time To Live (TTL):** This field indicates the number of gateways or routers a packet can go through; this value is set by the sender (32 or 64), and is decremented by one every time a router handles the datagram. If this field becomes zero, the datagram is thrown away.

**Protocol type:** The number in the field identifies the high-level protocol that generates this datagram or allows the destination IP to pass the datagram on to the required protocol. For example, ICMP is 1, UDP is 17, and TCP is 6.

**Header checksum:** This field is the checksum of the header (not the data field). The checksum is the sum of the one's complement of the 16-bit word of the header.

**Sending address:** This is the IP address of the source.

**Destination address:** This is the IP address of the destination.

### Maximum Transfer Unit (MTU)

**Maximum Transfer Unit (MTU)** is the largest frame length that can be sent on a given physical medium. There is a limit on the frame size. For example, 802.3's maximum frame size is 1500 bytes. If the datagram is larger than the MTU, the datagram is fragmented into several frames, each less than the MTU. Table 16.3 shows MTU values for a few common network types.

**TABLE 16.3** Network Types with MPU Values

| Network Type | MTU (Bytes) |
|---|---|
| 4-Mbps Token Ring | 4464 |
| 16-Mbps Token Ring | 17914 |
| FDDI | 4352 |
| Ethernet | 1500 |
| X.25 | 576 |
| Point-to-Point | 296 |

### Address Resolution Protocol (ARP)

The interior gateway must have the physical address of the host connected to one of its local networks. The gateway must record both the IP address and its physical address. If the gateway does not have the physical address of the host, it will send an **Address Resolution Protocol (ARP)** packet to get the host's physical address. The host will respond through the RARP protocol. Figure 16.11 shows an ARP packet format.

ARP is used by the interior gateway to find the physical address of the target station on the same network as the IP address of the network. Each station has a table located in cache, called the *ARP cache table*. If station A needs the physical address of B (as shown in Figure 16.12), station A will broadcast the ARP message on the network, and any station matching its IP address with the IP address in the ARP packet will accept the ARP packet. Then it responds to station A with RARP, which contains the physical address of station B.

**FIGURE 16.11**
ARP packet format

| Hardware Type 16 bits | |
|---|---|
| Protocol Type 16 bits | |
| HLEN Hardware Address Length 8 bits | PLEN IP Address Length 8 bits |
| Operation Code 16 bits | |
| ARP Request = 1<br>RARP Request = 3 | ARP Response = 2<br>RARP Response = 4 |
| Sender Hardware Address 48 bits | |
| Sender IP Address 32 bits | |
| Target Hardware Address 48 bits | |
| Target IP Address 32 bits | |

**FIGURE 16.12**
ARP and RARP architecture

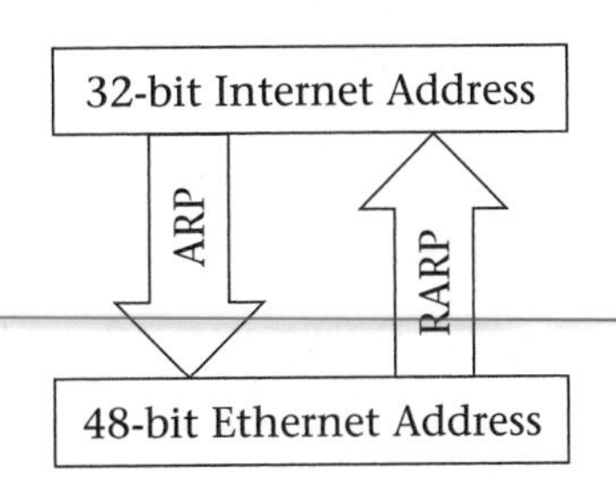

The following describes the function of each field in an ARP packet format:

**Hardware type:** Hardware type identifies the type of hardware interface and the following are some of the hardware types:

| Type | Description |
|---|---|
| 1 | Ethernet |
| 2 | IEEE 802.3 |
| 3 | X.25 |
| 4 | Token Ring |

**Protocol type:** Protocol type identifies the type of protocol the sending device is using. For example, protocol type 0800H is used for IP.

**HELN:** Hardware Address Length (HELN) in bytes (means 6 * 8 = 48 bits is the size of hardware address)

**PLEN:** Length of IP address

**Operation code:** Indicates whether the datagram is an ARP request or an ARP response:

1 = ARP request
2 = ARP response
3 = RARP request
4 = RARP response

## 16.6 IPv4 Addressing

An IP address is a 32-bit number that forms a unique address for each host connected to the Internet. No two hosts can have the same IP address. The assignment and maintenance of IP addressing is maintained by InterNIC.

An IP address is written in dotted decimal ($Base_{10}$) notation and is represented by four 8-bit binary numbers with the range of 0 to 255 ($4 \times 8 = 32$ bits).

| | | | |
|---|---|---|---|
| Binary ($Base_2$) | 00000000 | to | 11111111 |
| Decimal ($Base_{10}$) | 0 | to | 255 |

IP addresses are organized into five classes.

**Class A IP Address** The Class A IP address is used for organizations with a large number of users connected to the Internet and a small number of networks.

| | 7 bits | 24 bits |
|---|---|---|
| 0 | NET ID | HOST ID |

Class A IP address format

The first most significant bit of a Class A IP address is zero.

The network ID is 7 bits: $2^7$ is 128.

The host ID is 24 bits: $2^{24} = 16 \times 10^6$ nodes.

The range of a network ID is from 0 to 127. The numbers 0 and 127 are reserved.

The range of Class A addresses is from 0.0.0.0 to 127.255.255.255 in dotted decimal.

**Class B IP Address** A Class B IP address is used for medium-sized networks having more than 255 hosts.

The first two bits of a Class B address are 1 and 0.

The network ID is 14 bits and host ID is 16 bits.

With a Class B address we can have $2^{14}$ (16,384) networks and each network can have $2^{16}$ (65,536) hosts or nodes.

The range of a Class B network ID is from 128.0 to 191.255.

| | | 14 bits | 16 bits |
|---|---|---|---|
| 1 | 0 | NET ID | HOST ID |

Class B IP address format

**Class C IP Address** A Class C IP address is used for networks with a small number of hosts (those networks whose number of hosts does not exceed 255).

The first 3 bits of a Class C address are 1, 1, and 0.

Twenty-one bits are used for network ID, and 8 bits are used for host ID.

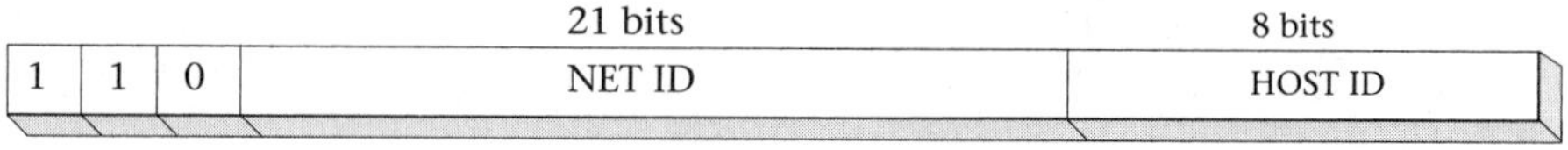

| 1 | 1 | 0 | NET ID (21 bits) | HOST ID (8 bits) |
|---|---|---|---|---|

Class C IP address format

A Class C IP address can handle $2^{21}$ networks. Each network can have 256 host IDs. The range of a Class C network ID is from 192.0.0 to 223.255.255. The IP address of 192.0.2.1 was never assigned and is used for *test* purposes only.

**Class D IP Address** A Class D IP address is reserved for multicasting. In multicasting, a packet is sent to a group of hosts.

The range of a Class D network ID is from 224.0.0.0 to 239.255.255.255.

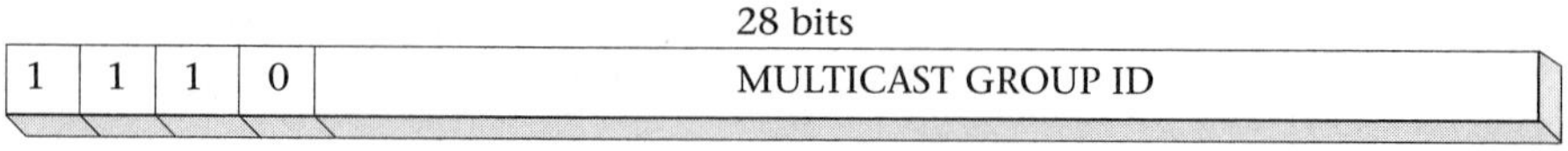

| 1 | 1 | 1 | 0 | MULTICAST GROUP ID (28 bits) |
|---|---|---|---|---|

Class D IP address format

**Class E IP Address** A Class E IP address is reserved for research and the range of Class E addresses is from 240.0.0.0 to 247.255.255.255.

**Loopback IP Address** The last address of each class is used as a **loopback address** for testing; the loopback address is used on a computer to communicate with another process on the same computer. The loopback addresses are as follows:

| | |
|---|---|
| Class A | 127.0.0.1 |
| Class B | 191.255.0.0 |
| Class C | 223.255.255.0 |

**Network Addresses** The host portion of a network address is set to zero. For example, 129.49.0.0 is a **network address,** but it is not a node address. No node is assigned to 0.0.

**Broadcast Addresses** The host portion is set to all ones in a broadcast. A packet with a broadcast address is sent to every node in the network. For example, address 129.49.255.255 is a **broadcast address.**

## 16.7 Assigning IP Addresses

Most universities will have networks connected to the Internet, as shown in Figure 16.13. The network administrator must contact InterNIC and obtain an IP address for the university. The InterNIC assigns a Class B address with network ID of 129.47 (the first two bytes of the IP address). The network administrator requires two bytes; from these two bytes he/she must decide how many bits are needed for each subnetwork. The answer is determined by how the network is growing and the future needs of the university. In the example on page 290, we use one byte to represent our subnetwork address.

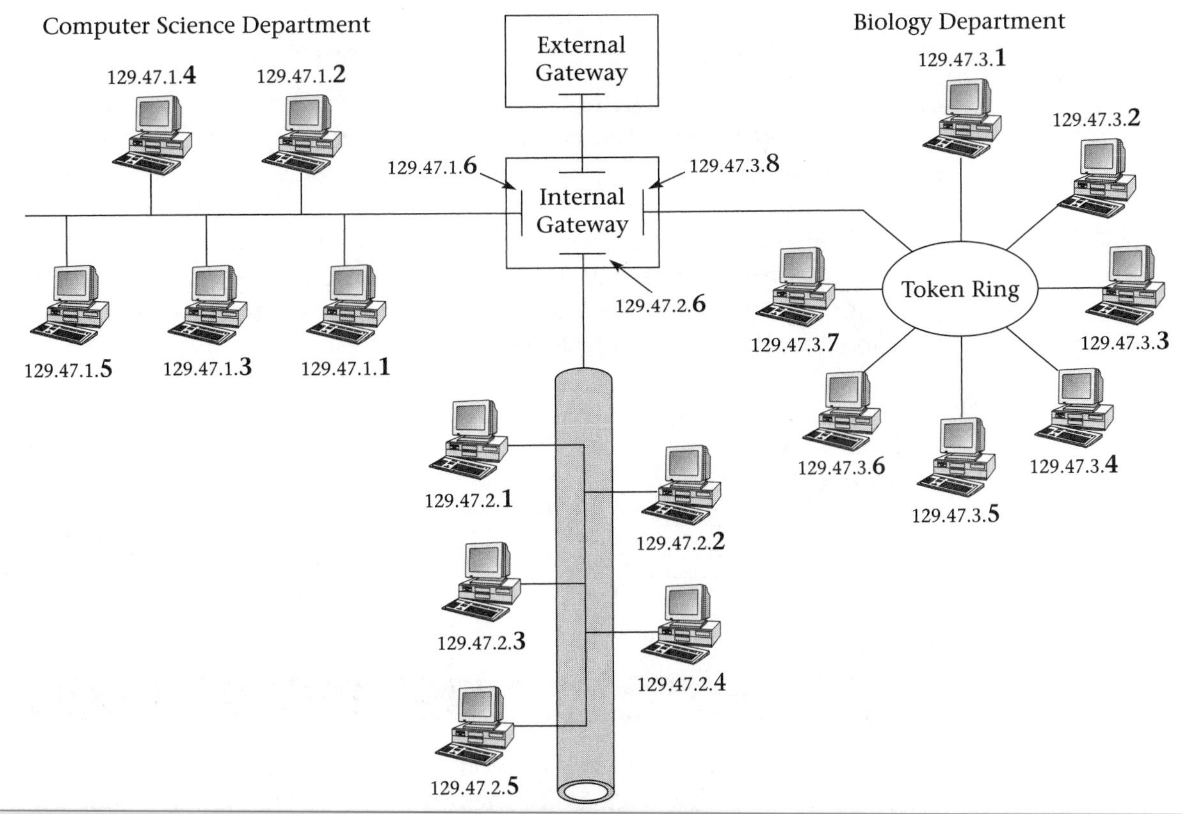


**FIGURE 16.13**
A network with three LANs and one gateway

| ←2 bytes→ | ←1 byte→ | ←1 byte→ |
|---|---|---|
| 129.47 | Sub ID | Host ID |

The subnetwork ID is eight bits, implying that we can have 256 networks in the university with 255 nodes each.

**Subnetwork ID** In Figure 16.13 you can see that the Computer Science network ID is 129.47.1 and the host's IDs are assigned 1, 2, 3, 4, and 5 (shown in bold type).

| | |
|---|---|
| 01 | Computer Science Department |
| 02 | Mathematics Department |
| 03 | Biology Department |
| 04 to 255 | reserved for future use |

The gateway uses the same NIC that the network is connected to. If the network is Ethernet, the interface card inside the gateway will be an Ethernet card.

**Address Mask** An **address mask** is used to define how many bits of an IP address are allocated for the host ID. It is used to separate the network address from a host ID. In the preceding example, the least significant byte is used for a host ID, as shown in Figure 16.14.

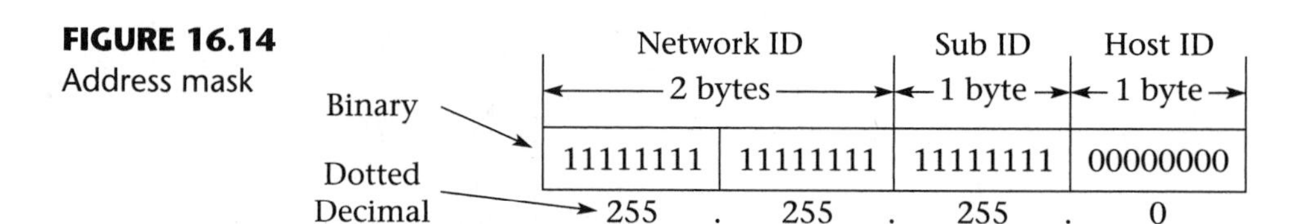


**FIGURE 16.14** Address mask

## 16.8 Demultiplexing Information

Figure 16.15 illustrates the general block diagram of Internet hardware and protocols. The packets (in the form of electrical signals) come to the Physical layer of the Network Interface Card. The Physical layer changes the signal to the bits and passes it to the MAC sublayer. The MCA sublayer takes off its header (preamble, SFD, SA, and DA) and passes it to

LLC (Logical Link Control). The LLC checks the type field. If the type field is 0800H, the pack is an IPv4 datagram and is passed to IP.

For IPv6 the type field would be 86DD H (hexadecimal). IP looks at the 8-bit protocol field (TCP = 6, UDP = 17, ICMP = 1, and IGMP = 6). IP will take off its header and pass the data to one of the these protocols, depending on the protocol number. Assuming the data is passed to TCP, TCP will look at the port number and pass it to the Application layer.

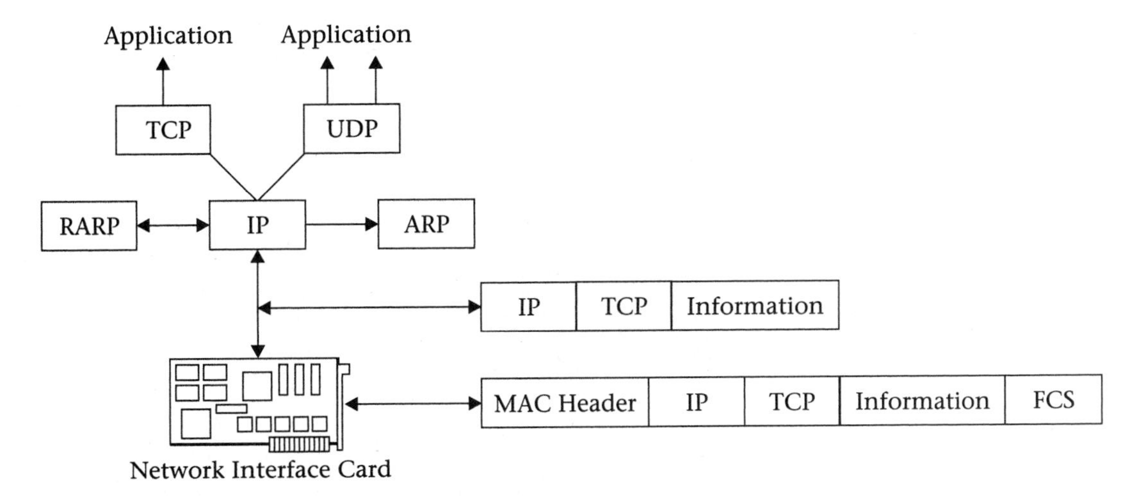


**FIGURE 16.15**
Demultiplexing of information

## 16.9 TCP Connection and Disconnection

**TCP Connection** TCP first establishes the connection with the destination before any information is transmitted, as shown in Figure 16.16. A connection is set up as follows:

1. The source sends a packet to the destination by setting SYN = 1 in the TCP header and setting a sequence number in TCP packet format (assuming the sequence number is $x$).
2. The destination will respond to the source by setting the SYN bit to 1, the acknowledge field to $x + 1$, and the sequence number to $y$.
3. The source will start transmitting data.

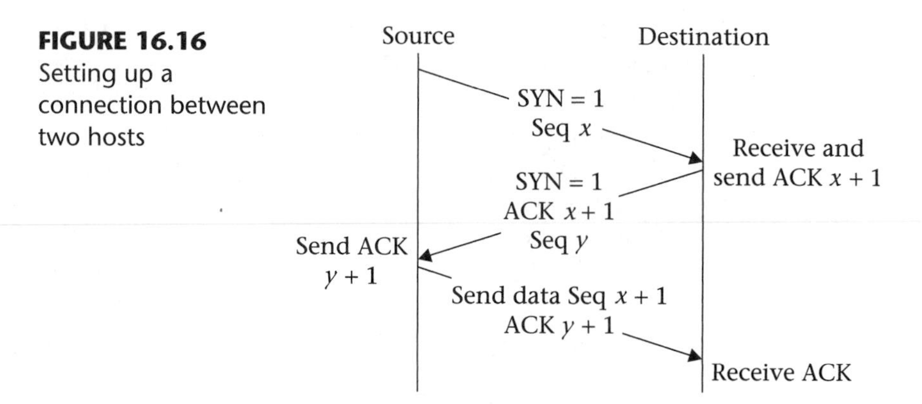


**FIGURE 16.16** Setting up a connection between two hosts

**TCP Disconnection** When the source sends the last packet to the destination, the source sets FIN to 1 to inform the destination that this is the last packet. The destination acknowledges the last packet and sets the FIN to 1 to inform the source that destination does not have any more packets to send. The source sends a packet with RS set to 1 and the destination responds to the source with a packet whose RS bit is set to 1. Figure 16.17 shows the disconnection process.

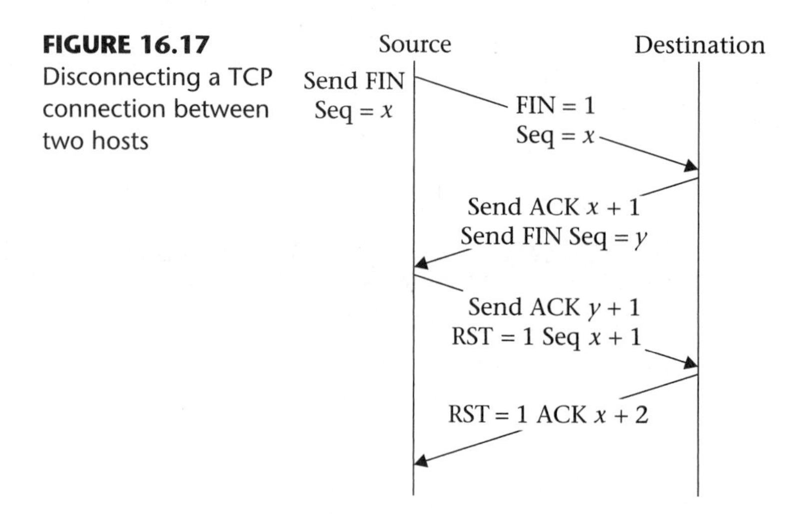


**FIGURE 16.17** Disconnecting a TCP connection between two hosts

## 16.10 Internet Protocol Version 6 (IPv6)

Due to the growth of the Internet and the address limitations of IPv4, in 1995, the Internet Engineering Task Force (IETF) approved **Internet Protocol version 6 (IPv6)**. The limitations that led to Internet Protocol Version 6 are discussed next.

The IPv4 address size is 32 bits and can connect up to $2^{32}$ = 4 billion users to the Internet. The IPv4 address field is divided into two parts, network address (Network ID) and host address (Host ID). Once a network number is assigned to an organization, the organization might not use all host IDs in the host ID field; this means that all IPv4 addresses might not be used completely. Also, the number of networks connected to the exterior gateway increases rapidly and this causes the routing table to become large. When the routing table becomes large, it takes more time to search through the table.

The IPv6 protocol reduces the size of the routing table in exterior gateways because IPv6 uses a hierarchical scheme to define an IP address. IPv6 has the following features:

- expanded addressing
- simplified header format
- support extension
- flow labeling
- authentication and privacy

## IPv6 Structure

IPv6 is divided into two parts: basic header and extension header. The first 40 bytes of the header are called the *basic header*, as shown in Figure 16.18. IPv6 addresses are 128 bits in length. Addresses are assigned to individual interfaces on the nodes rather than to the nodes themselves. A single interface may have multiple addresses. The fields shown in Figure 16.18 are as follows:

**Version:** Version field is 4 bits; and it is 0110, which is 6 in decimal.

**Priority:** This field defines the priority of an IP datagram. It is used for congestion control. For example, email has a priority of 2, and interactive traffic has a priority of 6.

**Flow label:** The flow label is 24 bits and it is used for labeling packets that belong to particular traffic that the sender requests for special handling, such as real-time traffic.

**Payload length:** 16 bits. Indicates the number of bytes in payload fields.

**Next header:** 8 bits. Determines which header follows the basic IP header.

**Hop limit:** 8 bits. The function of this field is the same as TTL for IPv4.

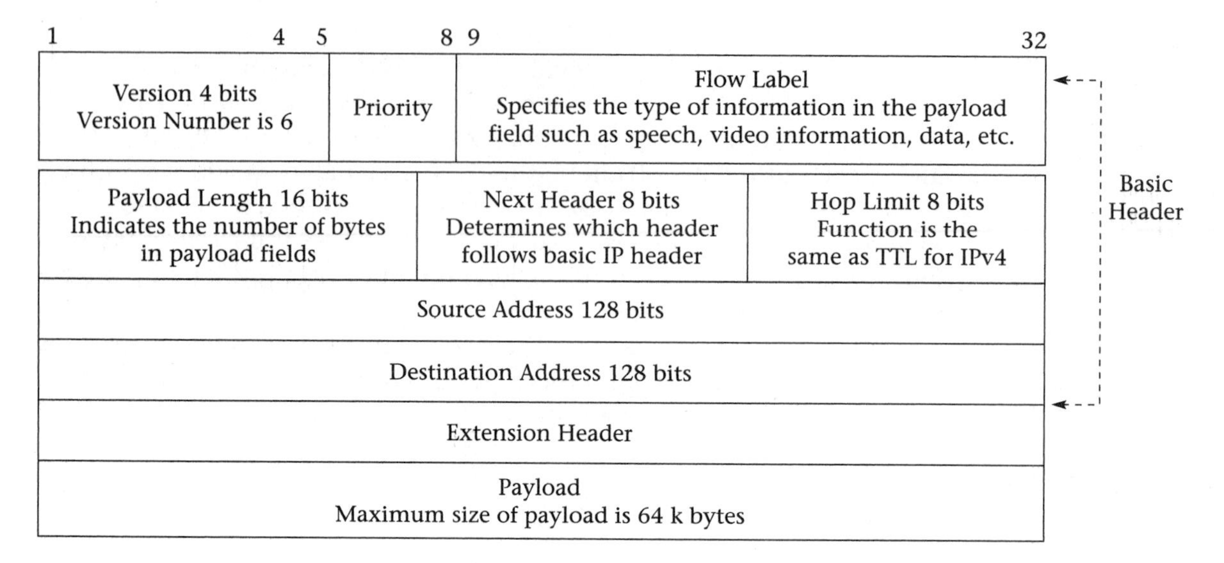


**FIGURE 16.18**
IPv6 packet format

**IP extension header:** Figure 16.19 shows an IP datagram with several extension headers and some of the header values. These values are also listed in Table 16.4. In Figure 16.19, the first value of the Next Header is 0, which determines that the following header is Hop-by-Hop Options. The second value of Next Header is 43, which determines that the following header is Routing Information. The third value of Next Header is 6, which identifies the third header as TCP.

**TABLE 16.4** Header Values for Figure 16.19

| Value | Header Function |
|---|---|
| 0 | Hop-by-Hop Option Header |
| 1 | IP |
| 6 | TCP |
| 17 | UDP |
| 43 | Routing Header |
| 50 | Encryption Security Payload |

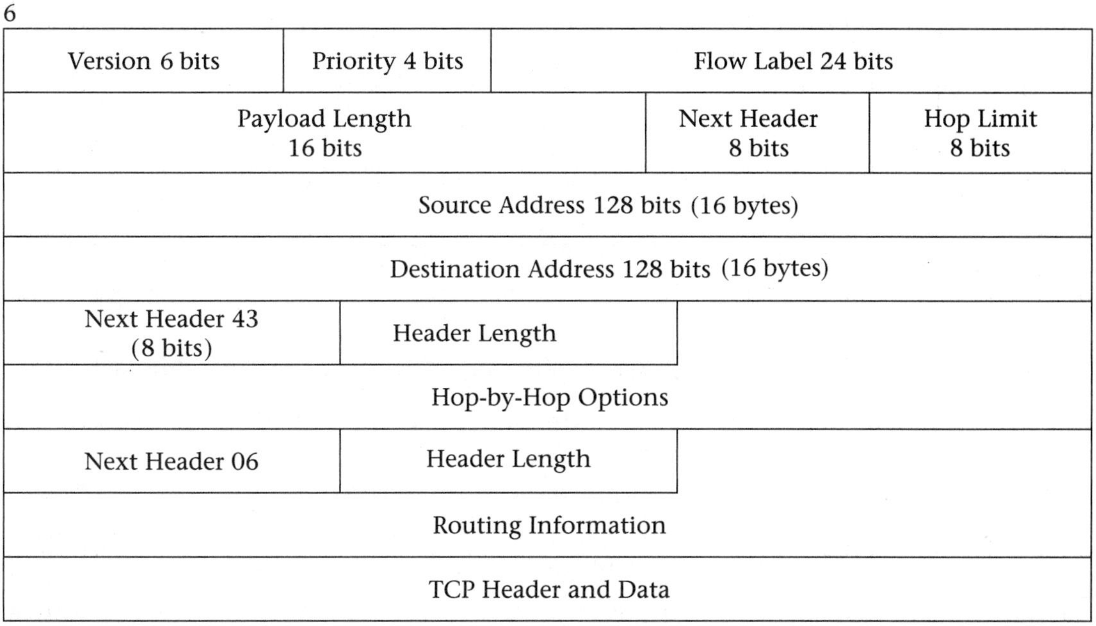

6

<table>
<tr><td>Version 6 bits</td><td>Priority 4 bits</td><td colspan="3">Flow Label 24 bits</td></tr>
<tr><td colspan="3">Payload Length<br>16 bits</td><td>Next Header<br>8 bits</td><td>Hop Limit<br>8 bits</td></tr>
<tr><td colspan="5">Source Address 128 bits (16 bytes)</td></tr>
<tr><td colspan="5">Destination Address 128 bits (16 bytes)</td></tr>
<tr><td>Next Header 43<br>(8 bits)</td><td colspan="2">Header Length</td><td colspan="2"></td></tr>
<tr><td colspan="5">Hop-by-Hop Options</td></tr>
<tr><td>Next Header 06</td><td colspan="2">Header Length</td><td colspan="2"></td></tr>
<tr><td colspan="5">Routing Information</td></tr>
<tr><td colspan="5">TCP Header and Data</td></tr>
</table>

**FIGURE 16.19**
IP datagram with several extension headers

## IPv6 Address Architecture

***IPv6 Address Representation.*** The IPv6 address is 128 bits. It is divided into eight fields of 16 bits. Each field is represented in hexadecimal form. In general, the IPv6 address is represented as:

Y: Y: Y: Y: Y: Y: Y: Y

Y is 4 digits in hexadecimal or 16 bits in binary. For example:

FE45:436A:12DF:4563:879E: 0008:0000:1232

In the preceding address, the leading zero of each field can be skipped. For example, the field containing 0008 can be written as 8 and the field containing all zeros can be represented by a double colon. Only one double colon in each address is allowed. Therefore, the address can be represented as:

FE45:436A:12DF:4563:879E: 8::1232

***Addressing Format.*** The specific type of addresses of IPv6 is indicated by the most significant bits. It is variable in length, and it is called the *format prefix*. Table 16.5 shows some of the format prefixes.

**TABLE 16.5** Some Prefix Values for IPv6

| Prefix Binary Value | Allocation |
|---|---|
| 0000 0000 | Reserved |
| 0000 0010 | Reserved for IPX |
| 011 | Provider-Based Unicast Address |
| 1111 1111 | Multicast Address |

***IPv6 Address Types.*** IPv6 supports three types of addresses:

1. **Unicast address:** This type of address is used for a single interface.
2. **Anycast address:** This type of address is used for more than one interface. A packet sent with an anycast address is delivered to only one of the interfaces, the one that is nearest to the source.
3. **Multicast address**: This type of address is used for a set of interfaces. A packet with a **multicast address** will transmit to all interfaces.

***IPv6 Unicast Address.*** One of the most useful addresses is the provider-based **unicast address.** Its format is shown in Figure 16.20. This type of address is used for Internet provider services to an organization.

| 3 bits | 5 bits | 16 bits | 16 bits | 8 bits | 32 bits | 48 bits |
|---|---|---|---|---|---|---|
| Format Prefix | Registry ID | Provider ID | Subscriber Type | Subscriber ID | Subnetwork ID | Interfact ID |

**FIGURE 16.20**
Unicast address format

The following describes the function of each field of an IPv6 address:

**Format prefix:** The prefix for the provider-based unicast address is 011.

**Registry ID:** Identifies the registry that assigns the provider portion of the unicast address.

**Provider ID:** Identifies the specific provider that is assigned to the subscriber of the address.

**Subscriber type:** Identifies the type of organization (education, corporation, etc.).

**Subscriber ID:** Identifies the specific subscriber from multiple subscribers.

**Subnetwork ID:** Identifies the network.

**Interface ID:** Identifies the interface. It is an IEEE 802 MAC address.

**Loopback address:** The unicast address 0:0:0:0:0:0:0:1 is used by a node to send an IPv6 datagram to itself.

***IPv6 Address with IPv4.*** Figure 16.21 shows an IPv6 address format that embeds IPv4 into the last 32 bits of IPv6. Another type of IPv6 unicast address used for nodes that do not support IPv6 addressing is shown in Figure 16.22.

| 80 bits | 16 bits | 32 bits |
|---|---|---|
| 0000..............0000 | 0000 | IPv4 Address |

**FIGURE 16.21**
IPv4 embedded into IPv6

| 80 bits | 16 bits | 32 bits |
|---|---|---|
| 0000..............0000 | FFFF | IPv4 Address |

**FIGURE 16.22**
IPv4 embedded into IPv6 for a station that does not support IPv6

## 16.11 Internet II

**Very High Performance Backbone Network Service (vBNS)** is a cooperative project between MCI and the National Science Foundation (NSF) to provide a high-bandwidth network for research in the Internet technologies.

The National Science Foundation and MCI Telecommunication Corp jointly provide vBNS to the following five NSF Supercomputer Centers (SCCs):

- Cornell Theory Center (CTC)
- National Center for Atmospheric Research (NCAR)
- National Center for Supercomputing Applications (NCSA)

- Pittsburgh Supercomputing Center (PSC)
- San Diego Supercomputing Center (SDSC)

vBNS provides service to these centers and other institutions for research and development of new technology for the Internet. These institutions are connected to vBNS by high-capacity interconnection points called **Gigabit Capacity Point-of-Presence (GigaPoP).** vBNS use advanced technology such as ATM and SONET for transmitting video, voice, and data. It operates at a speed of 622 Mbps using OC-12, as shown in Figure 16.23.

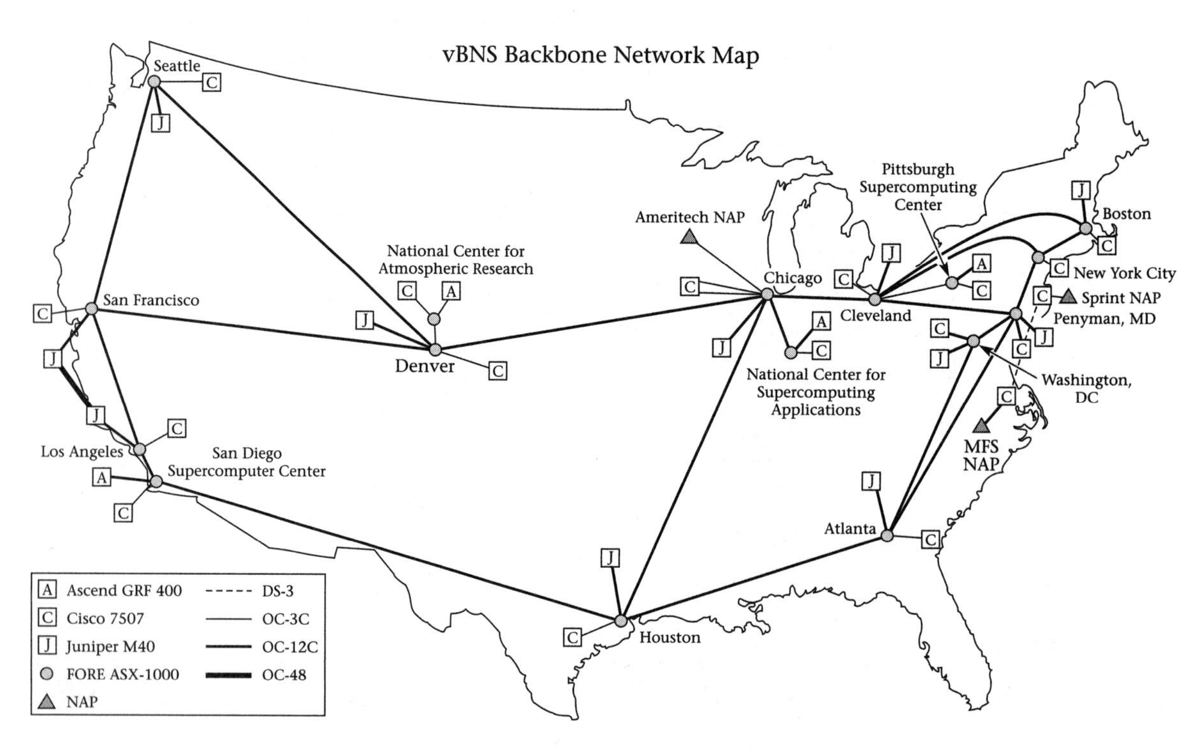


**FIGURE 16.23**
vBNS backbone network map

## Summary

- The Internet is a group of networks connected together through gateways and routers.
- The Internet uses a set of protocols for communications called *Transmission Control Protocol and Internet Protocol (TCP/IP).*
- Some applications used by the Internet are email, Telnet, File Transfer protocol (FTP), and Hypertext Transfer protocol (HTTP).
- An email address is made up of a user name and a mail server, such as Elahia1@southernct.edu.
- Telnet is one of the most useful Internet applications. It enables one computer to do a remote login to another computer.
- FTP is a protocol used for transferring a file from/to a remote computer without logging into the remote computer.
- HTTP is an advanced file retrieval protocol that can access distributed documents on the Web.
- TCP provides reliable service between hosts.
- IP provides packet delivery between the hosts.
- IPv4 is 32 bits and each byte is represented by a decimal number. They are divided into classes A, B, C, D, and E.
- The range of Class A IP addresses is from 0.0.0.0 to 127.255.255; the Network IDs of 0 and 127 are reserved.
- The range of Class B IP addresses is from 128.0.0.0 to 191.255.255.255.
- The range of Class C IP addresses is from 192.0.0.0 to 223.255.255.
- The Class D IP addresses are reserved for multicasting and Class E IP addresses are reserved for future use.
- DNS is a distributed database containing a domain name and the corresponding IP address. Top domain names are edu, gov, com, mil, org, net, aero, coop, biz, name, and int.
- UDP provides unreliable service between the hosts. Applications of UDP are Trivial File Transfer Protocol (TFTP) and Remote Procedure Call (RCP).
- The TCP header is 20 bytes and the IP header is 20 bytes.
- IPv6 or IP next generation address size is 128 bits.
- Internet II is used for research and for developing new technology for the Internet.

- Internet II uses Very High Performance Backbone Network Service (vBNS) for its backbone.
- vBNS operates at the speed of 622 Mbps using OC-12.

## Key Terms

Address Mask
Address Resolution Protocol (ARP)
Advanced Research Project Agency (ARPA)
Anonymous FTP
ARPANET
Broadcast Address
Domain Name
Gigabit Capacity Point-of-Presence (GigaPoP)
Internet Protocol (IP)
Internet Protocol Version 6 (IPv6)
Loopback Address
Maximum Transfer Unit (MTU)
Multicast Address
Network Address
NSFNET
Port Number
Remote Login
Telnet
Transmission Control Protocol (TCP)
Unicast Address
User Datagram Protocol (UDP)
Very High Performance Backbone Network Service (vBNS)

## Review Questions

### • Multiple Choice Questions

1. The Internet is a collection of ________.
   a. servers
   b. applications
   c. networks
   d. routers
2. The Internet uses the ________ protocol.
   a. X.25
   b. NWLink
   c. TCP/IP
   d. Window NT
3. Of the following, ________ is not an application of the Internet.
   a. email
   b. FTP
   c. WWW
   d. antivirus software
4. __________ provides packet delivery between networks.
   a. IP
   b. TCP
   c. X.25
   d. ARP

5. The Transport layer of TCP/IP consists of UDP and ____________.
   a. TCP
   b. IP
   c. a and b
   d. ICMP
6. ____________ performs reliable delivery of data.
   a. IP
   b. UDP
   c. TCP
   d. RARP
7. ________ organization ratifies standards for the Internet.
   a. IEEE
   b. ITU
   c. IETF
   d. EIA
8. The type of switching used in the Internet is ____________.
   a. virtual circuit
   b. packet switching
   c. circuit switching
   d. message switching
9. The protocol used for unreliable communication is ________.
   a. TCP
   b. UDP
   c. IP
   d. ARP
10. Telnet uses ____________ protocol for remote login.
    a. UDP
    b. TCP
    c. IP
    d. FTP
11. A modem uses ________ protocol to connect a computer to the Internet.
    a. TCP/IP
    b. UDP
    c. PPP
    d. ARP
12. Telnet enables a user to _______.
    a. transfer a file
    b. send email
    c. log in remotely
    d. transfer mail
13. The number of bits in an IPv4 address is ________.
    a. 24 bits
    b. 32 bits
    c. 48 bits
    d. 128 bits
14. The application of a loopback address is ________.
    a. reserved by Internet authority
    b. used for testing
    c. used for broadcast address
    d. used for unicast
15. Of the following addresses, ________ is a Class C broadcast address.
    a. 192.205.205.255
    b. 192.205.205.205
    c. 192.205.205.00
    d. 192.205.255.2555

16. ______ protocol is used for the Web.
    a. TCP/IP
    b. HTPP
    c. UDP
    d. ARP
17. TCP is used for______.
    a. reliable communication
    b. unreliable communication
    c. connection-oriented communication
    d. none of the above
18. The function of the source and destination port in a TCP header is _____.
    a. used to identify the source and destination host on the network.
    b. used to identify the application of the source protocol and the application of the destination protocol.
    c. used to identify the source protocol and the destination protocol.
    d. none of the above
19. The function of the Window Size field in a TCP header is that the ______.
    a. source reports its buffer size to the destination.
    b. source reports the number of packets received from the destination.
    c. source reports the size of its cache memory to the destination.
    d. source reports an error in the packet.
20. The function of Time To Live (TTL) in a TCP header is to ______.
    a. hold the time of day.
    b. define the number of routers a datagram can pass.
    c. define the transmission time of a datagram between the source and destination.
    d. define the number of words in a packet.
21. _________ protocol is used to set up a connection between source and destination.
    a. UDP
    b. TCP
    c. IP
    d. ARP
22. IPv6 is _______ number of bits.
    a. 32
    b. 48
    c. 64
    d. 128
23. The Internet routes a datagram from one gateway to another base on the datagram's _____.
    a. IP address
    b. MAC address
    c. port address
    d. none of the above

24. The function of an IP subnet mask is that ______.
    a. The IP subnet mask represents bits in the network portion of an IP address.
    b. The IP subnet represents the bits in the host portion of an IP address.
    c. a and b
    d. none of the above

## • Short Answer Questions

1. What is the markup language used to create files for the Web?
2. What is the protocol used by the Web to transfer a file?
3. What is the protocol used to access the Internet by a modem?
4. List the DNS indicators.
5. List the protocols in the Transport level.
6. List the protocols in the Internet level.
7. What is the size of an IPv4 address?
8. What is the size of an IP header?
9. Explain the function of IP.
10. Explain the function of TCP.
11. What is the size of a TCP header?
12. What is the function of the TTL field in an IP header?
13. What is the size of a UDP header?
14. List the IP address classes.
15. If an organization uses the third byte of an IP address for a subnet, how many subnets can be assigned to the Class B address?
16. What is the name of the organization that assigns IP addresses?
17. What is the function of ARP?
18. Show the IP frame format.
19. What is the current version of IP?
20. What is function of the TTL field in an IP packet?
21. What is the size of an IPv6 address?
22. Convert the following IP address from hexadecimal to dotted decimal representation, and find the class type of each IP address.
    a. 46EF3A94
    b. 23446FEC
23. List some Internet applications.

24. In the following email address, identify username and mail server address:
    Elahi@Xycorp.com
25. List two application protocols for UDP.
26. List three application protocols for TCP.
27. Show the TCP/IP reference model.
28. What is the function of ICMP?
29. What does DNS stand for? Describe its function.
30. Explain the function of Telnet.
31. What is a port number?
32. What is an MTU?
33. Identify the class of the following IP addresses:
    a. 129.234.12.08
    b. 117.243.56.89
    c. 92.92.92.9
34. How many bits is IPv6?
35. Assume the Internet Network Information Center assigns you the Class B IP address: 172.200.0.0.
    a. How many bits do you use from the host ID for a 128 subnet ID?
    b. How many host IDs can be generated for each subnet ID?
    c. What is the subnet mask ID?
36. There are two computers, A and B, with IP addresses of 174.20.45.37 and 174. 20.67.45. If these two computers have a subnet mask ID of 255.255.0.0, can you determine if these two computers are located in the same network?

CHAPTER

# Internet Protocols (Part II) and MPLS

OUTLINE



OBJECTIVES

After completing this chapter, you should be able to:

- Define the components of DHCP
- Explain the application of DHCP and its operation
- Understand PPP and its application
- Define the point-to-point connection operation
- Show the IP address of a client and an ISP server
- Explain the application of a VPN
- Explain the application of IPsec
- Explain the operation of MPLS

INTRODUCTION

Manually assigning IP addresses to host computers in a large network is very time consuming. To overcome this problem, TCP/IP offers **Dynamic Host**

**Configuration Protocol (DHCP),** which is an extension of the boot protocol. DHCP uses the UDP protocol to communicate with a DHCP server. The client broadcasts a packet with the IP address 255.255.255.255 (broadcast address). The broadcasted packet contains the hardware address of the client, and the DHCP server responds to this request.

## 17.1 Dynamic Host Configuration Protocol (DHCP)

**DHCP Components**

To have DHCP service on a network requires three types of software: DHCP client, DHCP server, and DHCP relay agent.

***DHCP Client.*** Most Network Operating Systems (NOS) offer DHCP software for clients, such as Windows NT and 2000. The DHCP client software enables the client workstation to obtain its IP address from the DHCP server automatically. Clients broadcast a packet to the network that has the DHCP server in its broadcast domain (that is, in the same segment).

***DHCP Server.*** The DHCP server holds a range of IP addresses and responds to any request made by a DHCP client. Note that the DHCP server and the client host must be in the same broadcast domain; otherwise the DHCP relay agent software is required.

***DHCP Relay Agent.*** When the DHCP server is not located in a broadcast domain of the client station, the router that the client station is connected to requires the **DHCP relay agent** software, as shown in Figure 17.1.

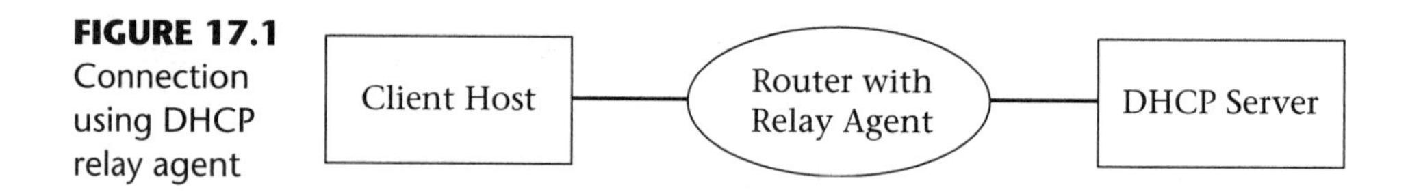


**FIGURE 17.1** Connection using DHCP relay agent

The function of the DHCP relay agent is to accept the broadcast packet from the DHCP client and send the packet to the DHCP server. The relay agent changes the broadcast packet to a unicast address by using its own IP address and then sends it to the DHCP server. The DHCP server responds to the DHCP relay agent. The DHCP relay agent then forwards this response to the DHCP client. The DHCP server supports three methods to allocate IP addresses to a client host:

1. **Automatic allocation:** DHCP assigns a permanent IP address to the host.
2. **Dynamic allocation:** DHCP assigns an IP address to a host for a limited period of the time (this time is called *lease time*). If the client does not need the IP address, the DHCP server can reuse this IP address and assign it to another host. This method is used by Internet Service Providers to assign IP addresses to their clients in order to be connected to the Internet temporarily.
3. **Manual allocation:** The network administrator assigns an IP address to the host and the DHCP server transfers that IP address to the client.

**DHCP Operation** The following steps describe the operation of DHCP, as shown in Figure 17.2:

1. The DHCP client does not have an IP address and it broadcasts a packet on the network requesting an IP address from the DHCP server. This packet is called a *DHCP Discovery* packet.
2. The DHCP server responds to the DHCP Discovery packet by sending a *DHCP Offer* packet to the client. The DHCP Offer packet includes the client IP address, the address mask, and the lease time (the amount of the time that the client can hold this address). In the middle of the lease time, the client host sends a renewal packet to the DHCP server to find out if it can keep the same IP address for the next lease time.
3. When the client host has received a DHCP Offer packet from the DHCP server, it can accept or reject the address offered. The client sends a packet to the DHCP server to inform either the acceptance or rejection of the IP address. This packet is called the *DHCP Request* packet.
4. The DHCP server sends a DHCP ACK to the client in response to the client's DHCP Request and informs the client of the completion of the DHCP process.
5. The client receives the DHCP ACK packet with configuration from the DHCP server. Now the client is configured. If a client receives a DHCP NACK (negative acknowledgment from the DHCP server), then the client host cannot use this IP address.

6. If the client host does not need an IP address, it will send a *DHCP Release* packet to the DHCP server to release the IP address.

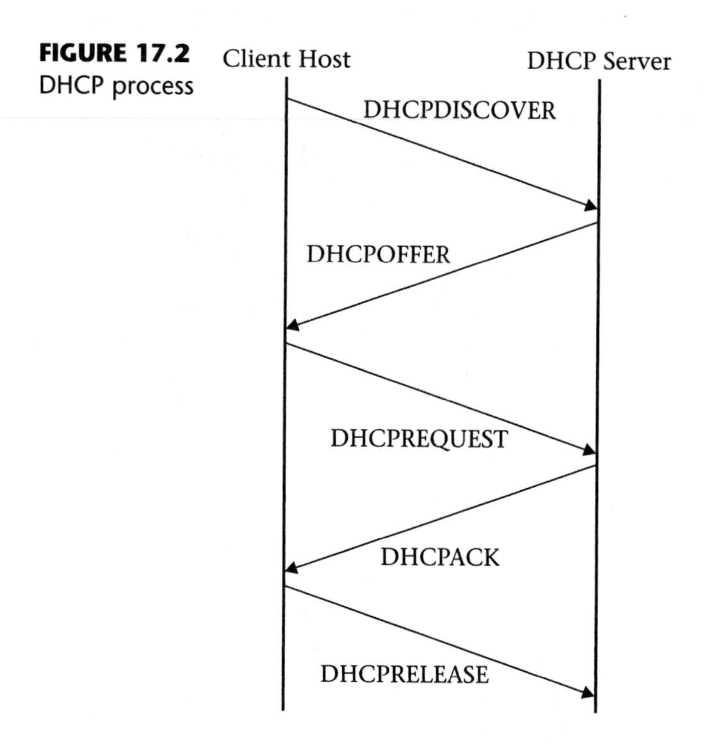


**FIGURE 17.2** DHCP process

***DHCP Packet Format.*** DHCP is an application protocol for UDP. Figure 17.3 shows the payload of an IP packet as a DHCP; Figure 17.4 shows the DHCP packet format.

**FIGURE 17.3** IP packet with DHCP

| IP Header | UDP Header | DHCP Packet |
|---|---|---|

The following describes the function of each field in the DHCP packet format:

**OP code:** The Op code indicates a request from a client or a reply to a request (1 for request, 2 for reply).

| 1 | | | 32 |
|---|---|---|---|
| Code 8 bits | Hardware Type 8 bits | Hardware Length 8 bits | Hop Count 8 bits |
| Transaction ID 32 bits | | | |
| Number of Seconds 16 bits | | Unused 16 bits | |
| Client IP Address 4 bytes | | | |
| Machine IP Address 4 bytes | | | |
| Server IP Address 4 bytes | | | |
| Gateway IP Address 4 bytes | | | |
| Client MAC Address 6 bytes | | | |
| Server Host Name up to 64 bytes | | | |
| Boot File Name up to 1284 bytes | | | |
| Vendor-Specification Information up to 64 bytes | | | |

**FIGURE 17.4**
DHCP packet format

**Hardware type:** The hardware type indicates the type of network card being used (such as IEEE 802.3 or Token Ring).

**Hardware length:** The hardware length indicates the size of the hardware address or MAC address (6 bytes).

**Hops:** The hops field indicates the number of hops a packet can make on route to the destination. The maximum number of hops is 3.

**Transaction ID:** A random number set by the client, used by the client and the server to coordinate messages and responses.

**Number of seconds:** The number of seconds is set by the client. The secondary server does not respond until this time has expired.

**Client IP address:** If the client does not have an IP address, this field will be set to 0.0.0.0.

**Server IP address:** The server IP address is set by the server.

**Router IP:** The router IP is set by the forwarding router.

**Client hardware:** The client address is set by the client and is used by the server to identify which client the request came from.

**Server host name:** Optional.

**Boot file name:** The client can leave this field null or indicate the type of the boot file.

**Vendor specification:** This field is used for various extensions of the bootstrap.

**UDP header:** The UDP header contains source and destination port numbers. The BOOTP uses two reserved port numbers. Port number 68 is used for the client and port number 67 is used for the server.

## 17.2 Internet Control Message Protocol (ICMP)

**Internet Control Message Protocol (ICMP)** is used to report error messages during packet delivery by an Internet Protocol such as protocol unreachable, network unreachable, network congestion, packet is too large, and announcing timeouts (TTL field drops to zero). ICMP is also used by the network administration for diagnostic purposes such as *Ping* and *Tracert* commands. ICMP is an Internet Layer protocol packet and ICMP packets are transported by IP to the destination. ICMP messages can be divided into two groups: error messages and information messages. Figure 17.5 shows the ICMP packet format.

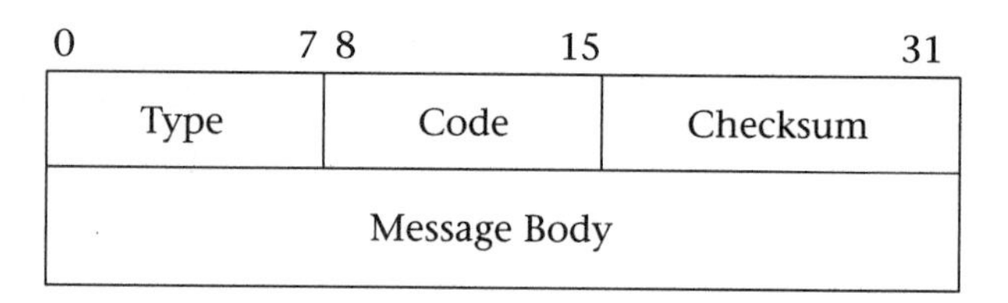


**FIGURE 17.5** ICMP packet format

The following describes the function of each field of the ICMP packet format:

**Type (8 bits):** Indicates type of message. Numbers between 0 and 127 indicate error messages, and numbers larger than 127 represent information messages. Table 17.1 shows the type values and their meanings.

**Code (8bits):** Provides additional information about the message type. For example, for a type value of 1 (destination unreachable), the code number specifies more detailed information as shown in Table 17.2.

**Checksum:** This field is used for error detection of ICM headers and messages.

**TABLE 17.1** Type Values and Meaning

| Type Value | Meaning |
|---|---|
| 1 | Destination unreachable |
| 2 | Packet is too large |
| 3 | Time exceeded |
| 4 | Parameter problem |
| 128 | Echo request |
| 129 | Echo response |

**TABLE 17.2** Error Messages and Associated Code Values

| Code | Definition |
|---|---|
| 0 | No route to destination |
| 1 | Communication with destination prohibited |
| 2 | No route to destination |
| 3 | Not assigned |
| 4 | Address unreachable |
| 129 | Port unreachable |

**Packet too large:** Consider Figure 17.6 in which two networks are connected by a router. Network A is a Token Ring with MTU of 18000 bytes and Network B is an Ethernet with MTU of 1500 bytes. If station A sends an IP packet to station B, station A sends the packet to the router and the router examines the DF bit (do not fragment) in the IP header. If this bit is set, indicating that the packet is not to be fragmented, then the router is not able to send the packet to station B, and the router discards the packet. The router then sends an error message to station A indicating that the packet is too large and the MTU for the packet should be 1500 bytes. Station A sends a new packet with the MTU of 1500 bytes to the router. This is called *Path MTU discovery*. Figure 17.7 shows the ICMP packet format for path MTU discovery.

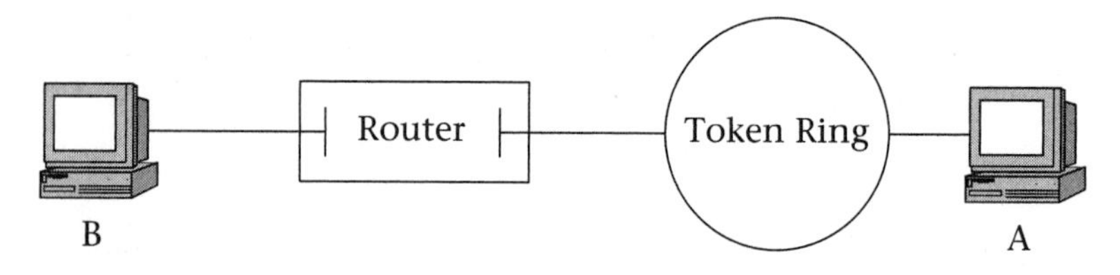


**FIGURE 17.6**
Two networks with different MTUs are connected via a router

**FIGURE 17.7**
ICMP packet format for MTU discovery

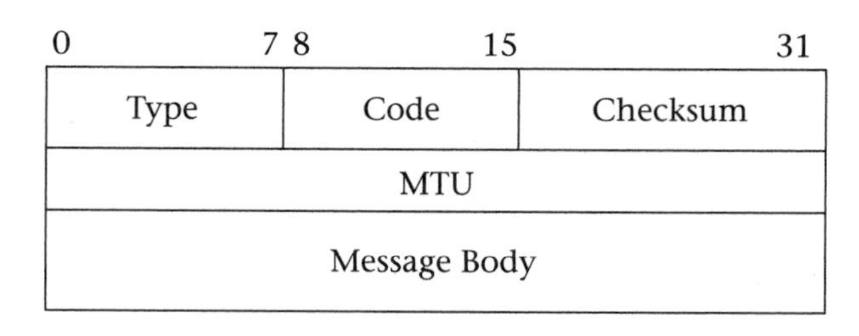


In Figure 17.7, the type field is equal to 2, the code field is zero, and the MTU field is the maximum transmission unit of the next-hop link.

**Time exceeded:** When an IP packet travels through the Internet, the TTL field decrements by one each time it passes a gateway. When the TTL value becomes zero, the gateway sends the ICM message of time exceeded.

**Parameter problem:** When the router discovers an error in an IP packet, it discards the packet and sends a parameter problem to the source.

**Echo request and response:** An ICMP echo-request message, which is generated by the *ping* command, is sent by any host to test node reachable across an Internetwork. The ICMP echo-reply message indicates that the node can be successfully reached.

## 17.3 Point-to-Point Protocol (PPP)

**Point-to-Point Protocol (PPP)** provides a standard method for transmitting IP packets over a serial link, such as a modem. PPP encapsulates an IP packet, and then transports it over a point-to-point link. The functions of PPP are establishing link configurations, configuring link quality testing, and managing IP addresses. Figure 17.8 shows the point-to-point protocol architecture. In this figure, the Data Link layer is divided into two sublayers: Network Control Protocol (NCP) and Link Control Protocol (LCP).

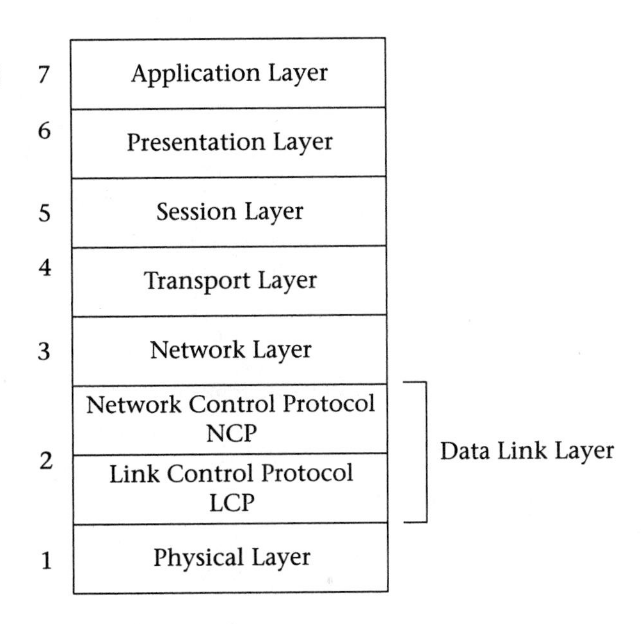


**FIGURE 17.8**
Point-to-point protocol architecture

**Network Control Protocol (NCP):** The basic functions of the **Network Control Protocol (NCP)** are to establish a link and configure different network protocols. PPP supports a family of NCPs such as IP Control Protocol (IPCP), AppleTalk Control Protocol, and IPX Control Protocol.

**Link Control Protocol:** The functions of LCP are to establish connections, configure connections, and negotiate link parameters such as authentication methods, Maximum Receiving Unit (MRU), and termination of the link. Also, the LCP periodically tests the link.

## PPP Connection Operation

Consider a source A that is trying to make a connection to destination B. A and B are connected by a point-to-point link such as a PC that is connected to an ISP via a modem. The following states describe the PPP connection operations:

1. **Dead state:** At this state the link is down and the source requests a connection to the destination, such as a PC dialing a server. When a physical connection is established, then it goes to the establishment state.
2. **Establishment state:** The source negotiates parameters of the link by sending LCP packets to the destination; these parameters are the Maximum Receive Unit (MRU), compression of certain fields of the PPP, and authentication protocol. If negotiations are completed successfully, then it goes to the next state for authentication. If it fails it goes to step 6.
3. **Authentication state:** LCP supports Password Authentication Protocol (PAP) and Challenge Handshake Authentication Protocol (CHAP).

   i. **Password Authentication Protocol (PAP):** The source transmits the name and password to the remote server for verification.

   ii. **Challenge Handshake Authentication Protocol (CHAP):** CHAP uses three-way handshaking and it is more reliable than PAP. CHAP performs the following steps for authentication of a user:

      a) The server sends a random number to the user's station.

      b) The user's station generates a hash value from the random number. The user's password and username are then sent along with that hash value to the server.

      c) The server generates a hash value of the random number. If this hash value is same as the hash value that was sent by the user, then the user is authenticated.

When authentication is complete it goes to the network state; otherwise it goes to step 6.

4. **Network state:** At this state, the NCP packets are exchanged between two end points to configure the Network layer. When the NCP configurations are complete, then it goes to an open state.
5. **Open state:** At this state, IP packets are transported between links.

6. **Termination state:** The NCP packet is used to terminate the link during this state.
7. The link is disconnected.

***Physical Layer.*** PPP is able to operate any DTE/DCE interface, such as EIA 232-C and EIA-422.

**PPP Packet Format**

Figure 17.9 shows the PPP packet format, which uses HDLC packet structure. The start and end flags are designated by 7E in hex; the address field is always FF in hex; and the control byte is 03.

| Flag 7E | Address FF | Control 03 for PPP | Protocol Type | Information | FCS | Flag 7E |
|---|---|---|---|---|---|---|
| 1 | 1 | 1 | 2 | 0-1500 | 2 | 1 Byte |

**FIGURE 17.9**
Point-to-point packet format

The following describes the function of each field in a PPP packet format:

***Protocol Type.*** The protocol type's value defines the type of packet, as shown in Table 17.3.

**TABLE 17.3** Protocol Numbers and Packet Types

| Protocol Number | Packet Type |
|---|---|
| C021 | LCP |
| C023 | PAP |
| 8021 | IPCP |

***Link Configuration Packets (LCP).*** There are three types of LCP packets, as shown in Table 17.4: link configuration, link termination, and link monitor. When the protocol type is C021 (Hex), then the packet is an LCP packet and it is used for link configuration, termination, and link monitor. Figure 17.10 shows the packet format for LCP.

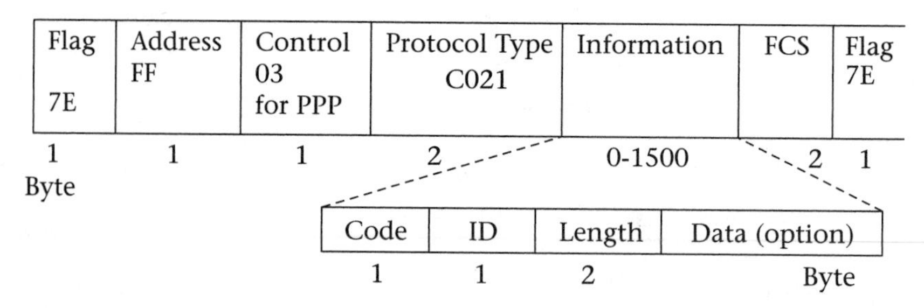


**FIGURE 17.10**
LCP packet format

Where:

**Code:** The code field is 8 bits and it defines the type of LCP packet, as shown in Table 17.4.

**ID:** The identification is 8 bits and it is used for matching packet requests and responses.

**Length:** The length field is 16 bits and it defines the number of bytes in an LCP packet.

**TABLE 17.4** LCP Packet Types and Codes

| Code | LCP Packet Type | Comments |
|---|---|---|
| 1 | Configuration Request | Link Configuration Packet |
| 2 | Configuration ACK | Link Configuration Packet |
| 3 | Configuration NACK | Link Configuration Packet |
| 4 | Configuration Reject | Link Configuration Packet |
| 5 | Terminate Request | Link Termination Packet |
| 6 | Termination ACK | Link Termination Packet |
| 7 | Code Reject | Link Termination Packet |
| 8 | Protocol Reject | Link Termination Packet |
| 9 | Echo Request | Link Monitor |
| A | Echo Reply | Link Monitor |

***Network Control Protocol (NCP).*** Network control protocol is used for configuring the Network layer. When the protocol type is 8021 (Hex), this signifies IP Control Protocol (IPCP). IPCP packets are used to configure the IP layer in the Network layer. IPCP packets are similar to LCP packets.

***IP Address.*** The ISP server dynamically assigns an IP address to a client's PC. To look up the IP address of your computer (if you are using MS Windows or XP) you should refer to the TCP/IP in your network connection in the Control Panel folder. Figure 17.11 shows the IP address of a client and an ISP.

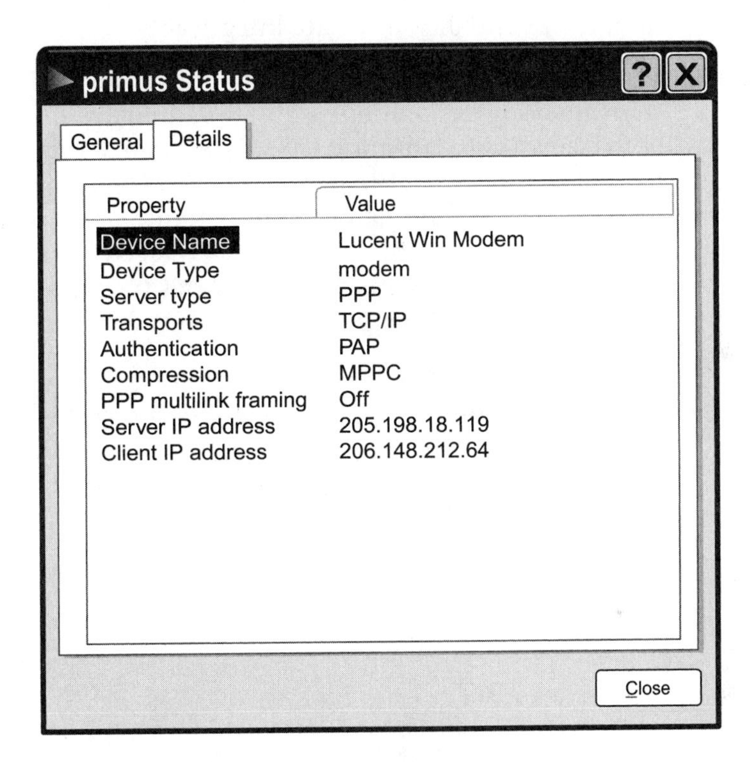


**FIGURE 17.11**
IP address of a client and server using PPP

## 17.4 Virtual Private Network (VPN)

Large organizations and corporations have multiple sites in different locations and would often like to connect LANs together. One solution is to lease a data communication line and connect the LANs together, thus creating a private network. A corporation with one hundred offices in different locations throughout a country must lease one hundred lines to connect its LANs together. This method is not cost effective. In order to reduce the cost of leasing private lines, a **Virtual Private Network (VPN)** can be employed over the Internet.

Figure 17.12 shows two networks that are connected through the Internet using VPN devices. The function of a VPN device is to make a secure communication between two LANs by establishing a tunnel between its two endpoints. The disadvantage of using the Internet as a communication channel between corporations is poor security. Therefore, the VPN needs to provide security components, such as authentication, confidentiality, and data integrity.

**Tunneling**

**Tunneling** is the process of placing one packet inside another packet for transmission over a public network. In Figure 17.12, it is assumed that both networks are running IPX. If station A wants to send an IPX packet to station B, it sends the packet to the VPN device A first, and then the VPN device A encapsulates the packet into an IP packet and transmits it over the Internet. When VPN device B receives the IP packet, it discards the IP header and sends the IPX packet to station B. The following are some of the tunneling protocols:

- **Point-to-Point Tunneling Protocol (PPTP):** PPTP was developed by the PPTP forum. PPTP supports 40- and 128-bit encryption.
- **Layer-2 Forwarding (L2F):** L2F was developed by CISCO and uses any authentication method.
- **Layer-2 Tunneling Protocol (L2FT):** L2FT was developed by CISCO and IETF. It is a combination of L2F and Point-to-Point Tunnel Protocol.

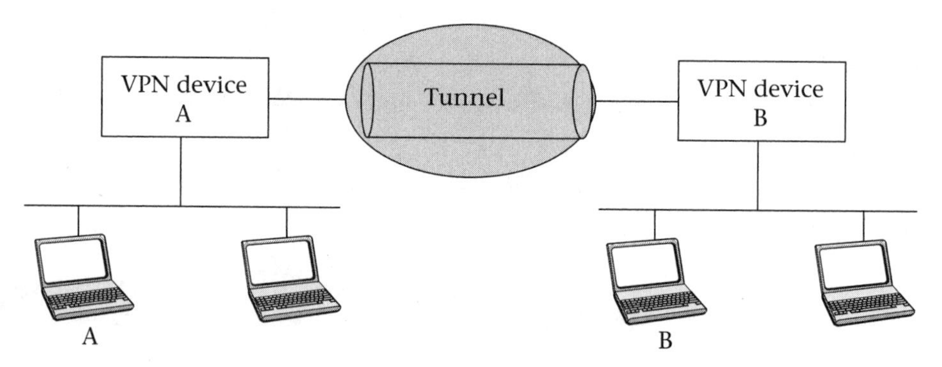


**FIGURE 17.12**
Virtual Private Network (VPN)

## 17.5 IP Security (IPsec) Protocol

In order to protect information from attacks in VPNs, the IETF developed the IPsec Protocol for integrity, authentication, and confidentiality of data over IP networks. The **IP Security (IPsec) Protocol** is an open standard and does not define any specific protocol for authentication and encryption. IPsec is becoming increasingly more popular than the other protocols. IPsec is made of two protocols: **Authentication Header (AH) Protocol** and **Encapsulation Security Protocol (ESP).**

## Authentication Header Protocol

The Authentication Header Protocol (AH) is used for identification of a user or application and it can operate in either tunneling mode or transport mode. AH does not provide any encryption. Figure 17.13 shows the AH packet format.

**FIGURE 17.13**
AH packet format of IPsec

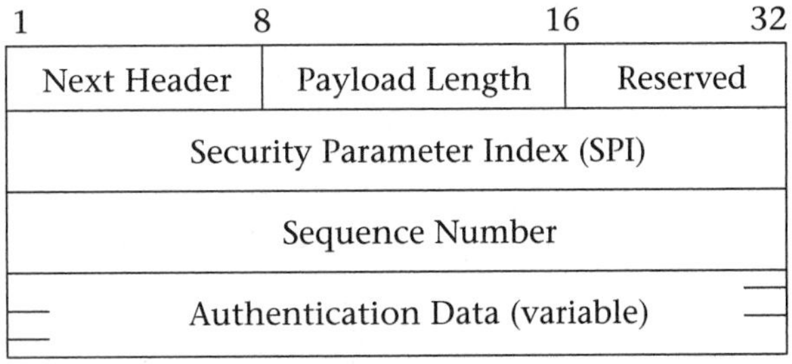

| 1 — 8 | 8 — 16 | 16 — 32 |
|---|---|---|
| Next Header | Payload Length | Reserved |
| Security Parameter Index (SPI) | | |
| Sequence Number | | |
| Authentication Data (variable) | | |

The following describes the function of each field of the authentication header in Figure 17.13:

**Next header:** Identifies the next header (such as TCP).

**Payload length:** Number of 32-bit words in AH.

**Security Parameter Index (SPI):** The SPI informs the receiving station of the location of the security information, such as the decryption key.

**Sequence number:** Each packet has a unique number. If an intruder copies a packet and resends it with the same sequence number, then the receiver discards the packet.

**Authentication data:** This is the digital signature of the payload.

Figure 17.14 shows the AH location in an IPsec packet in transport mode. In the IP header, the protocol type is 50, thus indicating that the next header is an IPsec header.

Figure 17.15 shows the location of AH in the IPsec packet using tunneling mode. In tunneling mode, the entire packet is encapsulated into another IP packet and AH includes the entire IP packet.

**FIGURE 17.14**
Authentication header using transport mode

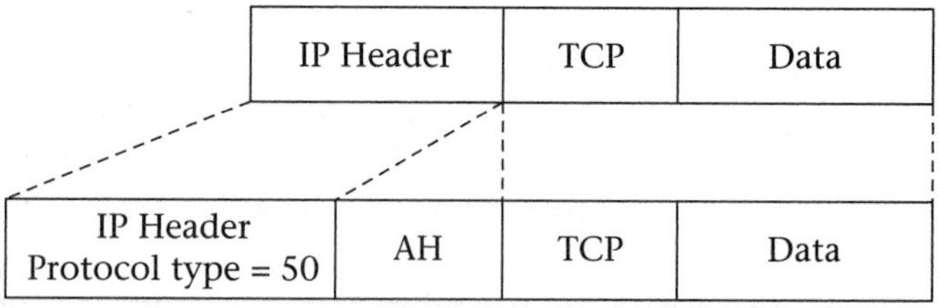

| IP Header | TCP | Data |
|---|---|---|

| IP Header Protocol type = 50 | AH | TCP | Data |
|---|---|---|---|

**FIGURE 17.15**
Authentication header using tunneling mode

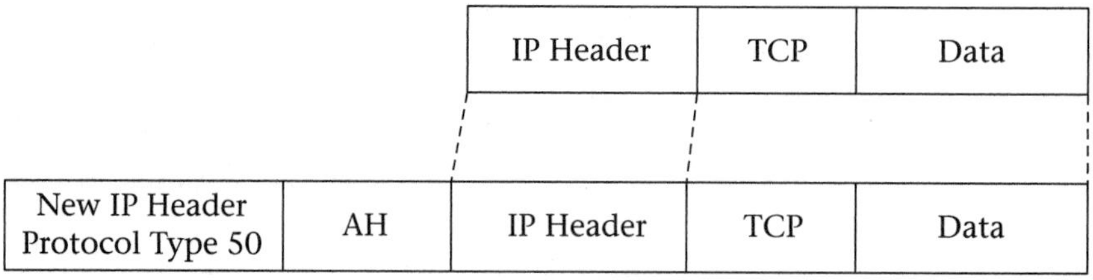

| IP Header | TCP | Data |
|---|---|---|

| New IP Header Protocol Type 50 | AH | IP Header | TCP | Data |
|---|---|---|---|---|

**Encapsulation Security Packet (ESP)**

The Encapsulation Security Packet performs encryption and authentication of data. Figure 17.16 shows an ESP packet format.

**FIGURE 17.16**
ESP packet format

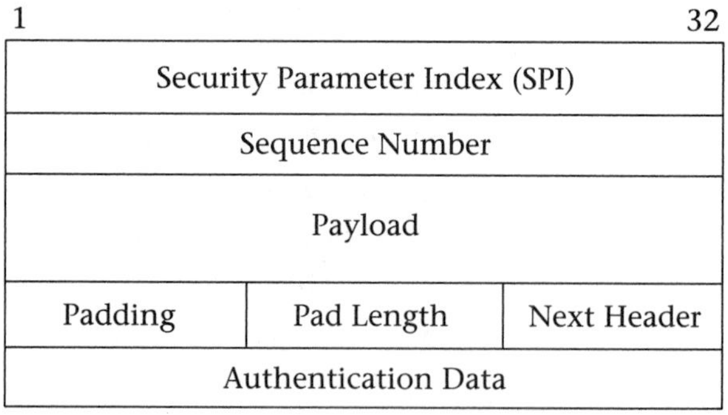

1 ... 32

| Security Parameter Index (SPI) | | |
|---|---|---|
| Sequence Number | | |
| Payload | | |
| Padding | Pad Length | Next Header |
| Authentication Data | | |

ESP can operate in tunneling mode or in transport mode. Figure 17.17 shows the ESP format using transport mode. In transport mode, only the TCP header and data fields are encrypted. The authentication data is a digital signature of the data, padding, TCP header, and ESP header. As shown in Figure 17.17, the protocol type in the IP header is 50, thus indicating that the next protocol is ESP. Figure 17.18 shows ESP using tunneling mode. In tunneling mode, the entire IP packet is encrypted and has the entire ESP header encapsulated into the IP packet for transmission.

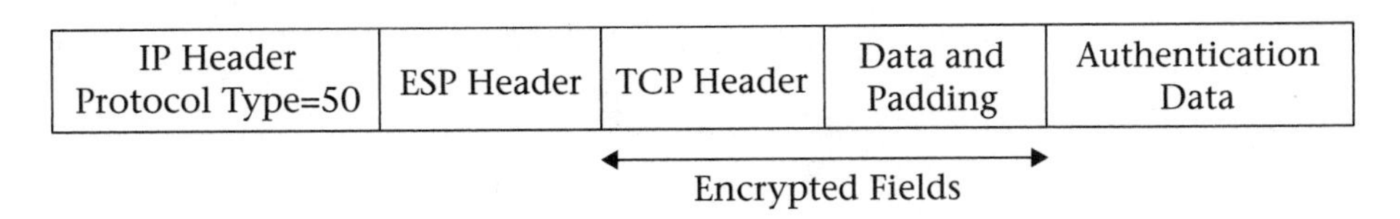


**FIGURE 17.17**
ESP using transport mode

| New IP Header Protocol Type 50 | ESP Header | IP Header | TCP Header | Data and Padding | Authentication Data |
|---|---|---|---|---|---|

Encrypted Fields

**FIGURE 17.18**
ESP using tunneling mode

## 17.6 Routing

The function of a **router** is to determine the path for transporting information (packets) through the Internet. Furthermore, routing is the method of moving packets from one network to another network. Figure 17.19 shows four networks connected together through the routers R1, R2, R3, and R4. Consider the following questions when host A wants to send a packet to host B:

1. How does host A know that host B is connected to router R3?
2. What is the best route for sending information from host A to host B? Host A can send packets to host B by using three different paths. Host A can send packets to host B using the path R1–R2–R3 or R1–R4–R3 or R1–R2–R4–R3. The router determines the best route for sending information by using a routing algorithm to build a routing table.

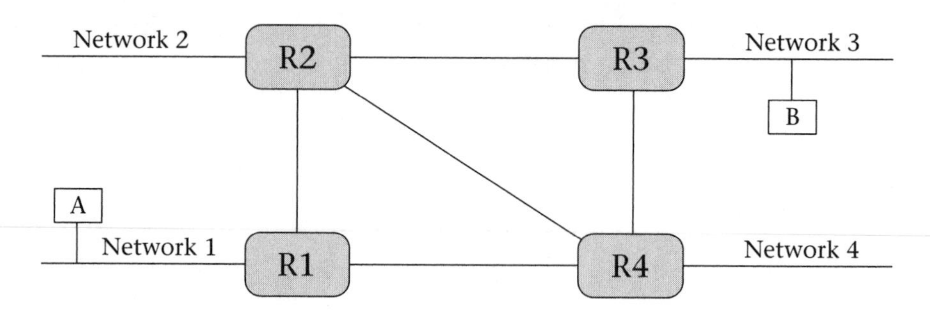


**FIGURE 17.19**
Networks connected together by four routers

**Link Characteristics**

Routers use different metrics (link characteristics) to find the best route for sending a packet to the destination. The metrics are as follows:

**Hop count:** The hop count measures the number of hops that a packet must go through in order to get to the destination.

**Throughput:** The data rate of the link.

**Communication cost:** The cost of transmitting information from the source to the destination.

**Delay:** Measures the amount of time it takes a packet to travel from the source to the destination.

**Configuring the Routing Table**

There are two ways to configure the routing table: with **dynamic routing** or **static routing.**

***Dynamic Routing.*** In dynamic routing, the routers use a routing protocol to build their routing tables. If there is a change in the network configuration, a dynamic routing protocol broadcasts the changes to all the routers in the network in order to update all routing tables. Some of the most popular dynamic routing protocols are as follows:

- Routing Information Protocol (RIP)
- Open Short Path First Protocol (OSPF)
- Interior Gateway Routing Protocol (IGRP)

***Static Routing.*** Static routing tables are configured manually by the network administrator. Static routing is often used for small networks. The problem with this type of routing is that if there is a change in network topology or a network link failure, all the routing tables need to be manually updated.

Figure 17.20 shows a network with three routers: A, B, and C. Router A has an Ethernet link E0 and one serial link S0. Router B has two serial links S0 and S1 and one Ethernet link E0. Router C has one serial link S0 and one Ethernet link E0. Class B IP addresses (180.160.0.0) are used to assign IP address to each host.

As a network administrator, you would need to assign an IP address to each network and to each host.

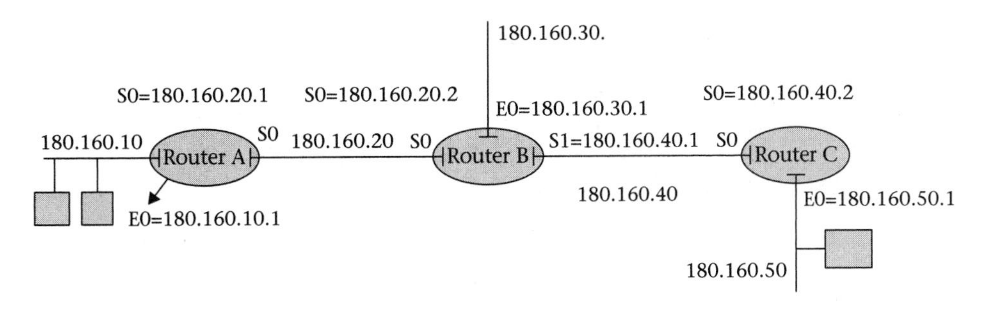


**FIGURE 17.20**
Networks connected together by three routers

The following IP addresses are assigned to the following networks:

180.160.10 the network connected to the Ethernet port of router A
180.160.20 the serial link between router A and router B
180.160.30 the Ethernet network connected to router B
180.160.40 the serial link between routers B and C
180.160.50 the Ethernet network connected to router C

In static routing, the routing table is built and updated manually, and the IP routing table has the following fields:

A. IP address of remote network
B. Subnet mask
C. IP address of default gateway

Tables 17.5, 17.6, and 17.7 show the routing tables for routers A, B, and C.

**TABLE 17.5** Routing Table for Router A

| Remote Network | Subnet Mask | Default Gateway |
|---|---|---|
| 180.160.50 | 255.255.255 | 180.160.20.2 |
| 180.160.40 | 255.255.255 | 180.160.20.2 |
| 180.160.30 | 255.255.255 | 180.160.20.2 |

Router A can access the other networks through the S0 link of router B. Therefore, S0 is the default gateway for router A.

**TABLE 17.6** Routing Table for Router B

| Remote Network | Subnet Mask | Default Gateway |
|---|---|---|
| 180.160.50 | 255.255.255 | 180.160.40.2 (S0 of router C) |
| 180.160.10 | 255.255.255 | 180.160.20.1 (S0 of router A) |

**TABLE 17.7** Routing Table for Router C

| Remote Network | Subnet Mask | Default Gateway |
|---|---|---|
| 180.160.10 | 255.255.255 | 180.160.40.1 (S1 of router B) |
| 180.160.120 | 255.255.255 | 180.160.40.1 (S1 of router B) |
| 180.160.30 | 255.255.255 | 180.160.40.1 (S1 of router B) |

## 17.7 Networking Diagnostic Commands

TCP/IP developed some networking commands for network diagnostics, including *ping, tracert, IPconfig, netstate, arp,* and *nslookup*. The following command descriptions are based on the Microsoft Windows OS.

### PING Command

The PING command is one of the most useful tools for debugging networks. The PING (Packet Internet Groper) command is an ICMP echo process and it is used for checking whether the destination host is reachable. The PING command comes with different options. PING will send an ICMP request echo to the destination and it expects an ICMP echo packet from the destination. The PING command transmits 32 bytes to the destination and receives 32 bytes from the destination. PING also displays the time it took to transmit and receive 32 bytes of data. The following shows an example using the PING command:

C:\>**ping yahoo.com**

Pinging yahoo.com [66.218.71.198] with 32 bytes of data:
Reply from 66.218.71.198: bytes = 32 time = 101ms TTL = 237
Reply from 66.218.71.198: bytes = 32 time = 130ms TTL = 237
Reply from 66.218.71.198: bytes = 32 time = 111ms TTL = 237
Reply from 66.218.71.198: bytes = 32 time = 140ms TTL = 237

Ping statistics for 66.218.71.198:
Packets: Sent = 4, Received = 4, Lost = 0 (0% loss),
Approximate round trip times in milliseconds:
Minimum = 101ms, Maximum = 140ms, Average = 120ms

## ARP Command

If a router receives a packet that does not have the hardware address of the packet in its arp cache table, the router sends a broadcast packet to the hosts on the network and requests the hardware address of the host. All the hosts in the network accept the packet and compare the IP address of the packet with their own IP address. In the event that both IP addresses are the same, the host will respond to the router with a RARP that contains the hardware address of the host.

ARP (Address Resolution Protocol) tables contain both the IP address and the MAC address of a computer. The following is the list of ARP commands:

| | |
|---|---|
| arp -a | Display the contents of the ARP table |
| arp -c | ARP table (do not use this command) |
| arp -d | Delete an entry with an IP address |
| arp -s | Add an entry with a MAC address |

The following shows an example using the ARP command:

C:\>**arp -a**

Interface: 10.72.8.6 on Interface 0x1000003

| Internet Address | Physical Address | Type |
|---|---|---|
| 10.72.11.254 | 00-d0-00-38-18-00 | dynamic |

### IPconfig Command

IPconfig displays network settings, the physical address, the IP address, the subnet mask of the host, and the IP address of the default gateway. The following shows an example using the IPconfig command:

C:\>**ipconfig/all**

Windows 2000 IP Configuration

Host Name . . . . . . . . . . . . : CSCAXE055689
Primary DNS Suffix . . . . . . . : scsu.southernct.edu
Node Type . . . . . . . . . . . . : Hybrid
IP Routing Enabled. . . . . . . . : No
WINS Proxy Enabled. . . . . . . . : No
DNS Suffix Search List. . . . . . : scsu.southernct.edu

### Tracert Command

Tracert shows the path of a packet from source to destination and the number of gateways the packet travels through. The following shows an example using the tracert command:

H:\>**tracert yahoo.com**

Tracing route to yahoo.com [66.218.71.198]
over a maximum of 30 hops:

1 <10 ms <10 ms <10 ms 10.72.11.254
2 <10 ms <10 ms <10 ms enpbx1-gi92-je1a1.owlnet [10.95.254.233]
3 20 ms 10 ms <10 ms brick-s1p2c0-enpbx1.owlnet [10.95.254.193]
.
.
15 80 ms 90 ms 90 ms unknown.Level3.net [64.152.69.30]
16 80 ms 91 ms 90 ms w1.rc.vip.scd.yahoo.com [66.218.71.198]

### Netstat Command

The netstat command displays information about your network configuration. The netstate command comes with the following options:

| | |
|---|---|
| netstat -n | displays information on the NIC of your computer |
| netstat -r | displays the IP routing table |
| netstat -a | displays information on TCP and UDP ports |
| netstat -s | displays operational statistics of network protocols |

The following shows an example using the netstat command:

H:\>**netstat -r**

Route Table
Interface List
0×1 .......................... MS TCP Loopback interface
0×1000003 ...00 b0 d0 07 10 b8 ...... 3Com EtherLink PCI
Active Routes:

| Network Destination | Netmask | Gateway | Interface | Metric |
|---|---|---|---|---|
| 0.0.0.0 | 0.0.0.0 | 10.72.11.254 | 10.72.8.6 | 1 |
| 10.72.8.0 | 255.255.252.0 | 10.72.8.6 | 10.72.8.6 | 1 |
| 10.72.8.6 | 255.255.255.255 | 127.0.0.1 | 127.0.0.1 | 1 |
| 10.255.255.255 | 255.255.255.255 | 10.72.8.6 | 10.72.8.6 | 1 |
| 127.0.0.0 | 255.0.0.0 | 127.0.0.1 | 127.0.0.1 | 1 |
| 224.0.0.0 | 224.0.0.0 | 10.72.8.6 | 10.72.8.6 | 1 |
| 255.255.255.255 | 255.255.255.255 | 10.72.8.6 | 10.72.8.6 | 1 |
| Default Gateway: | 10.72.11.254 | | | |
| Persistent Routes: | None | | | |

## Nslookup Command

The function of DNS is to resolve hosts with a given IP address. The DNS service accepts requests from a DNS client to resolve names with an IP address.

The nslookup command allows you to match a machine's Internet domain name (www.ibm.com) with its IP address (129.248.150.51) or vice versa. The following shows an example using the nslookup command:

H:\>**nslookup 129.42.17.99**

Server: scsudc01.scsu.southernct.edu
Address: 10.64.24.19
Name: www.ibm.com
Address: 129.42.17.99

H:\>**nslookup ibm.com**

Server: scsudc01.scsu.southernct.edu
Address: 10.64.24.19
Non-authoritative answer:
Name: ibm.com
Addresses: 129.42.17.99, 129.42.18.99, 129.42.19.99, 129.42.16.99

## 17.8 Multi-Protocol Label Switching (MPLS)

**Multi-Protocol Label Switching (MPLS)** was developed to overcome the deficiency of IP packet delivery over private and public networks. An IP network is connectionless and when an IP packet is received in a router, the router uses its routing table and the IP address of the packet to find the next hop. The IP address does not use a fixed size such as class A, B, C, or D; therefore, it takes time to find the next hop. MPLS defines a method for fast packet forwarding over an IP network. MPLS is an independent protocol and it works with multiple protocols such as IP, ATM, and frame relay. MPLS adds a label or tag to an IP packet and it uses this label to find the path or next hop. Because the length of the label is fixed, it becomes faster to look up the table than rely on IP routing. Routers on an IP network use this label to find the next hop. The operation of MPLS is similar to cell switching in an ATM network and frame relay. MPLS performs the same function as a router but with higher performance. MPLS can be used for VPN, Transport layer (layer 2), and connection-oriented service.

### MPLS Components and Operation

Figure 17.21 shows an MPLS network that consists of the following components:

***Label Edge Router (LER).*** LER is located between the MPLS network and the users' networks, as shown in Figure 17.21. LER performs the following functions:

- When LER receives an IP packet from the user's network, it adds its label to the IP packet and uses its routing table to forward the IP packet based on the label value to the next label switch router.
- When LER wants to transmit a packet to the user's network, it removes the label field from the IP address and transmits the IP packet to the user's network.

***Label Switch Router (LSR).*** The function of LSR is to examine the IP packet based on its label and replace the label with a new label, which identifies the next hop, and then forwards the packet to the next hop.

In Figure 17.21, network A transmits two IP packets to LER1. LER1 uses its routing table and adds a label to each packet and then forwards the packets to the next LSR. The LSR2 examines each packet's label and strips each packet of its label and inserts a new label and forwards each packet to the appropriate port for the next hop. An IP packet with label 40 is transmitted to LER4 and an IP packet with label 60 is transmitted to

LSR3. LER4 examines the label of the IP packet and finds out that packet must be transmitted to user network B. LER4 removes the label from the IP packet and forwards the packet to user network B.

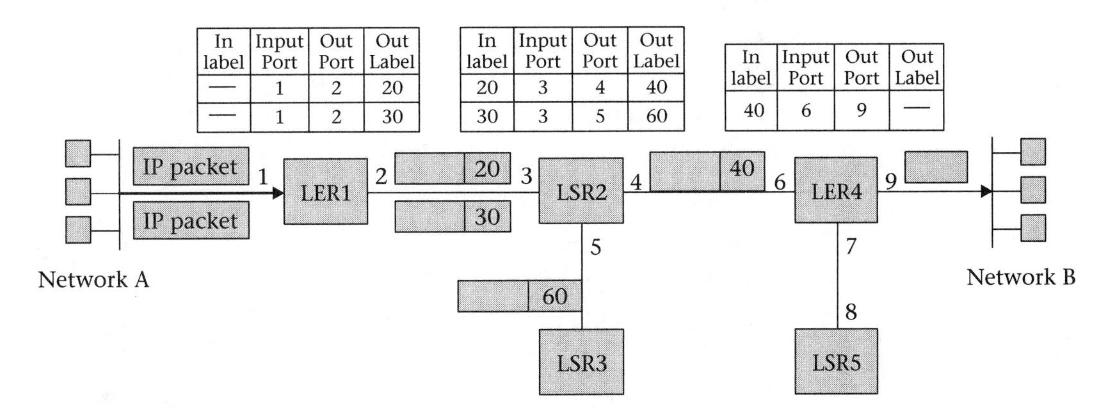


**FIGURE 17.21**
MPLS network

***Forward Equivalence Class (FEC).*** This is a group of IP packets that are forwarded in the same manner over the same path. Therefore, a group of packets with same FEC can assign the same labels throughout the path. Considering that multiple IP packets with the same forwarding characteristics are transmitted by network A to network B, the LER1 assigns these IP packets the same labels throughout the path. The packets with the same FEC will use the same path. This method reduces the size of the routing table.

MPLS defines only the forwarding methods and other protocols that are needed to perform routing and signaling protocols. The routing protocol distributes the networking topology and configures the routing table by using an IP routing protocol such as OSPF, BGP, or RIP. Signaling protocols are used to inform the routers which label and link are to be used by the switch for each label switching path. The MPLS advantages are as follows:

- The path from the source to the destination can be identified in advance.
- MPLS can select network paths in order to have a balanced load in the network.
- MPLS can provide a specific path for the data.

- MPLS offers Quality of Service (QoS) by choosing the specific path in order to provide bandwidth to the application, resulting in less delay and less packet loss.

***MPLS Label.*** An MPLS label is inserted in the frame and the location of the label depends on the layer 2 technology. In ATM, the label is VPI/VCI and in frame relay, it is DLCI filed. In IP packets the label is inserted between layer 2 and layer 3, as shown in Figure 17.22. The MPLS label field is 32 bits and the following describes the function of each field:

**Label:** This is 20 bits.

**CoS (Class of Service):** This field is 3 bits and is used in queuing and discarding packets while traveling through the network.

**S (Stack):** This field is one bit and is used for having multiple MPLS labels.

**TTL (Time-to-Live):** This field is 8 bits and works similar to the TTL field in IPv4.

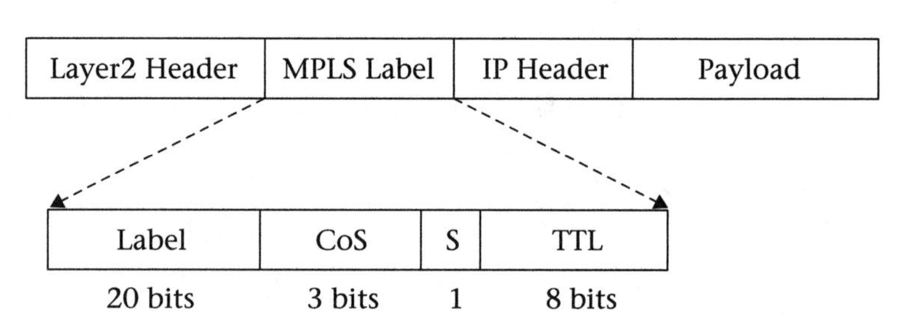


**FIGURE 17.22** MPLS label format

## Summary

- Dynamic Host Configuration Protocol (DHCP) is used to assign IP addresses to a host dynamically.
- The components of DHCP are DHCP client, DHCP server, and DHC relay agent.
- A DHCP server supports automatic allocation, dynamic allocation, and manual allocation for assigning IP addresses to clients.
- Point-to-Point Protocol (PPP) provides standards for transmitting IP packets over serial connections (modem).
- Virtual Private Networks (VPN) are used to connect networks of an organization that are located in different locations, using the Internet.
- VPNs use tunneling to transfer packets between networks of a VPN.
- Tunneling is a process that places one packet inside another packet for transmission.
- IPsec is a tunneling protocol.
- IPsec is used to protect information from an attack on a VPN.
- The function of a router is to determine the path for transmitted packets through the Internetworking.
- In a router the path of information is determined by the routing table.
- The routing table can be configured in two ways: static routing or dynamic routing.
- A static routing table is configured manually and a dynamic routing table is configured automatically.
- MPLS uses label switching to route the information.

## Key Terms

Authentication Header (AH) Protocol

DHCP Relay Agent

Dynamic Host Configuration Protocol (DHCP)

Dynamic Routing

Encapsulation Security Protocol (ESP)

Internet Control Message Protocol (ICMP)

IP Security (IPsec) Protocol

Muliti-Protocol Label Switching (MPLS)

Network Control Protocol (NCP)

Point-to-Point Protocol (PPP)

Router

Routing Information Protocol (RIP)
Static Routing
Tunneling
Virtual Private Network (VPN)

## Review Questions

- **Multiple Choice Questions**

1. DHCP is used to assign ______ to a client.
   a. IP addresses automatically
   b. MAC addresses automatically
   c. IP addresses manually
   d. MAC addresses manually
2. A ______ holds a range of IP addresses.
   a. client server
   b. DHCP server
   c. router
   d. relay agent
3. The function of a relay agent is to ______.
   a. pass the broadcast message
   b. block the broadcast message
   c. change the broadcast message to unicast
   d. change the unicast message to multicast
4. PPP is used for ______.
   a. an Ethernet connection
   b. a serial connection
   c. a VPN
   d. a VLAN
5. A VPN is used to connect networks that are ______.
   a. located in the same building
   b. located in a campus
   c. located in different locations
   d. located in the same room
6. VPNs use ______ to connect networks.
   a. leased lines
   b. modems
   c. the Internet
   d. public networks

7. IPsec is used for ______.
   a. Ethernets
   b. the Internet
   c. VPNs
   d. VLANs
8. IPsec was developed by ______.
   a. ITU
   b. IETF
   c. IEEE
   d. EIA
9. The function of a router is to ______.
   a. find destination IP addresses
   b. find destination MAC addresses
   c. find the path to the destination for transporting packets
   d. none of the above
10. Static routing tables are updated by ______.
    a. a server
    b. the network administrator
    c. an automatic method
    d. none of the above
11. Dynamic routing tables are updated by ______.
    a. a server
    b. the network administrator
    c. an automatic method
    d. none of the above
12. MPLS uses a/an ______ for routing of packets.
    a. label
    b. IP address
    c. a and b
    d. MAC address

## • Short Answer Questions

1. What is the application of DHCP?
2. List the component of DHCP.
3. List DHCP IP address allocation methods.
4. Explain a DHCP dynamic allocation IP address.
5. What does PPP stand for?
6. What is the application of PPP?

7. Show the PPP layered architecture.
8. What is the function of Network Control Protocol?
9. What are the functions of Link Control Protocol?
10. Explain Password Authentication Protocol (PAP).
11. Explain Challenge Handshake Authentication (CHAP).
12. Find the IP address of your home computer and the IP address of the ISP when your computer is connected to your ISP.
13. In question #12 determine the type of authentication used.
14. What does VPN stand for?
15. Describe tunneling and its application.
16. What are the functions of IP security?
17. What is the function of a router?
18. List the link characteristics.
19. Explain static routing and dynamic routing.
20. Send a ping command to host www.sun.com.
21. Send a ping to a non-existing host.
22. Send a ping to the broadcast address 192.0.1.255.
23. In the following figure, show the network of an organization. As a network administrator:
    A. Assign an IP address to each network and interface using a class B address.
    B. Show the routing table for each router.

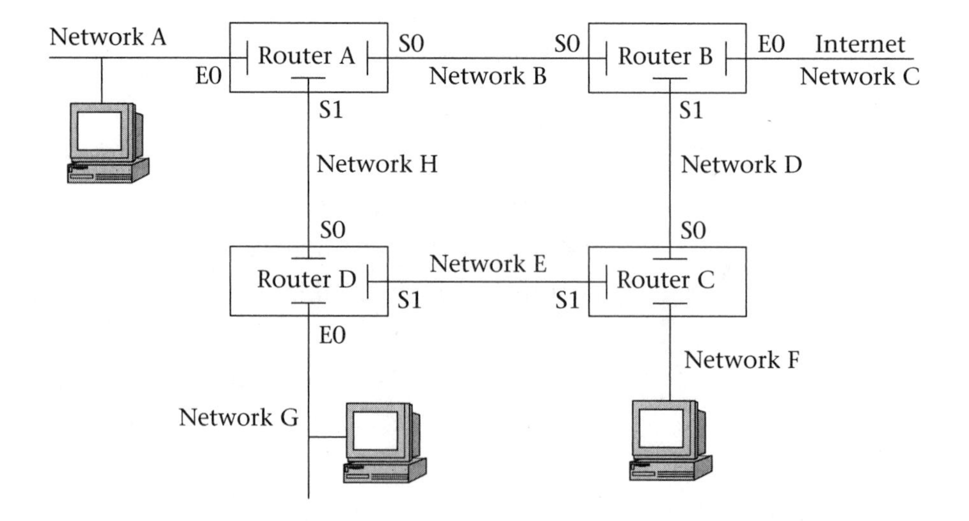

CHAPTER 18

# Wireless Local Area Network (WLAN)

OUTLINE



OBJECTIVES

After completing this chapter, you should able to:

- Discuss the applications and advantages of a Wireless LAN (WLAN)
- Understand WLAN technology
- Describe the applications of the ISM and UNII bands
- Explain the operation of Physical layers for a WLAN

- Explain the access methods for WLANs
- Distinguish between different types of IEEE 802.11
- Discuss WLAN security

## INTRODUCTION

The **Wireless Local Area Network (WLAN)** or IEEE 802.11 is a LAN technology that enables users to access an organization's network from any location inside the organization without any physical connection to the organization's network. WLAN uses radio frequency or infrared waves as transmission media. The WLAN is the next generation of campus network. Students are able to connect their laptops to the campus network from any location inside the campus. In hospitals WLAN facilitates access to patients' files from any site in the hospital. Likewise, WLANs are used in warehouses and workshops. The following are some of the advantages of wireless LAN over wired LANs:

1. Wireless LANs can be used in places where wiring is impossible.
2. Wireless LANs can be expanded without any rewiring.
3. Wireless LANs provide mobility; that is, users can move their computers anywhere inside the organization.
4. Wireless LANs support roaming; users move around with their laptops without interrupting their connections.
5. Wireless LANs are cost effective as they make movement from one location to another possible without the expense of connecting wires.

## 18.1 WLAN Components

Wireless LANs have three main components: a WLAN Network Interface card (NIC), an access point, and a Network Operating System (NOS).

### WLAN Network Interface Card (NIC)

The WLAN NIC is the interface between the user's system and an access point. A WLAN NIC is shown in Figure. 18.1.

**FIGURE 18.1**
WLAN NIC for a laptop computer (Courtesy of Cisco Corp.)

## Access Point (AP)

An **access point (AP)** is a wireless hub. It is connected to a wired LAN and provides coordination between users. Figure 18.2 shows an AP. An AP is made of three components: an antenna, a receiver, and a transmitter.

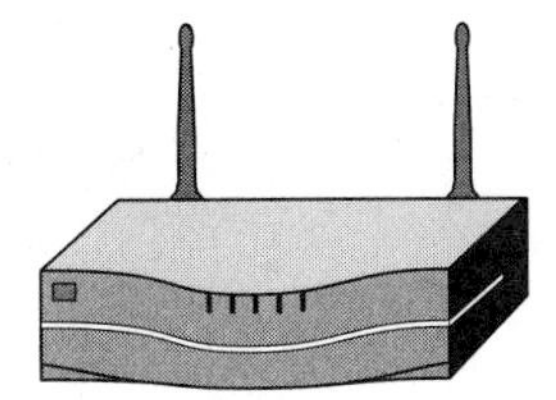

**FIGURE 18.2**
A wireless hub serves as the access point (Courtesy Dlink Corp.)

***Antenna.*** An antenna is a conductor that it is used to radiate and to receive electromagnetic waves. The anntena can be characterized by directionality and gain.

Directionality refers to the direction of the **Radio Frequency (RF)** signal transmitted by the antenna. The two types are omni-directional antenna that transmits the RF signal through 360 degrees, and directional antenna that transmits in a specific direction, as shown in Figure 18.3.

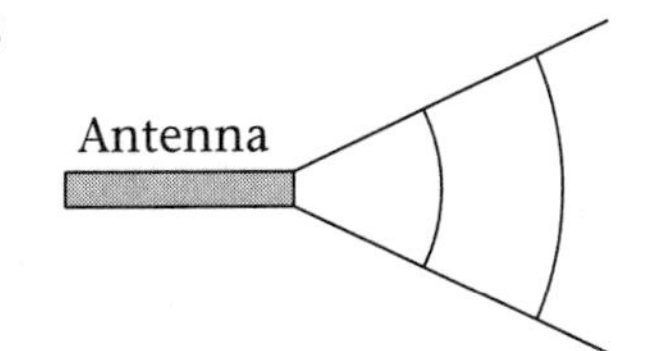


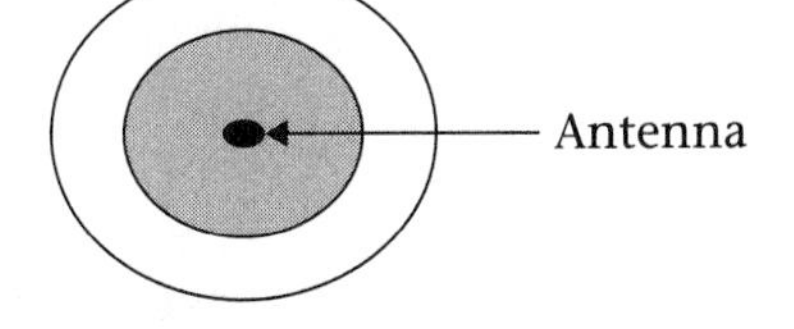


**FIGURE 18.3**
Directional and omni-directional antennas

***Antenna Gain.*** The antenna gain is measured in dBi. dB stands for *decibel* and *i* stands for *isotropic*. An isotropic antenna is an ideal antenna that transmits the RF signal in all directions equally; but real antennas do not transmit RF signals in all directions. The gain of antennas is given by Equation 18.1:

$$G = Pa/Pi \qquad (18.1)$$

Where:

*G:* antenna gain
*Pi:* power density of isotropic anntena at the same distance
*Pa:* power density of real antenna in specific direction and distance

Pi is defined by Equation 18.2:

$$Pi = Pt/4Pi\, r^2 \tag{18.2}$$

Where:

*Pi:* power density of an isotropic anntena
*P*t: transmitted power in watts
*r:* distance from the antenna in meters

**Network Operating System**

Most operating systems come with a wireless NOS.

## 18.2 WLAN Topologies

WLAN topologies can be classified as either managed wireless networks or unmanaged wireless networks.

**Managed Wireless Networks**

In a managed wireless network topology, the AP coordinates the communications between users. Topologies for managed wireless networks are **Basic Service Set (BSS)** topology and **Extended Service Set (ESS)**.

***Basic Service Set (BSS).*** BSS is composed of a group of workstations that can access each other and the wired LAN through an AP, as shown in Figure 18.4. The function of the AP is to coordinate communication between wireless LAN clients and the wired LAN. Also, the AP receives information from clients and retransmits that information to the hub. The AP works as a bridge between a wireless LAN and a wired LAN. The area covered by an AP is called a **cell**. As shown in Figure 18.4, any station inside the cell can access the AP. The access point coverage varies and depends on the manufacturer's design of the wireless LAN product, transmitter power, and the environment in which the WLAN operates (indoors or outdoors). When the number of clients in a cell or area of coverage increases, multiple access points will be added to the wireless LAN, which leads to an Extended Service Set (ESS) topology.

**FIGURE 18.4**
BSS topology

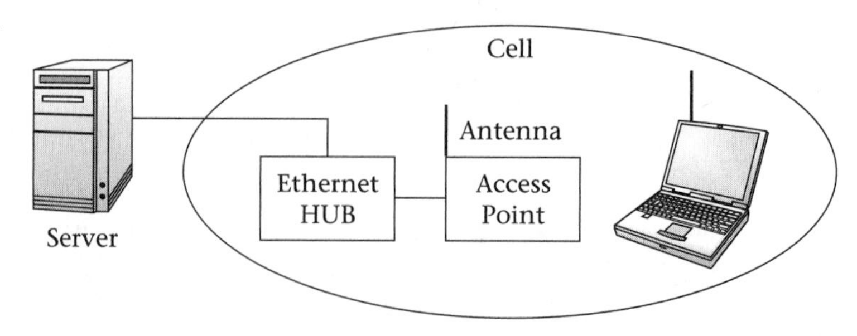

***Extended Service Set (ESS).*** ESS is used to expand a wireless LAN. Figure 18.5 shows an ESS with a number of APs. The APs are connected to a Distribution System (DS). The functions of the DS are to form extended service and provide distribution services between access points. The DS basically acts as a bridge for access points.

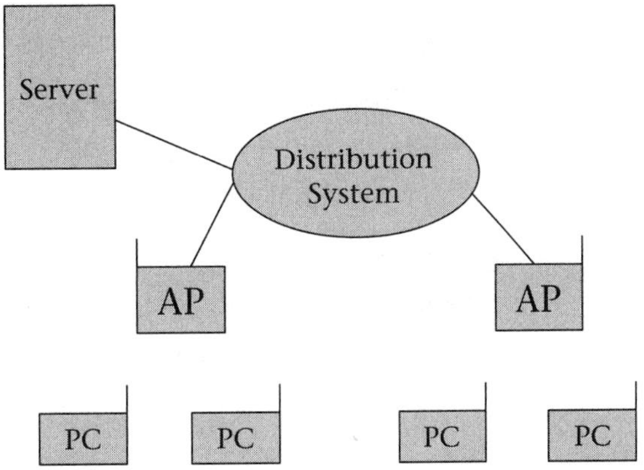

**FIGURE 18.5**
ESS topology

Server
Distribution System
AP
AP
PC
PC
PC
PC

#### Unmanaged Wireless Networks

The topology for an unmanaged network is called **Ad-Hoc**. In Ad-Hoc topology, the LAN is made of wireless devices without any access point. In this topology each device communicates directly with other devices, as shown in Figure 18.6.

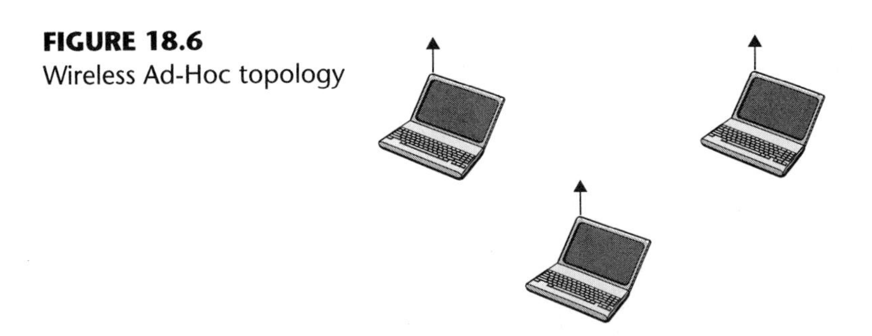

**FIGURE 18.6**
Wireless Ad-Hoc topology

## 18.3 Wireless LAN Technology

Two types of technologies are used for transmission of information in WLANs: Infrared (IR) and Radio Frequency (RF).

#### Infrared Technology

Infrared technology is suitable for indoor WLANs because infrared rays cannot penetrate through walls, ceilings, or other obstacles. In

**infrared (IR)** technology, the transmitter and receiver should see each other (be in line of sight) just like the remote control of a television set. In an environment where there are obstacles, for example, buildings, walls, and so forth, between the transmitter and a receiver, the transmitter may use diffused IR. However, most WLANs use RF technology.

### Radio Frequency Technology

There are two types of RF signals in use for transmission of information: Narrowband and Spread Spectrum.

***Narrowband Signal.*** **Narrowband signal** refers to a signal with a narrow spectrum, as shown in Figure 18.7. In narrowband, information is transmitted at a specific frequency, such as AM or FM radio waves.

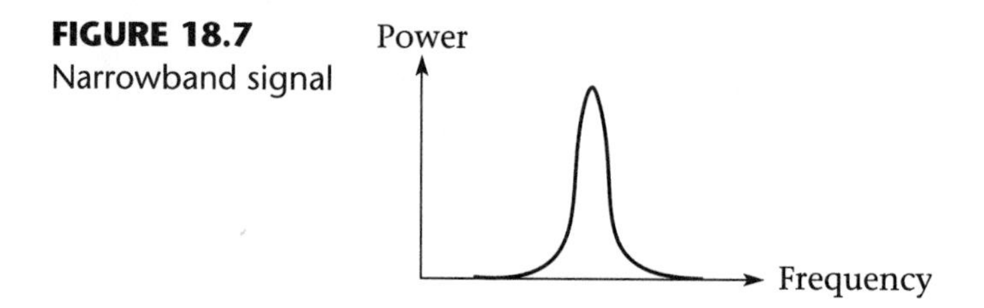


**FIGURE 18.7**
Narrowband signal

***Spread Spectrum.*** In **spread spectrum** technology, information is transmitted over a range of frequencies, as shown in Figure 18.8. Spread spectrum is the most popular technology used for WLAN.

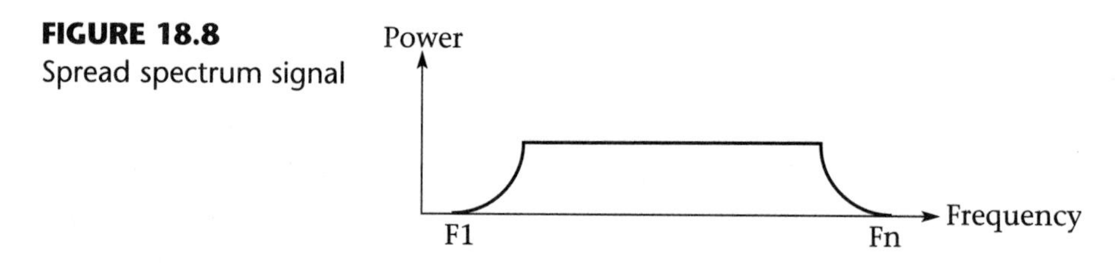


**FIGURE 18.8**
Spread spectrum signal

There are certain advantages to using the spread spectrum band over the narrowband:

- In spread spectrum technology, information is transmitted at different frequencies.
- It is hard to jam a spread spectrum signal; the signal cannot be disrupted by other signals.
- Interception of spread spectrum signals is more difficult than interception of a narrowband signal.
- Noise is less disruptive in spread spectrum signals than in narrowband signals.

## 18.4 WLAN Standards (IEEE 802.11)

The IEEE 802.11 committee has approved several standards for WLAN. The standards define functions of the Medium Access Control (MAC layer) and the Physical layer. Table 18.1 shows the Physical layer and Data Link layer for various WLAN standards.

**TABLE 18.1** IEEE 802.11 Standards and Their Data Rates

| Logical Link Control | | | | |
|---|---|---|---|---|
| MAC Layer<br>CSMA/CA and Point Coordination Function | | | | **Data Link Layer** |
| IEEE 802.11<br>DSSS and FHSS<br>1 and 2 Mbps | IEEE 802.11b<br>HR-DSSS<br>1, 2, 5.5, and 11 Mbps | IEEE 802.11a<br>OFDM<br>6, 12, 18, 24, 36, 45, and 54 Mbps | IEEE 802.11g<br>OFDM<br>6, 12, 18, 24, 36, 45, and 54 Mbps | **Physical Layer** |
| **ISM band** | **ISM band** | **U-NII band** | **ISM band** | |

## 18.5 Wireless LAN Physical Layers

In general, the Physical layer performs the following functions:

- Modulation and encoding: Information is modulated and then transmitted to the destination.
- Supports multiple data rate.
- Senses the channel to see if it is clear (carrier sense).
- Transmits and receives information.

### IEEE 802.11 Physical Layer

The IEEE 802.11 standard operates in the ISM band and is designed for data rates of 1 Mbps and 2 Mbps. IEEE 802.11 defines two types of radio frequency for data transmission: **Frequency Hopping Spread Spectrum (FHSS)** and **Direct Sequence Spread Spectrum (DSSS)**.

***ISM Band.*** The Federal Communications Commission (FCC) allocates separate ranges of frequencies to radio stations, TV stations, telephone companies, and navigation and the military agencies. The FCC also allocates a band of frequencies called the **Industrial, Scientific, and Medical Band (ISMB)** for industrial, research, and medical applications. The use of the ISM band does not require a license from the FCC (with power of transmission up to one watt). Figure 18.9 shows the ISM band.

902 MHz 928 MHz 2.4 GHz 2.48 GHz 5.725 GHz 5.85 GHz

| Industrial Band<br>I-band | | Scientific Band<br>S-band | | Medical Band<br>M-band |
|---|---|---|---|---|

**FIGURE 18.9**
ISM band

***Frequency Hopping Spread Spectrum (FHSS).*** The IEEE 802.11 standard recommends the scientific band (2.4 GHz to 2.483 GHz) of the ISM band for WLAN. This band is divided into 79 channels of 1-MHz each. The transmitter sends each part of its information on a different channel. Figure 18.10 shows the frequency hopping spectrum.

The order of the channels used by the transmitter to transmit information to the receiver is predefined and the receiver knows the order of the incoming channels. For example, the transmitter may use a hop pattern of 3, 6, 5, 7, and 2 for transmitting information. The hop sequence can be selected during the installation of the WLAN. The FCC requires that a transmitter spend a maximum of 400 ms in each frequency for the transmission of data (this time is called the *Dwell time*) and use 75 hop patterns (each hop is one channel). The FCC also requires that the maximum power for the transmitter in the United States should not exceed one watt.

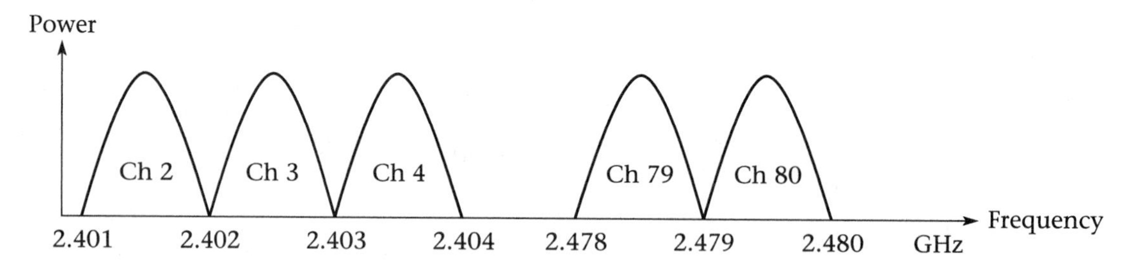


**FIGURE 18.10**
Frequency hopping spread spectrum

FHSS is more immune to noise because information is transmitted at different channels. In FHSS, if one channel is noisy, it can retransmit information on another channel. Table 18.2 shows FHSS channels and frequencies.

**TABLE 18.2** FSSH Channels and Frequencies

| Channel Number | Frequency in GHz |
|---|---|
| 2 | 2.402 |
| 3 | 2.403 |
| 4 | 2.404 |
| . | . |
| . | . |
| . | . |
| 80 | 2.480 |

***Direct Sequence Spread Spectrum.*** In DSSS, before transmission each bit of information is broken down to a pattern of bits called a **chip**.

For generating chip bits, each information bit is exclusive-OR'ed with pseudo random code, as shown in Figure 18.11. The output of the exclusive-OR for each data bit is called *chip bits*, which are modulated and then transmitted. This method creates a higher modulation rate because the transmitter transmits the chip bit over a larger frequency spectrum. Figure 18.12 shows the transmission section of the Physical layer. The receiver uses the same pseudo random code to decode the original data. A larger chip sequence generates a larger frequency band. IEEE 802.11 recommends 11 bits for each chip.

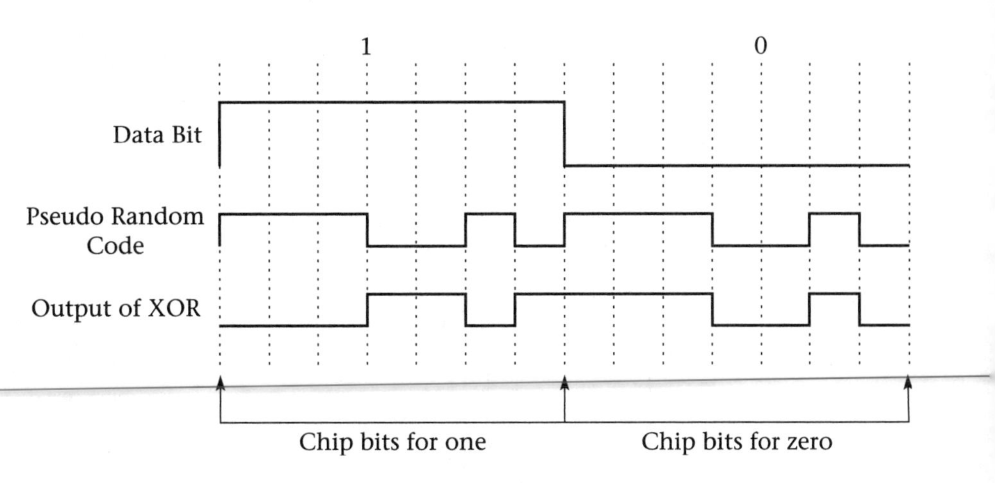


**FIGURE 18.11**
Generation chips

The DSSS supports two types of modulation: Differential Binary Phase Shift Keying (DBPSK), which is used for a data rate of 1 Mbps, and Differential Quadrature Phase Shift Keying (DQPSK), which is used for data rates of 2 Mbps.

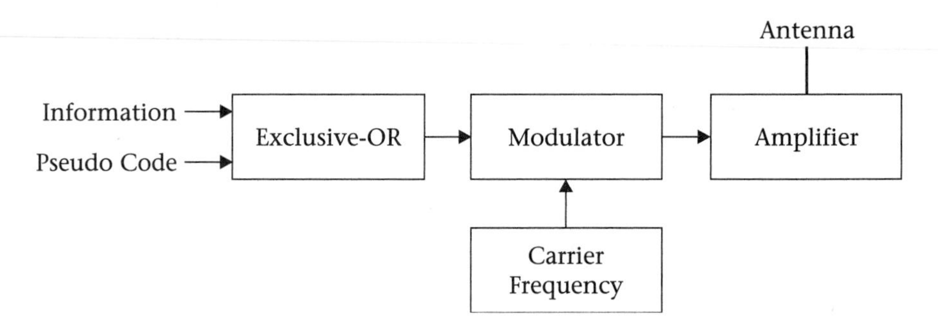


**FIGURE 18.12**
Transmission section of Physical layer

## IEEE 802.11b Physical Layer

The IEEE 802.11b standard extends the DSSS Physical layer of 802.11 to provide higher data rates of 5.5 Mbps and 11 Mbps. 802.11b uses **Complementary Code Keying (CCK)** to support the two new data rates: 5.5 Mbps and 11 Mbps, in addition to 1 Mbps and 2 Mbps. CCK converts the information to sixty-four 8-bit code words for transmission. These code words have unique characteristics so that the receiver can distinguish them from each other, even if noise and multi-path interference changes these code words. In order to transmit information at rates of 5.5 Mbps, the Physical layer uses CCK to transmit each 4 bits per signal with a transmitting frequency of 1.375 M symbols per second. To obtain 11 Mbps, the Physical layer transmits 8 bits per signal with 1.375 M signals/second. IEEE 802.11b data rates are shown in Table 18.3.

**TABLE 18.3** IEEE 802.11 Data Rates and Modulations

| Data Rate Mpbs | Code Length | Modulation Method | Signal Rate | Bits/Signal |
|---|---|---|---|---|
| 1 | 11 chip bits | BPSK | 1 Mbps | 1 |
| 2 | 11 chip bits | QPSK | 1 Mbps | 2 |
| 5.5 | 8 CCK | QPSK | 1.375 M | 4 |
| 11 | 8 CCK | QPSK | 1.375 M | 8 |

***IEEE 802.11b Channels.*** The IEEE 802.11b standard defines 11 channels that may be used. Each channel is represented by its center frequency, as shown in Table 18.4. The center frequency of each channel is separated from adjacent channels by 5 MHz. The bandwidth of each channel is 16 MHz, and using adjacent channels will cause interference. IEEE 802.11b supports three non-overlapping channels: 1, 6, and 11 to overcome interference.

**TABLE 18.4** IEEE 802.11b Channels and Frequencies

| Channel Number | Center Frequency (MHz) |
|---|---|
| 11 | 2462 |
| 10 | 2457 |
| 9 | 2452 |
| 8 | 2447 |
| 7 | 2442 |
| 6 | 2437 |
| 5 | 2432 |
| 4 | 2427 |
| 3 | 2422 |
| 2 | 2417 |
| 1 | 2412 |

To avoid overlapping of channels, the most common channels that are used are channels 1, 6, and 11. Figure 18.13 shows an arrangement of APs in a three-story building, where each row represents one floor with its access point and channel number.

| AP(1) | AP(6) | AP(11) |
|---|---|---|
| AP(11) | AP(1) | AP(6) |
| AP(6) | AP(11) | AP(1) |

**FIGURE 18.13**
Locations of access points in a three-story building

## IEEE 802.11a Physical Layer

IEEE 802.11a is an enhancement of IEEE 802.11 and uses DSSS technology. IEEE 802.11a operates in **Unlicensed National Information Infrastructure Band (U-NII).** A U-NII band consists of three 100-Mhz frequency bands, as shown in Figure 18.14. The Physical layer of IEEE 802.11a uses **Orthogonal Frequency Division Multiplexing (OFDM)** for transmitting data at higher rates. IEEE 802.11a offers data rates of 6, 9, 12, 18, 24, 36, 48, and 58 Mbps. The Physical layer can use any of the BPSK, QPSK, 16QAM, and 64QAM for modulation depending on the data rate. The frequency of operation is made of twelve 20-Mhz channels, as shown in Table 18.5.

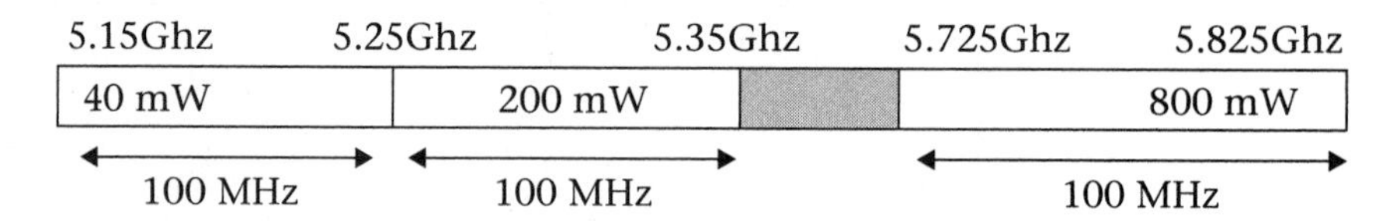


**FIGURE 18.14**
Frequency of operation for U-NII band

**TABLE 18.5** IEEE 802.11a Channels and Transmission Powers

| Frequency (kHz) | Channel Number | Maximum Output Power (mW) |
|---|---|---|
| 51.6–51.80 | 1 | 50 Low band |
| 51.80–52.00 | 2 | 50 Low band |
| 52.00–52.20 | 3 | 50 Low band |
| 52.20–52.40 | 4 | 50 Low band |
| 52.40–52.60 | 5 | 250 Middle band |
| 52.60–52.80 | 6 | 250 Middle band |
| 52.80–53.00 | 7 | 250 Middle band |
| 53.00–53.20 | 8 | 250 Middle band |
| 57.25–57.47 | 9 | 1000 High band |
| 57.45–57.65 | 10 | 1000 High band |
| 57.65–57.85 | 11 | 1000 High band |
| 57.85–58.05 | 12 | 1000 High band |

IEEE 802.11a offers 12 channels and each channel has a band of 20 MHz. Each channel is divided into 52 subchannels of 300 kHz each. OFDM uses 48 channels for transmission of information in parallel and 4 channels are used for pilot data. IEEE 802.11a highband is used for building-to-building applications and the low and middle bands are used for in-building applications. The IEEE 802.11a supports Forward Error Correction (FEC). This means that the transmitter sends the second copy of the information to the destination in case part of the primary information was lost. The receiver can then recover any lost data.

***Comparing IEEE 802.11a and IEEE 802.11b.*** One of the disadvantages of IEEE 802.11a is that OFDM consumes more power than IEEE 802.11b, therefore reducing the life of a notebook battery. 802.11a and 802.11b are also incompatible. IEEE 802.11b operates in ISM band, whereas IEEE 802.11a operates in the U-NII band. IEEE 802.11b covers an area with a diameter of 100 meters whereas IEEE 802.11a covers a 50-meter diameter area. This means that using IEEE 802.11a, a customer needs to install twice as many access points to ensure the same coverage area with IEEE 802.11b. IEEE 802.11a supports eight non-overlapping channels, which means it can use eight access points and support 512 users in one area. IEEE 802.11b supports three non-overlapping channels, which means it can use three access points in one area with the maximum of 192 users.

### IEEE 802.11g Physical Layer

IEEE 802.11a and IEEE 802.11b define different standards, which are not compatible with each other. IEEE 802.11b operates at 2.4 GHz and transmits data at the rate of 11 Mbps using DSSS technology, whereas IEEE 802.11a uses OFDM. IEEE.80211g operates in 2.4 GHz using DSSS and ODFM for transmission of information.

Table 18.6 shows the characteristics of IEEE 802.11 a, b, and g.

**TABLE 18.6** Characteristics of IEEE 802.11 a, b, and g

| | 802.11b | 802.11g | 802.11a |
|---|---|---|---|
| Available RF Channel | 3 non-overlapping | 3 non-overlaping | 8 non-overlapping |
| Frequency Band | 2.4 GHz | 2.4 GHz | 5 GHz |
| Maximun Data Rate | 11 Mbps | 54 Mbps | 54 Mbps |
| Rate/Typical Range | 100 ft at 11 Mbps<br>300 ft at 1 Mbps | 50 ft at 54 Mbps<br>150 ft at 11 Mbps | 40 ft at 54 Mbps<br>300 ft at 6 Mbps |

## 18.6 Physical Layer Architecture

Figure 18.15 shows the Physical layer architecture of the IEEE 802.11 familly. The Physical layer is divided into two sublayers: the Physical Layer Convergence Procedure (PLCP) and the Physical Medium Dependent.

**FIGURE 18.15**
Physical layer architecture

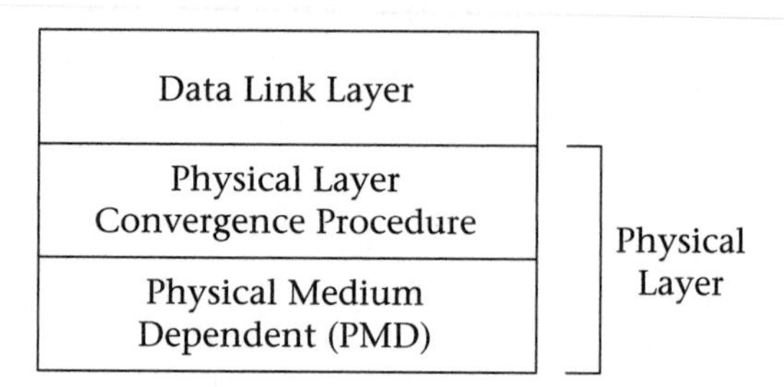


The MAC layer transfers a frame to PLCP. The PLCP adds its header to the MAC frame and transmits the frame to PMD for transmission. The IEEE 802.11 defines frequency hopping and DSS for the Physical layer. Figure 18.16 shows the PLCP for DSSS and Figure 18.17 shows the PLCP frame format for FHSS.

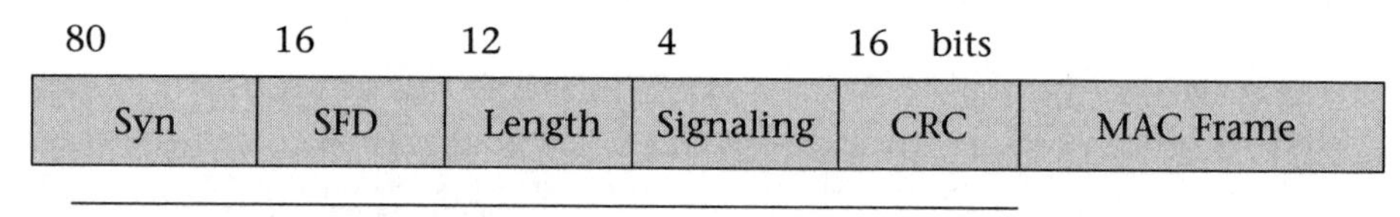


**FIGURE 18.16**
PLCP frame format for DSSS

| 128 | 16 | 8 | 8 | 16 | 16 bits | |
|---|---|---|---|---|---|---|
| Syn | SFD | Signaling | Service | Length | CRC | MAC Frame |

**FIGURE 18.17**
PLCP for FHSS

The following describes the function of each field in the PLCP frame format:

**Sync field:** Sync field is used for synchronization and it is 80 bits of alternating 0's and 1's.

**SFD field:** SFD field is 00001100 10111101.

**Length field:** Defines the length of PLCP in bytes.

**Signaling:** This field indicates to the Physical layer the modulation type that must be used for transmission of the frame. The data rate is calculated as follows:

$$\text{Data Rate} = \text{Value in signal field} * 100 \text{ Kbps}$$

The value of the signaling field and its data rate is shown in Table 18.7.

**TABLE 18.7** Value of Signaling Field and Data Rate

| Value of Signal Field in Hex | Data Rate in Mbps |
|---|---|
| 0A | 1 |
| 14 | 2 |
| 37 | 5.5 |
| 6E | 11 |

**CRC-16:** Used for error detection in the PLCP header.

**Service field:** This field is reserved.

IEEE 802.11b has two types of PLCP preamble headers: short PLCP preamble headers and long PLCP preamble headers. The short PLCP preamble header has 56 bits for the synchronization field and the long PLCP preamble header has a 128-bit synchronization header.

**Interframe Space**

In general, the interframe space enables the receiver to complete the frame before the next frame comes. The IEEE 802.11 defines three types of Interframe Spaces (IFS) among the frames transmitted between source and destination:

1. Short Interframe Space (SIFS): This interframe space is used for an immediate response such as ACK, CTS, and RTS.
2. Distribution Coordination Function Interframe Gaps Space (DIFS): DIFS is used for the spacing of data frames.
3. Point Coordination Function Interframe Space (PIFS): This interval is used for the point condition access method and the gap is used for polling of the client. The client should respond after this time.

## 18.7 WLAN Medium Access Control (MAC)

The Medium Access Control (MAC) layer performs the following functions:

- Supports multiple Physical layers
- Supports access control
- Fragmentation of frame
- Roaming

IEEE 802.11a defines distribution coordination function (Carrier Sense Multiple Access with Collision Avoidance CSMA/CA) and Point Coordination Function as methods for a station to access Wireless LANs.

**Carrier Sense Multiple Access with Collision Avoidance**

**Carrier Sense Multiple Access with Collision Avoidance (CSMA/CA)** is similar to CSMA/CD. In CSMA/CA when a station wants to transmit a frame, first it listens to the medium. If there is no traffic, it continues to wait for a short interframe space and if there is still no traffic on the medium, then the station will start transmitting. Otherwise, it has to wait for the medium to become clear. Figure18.18 shows a CSMA/CA flowchart operation.

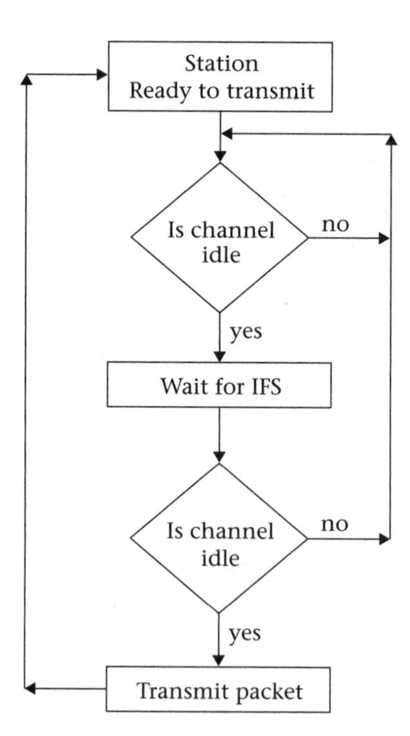


**FIGURE 18.18**
CSMA/CA flowchart

In order to reduce the probability of two stations transmitting simultaneously because they cannot hear each other, the standard defines a virtual carrier sense mechanism.

***Virtual Carrier Sense.*** Figure 18.19 shows two stations and one access point. Stations B and C are covered by an access point, but station B cannot cover station C. While the access point communicates with station B, C can listen to the AP and find out whether the medium is clear. Station C waits for the medium to be clear. When B is transmitting to the AP, C is not in the range to detect that B is transmitting to the AP. Therefore, C will see the medium as clear and start transmission. B and C will be transmitting to AP at the same time and cause a collision. Station C is a hidden station and there is therefore no physical connection to detect this collision. The following steps describe CSMA/CA operation and Figure 18.20 shows the CSMA/CA process:

1. Station B wants to transmit to the AP. It senses the medium, and if the medium is clear, it sends a short message to the AP, which is called the *Request to Send (RST)*. This message contains destination, source address, and size of the data to be transmitted.
2. If the AP is ready to communicate with B, the AP will send a *Clear to Send (CTS)* frame to B; otherwise it will send a busy frame. This signal can be detected by station C and is taken as a busy medium. Station B receives a CTS signal and then transmits its frame. The receiver acknowledges each frame transmitted by B.

**FIGURE 18.19**
CSMA/CA for WLAN

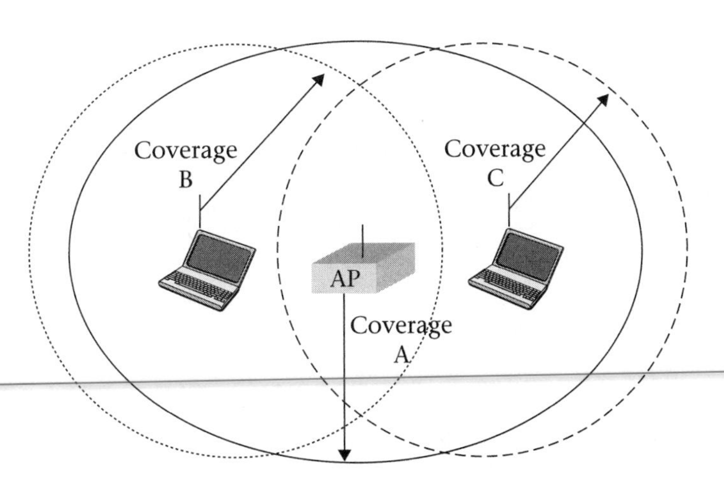

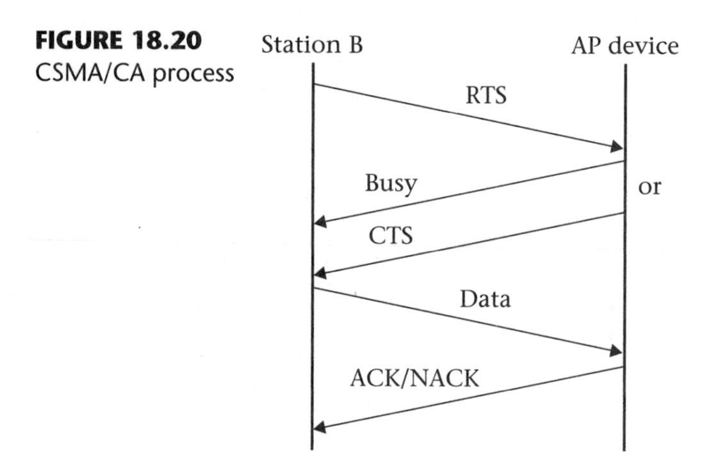


**FIGURE 18.20**
CSMA/CA process

**Point Coordination Function (PCF) Access Method** IEEE 802.11 provides the point coordination function as an optional access method. The AP polls each user for transmission. In the PCF method, the AP listens to the medium; if there is no traffic in the medium, it sends a **beacon frame** to all users indicating that the PCF method will be used for the access method and all users must stop transmission. Then the AP sends a poll frame to a specific user. If the user has any data, it will transmit the data; otherwise, the user will send a null frame to the AP.

## 18.8 MAC Frame Format

IEEE 802.11 defines three types of frames for WLAN: management frame, control frame, and data frame. Figure 18.21 shows MAC frame format for IEEE 802.11.

| 2 | 2 | 6 | 6 | 6 | 2 | 6 | 0-2312 | 4 bytes |
|---|---|---|---|---|---|---|---|---|
| Frame Control | Duration/ID | Address 1 | Address 2 | Address 3 | Sequence number | Address 4 | Information | CRC |

**FIGURE 18.21**
MAC frame for IEEE 802.11

The following describes the function of each field in the MAC frame of IEEE 802.11:

**Frame control:** The frame control field is 2 bytes and defines the types of frames, as shown in Figure 18.22.

| bits 2 | 2 | 4 | 1 | 1 | 1 | 1 | 1 | 1 | 1 | 1 |
|---|---|---|---|---|---|---|---|---|---|---|
| Protocol | Type | Subtype | To DS | From DS | More Fragment | Retry | Power Man | More Data | WEP | RS |

**FIGURE 18.22**
Control fields of MAC frame

The following describe the function of each field of Figure 18.22:

**Protocol:** Defines the protocol version. The current version is zero.

**Type:** Defines the type of the frame. 00 means management frame, 01 means control frame, and 10 means data frame.

**Subtype:** Defines the subframe in each type; there are several subtypes in each management frame.

**To DS and From DS Fields:** To DS and From DS define the direction of the frame and also the function of address fields in Figure 18.21. Table 18.8 describes the function of To DS, From DS, and address fields (Address 1, Address 2, Address 3, and Address 4 fields).

**TABLE 18.8** To DS and From DS

| To DS | From DS | Address 1 | Address 2 | Address 3 | Address 4 | Comments |
|---|---|---|---|---|---|---|
| 0 | 0 | DA | SA | BSS ID | Unused | Transferring a frame between two stations located in the same BSS or cell |
| 0 | 1 | DA | BSS ID | SA | Unused | Transferring a frame from DS to a station in a cell |
| 1 | 0 | BSS ID | SA | DA | Unused | Transferring a frame from a station to DS |
| 1 | 1 | Receiver Address | Transmitter Address | DA | SA | This is used for a wireless distribution system |

In Table 18.8, DA is the destination address (MAC address) and SA is the source address.

**More fragment:** The fragment field set to 1 indicates that more frames, belonging to the same application, are coming. This field set to zero indicates to the destination that the current frame is the last frame.

**Power management:** This bit set to 1 indicates that the transmitter is operating under power management.

**Wired Equipment Privacy (WEP):** This bit set to 1 indicates that a cryptographic algorithm has changed the information.

**Retry:** Retry set to1 indicates that this is a copy of the previous frame.

**RS:** Reserved bit.

**Duration/ID:** This field contains the duration value Network Allocation Vector (NAV) to inform other stations how long it will take for the source to complete its transmission (in microseconds). The other stations use the NAV value to defer their transmissions.

**Sequence number:** This field is divided into two fields: a 4-bit field and a 12-bit field. The first four bits indicate a fragment number and the 12-bit field indicates the sequence number of the frame.

## 18.9 WLAN Frame Types

The type field in Figure 18.22 defines the type of information in the frame. The IEEE defines three types of frame: management frames, control frames, and data frames.

### Management Frames

The management frame is used by a station to make a connection to the AP, to disconnect the station from the AP, and for timing and synchronization. Some of the management frames are as follows:

- **Association request** (subtype = 0000): Client sends a request to the frame for joining a BSS network.
- **Association response** (subtype = 0001): AP responds to the client request as to whether the AP is accepting this request.
- **Reassociation** (subtype = 0010): Sending a frame by a client when moving from one BSS to another.
- **Reassociation response** (subtype = 0011): In response to reassociation, AP sends a reassociation frame to the client on whether it accepts the client joining the AP to BSS.
- **Disassociation** (1010): Used by a client to terminate its association with BSS.
- **Beacon frame** (subtype = 1000): A beacon frame is transmited by the access point periodically to inform other wireless stations of its presence. This frame contains the following information:

The beacon interval that is used by the station to know when to wake up for the next beacon

Timestamp that is used for synchronization between the access point and the wireless station

Types of signaling, such as FHSS and DSSS

Supported data rates

- **Probe request frame (subtype = 0101):** Station requesting information from another station.

**Control Frames** Control frames are used for flow control such as positive acknowledgement (ACK), RTS (subtype = 1011), and CTS (subtype = 1100).

**Power Management** Most wireless LAN clients are laptop computers, and power saving is an important factor for a wireless client. Therefore, IEEE 802.11 defines two power management modes: Active or Continous Aware Mode (CAM), and Power Save Polling (PSP) "sleep" mode.

**Active Mode or Continous Aware Mode (CAM):** In this mode the wireless client uses full power.

**Power Save Polling (sleep):** In this mode the wireless client goes to sleep, meaning that the client uses less power and turns off power to the display, disks, and other peripherals that are not needed.

In PSP mode the following operations take place:

The client sends a frame to the AP informing the AP that the client is going to sleep.

The access point records the stations that are asleep.

The access point buffers the sleeping clients' frames.

The clients that are asleep continuously receive beacon frames.

The access point sends beacon frames with traffic indication map (TIM), which informs clients that were asleep that they have frames in the AP.

The clients that have frames in the AP switch their power mode to active mode.

The client sends a request for its frames.

**Data Frames** Data frames are used for transfer of information.

## 18.10 Station Joining a Basic Service Set (BSS)

For a client to associate with an access point, it must perform the probe phase, authentication phase, and association phase. The following describes each phase in detail:

**Probe phase:** When a station wants to join a BSS, the station needs to get synchronization information from the AP (clock value of AP). This can be accomplished in two ways: by passive scanning and by active scanning.

**Passive scanning:** In this method the station can receive a beacon frame, which is being sent out periodically by the access point. This beacon frame contains synchronization information.

**Active scanning:** In this method the station transmits a probe request frame to locate an access point and waits for a probe response; the probe response frame contains the synchronization clock.

**Association phase:** If the authentication phase is completed successfully, the station will send an association request packet to the access point. The access point adds the station to its association table. A station can associate with only one access point at a time.

The authentication phase is covered in the WLAN security (Section 18.14).

## 18.11 Roaming

When a station is moving from one cell to another cell without losing connection, it is called **roaming**. In WLAN, moving from one cell to another must be performed between the packet transmissions, meaning that the packet must be transmitted completely before moving to another cell.

## 18.12 Wi-Fi Certification

There are many manufacturers of wireless equipment and these devices should be interoperable. Therefore the Wireless Ethernet Alliance (WECA) offers certification for wireless equipment interoperability, referred to as **Wireless Fidelity (Wi-Fi)**.

## 18.13 WLAN Signal Distortion

The RF signal can be distorted while going through physical obstacles such as a wall, a ceiling, or by multi-path fading.

**Multi-Path Fading**

**Multi-path fading** occurs when a communication signal transmitted through the air takes multiple paths to reach the destination, as shown in Figure 18.23. The receiver receives multiple signals with different delays and amplitudes. When the delay of the transmitted signal increases, it will distort the signal, causing data to be corrupted and reducing data throughput.

**FIGURE 18.23**
Multi-path fading

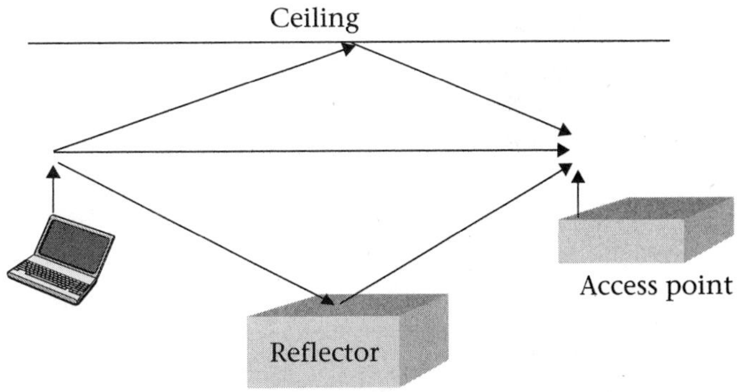

## 18.14 WLAN Security

A WLAN is less secure than a wired LAN. In a wired LAN, anyone who wants access to the network must make a physical connection. In WLANs, users can access the AP if the AP signal is detectable. The following methods are used for security of WLANs:

- Service Set Identifier
- MAC Address Filtering
- Wired Equivalent Privacy
- Authentication

**Service Set Identifier (SSID)**

In WLAN a **Service Set Identifier (SSID)** is assigned to each AP and the client computer must know the SSID of the AP in order to communicate with the wireless network. In a building with several APs the client computer must be programmed with multiple SSIDs. The SSID can be any alphanumeric character from 2 to 32 characters. The SSID helps client stations to identify multiple access points.

**MAC Filtering** In this method, each AP can be programmed with a list of the MAC addresses of clients that are allowed access. If a client's MAC address is not on the list, then the AP denies the access.

**Wired Equivalent Privacy (WEP)** The IEEE 802.11 includes **Wired Equivalent Privacy (WEP)** in the MAC layer of WLAN in order to protect wireless communication. WEP uses a secret key that is shared between the WLAN access point and the users. The transmitter encrypts the data before transmission, as shown in Figure 18.24. At the transmitter side, a 24-bit initialization vector (IV) is appended to the secret key, and it is used as input to an RC4 algorithm to generate a keystream for encryption.

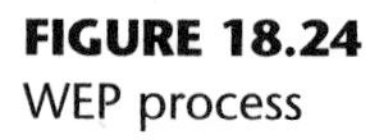

**FIGURE 18.24**
WEP process

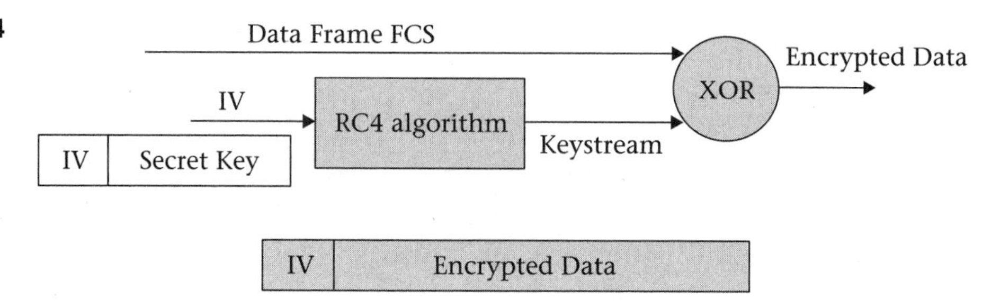


The size of the keystream is equal to the size of the frame. The keystream is exclusive-OR'ed with the data frame including FCS, resulting in encrypted data. The transmitter appends IV to the encrypted data and transmits to the receiver. At the receiver side, the reciever uses the IV with the secret key to generate the keystream in the same method as the transmitter side. The receiver uses this keystream and XOR's it with encrypted data, which results in decrypted data. Therefore, different values for the IV can be selected to generate different keystreams. If a secret key of 40 bits with a 24-bit IV results in a 64-WEP key, then a 104-bit secret key will result in a 128-bit WEP key.

**Authentication** **Authentication** is a process in which the AP is used to identify the WLAN client before giving permission for connection. The client sends an authentication frame and the access point responds by accepting the client or denying the client for connection. The authentication process might be done by the access point, or the AP might pass the client request to a RADIUS server for processing the client request based on a list of criteria that was set by the network administrator. IEEE 802.11 defines two types of authentication: open system authentication and shared key authentication.

***Open System Authentication.*** This is the simplest method of authentication for a WLAN. The access point and client set their own criteria for authentication. If a client has the SSID of the access point, the client sends an authentication frame with its secret key to the access point. If the secret key of the client matches the secret key of the access point, then the AP sends a positive response to the client and the client becomes associated with the AP.

***Shared Key Authentication.*** Figure 18.25 shows the shared key authentication process. In shared key authentication, the following steps take place:

1. Client sends a request to the AP for authentication.
2. AP sends a plain text (non-encrypted text) to clients.
3. Client uses its secret key with an initialization vector (IV) to generate a keystream, and then the client uses this keystream and encrypts the plain text.
4. Client sends encrpted text with IV to the AP.
5. AP uses the IV and its secret key to generate keystream, and then keystream decrypts the encypted text transmitted by the client.
6. If the encrypted text is the same as the plain text, then the AP responds with positive acknowledgement for the client to be associated with the access point; otherwise it sends a negative response and denies authentication of the client.

In shared key authentication the clients and AP use the same key, and it is easy for the unauthorized client to obtain the shared key by using special sniffer software, such as AirSnort or WEPcrak. Therefore, the network administrator needs to change the shared secret key frequently. In a large network, this is hard to do. To overcome this problem, IEEE 802.1x offers Extensible Authentication Protocol (EAP), which is covered in the network security chapter (Chapter 23).

**FIGURE 18.25** Shared authentication process

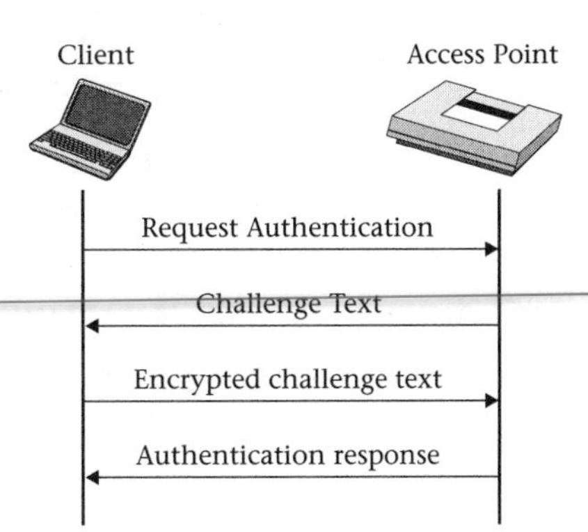

## Summary

- The standard for wireless LANs is the IEEE 802.11 series.
- Components of WLANs are WLAN NIC, transmitter, AP device, and NOS.
- WLANs use RF and IR signals for transmitting information.
- The IR ray mode of transmission is used indoors and cannot penetrate through obstacles.
- IEEE 802.11 defines Frequency Hopping Spread Spectrum (FHSS) and Direct Sequence Spread Spectrun (DSSS) for the Physical layer using RF signals.
- IEEE 802.11 defines Carrier Sense Multiple Access with Collision Avoidance (CSMA/CA) for the access method.
- WLANs use ISM-band and U-NII band.
- WLAN topologies are managed wireless network and unmanaged wireless network.
- Managed wireless network topologies are Basic Service Set (BSS) and Extended Service Set (ESS).
- Wireless unmanaged networks are called *ad-hoc networks*.
- Narrow band signal refers to a signal with a narrow spectrum.
- Spread spectrum signal refers to a signal with a range of frequencies.
- IEEE 802.11, IEEE 802.11b, and IEEE 802.11g operate in the ISM band.
- The data rates for IEEE 802.11b are 1, 2, 5.5, and 11 Mbps.
- IEEE 802.11b offers three non-overlapping channels: 1, 6, and 11.
- IEEE 802.11b uses complementary code keying for transmission of information.
- The IEEE 802.11 family uses CSMA/CA and point of coordination function (PCF) for access methods.
- IEEE 802.11g operates in 2.4 GHz and uses DSSS and OFDM for transmission of information.
- The wireless alliance offers certification for wireless equipment interoperability and it is refered to as *Wi-Fi.*
- WLAN uses SSID, MAC filtering, WEP, and authentication for security.
- IEEE 802.11 defines open system authentication and shared key authentication.

- The most popular authentication protocols are EAP and EAPP.
- IEEE 802.1x is an authentication standard for wireless and wired LAN.

## Key Terms

Access Point (AP)
Ad-Hoc
Authentication
Basic Service Set (BSS)
Beacon Frame
Carrier Sense Multiple Access with Collision Avoidance (CSMA/CA)
Cell
Chip
Complementary Code Keying (CCK)
Direct Sequence Spread Spectrum (DSSS)
Extended Service Set (ESS)
Frequency Hopping Spread Spectrum (FHSS)
Industrial, Scientific, and Medical Band (ISMB)
Infrared (IR)
Multi-Path Fading
Narrowband Signal
Orthogonal Frequency Division Multiplexing (OFDM)
Radio Frequency (RF)
Roaming
Service Set Identifier (SSID)
Spread Spectrum
Unlicensed National Information Infrastructure Band (U-NII)
Wired Equivalent Privacy (WEP)
Wireless Fidelity (Wi-Fi)
Wireless Local Area Network (WLAN)

## Review Questions

- **Multiple Choice Questions**

1. The transmission media for WLAN is ______.
   a. air
   b. cable
   c. optical cable
   d. UTP
2. The maximum transmission power for WLAN is ______.
   a. 2 watts
   b. 1 watts
   c. 3 watts
   d. 10 watts

3. The access method for WLAN is ______.
   a. SMA/CD
   b. SMA/CA
   c. token passing
   d. full-duplex
4. An advantage of spread spectrum signal over narrowband signals is that ______
   a. spread spectrum has more power.
   b. spread spectrum signal uses a range of frequency.
   c. narrowband signal uses a range of a frequency.
   d. spread spectrum uses a single frequency.
5. The area covered by an access point is called a ______.
   a. frame
   b. token
   c. cell
   d. chip
6. Of following technologies, ______ is used for WLAN.
   a. infrared
   b. radio frequency
   c. a and b
   d. digital signal
7. The IEEE standard for WLAN is ______.
   a. IEEE 802.10
   b. IEEE 802.11
   c. IEEE 802.12
   d. IEEE 802.13
8. EEE 802.11b offers data rates of ______.
   a. 1 and 2 Mbps
   b. 1, 2, and 11 Mbps
   c. 54 Mbps
   d. 11 and 45 Mbps
9. The non-overlapping channels for IEEE 802.11b are ______.
   a. 1, 10, and 11
   b. 1, 6, and 11
   c. 1, 4, and 10
   d. 3, 4, and 5
10. IEEE 802.11b operates in the ______ band.
    a. U-NII
    b. ISM
    c. B
    d. C

11. IEEE 802.11a operates in the ______ band.
    a. U-NII
    b. ISM
    c. B
    d. C
12. IEEE 802.11g uses ______ for transmitting information.
    a. DSSS and FHS
    b. DSSS and OFDM
    c. DSSS and CCK
    d. FHS and OFDM
13. A beacon frame is transmitted by ______.
    a. client station
    b. access point
    c. client and access point
    d. distributed system
14. DSSS uses ______ chip bits.
    a. 11
    b. 12
    c. 15
    d. 20
15. IEEE 802.11b uses ______ technology for transmission of information.
    a. CCK
    b. OFDM
    c. FHS
    d. HR-DSSS

## • Short Answer Questions

1. What does WLAN stand for?
2. What are the components of WLAN?
3. What are the WLAN topologies?
4. What is a cell?
5. What is an AD-Hoc topology?
6. Explain narrowband signal and spread spectrum signal.
7. What is the function of an access point device in WLAN?
8. What are the advantages of a spread spectrum signal over a narrowband signal?
9. What are IEEE standards for WLANs and their data rates?

10. What does OFDM stand for?
11. Explain OFDM operation.
12. What does DSSS stand for?
13. Explain DSSS operation.
14. What are the data rates for IEEE 802.11b?
15. How many non-overlapping channels are offered by IEEE 802.11b?
16. What is the maximum data rate for IEEE 802.11a?
17. What is the range of frequencies for U-NII band?
18. What are the types of access methods for WLANs?
19. What are the types of frame formats for WLANs?
20. What is the function of the association request frame?
21. Which devices transmit beacon frames?
22. What is the service set identifier?
23. What does CSMA/CD stand for?
24. What does WEP stand for?
25. What is the function of Wi-Fi?
26. What are the causes of signal distortion?
29. Explain multi-path fading.
28. Explain FHSS operation.
29. What is the maximum transmitter power for WLAN?
30. Explain access methods for WLAN.
31. What does ISM stand for?

C H A P T E R

# Bluetooth Technology

**OUTLINE**



**OBJECTIVES**

After completing this chapter, you should able to:

- Describe the application of Bluetooth technology
- Discuss the Bluetooth topology
- Explain the Bluetooth protocol architecture
- Discuss the Bluetooth packet format
- Explain the function of each layer of the Bluetooth protocol architecture

**INTRODUCTION**

In 1994, Ericsson Mobile Communication of Sweden began research to investigate the use of radio frequency for communication between mobile phones, wireless devices, and PCs. They concluded that all portable devices can be linked using radio frequencies. **Bluetooth** is named after the Danish King who ruled from 940 to 981 AD and who is remembered for unifying Denmark and Sweden.

Ericsson, IBM, Nokia, and Toshiba formed the Bluetooth Special Interest Group (SIG) in 1998; and Motorola, Microsoft, Lucent, and 3Com joined in 1999. Bluetooth is intended for devices to be connected wirelessly to each other over a maximum distance of 10 meters. The applications of Bluetooth are as follows:

- Connecting keyboards, mice, and other peripherals to a PC without using wires.
- Enabling mobile phones to communicate with laptops and Personal Digital Assistants (PDA) wirelessly. For example, a laptop computer can access the Internet in any location by using a mobile phone that has modem capability.

## 19.1 Bluetooth Topology

Bluetooth offers two types of topologies: **Piconet** and **Scatternet.**

**Piconet Topology**

The basic topology of Bluetooth is called **Piconet.** Piconet is a collection of up to seven devices, which are connected in an **Ad Hoc** fashion as shown in Figure19.1. In this figure, one of the devices will be the master, and the others are slaves. The function of the master is to set up a hopping pattern to the slaves, and assigning time slots to the slaves for transmission. The slaves communicate only with the master and each master can be connected to seven slaves simultaneously or 200 inactive slaves (parked slaves) in a Piconet topology. Each device in Piconet topology can become either a master or a slave. In Piconet topology the active devices are assigned a 3-bit number (from 1 to 7) that is called the **Active Member Address (AMA);** and devices that are parked are assigned an 8-bit number called the **Parked Member Address (PMA).**

**Scatternet Topology**

When a group of Piconets communicate with each other, it is called a **Scatternet.** Figure 19.2 shows a diagram of Scatternet topology. Two

**FIGURE 19.1**
Piconet topology

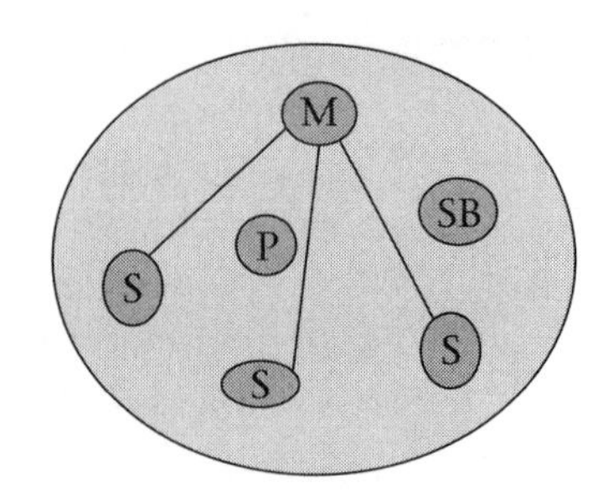


M: Master station
S: Slave station
P: Parked station (inactive)
SB: Standby station

**FIGURE 19.2**
Scatternet topology

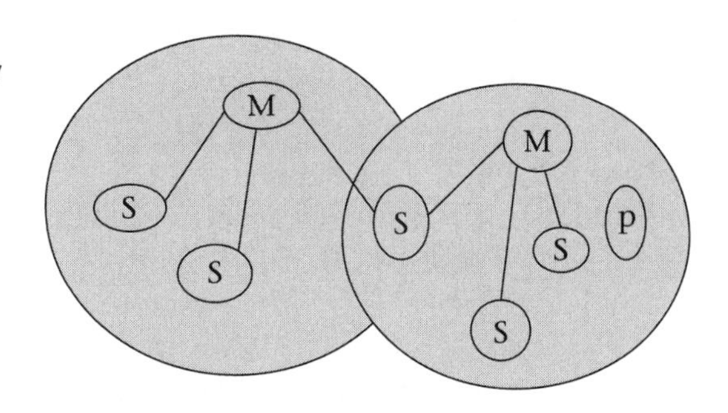


Piconets can create a Scatternet if one slave located in both Piconets acts as a bridge between the two Piconets.

## 19.2 Bluetooth Protocol Architecture

Figure 19.3 shows the Bluetooth protocol architecture; it is made of numerous layers and protocols that do not follow the TCP/IP protocol or the OSI model.

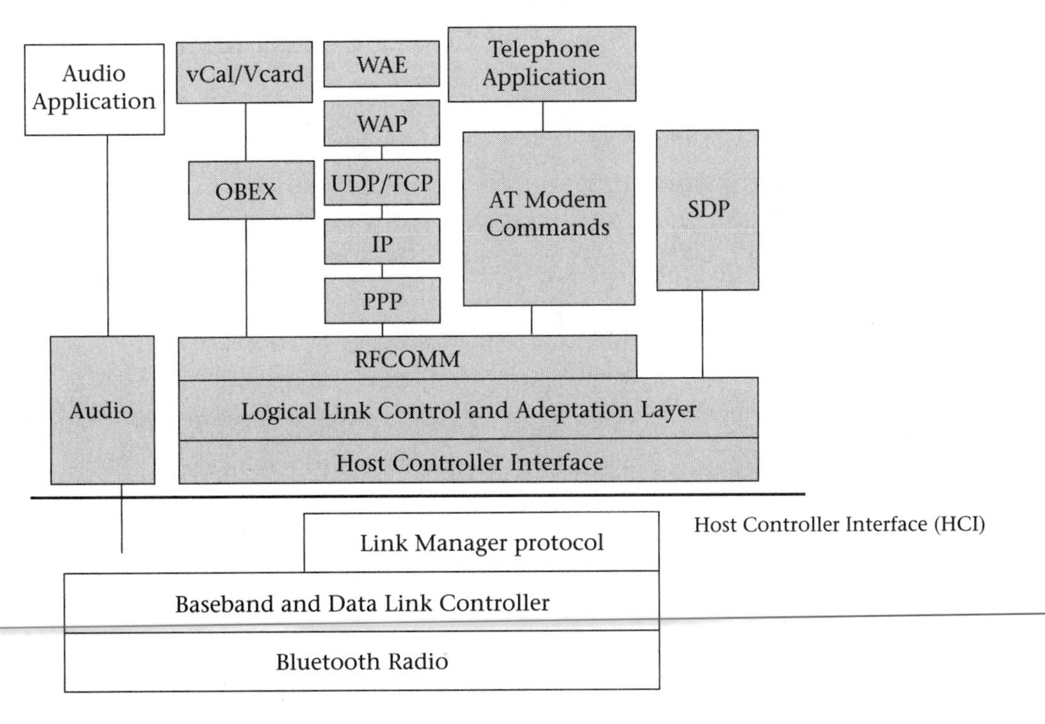


**FIGURE 19.3**
Bluetooth protocol architecture

Next, we describe the function of each layer of the Bluetooth protocol architecture.

**Bluetooth Radio Layer**

Bluetooth operates in the ISM band in the 2.40–2.480-GHz range. The ISM frequency spectrum is divided into 79 channels of 1 MHz each. Bluetooth uses frequency shift keying for modulation; each signal therefore represents one bit, which results in a data rate of 1 Mbps. Bluetooth uses FHSS with Time Division Duplexing (TDD) for communication between slave and master. The hopping rate is 16,000 hops per second, and the time between two hops, called the *interval time,* is 625 microseconds (time slot). Bluetooth devices can operate at three different power levels; Table 19.1 shows Bluetooth transmission power and distance between slaves and master.

**TABLE 19.1** Bluetooth Transmission Power

| Power Class | Max Power | Min Power | Distance from Master |
|---|---|---|---|
| Class 1 | 100 mW | 1 mW | 100 meters |
| Class 2 | 2.5 | 0.25 mw | 10 meters |
| Class 3 | N/A | 1 mW | N/A |

**Baseband and Link Control**

The functions of the Baseband and Link Control layers are to manage the radio layer operation and physical channels. The Baseband and Link Control layers perform the following functions:

1. Accepting bits from the Radio layer and forming a frame. Sending bits to the Radio layer for transmission
2. Managing physical channels
3. Managing physical addressing
4. Error control
5. Packet sequence numbering
6. Payload encryption
7. Packet ACK/NAK
8. FEC decoding
9. Enabling the Bluetooth Radio layer to form Piconet with other Bluetooth devices
10. Controlling frequency hopping and synchronization

11. Managing Synchronous Connection Oriented (SCO) and Asynchronous Connectionless Links (ACL).

### Link Controller

The function of the **link controller** is to establish a connection between the master and the slave and to determine the state of a station. A station can be in standby state or connection state. In a connection state the slave and master exchange information, and the standby state is the low-power consumption mode.

***Connection Setup.*** In order for a Bluetooth device (master) to make a connection to another device, the device must be in standby mode.

***Standby mode.*** A device in **standby mode** wakes up every 1.28 seconds and listens for pages or inquiry messages.

***Page message.*** The page message is sent over 32 different frequencies; the first page message is sent 128 times over the first 16 frequencies, and if there is no response from any device, the master sends another page message 128 times over the next 16 frequencies.

***Inquiry message.*** If the address of the device is unknown, then an inquiry message is needed before paging. Figure 19.4 shows the process that takes place for two devices to establish a connection.

**FIGURE 19.4**
Bluetooth connection process

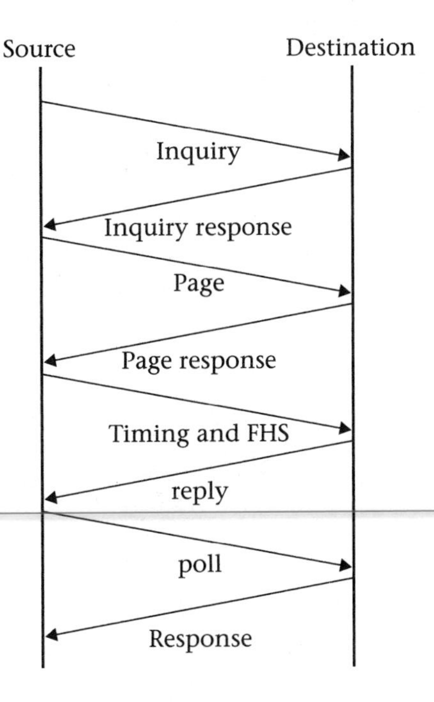

The following steps describe the connection process shown in Figure 19.4:

1. The source sends an inquiry packet that includes its address and clock to all the devices in its range.
2. The destination receives the inquiry packet, and responds with an inquiry response packet. This response contains the address and clock of the destination.
3. The source sends a page packet to the destination and becomes the master.
4. The destination responds with a page packet and becomes the slave.
5. The master sends the FHS to the slave.
6. The slave responds to the FHS packet.
7. The master sends a poll packet to verify that the slave has switched to the frequency hopping pattern and timing of the master.
8. The slave responds to the master poll. They are now ready to exchange information.

### Link Management Protocol (LMP)

The following describes the functions of **Link Management Protocol (LMP)**:

1. To set up a link between Bluetooth devices.
2. To negotiate with other devices for selecting a packet size of 1 slot, 3 slots, or 5 slots for data transmission.
3. To manage power operations of Bluetooth devices such as class 1, class 2, or class 3.
4. Authentication.
5. Encryption.
6. To switch the role of slave to master and vice versa.
7. To close a connection between the master and slave.

### Logical Link Control and Adaptation Protocol (L2CAP)

The following describes the functions of L2CAP:

1. To create a logical link between devices.
2. To terminate a logical link.
3. Negotiating the quality of service between Bluetooth devices.

4. Multiplexing of multiple channels using a single Bluetooth link. This means that multiple protocols such as TCP/IP, OXEB, and SDP can be used simultaneously.
5. Segmentation of data packets that exceed the maximum transfer unit of the Bluetooth payload (2755 bytes).
6. Reassembly of incoming packets in the appropriate order.

**RFCOM**

The **RFCOM** protocol emulates the RS-232 serial protocol, which is used for transmitting characters to and from applications that are using the serial port for communication.

**Service Discovery Protocol (SDP)**

The **Service Discovery Protocol (SDP)** is used by a Bluetooth client to search for a service, or by a Bluetooth server to respond to an SDP request.

**Object Exchange Protocol (OBEX)**

The **Object Exchange Protocol (OBEX)** provides simple file transfers between mobile devices such as sending business cards, personnel calendar entries, and scheduling information.

**Host Controller Interface (HCI)**

The function of the **Host Controller Interface (HCI)** is to act as the interface between the higher layer of Bluetooth and the lower layer.

## 19.3 Bluetooth Physical Links

The Bluetooth device supports two types of connections: **Synchronous Connection Oriented** (SCO) and **Asynchronous Connectionless** (ACL).

**Synchronous Connection Oriented (SCO)**

The **Synchronous Connection Oriented (SCO)** link is used for voice traffic that requires a constant bit rate. The master reserves time slots for transmission, and in this type of connection the master establishes a connection to the slave by transmitting a Link Management (LM) packet to the slave. The LM packet contains reserved time slots. Slaves then respond to the master with a time slot that was reserved. A master can support up to three SCO links simultaneously with single or multiple slaves, and a slave can support up to three SCO links simultaneously to the same master. SCO uses FEC for error corrections, so there is no need for retransmission of packets. Bluetooth specifies three types of High quality Voice packet (HV): HV1, HV2, and HV3. HV1 carries 10 bytes, which are protected by

1/3 FEC; HV2 packet carries 20 bytes, which are protected by 2/3 FEC; and HV3 packet carries 30 bytes, which are not protected by FEC.

**Asynchronous Connectionless (ACL)**

**Asynchronous Connectionless (ACL)** is used for data traffic with a maximum data rate of 721 Kbps. It can carry data in 1-, 3-, or 5-slot packets. ACL links are used for point-to-multi-point connections between master and slaves, but only one ACL connection is allowed per slave. ACL uses FEC for error correction.

**Device Addressing**

Bluetooth uses three different addressing schemes:

1. Bluetooth device address or hardware address: a unique 48-bit address that is assigned to each device.
2. Active Member: a 3-bit number that specifies the slave ID in a Piconet.
3. Parked Member's Address: an 8-bit number that is assigned to each Bluetooth slave that is in the parked state in a Piconet.

## 19.4 Bluetooth Error Correction

Bluetooth uses 1/3 FEC, 2/3 FEC, and Automatic Repeat Request (ARQ) for error control. The 1/3 FEC means that each data bit is repeated three times, and 2/3 FEC means that every two data bits change to three bits. In the ARQ method, the transmitter keeps transmitting the same packet until it receives an acknowledge packet from the receiver or until the transmission timeout is exceeded.

## 19.5 Bluetooth Packet Types

The function of baseband is to accept bits from the Radio layer and make a packet. The Baseband layer defines 13 different types of packet. The master defines a series of slots of 625 microseconds each. The master transmits at even slots and the slave transmits at odd slots using time division multiplexing. Bluetooth defines 1-, 3-, and 5-slot packets and multi-slot packets are used for higher data rates. Figures 19.5a, b, and c show 1-, 3-, and 5-slot frames.

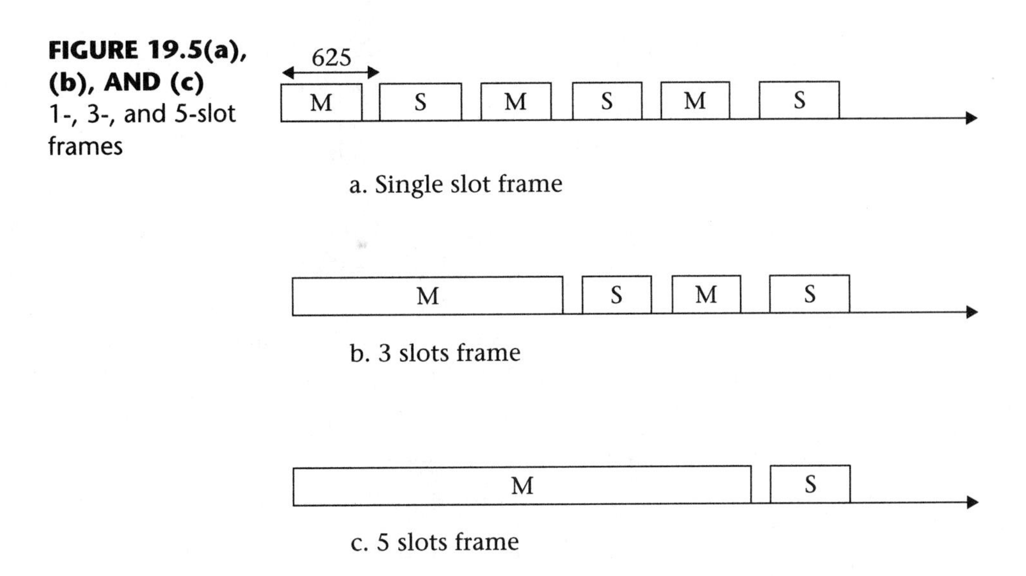


**FIGURE 19.5(a), (b), AND (c)** 1-, 3-, and 5-slot frames

## Baseband Layer Packet Format

Figure 19.6 shows the baseband packet format. The packet is divided into three fields: access code, header, and payload.

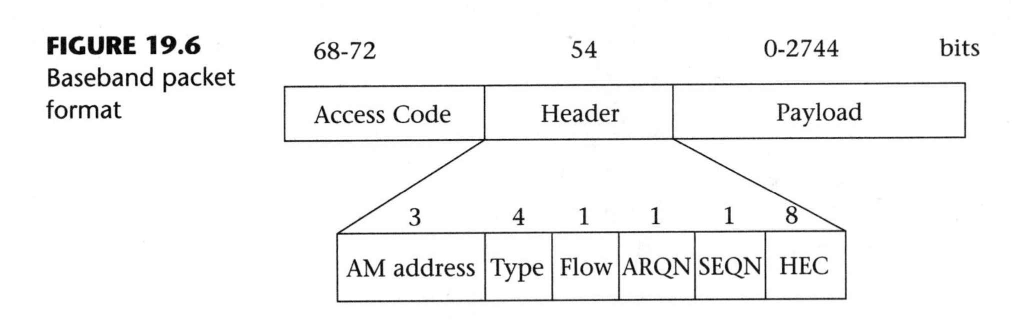


**FIGURE 19.6** Baseband packet format

The following describes the function of each field of the Baseband packet format:

**Access code:** The access code is used for synchronization and the identification of the master. If the slave is located within range of two masters by an access code such as Scattrenet, then the slave is able to identify the master.

**Packet header:** The packet header is actually 18 bits but it uses FEC 1/3; FEC 1/3 repeats the header three times.

**AM address:** The active member address represents the active address of the slave for which the packet is intended. This address is temporarily assigned to the slave by the master in a Piconet topology. The address is 3 bits; address 000 is used by the master to broadcast a packet to all the slaves in a Piconet.

**Type:** The type field determines the type of the packet; the packet may carry control information as well as synchronous and asynchronous data.

**Flow:** The flow bit is used by the slave to indicate to the master that its buffer is full.

**ARQN:** Used for acknowledgment of a packet.

**SEQN:** The sequence number is used for retransmission of a packet.

**HEC:** Header Error Check (HEC) is used for header protection.

**Payload:** The maximum payload is 343 bytes.

## Summary

- Bluetooth technology is used to connect keyboards, mice, and other peripherals to a PC without using wires.
- Bluetooth uses Piconet and Scatternet topologies.
- Master devices control the operation of Piconet.
- Masters set the hopping pattern.
- In a Piconet, a station can be a master station, slave station, parked station, or standby station.
- Bluetooth operates in the ISM frequency spectrum.
- Bluetooth uses FHSS with Time Division Multiplexing (TDM) for communications between master and slave.
- Bluetooth supports Synchronous Connection Oriented (SCO) and Asynchronous Connectionless Links (ACL).
- Bluetooth supports 1/3 FEC, 2/3 FEC, and ARQ for error correction.
- Bluetooth devices use four different addresses: active member address, parked member address, hardware address, and access request address.
- Bluetooth supports single-slot, 3-slot, and 5-slot frames.

## Key Terms

Active Member Address (AMA)
Ad Hoc
Asynchronous Connectionless (ACL)
Bluetooth
Host Controller Interface (HCI)
Link Controller
Link Management Protocol (LMP)
Object Exchange Protocol (OBEX)
Parked Member Address (PMA)
Piconet
RFCOM
Scatternet
Service Discovery Protocol (SDP)
Standby Mode
Synchronous Connection Oriented (SCO)

## Review Questions

- **Multiple Choice Questions**

1. Applications of Bluetooth technology are ______.
   a. connecting PCs together
   b. connecting mobile devices together
   c. connecting peripherals to PC without using wires
   d. a and b
2. Bluetooth topologies are ______.
   a. ring and bus
   b. star and fully connected
   c. Piconet and Scatternet
   d. none of the above
3. The maximum number of devices in a Piconet topology is ______.
   a. 2
   b. 4
   c. 5
   d. 7
4. When a group of Piconets communicates with each other, it is called ______ topology.
   a. bus
   b. ring
   c. Scatternet
   d. Piconet

5. The device that manages a Piconet topology is called a ______.
   a. slave
   b. master
   c. standby station
   d. parked station
6. Bluetooth uses ______ channels.
   a. 25
   b. 45
   c. 79
   d. 85
7. The bandwidth of each channel in Bluetooth technology is ______.
   a. 5 MHz
   b. 1 MHz
   c. 2 Mhz
   d. 10 Mhz
8. Bluetooth operates in the ______ band.
   a. UNII
   b. ISM
   c. VHF
   d. UHF
9. Bluetooth uses ______ for communication between the master and slave.
   a. DSSS
   b. FHSS
   d. FDM
   d WDM
10. Blue tooth offers ______.
    a. connection oriented and connectionless oriented
    b. synchronous connections and asynchronous connectionless
    c. character oriented and bit oriented
    d. none of the above
11. Bluetooth synchronous connections are used for ______.
    a. information with constant bit rate
    b. information with variable bit rate
    c. data information
    d. none of the above

12. Bluetooth technology uses ______ for error corrections.
    a. parity bit and CRC
    b. parity bit and FEC
    c. FEC and ARQ
    d. hamming code and FEC
13. In Bluetooth technology, asynchronous connectionless is used for transmission of ______.
    a. voice
    b. data
    c. video
    d. images
14. A Bluetooth device address is ______ bits.
    a. 24
    b. 48
    c. 32
    d. 64
15. A Bluetooth active member address is ______ bits.
    a. 3
    b. 7
    c. 12
    d. 32
16. The function of the Link layer is ______.
    a. to transmit information
    b. to establish a connection between the master and slave
    c. error detection
    d. to receive information from the Physical layer

## • Short Answer Questions

1. What are the applications of Bluetooth technology?
2. List the Bluetooth topologies.
3. Show Piconet topology.
4. Show Scatternet topology.
5. How many devices can be in one Piconet?
6. What is the function of a master in a Piconet?
7. How many slaves are in a Piconet?
8. List the types of device addressing used for Bluetooth.
9. Show Bluetooth protocol architecture.
10. What frequency spectrum is used for Bluetooth?

11. List the power transmissions for Bluetooth.
12. What are the functions of Baseband and Link layer?
13. What are the types of physical connections supported by Bluetooth?
14. What are the applications of Synchronous Connection Oriented (SCO) links?
15. How many SCO links are supported by the Physical layer?
16. What are the applications of Asynchronous Connectionless (ACL) links?
17. List the types of error correcting methods used by Bluetooth.
18. Show Bluetooth baseband packet.
19. What are the functions of the link controller?
20. List three functions of the Link Management protocol.
21. What is the function of RFCOM?
22. What is the function of SDP?

CHAPTER 20

# Wireless MAN

OUTLINE



OBJECTIVES

After completing this chapter, you should be able to:

- Identify and discuss the applications of wireless MANs
- Discuss the standards for wireless MANs
- Describe the topology and components of wireless MANs
- Explain the IEEE 802.16 protocol architecture
- Explain the IEEE 802.16 MAC layer functions
- Discuss the Physical layers of IEEE 802.16
- Discuss TDD and FDD operations
- Explain the Physical layers of IEEE 802.16a

INTRODUCTION

Wireless MAN is also known as **Broadband Wireless Access (BWA).** The standard for wireless MAN was approved by the IEEE 802.16 committee in April 2002. Some of the applications of wireless MAN are as follows:

- Providing links between campus buildings and dormitories
- Providing links between hospitals within a city for sharing patient information

- Providing links between airport buildings and railway terminals
- Providing links for police and fire departments

Wireless MAN can be used by ISPs to provide Internet access to buildings and offices, replacing cable and DSL modems. Figure 20.1 shows the application of BWA. In this figure, there are three types of stations: **Base Station (BS), Subscriber Station (SS),** and repeater. The base station is connected to the core network of an organization or public network such as the Internet. The subscriber station serves users inside the building. In Figure 20.1, all networks are connected to the BS without any cabling. The advantage of BWA is that it is easy to install and less expensive than wired networks.

## 20.1 Wireless MAN Topology

IEEE 802.16 supports **Point-to-Multi-Point (PMP)** and optional **mesh topology.** In PMP topology, the base station transmits data to all of the subscriber stations in its cell, as shown in Figure 20.1. The transmission from the BS to the SS is called *downlink transmission* and transmission from subscriber to the BS is called *uplink transmission.* Figure 20.1 shows point-to-multi-point topology.

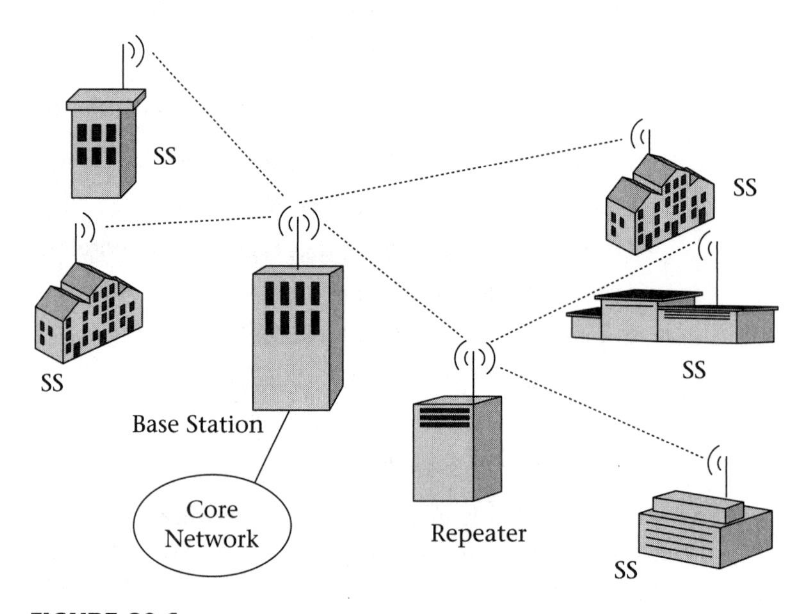


**FIGURE 20.1**
Point-to-multi-point wireless MAN topology

## 20.2 Wireless MAN Protocol Architecture

IEEE 802.16 defines the MAC and Physical layer standards for wireless MAN. Figure 20.2 shows the protocol architecture for wireless MAN. The IEEE defines two types of Physical layers: **IEEE 802.16** and **IEEE 802.16a.**

**FIGURE 20.2**
Protocol architecture of wireless MAN

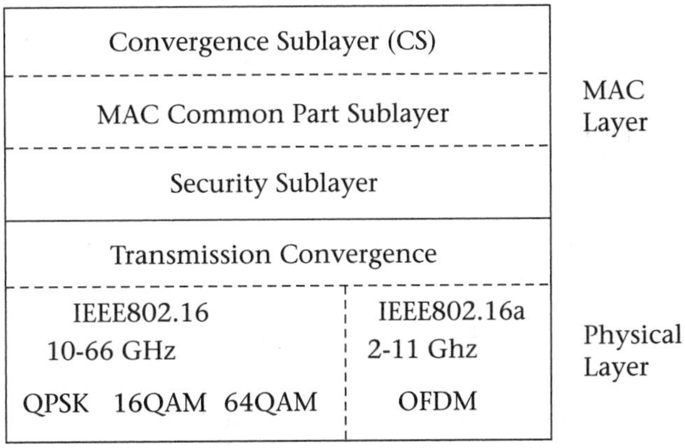

| | | |
|---|---|---|
| Convergence Sublayer (CS) | | MAC Layer |
| MAC Common Part Sublayer | | |
| Security Sublayer | | |
| Transmission Convergence | | Physical Layer |
| IEEE802.16<br>10-66 GHz<br>QPSK 16QAM 64QAM | IEEE802.16a<br>2-11 Ghz<br>OFDM | |

## 20.3 Wireless MAN MAC Layers

As shown in Figure 20.2, the MAC layer supports two Physical layers. The MAC layer is divided into the three sublayers: the Convergence sublayer, the Common Part sublayer, and the Security sublayer.

**Convergence Sublayer (CS)** IEEE 802.16 is designed to carry various types of Protocol Data Units (PDU) such as ATM, IPv4, IPv6, and Ethernet. The function of the CS is to accept information from the upper layer (LLC) and convert it to the IEEE 802.16 MAC data format as well as convert 802.16 MAC data to other data protocol units, such as ATM or IP.

**MAC Common Part Sublayer (CPS)** The IEEE 802.16 MAC layer uses a connection oriented PMP transmission. The BS has an antenna with multiple sectors and it is able to transmit data to all subscriber stations simultaneously without any coordination. The BS is capable of transmitting multicast or broadcast messages to the subscriber stations.

The MAC Common Part sublayer performs the following functions:

- Connection request; the CS requests a connection from the MAC Common Part sublayer

- Connection confirmation
- Termination of connection
- Bandwidth allocation
- Selection of multiplexing type (TDD or FDD)

**Security Sublayer**

The Security sublayer performs authentication between the BS and SSs, encryption, and decryption. The Security sublayer is able to operate using multiple encryption algorithms. The Security sublayer uses X.509 certificates for authentication and 56-bit Data Encryption Standard (DES) for encryption.

**Addressing and Connections**

Each SS has a 48-bit MAC address and a 16-bit **Connection Identifier (CID)**, which is used to identify the type of connection. The following are the three types of connections between SSs and BS:

**Basic connection:** BS and SSs use basic connections to exchange short and urgent management messages such as SS basic capability requests and responses.

**Primary management connection:** BS and SSs use primary management connections to exchange longer management messages, such as privacy key messages for encryption and decryption.

**Secondary management connection:** BS and SSs use secondary management connections to transfer long messages such as TFTP, SNMP, and DHCP packets.

**Wireless MAN MAC Frame Format**

Figure 20.3 shows the MAC Protocol Data Unit (PDU) frame format for wireless MAN. The PDU consists of three fields: generic MAC header, payload (the size of the payload is a variable), and CRC (optional).

| Generic MAC Header | Payload (variable) | CRC (optional) |
|---|---|---|

**FIGURE 20.3**
MAC protocol data unit (PDU) frame format

***Generic MAC Header Format.*** The IEEE 802.16 defines two types of generic MAC headers, one for data and management frames, and another for additional bandwidth requests. Figure 20.4 shows a **generic MAC header** for data and management frames.

| 1 | | | | | | | 16 bit |
|---|---|---|---|---|---|---|---|
| HT (1) = 0 | EC (1) | Type (6) | RSV (1) | CI (1) | EKS (2) | RSV (1) | LEN (3) |
| LEN (8) | | | CID Msb (8) | | | | |
| CID Lsb (8) | | | HCS (8) | | | | |

**FIGURE 20.4**
Generic MAC header for data and management frames

The following describes the function of each field in Figure 20.4:

**HT = 0:** Signifies a generic MAC header (for data and management frames).

**EC (Encryption Control bit):** If EC = 0, then the payload is not encrypted; if EC = 1, then the payload is encrypted.

**Type:** This field indicates the type of payload.

**RSV:** Reserved bit.

**EKS (Encryption Key Sequence):** Identifies the encryption key.

**HCS (Header Check Sequence):** This field is used for header error detection.

**LEN (Length):** This field defines the number of bytes in a MAC PDU including the header. The size of LEN is 11 bits.

**CID (Connection Identification):** Identifies the connection.

**CI (CRC indicator):** If CI = 0, then no CRC is attached to the PDU; if CI = 1, then the CRC is attached to the PDU.

If HT = 1, then a generic MAC header is used for requesting additional bandwidth, as shown in Figure 20.5. The SS requests additional bandwidth from the BS station; then the BS grants bandwidth to the SS station; then the SS station shares the bandwidth with the users.

**FIGURE 20.5**
Bandwidth request header format

| HT (1) = 1 | EC (1) = 0 | Type (6) | BR msb (8) |
|---|---|---|---|
| BR lsb (8) | | | CID msb (8) |
| CID lsb (8) | | | HCS (8) |

The following describes the function of each field in Figure 20.5:

**HT = 1:** Indicates that a generic header is being used for requesting additional bandwidth.

**EC = 0:** Indicates no encryption.

**BR (Bandwidth Request):** Represents the number of bytes requested by the SS from the BS for uplink transmission.

**Connection Identification (CID):** Represents the uplink connection identification.

**HCS (Header Check Sequence):** Used for detecting errors located in the header.

## MAC Management Messages

The MAC layer defines 33 MAC management messages; some of the MAC messages are shown in Table 20.1. The frame format of the MAC message is shown in Figure 20.6. The management message type and message payload fields make the payload for the MAC PDU.

| Generic Header | Management Message Type | Message payload |
|---|---|---|

**FIGURE 20.6**
MAC message frame format

**TABLE 20.1** Some MAC Management Frames and Their Descriptions

| Type | Message Name | Description | Type of Transmission |
|---|---|---|---|
| 0 | UCD | Uplink Channel Descriptor | Broadcast |
| 1 | DCD | Downlink Channel Descriptor | Broadcast |
| 2 | DL-Map | Down Link Map Message | Broadcast |
| 3 | UL-Map | Uplink Map Message | Broadcast |
| 4 | RNG-REQ | Ranging Request | Basic |
| 5 | RNG-RSP | Ranging Response | Basic |
| 6 | REG-REQ | Registration Request | Primary Management |
| 7 | REG-RSP | Registration Response | Primary Management |

The following describes the functions of each management frame.

***Uplink Channel Descriptor (UCD).*** A UCD frame is transmitted by BS to SS to inform SS of its uplink channel characteristics. The frame contains an uplink channel ID (8 bits), the number of the mini-slot size (number of physical slots is determined by $2^m$ where m is between 0 and 7), and the uplink burst profile (RF channel, symbol rate, and the number of the active burst profile).

***Downlink Channel Descriptor (DCD).*** A DCD is transmitted by the BS to the SS in the form of a broadcast packet at a periodic interval to define downlink physical channel parameters, such as BS transmitter power and FDD or TDD frame duration.

***Downlink Map Message (DL-MAP).*** A DL-map message is transmitted by BS as a broadcast message to all subscriber stations. It contains the following information:

*Physical Synchronization:* The DL-map synchronization field contains frame duration code and frame numbers. IEEE 802.16 defines three frame durations: 1 ms, 2 ms, and 5 ms.

*BS Station ID*: The base station ID is 48 bits (Mac address).

*DL-Map Information Elements*: A DL-map information element contains the Downlink Interval Usage Code (DIUC), which is defined as the burst profile for downlinks and the starting physical slot of the frame. The DIUC is a number that identifies a particular burst profile that is used by the downlink to transmit a MAC PDU to the SS stations. A DL-MAP may contain several DIUCs for transmitting several packets to several SS stations. DIUC represents a number from 0 to 15 and each number identifies specific burst profiles. The downlink burst profile contains modulation type and FEC type. The DICU = 0 (downlink burst profile 1) should be stored in each SS and it is used by BS for transmission of control headers such as preamble, DL-map, and UL-map.

***Uplink Map Message (UL-MAP).*** The uplink message is transmitted by the BS station to SSs to allocate uplink channels to them. It contains an uplink Channel ID (8 bits), the number of UL-map information elements, and allocation start time for uplink transmission that is defined by the UL-map in the mini-slots. The UL-map information elements contain connection identifiers and an uplink interval usage code (UIUC).

***Ranging Request (RNG-REQ).*** The RNG-REQ is transmitted from the SS to the BS to request frequency adjustment, power adjustment, and time alignment.

***Ranging Response (RNG-RSP).*** The BS station transmits a ranging response message to ranging requests by SS. The ranging response contains power adjustment, frequency adjustment, and timing adjustment information.

## 20.4 Wireless MAN Physical Layers

IEEE approved two physical layers for Wireless MAN: IEEE 802.16 and IEE802.16a.

### IEEE 802.16 Physical Layer

The IEE 802.16 operates in the 10-66 GHz frequency spectrum, and it requires **Line-of-Sight (LOS)** for communication between BS and SSs. The Physical layer supports three types of modulation: QPSK, 16QAM, and 64QAM. The type of modulation can be assigned dynamically by the BS based on the distance between the BS and SS.

The maximum data rate for IEEE 802.16 is 134.4 Mbps and it operates in a 20-, 25-, or 28-MHz channel bandwidth. Table 20.2 shows data rates for a variety of channels.

**TABLE 20.2** Channel Bandwidth and Data Rates for IEEE 802.16

| Channel Bandwidth | Symbol Rate | QPSK Data Rate | 16QAM Data Rate | 64QAM Data Rate |
|---|---|---|---|---|
| 20 MHz | 16 Ms/s | 32 Mbps | 64 Mbps | 96 Mbps |
| 25 MHz | 20 Ms/s | 40 Mbps | 80 Mbps | 120 Mbps |
| 28 MHz | 22.4 Ms/s | 44.8 Mbps | 89.6 Mpbs | 134.4 Mbps |

The IEEE 802.16 Physical layer operates using **Time Division Duplexing (TDD)** and **Frequency Division Duplexing (FDD).**

### Time Division Duplex (TDD)

TDD uses a single channel for uplink and downlink transmission but TDD allocates time slots for uplink and downlink transmission. Each frame is divided into downlink and uplink subframes, as shown in Figure 20.7 and each subframe is divided into time slots. The BS and SS must transmit their information on the time slots that were allocated by the DL-MAP and UL-MAP frames. The BS transmits downlink subframes first and then SS responds with uplink subframes. TDD requires what is called *guard time* or *transmission gap* equal to the round trip of packets from BS to the SS. Guard time is used so the BS is able to switch from transmitting mode to receiving mode. The guard time is increased by increasing link distance. The TDD is suitable for variable-bit rate ("bursty traffic").

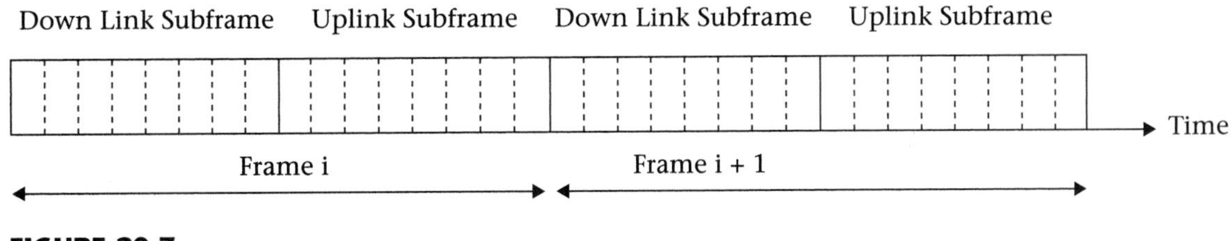


**FIGURE 20.7**
TDD frame

***TDD Subframes.*** Figure 20.8 shows the TDD frame format. The downlink subframe is transmitted to the SS stations, and then the SS stations respond to the downlink frame. The DL-map indicates the starting frame time slot for each frame transmitted from BS to each SS, and the UL-map indicates the starting slot for each frame transmitted by each SS to BS.

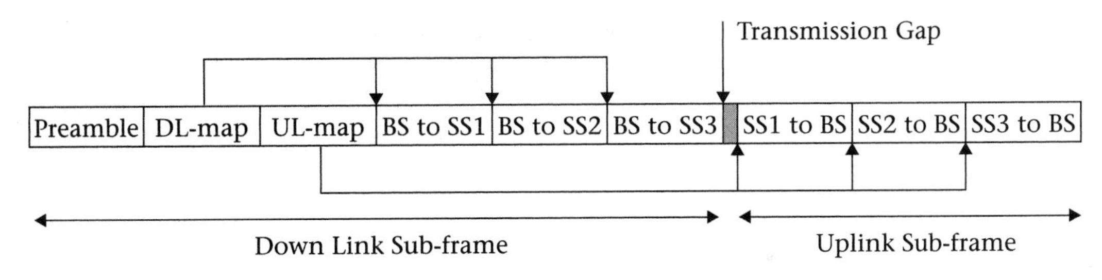


**FIGURE 20.8**
TDD frame structure

***Down Link Frame Format.*** Figure 20.9 shows the downlink frame format. The downlink frame uses Time Division Multiplexing (TDM) and transmits data at a specific time to each SS. The DL-map contains Data Interval Usage Code (DIUC) for each SS. The BS station transmits information to each SS based on the DIUC in the DL-map. In downlink transmission, the BS station transmits its packet to all SS stations and only one SS matches the address, which is the one that accepts the packet. During uplink, each SS has been allocated time slots by the BS and will transmit accordingly.

***Uplink Frame.*** An uplink frame is transmitted based on TDMA. The uplink channel is divided into a number of slots and the number of slots assigned depends on the type of frame (such as registration, connection, and user data). The UL-map contains the time slots for each SS transmission and also Uplink Usage Interval Code (UIUC) for each station.

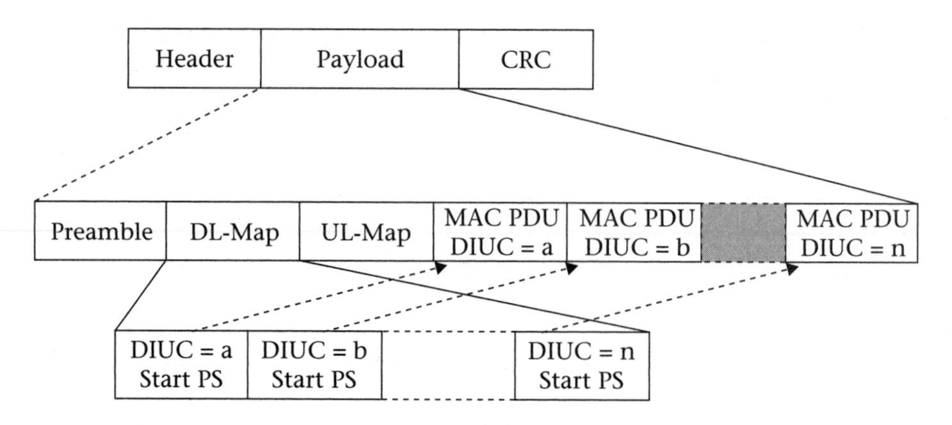


**FIGURE 20.9**
Downlink frame structure

### Frequency Division Duplex

Frequency Division Duplex (FDD) uses two channels for communication, one for downlink transmission (from BS to SS) and another channel for uplink transmission (from SS to BS). In the FDD method, the downlink and uplink use different frequencies for transmission and they can transmit frames at the same time. FDD supports both full-duplex and half-duplex transmission between BS and subscriber stations. Figure 20.10 shows the FDD frame structure.

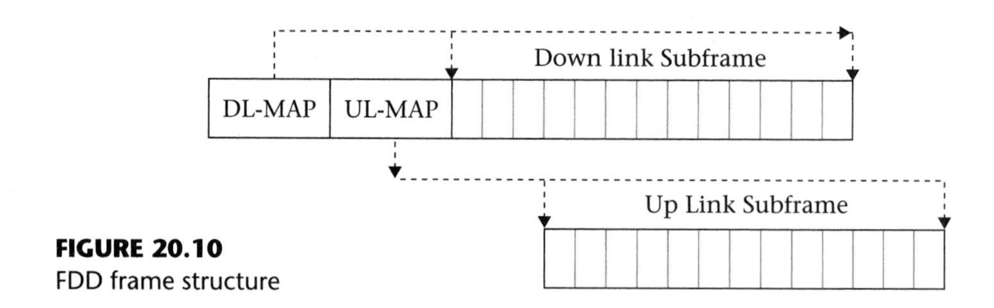


**FIGURE 20.10**
FDD frame structure

### Channel Access Method

A subscriber station performs the following functions in order to access the network:

- The subscriber station starts scanning its frequency list to find operating channel(s). Operating channels can also be manually programmed.
- The SS synchronizes itself with the BS using the preamble of the downlink frame transmitted periodically by the BS.

- The subscriber station obtains transmitted parameters using UCD that are periodically broadcasted by the base station.
- The subscriber station performs ranging. The SS sends a range request message to the BS, and then the BS measures the arrival time of the signal and also the signal power. The BS returns a range response frame in order to tell the SS to adjust its transmitter power.
- Performs registration.
- Negotiates basic capability.
- Establishes IP connectivity.

## 20.5 IEEE 802.16a Physical Layers

IEEE 802.16a is an extension of IEEE 802.16 approved in April 2003. The IEEE 802.16a Physical layer operates in the 2- to 11-GHz frequency spectrum and it supports **Non-Line-of-Sight (NLOS)** operation. IEEE 802.16a has a range of 30 to 50 miles and is able to transfer data at the rate 70 Mbps. It can be used for connecting hot spots where obstacles such as trees and tall buildings are presented. IEEE 802.16a can also be used for Wi-Fi hot spots and remote regions, such as rural areas.

The IEEE 802.16a supports three types of Physical layers: OFDM with 256 sub-carriers (supported by European corporations), OFDM with 2048 sub-carriers, and also a single carrier.

IEEE 802.16a supports PMP and mesh topology. Figure 20.11 shows mesh topology, which is used for communication between subscriber stations using NLOS offering Ad-Hoc networking between subscribers. In a mesh topology, each subscriber is called a *mesh node*. Nodes communicate with each other in non-line-of-sight. The 802.16a supports both FDD and TDD and also uses smart antennas. A *smart antenna* is made of multiple antenna elements and uses digital signal processing to combine multiple incoming signals created by multi-path fading and also automatically changes the direction of radiation in response to an incoming signal.

Table 20.3 shows the characteristics of IEEE 802.16 and IEEE 802.16a.

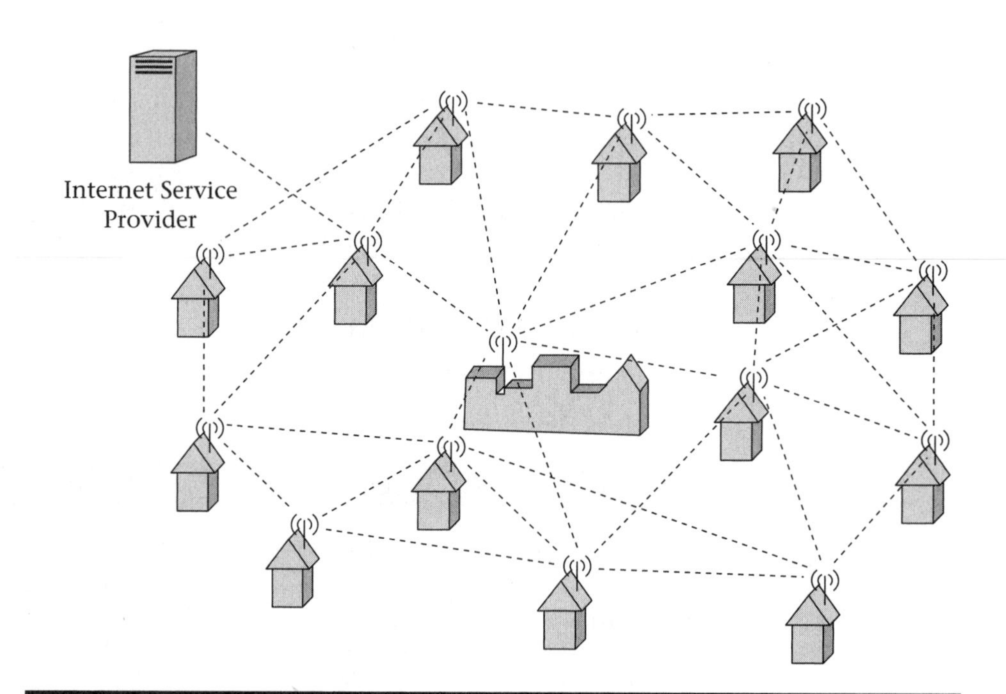


**FIGURE 20.11**
Mesh topology

**TABLE 20.3** Characteristics of IEEE 802.16 VS IEEE 802.16a

| Characteristics | IEEE 802.16 | IEEE 802.16a |
|---|---|---|
| Frequency Spectrum | 10–66 GHz | 2–11 GHz |
| Frequency Operation | Line-of-Sight | Non-Line-of-Sight |
| Maximum Bit Rate | 134 MHz | 75 MHz |
| Modulation Type | QPSK, 16QAM, 64QAM Single Carrier | QPSK, 16QAM, 64QAM, Single carrier, OFDM 256 Sub-carriers, OFDMA 2048 Sub-carriers |
| Downlink and Uplink Transmission | TDD/FDD | TDD/FDD |
| Channel Bandwidth | 20, 25, and 28 MHz | Between 1.25 and 20 MHz |
| Topology | PMP | PMP and Mesh |

## 20.6 WiMAX

WiMAX (Worldwide Interoperability for Microwave Access) is a non-profit forum and its members are manufacturers of broadband wireless products. The function of WiMAX is to promote certification for interoperability and compatibility of broadband wireless products. The member companies support IEEE 802.16 and ETSI (European Telecommunications Standards Institute) HiperMAN wireless standards.

## Summary

- Wireless MAN is also known as **Broadband Wireless Access (BWA).**
- Wireless MAN provides links between campus buildings and hospitals within a city for sharing patient information.
- Wireless Man defines three types of stations: **base station (BS), subscriber station (SS)**, and repeater.
- This base station is connected to the core network of an organization.
- The subscriber station (SS) serves users inside the building.
- IEEE 802.16 supports **Point-to-Multipoint (PMP)** and optional **mesh topology** transmission from subscriber to the BSs called **uplink transmission.**
- The IEEE defines two types of Physical layers: IEEE 802.16 and IEEE 802.16a.
- The MAC layer is divided into the three sublayers: the Convergence sublayer, the Common Part sublayer, and the Security sublayer.
- **Convergence sublayer (CS):** The IEEE 802.16 is designed to carry various types of Protocol Data Units (PDU) such as ATM, IPv4, IPv6, and Ethernet.
- **MAC Common Part sublayer (CPS):** The IEEE 802.16 MAC layer uses a connection oriented Point-to-Multipoint (PMP) transmission.
- **Security sublayer:** The Security sublayer performs authentication between the BS and SSs, encryption, and decryption.
- Each SS has a 48-bit MAC address and a 16-bit **Connection Identifier (CID).**
- **Basic connection:** BS and SSs use basic connections to exchange short and urgent management messages such as SS basic capability requests and responses.
- **Primary management connection:** BS and SSs use primary management connections to exchange longer management messages.
- **Secondary management connection:** BS and SSs use secondary management connections to transfer long messages such as TFTP, SNMP, and DHCP packets.
- The MAC layer defines 33 MAC management messages.
- Some of the MAC messages are: Uplink Channel Descriptor (UCD), Downlink Channel Descriptor (DCD), Downlink Map Message (DL-Map), and Uplink Map Message (UL-MAP).

- The IEEE 802.16 operates in the 10-66 GHz frequency spectrum, and it requires **Line-of-Sight (LOS)** for communication between BS and SSs.
- The IEEE 802.16 Physical layer operates using **Time Division Duplex (TDD)** and **Frequency Division Duplex (FDD).**
- IEEE 802.16a Physical layer operates in the 2- to 11-GHz frequency spectrum and it supports **Non-Line-of-Sight (NLOS)** operation.
- IEEE 802.16a has a range of 30 to 50 miles and is able to transfer data at the rate 70 Mbps.

## Key Terms

Base Station (BS)
Basic Connection
Broadband Wireless Access (BWA)
Connection Identifier (CID)
Frequency Division Duplexing (FDD)
Generic MAC Header
IEEE 802.16
IEEE 802.16a
Line-of-Sight (LOS)
Mesh Topology
Non-Line-of-Sight (NLOS)
Point-to-Multi-Point (PMP)
Primary Management Connection
Secondary Management Connection
Subscriber Station (SS)
Time Division Duplexing (TDD)

## Review Questions

- **Multiple Choice Questions**

1. The standard number for wireless MANs is ______.
   a. IEEE 802.11
   b. IEEE 802.16
   c. IEEE 802.14
   d. IEEE 802.14
2. IEEE 802.16 is used for ______.
   a. wireless LAN
   b. wireless MAN
   c. bluetooth
   d. none of the above

3. IEEE 802.16 topology is ______.
   a. bus
   b. ring
   c. point-to-multi-point
   d. star
4. IEEE 802.16 operates at ______.
   a. 2 to 11 GHz
   b. 10 to 40 GHz
   c. ISM band
   d. 10 to 66 GHz
5. IEEE 802.16 requires the BS and SS ______.
   a. to be close to each other
   b. to be line-of-sight
   c. to be far away
   d. none of the above
6. IEEE 802.16a operates at ______.
   a. 10 to 66 GHz
   b. 2 to 11 GHz
   c. UNII
   d. ISM
7. IEEE 802.16a requires the BS and SS ______.
   a. to be close to each other
   b. to be line-of-sight
   c. to be far away
   d. none of the above
8. The IEEE 802.16 MAC layer is divided to ______ sublayers.
   a. 2
   b. 3
   c. 4
   d. 5

## • Short Answer Questions

1. What is the IEEE standard number for wireless MAN?
2. What are the applications of wireless MAN?
3. What is the topology for IEEE 802.16?
4. List the MAC sublayers for IEEE802.16.
5. What is the frequency spectrum of IEEE 802.16?
6. What is the frequency spectrum of IEEE 802.16a?

7. What is difference between the IEEE 802.16a and the 802.16 Physical layers?
8. List the modulation types and how they are used by IEEE 802.16.
9. What are the types of connections that are used by IEEE 802.16 to connect BS to SS?
10. Show the MAC frame format of IEEE 802.16a.
11. List the four different management frames.
12. Show the TDD frame format.
13. Show the FDD frame format.
14. Explain what UIUC stands for and describe its application.
15. What is the function of a DCD frame?
16. What is the function of a DL-map frame?
17. What is the function of a UL-map frame?

CHAPTER 21

# Voice over Internet Protocol (VoIP)

OUTLINE



OBJECTIVES

After completing this chapter, you should be able to:

- Discuss the applications of VoIP
- Describe the factors that impact voice quality using VoIP
- Explain VoIP operation
- Discuss standards and protocols used for VoIP
- List the components of H.323

- List the components of SIP
- Show the SIP protocol architecture
- Show the H.323 protocol architecture
- Describe the H.323 connection setup between end users
- Describe the SIP connection setup between end users
- Calculate the minimum bandwidth requirement for VoIP

## INTRODUCTION

Public telephone systems are based on circuit switching networks, which allow real-time communication between users. **Voice over Internet Protocol (VoIP)** enables data networks such as the Internet, LANs, and WANs to be used for voice communication. Since VoIP reduces the cost of voice communication, it is in high demand by corporations with multiple locations. Corporations can avoid paying extra telephone charges by setting up a VoIP network for long distance communication between office locations. **Voice quality** is an important factor in the success of voice over data networks; it is imperative that VoIP offers the same quality as voice over the **Public Switch Telephone Network (PSTN).**

## 21.1 Voice Quality

There are three factors that impact voice quality over data networks: transmission delay, jitter, and packet loss.

### Transmission Delay (Latency)

**Transmission delay** is the time it takes a packet to travel from the source to the destination. For one-way transmission, delay between 150 and 250 msec is acceptable. This delay is generally caused by the following factors: propagation delay, storing and forwarding of packets in routers and gateways, compression at the source, and decompression at the destination. When a voice packet is received late at the destination, it will be discarded; therefore, the loss of a packet reduces the quality of the service. The latency can be reduced in private networks (Intranet, LAN) by adding Quality of Service (QoS) such as priority to the voice packets. However, the latency cannot be controlled for transmission of voice packets over the Internet.

### Jitter Delay

**Jitter delay** is the difference in arrival time between packets. It is preferable that the average arrival time between packets be constant. A variable delay is caused by congestion on networks and also by voice packets being sent over different paths. If voice packets are received by the destination at irregular times, distortions in the sound will occur.

**Packet Loss** Voice packets are transmitted over UDP; therefore, there is no guarantee that packets will reach their destinations. Packets may also be dropped by gateways when there is congestion in the network. When a packet is dropped, the gateway inserts a silence packet instead; this results in gaps in the conversation. VoIP can tolerate about a 2% packet loss.

## 21.2 Applications of VoIP

VoIP can be implemented on the Internet, a LAN, or a WAN. Currently, many corporations offer long distance calls using voice over Internet technology at one-fifth the price of using the PSTN. VoIP offers voice communication in three ways.

**Computer to Computer** In order to use a PC for voice communication, a microphone and special software are required at both ends. Voice packets use UDP for transmission over the Internet. Examples of VoIP software are Microsoft Net Meeting and Vocal Tec.

**Corporations with Multiple Sites** Figure 21.1 shows two PBXs (Public Branch Exchange) of a corporation that are located in New York and Paris. Each PBX is connected to the Internet by a device called a *VoIP gateway.* The function of a VoIP gateway is to bridge a voice network (PSTN or PBX) to a data network.

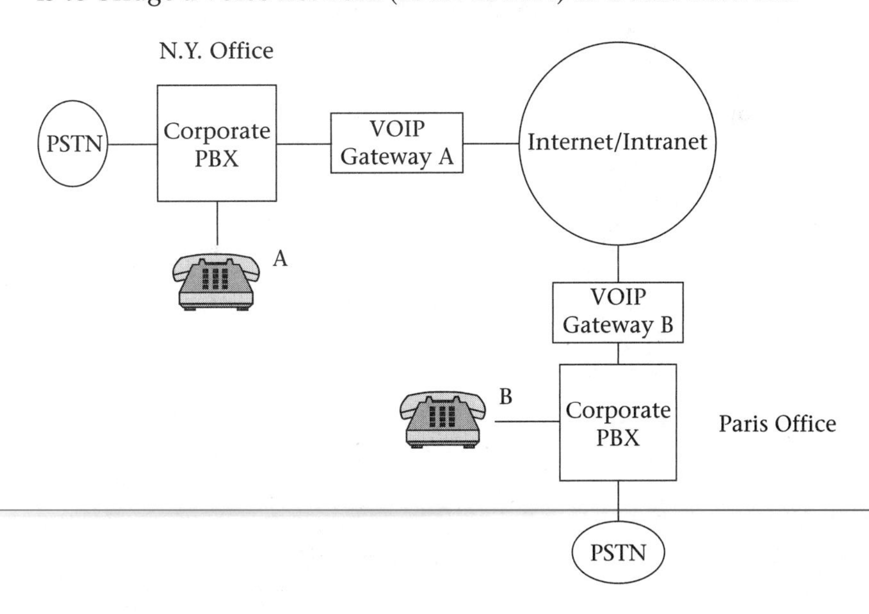


**FIGURE 21.1**
VoIP for corporations with multiple sites

In Figure 21.1, caller A accesses the local gateway A and dials the telephone number of the destination B. Gateway A uses the destination phone number to find the IP address of gateway B and makes a connection to gateway B via the Internet. Then, gateway A transfers the phone number of telephone B to gateway B by encapsulating the phone number into the IP packet. Next, gateway B dials phone B and makes a connection with phone B. Once a connection is made, both parties will start communication. Each gateway encapsulates several voice frames into one IP packet. The functions of a VoIP gateway are to establish call setup, make voice packets, and apply compression. The minimum bandwidth for an uncompressed voice packet is 64 kbps, but with voice compression this bandwidth can be reduced for more efficient transmission.

**Voice over Internet by Service Providers**

Currently, many corporations offer voice over Internet service to telephone users. In order to use the Internet for transmission of voice packets, a gateway is required. Figure 21.2 shows a block diagram of Internet telephony. The user in country A dials gateway A and gateway A prompts the user, requesting the phone number of the destination. The user then dials the phone number of the destination. Gateway A accepts this number and finds the IP address of destination gateway B. Gateway A makes a connection with gateway B and then passes the phone number of the destination to gateway B. Gateway B then dials the phone number and the parties start the conversation.

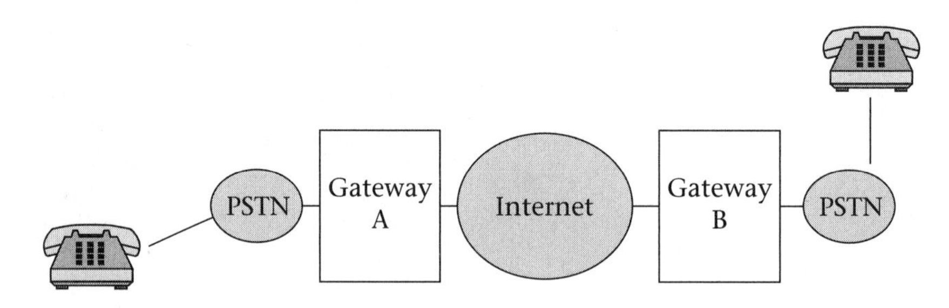


**FIGURE 21.2**
Voice over by Internet Service Provider (ISP)

**Voice over Intranet or LAN**

A corporation with multiple networks can use the data communication wiring for voice communication rather than using a PBX. The routers in a corporation network are capable of giving QoS to the voice packet.

## 21.3 VoIP Operation

Figure 21.3 shows the components and protocol for transmitting voice packets over a data network in which the following process will take place:

- **At the transmission side**

  1. The microphone accepts the voice signal and passes it to the Pulse Code Modulation (PCM) section.
  2. The PCM converts the voice signals to digital signals and passes the signals to the compression section.
  3. The compression section compresses the voice bits and forms a voice packet. The voice packet is then passed to the **Real-Time Protocol (RTP).**
  4. The RTP adds its header to the voice packet and passes the packet to UDP for transmission over the IP.

- **At the receiving side**

  1. The RTP passes its payload to the decompression section. The decompression section decompresses the voice packet and passes it to the analog-to-digital (A/D) converter.
  2. The A/D converter converts the voice packet to analog and then passes it to the speaker.

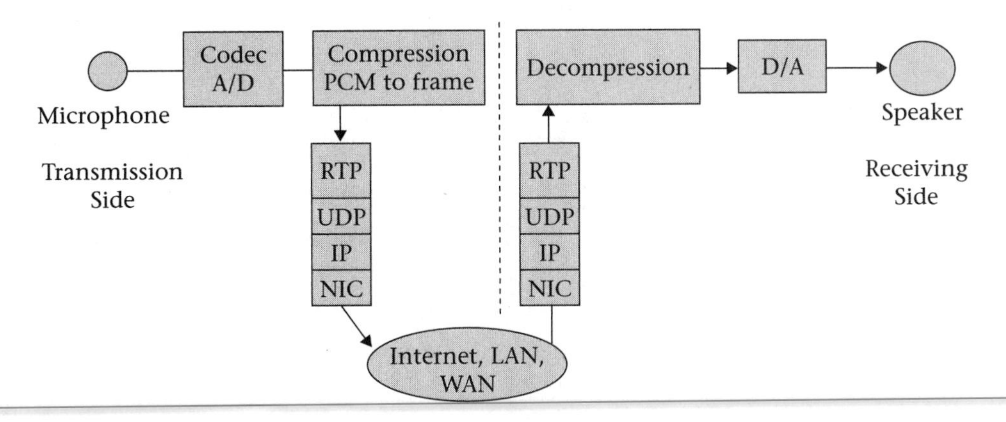


**FIGURE 21.3**
VoIP operation

## 21.4 VoIP Protocols and Standards

Before a voice packet is transmitted over a data network, a connection between the parties must exist. Currently, there are two protocols used for VoIP: **H.323** and **Session Initiation Protocol (SIP).** The functions of these protocols are to set up a connection, disconnect a connection, and handle call management. The SIP and H.323 protocols are used at the application level of the TCP/IP protocol.

## 21.5 H.323 Components

H.323 defines the four types of components used for point-to-point and point-to-multi-point communications, as shown in Figure 21.4:

1. H.323 terminal or endpoint
2. H.323 gateway
3. H.323 gatekeeper
4. Multi-point Conference Unit (MCU)

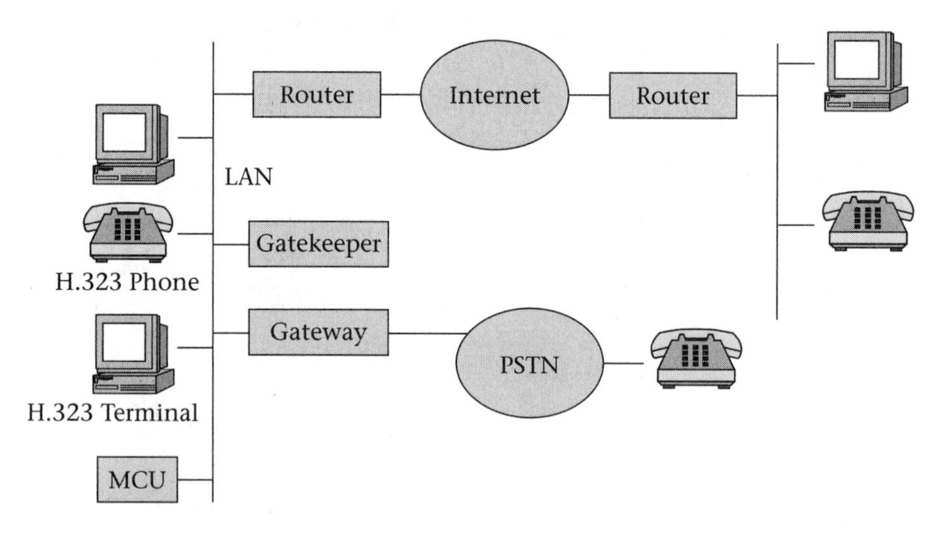


**FIGURE 21.4**
H.323 components

**H.323 Terminal:** The H.323 endpoint can be an H.323 terminal, a gateway, or an MCU. The terminal can be a PC with multimedia capability or an IP phone running H.323 protocol. The H.323 terminal must be able to originate a connection and disconnect a connection.

**H.323 Gateway:** A gateway converts voice and video packets into PSTN formats and vice versa.

**H.323 Gatekeeper:** A gatekeeper is an optional device that provides the following services to endpoints:

- **Address translation:** Translates aliases to IP addresses or phone numbers.
- **Gatekeeper discovery:** Upon boot-up of terminals, terminals can discover that their gatekeepers are using multicast addresses.

**Multi-Point Conferencing Unit (MCU):** The MCU is responsible for managing multi-point conferences (three or more parties involved in a conference call). The MCU initiates multi-point call setups. In multi-point conferencing, each terminal establishes a point-to-point connection with the MCU so that the MCU can control all communication.

## 21.6 H.323 Protocols

H.323 was published by the ITU in 1996 and is made up of multiple call control protocols, which define all aspects of call transmission. The H.323 is a standard used for transmission of real-time voice and video over a data network such as a LAN, a WAN, a MAN, and the Internet. H.323 is a combination of the following protocols that define the VoIP operation: H.245, H.225, H.261, and H.263. Figure 21.5 shows the H.323 protocol architecture.

<table>
<tr><td rowspan="2">RAS<br>Registration<br>Admission<br>Status<br>H.225</td><td>Video<br>Codecs<br><br>H.261<br>H.263</td><td>Audio<br>Codecs<br>G.711<br>G.722<br>G.723<br>G.729</td><td>Call Control<br>H.245</td><td>Call<br>Signaling<br>Q.931</td><td rowspan="2">Data<br>Protocols<br>T.120<br>T.125</td></tr>
<tr><td colspan="2">RTP/RTCP</td><td colspan="2">TPKT</td></tr>
<tr><td colspan="3">UDP</td><td colspan="3">TCP</td></tr>
<tr><td colspan="6">IP</td></tr>
<tr><td colspan="6">NIC</td></tr>
</table>

**FIGURE 21.5**
H.323 protocol architecture

### Audio Codec

The function of an **audio codec** at the transmission side is to accept audio signals from a microphone and convert the audio signals to digital using an A/D converter. The audio codec then compresses the digital bits (encode) and forms a voice packet. The function of the audio codec at the receiving side is to decompress (decode) the voice packets and convert them to audio signals using a D/A converter and send the analog signal to the speaker.

Voice compression is preformed by a device called a *Vocoder* (Voice Encoder/Decoder), which provides multiple types of voice compression. The type of compression is selected by negotiation between the source and destination gateways. The following are some of the voice compression standards:

| Standard | Data Rate |
|---|---|
| G.723 | 5.3 Kbps/6.3 Kbps |
| G.729 | 8 Kbps |
| G.711 | uncompressed 64 Kbps |

One method used to optimize voice quality is silence suppression. The function of a silent suppresser is to detect any gaps in speech and suppress those gaps. A pause indicates a gap in the voice at the receiving side.

### Video Codec

The function of a **video codec** at the transmitting side is to convert video signals (analog) from a camera to digital signals, compress the signals, create a video packet, and then pass the video packet to RTP. The function of a video codec at the receiving side is to decompress the video packet and convert the video signals to analog for video display. The ITU defines the H.261 and H.263 standards.

### H.225 Registration Admission and Status

**H.225** Registration Admission and Status (RAS) is used to carry messages to the gatekeeper for registration and admission of endpoints.

***Registration.*** When a terminal finds its gatekeeper, the terminal registers its IP address and its alias address with the gatekeeper. All H.323 endpoints are registered to a single gatekeeper.

***Admission.*** Another function of a gatekeeper is to permit or deny a call. The H.323 terminal sends an admission request to the gatekeeper. The admission request includes a destination address and the bandwidth required for the call. The gatekeeper might accept or reject the request based on the policies used by the gatekeeper such as link information and number of users currently using the link. The endpoints use RAS protocol to send messages to the gatekeeper for registration and status requests such as:

- Gatekeeper discovery
- Endpoint de-registration
- Admission request
- Registration of an endpoint with a gatekeeper
- Name registration
- Bandwidth request

**H.245 (Call Control)** H.245 is used for end-to-end control messaging such as flow control and opening and closing of a logical channel.

**T.125 (Data Protocol)** T.125 is used for data conferencing, such as text chat, file transfer, and image sharing.

**H.225/Q.931 Call Signaling** Q.931 is a signaling protocol for ISDN. H.225 uses Q.931 for **call signaling,** such as setting up a call and disconnecting a call between two endpoints.

**Real-Time Protocol (RTP)** Real-Time Protocol (RTP) is used for transporting audio and video packets over UDP. Figure 21. 6 shows the RTP format.

**FIGURE 21.6**
RTP packet format

| 2 | 1 | 1 | 4 | 1 | 7 | 16 bits |
|---|---|---|---|---|---|---|
| V | P | X | CC | M | PT | Sequence Number |
| Timestamp | | | | | | |
| SSRC | | | | | | |
| CSRC | | | | | | |
| Payload | | | | | | |

The following describes the function of each field in the RTP packet format:

**Version (V):** Defines the RTP version.

**Padding (P):** This field set to 1 means that extra bytes were padded to the payload. The last byte of the payload determines the number of the bytes that were padded to the payload; these bytes should be discarded.

**Extension (X):** This bit set to 1 means that that header is extended (for experimental use).

**Contributing Source Count (CC):** Used for multi-point call management.

**Mark (M):** This bit is to inform the receiver whether the packet is from a voice source or a video source. For voice application, this bit is set for the first packet following silent suppression. For video application, this bit is set for only the last packet of video frame.

**Payload type (PT):** Determines the type of payload.

**Sequence number:** The receiver uses this number to correct any packets that were received out of order or to detect any packet losses.

**Timestamp:** The timestamp depends on the payload. If the payload is a voice packet the timestamp is 8000, which is the sampling rate of the digitized voice. If the payload is a video packet, the timestamp is the clock rate for the video payload, which is 9000 Hz.

**Synchronization Source Identifiers (SSRC):** Used for multi-point calls.

**Contributing Source Identifiers (CSSRC):** Used for multi-point calls.

**Real-Time Control Protocol (RTCP):** RTCP provides a control mechanism for jitter delay and packet loss in RTP; it is used for end-to-end monitoring of data delivery. The endpoints use RTP to exchange packets that carry voice data and periodically they exchange RTCP packets to monitor the quality of data exchange.

## 21.7 H.323 Call Setup

Establishing a call between two H.323 endpoints requires two TCP connections. When a call is set up between the two endpoints, the endpoints use UDP for voice or video packet. One connection is for call setup using Q.931/H.225 and the other is for capability exchange between two endpoints. The RAS (H.225) is used for call setup between two endspoints and requires the following four phases:

1. **Endpoint registration:** Endpoints register with gatekeepers by using RAS commands, obtaining permission, and translating addresses such as phone numbers to IP numbers. RAS uses UDP port 1719. Some of the RAS commands are as follows:
    - ARQ: Admission Request
    - ACF: Admission Confirm
    - ARJ: Admission Reject
2. **Call signaling:** Call signaling is used to set up calls between two endpoints using TCP connection port 1720.

3. **Call control:** Call control is used for capability exchanges between endpoints such as determining master and slave, and determining the mode of operation. The endpoints exchange information about what type of media capability they have, such as G.711, G.723, H.261, and the characteristics of audio frames (number of samples per packet). Call control uses port 1024.
4. **Configure:** Configure an open logical channel for transmitting and receiving data, using UDP port 1024 with RTP.

Figure 21.7 shows the process that takes place to set up a call between endpoint A and endpoint B:

1. Endpoint A requests admission and the IP address of endpoint B from gatekeeper A by sending an ARQ message.
2. Gatekeeper A sends a Location Request (LRQ) to gatekeeper B to resolve the endpoint B alias address to an IP address.
3. Gatekeeper B sends a Location Confirm to gatekeeper A including the IP address of endpoint B.
4. Gatekeeper A sends an Admission Confirm with the IP address of endpoint B.
5. Endpoint A sends a call setup to endpoint B.
6. Endpoint B requests admission from gatekeeper B.
7. Gatekeeper B confirms the admission with an Admission Confirm Packet.
8. Endpoint B sends a connect to endpoint B.
9. Endpoint A and B exchange capabilities, deciding who is the master and the slave.
10. A logical channel opens.

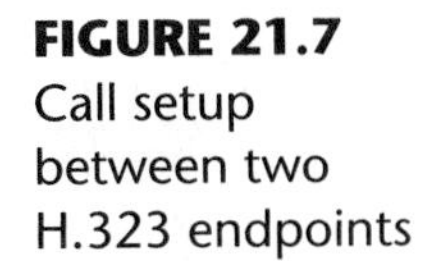

**FIGURE 21.7**
Call setup between two H.323 endpoints

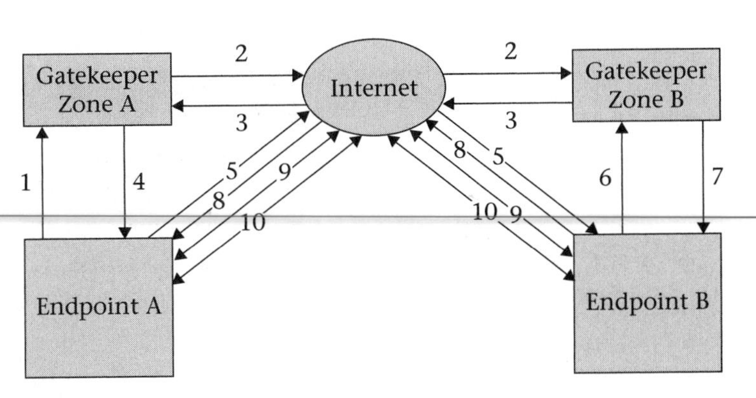

## 21.8 Session Initiation Protocol (SIP)

SIP is an Internet Engineering Task Force (IETF) Internet-based protocol designed for call setup and management between two or more endpoints. SIP is a signaling protocol used for real-time communication for VoIP such as Internet telephony and multimedia conferencing. SIP performs the following functions:

1. Sets up calls between users.
2. Disconnects a call between callers.
3. Determines the location of the destination.
4. Supports address resolution (converting phone numbers to IP addresses).
5. Determines whether the endpoint is available or not (busy).

Figure 21.8 shows SIP protocol architecture.

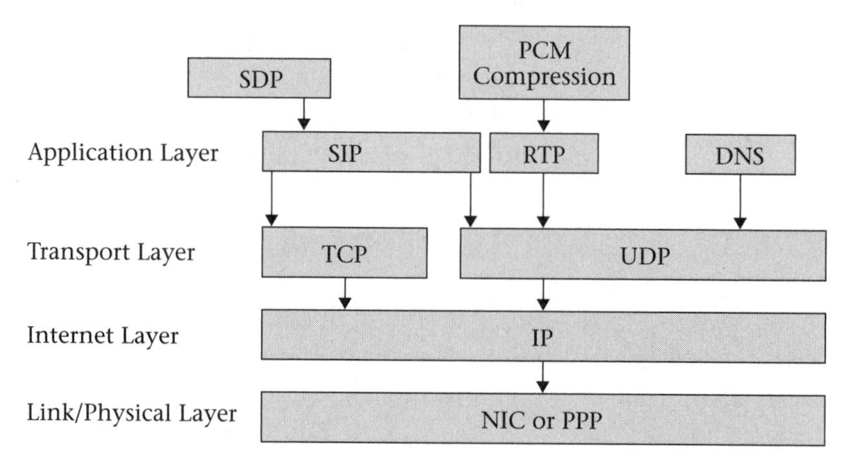


**FIGURE 21.8**
SIP protocol architecture

## 21.9 SIP Components

The SIP components are user agents, servers, and gateways. The following describes the function of each component.

**SIP User Agent or SIP Endpoint**

The **SIP user agent** can be an IP phone or a PC with SIP protocol. The User Agent (UA) should be able to send a SIP request and response. The UA works in client and server mode. The UA communicates with other user agents through a proxy server.

**Gateway** A gateway is a special device that connects the PSTN to the Internet.

**Server** SIP defines three types of servers: proxy server, redirector server, and registered server. These three servers can be implemented in a SIP proxy server.

***SIP Proxy Server.*** A **SIP proxy server** performs the following functions:

1. Accepts a user agent request and forwards the request to another user agent or server.
2. Accepts a response from a server or user agent and forwards it to the user agent.
3. If a proxy server does not have the IP address of the destination user agent, the proxy will contact a DNS server to obtain the IP address of the UA.
4. Requests a route on behalf of the user agent from a location server and also requests an IP address of the next proxy from a DNS server.

***Proxy Server Types.*** The SIP standard defines two types of SIP proxy servers: stateless proxy and stateful proxy:

***Stateless Proxy Server.*** The stateless proxy server receives a request from a UA, processes the request, and forwards the response to a UA (User Agent) or a server. The stateless proxy server does not keep any information (transactions) about the forwarded responses or requests. Therefore, if a response were lost due to congestion, the server would be unable to retransmit the response. The stateless proxy server is the simplest form of a proxy server.

***Stateful Proxy Server.*** The server also acts as client when it responds to requests and sends requests. The stateful proxy server keeps information about responses and requests. Therefore, if a packet were lost due to congestion, the server would be able to retransmit the packet.

***SIP Redirector Server.*** The function of a **SIP redirector server** is to accept requests and direct the client to contact alternate user agents (the same concept as when a secretary answers a phone and gives the caller another phone number).

The redirector uses 3XX code to respond to a request. Some of the codes are as follows:

301 The destination has moved permanently.

302 The destination has moved temporarily and the user is available at a different address.

305 The request source should contact the proxy server.

***Registrar and Location Server.*** The user agents register with a registrar server and the registrar server updates the location database (location server). The location database holds the address of the server that the UAs are connected to. Therefore, the proxy server can submit a client URL address to the location server and obtain an IP address of the user agent. The DNS server holds the IP addresses of the proxy servers.

## 21.10 SIP Request and Response Commands

SIP uses request and response commands to set up, change, or terminate a conversation between endpoints. A request is initiated by the client to the server and a response is initiated by the server to the clients. The SIP entities use special words for requests and call methods. The following defines the methods and their descriptions:

| Method | Description |
|---|---|
| INVITE: | used for inviting an endpoint for communication |
| BYE: | request for terminating a connection |
| ACK: | used for response to an invitation or for reliable communication between source and destination |
| REGISTER: | used by a user agent to register with a registrar server |
| CANCEL: | used for canceling a pending call |
| OPTIONS: | used for requesting information about call connection, such as bandwidth or compression methods |

### Response Codes

SIP uses codes for responding to a request. The response codes are classified as follows:

| Code | Description |
|---|---|
| Class 1XX | used to indicate progress, such as ringing and searching |
| Class 2XX | used to indicate success |
| Class 3XX | used for redirecting and forwarding |
| Class 4XX | client error |
| Class 5XX | server failure |
| Class 6XX | global failure (such as busy, not available) |

The following table shows some of the response codes and their descriptions:

| Code | Description |
|---|---|
| 100 | trying |
| 180 | ringing |
| 200 | OK |
| 301 | destination has moved permanently |
| 302 | destination has moved temporarily |
| 403 | not permitted |
| 480 | unavailable |
| 600 | busy |
| 603 | declined |

## 21.11 SIP Addressing

SIP uses Universal Resource Locators (URLs) for addressing. It is similar to email addressing, such as:

sip: elahi@southern.edu
sip: +1-800-555-2020@xyz.com; user located in a different network
sip: 25819@southernct.edu; user located in the same network

## 21.12 SIP Connection Operation

A connection is made between two UAs through one proxy server. Figure 21.9 shows two UAs and a proxy server. SIP uses the following process to set up a connection between user A and B:

1. The UA A sends a packet called an *invite* to the proxy sever.
2. The proxy server accepts the packet and sends it to UA B.
3. The Proxy server sends a code 100 to user A and user A waits.
4. Endpoint B accepts the invite packet and it starts ringing (UA B is ringing).

5. The proxy server passes the code 180 to UA A and A generates rings indicating that user B is ringing.
6. When user B picks up the phone, B sends an OK packet with code 200 to the proxy server.
7. The proxy server sends an OK packet with code 200 to user A, user A stops ringing, and a session for communication using RTP is established between A and B.

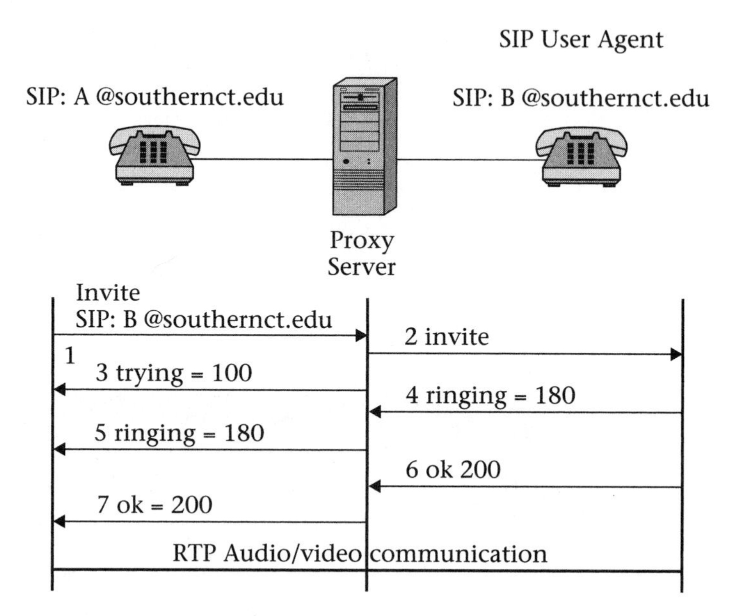


**FIGURE 21.9**
Connections between two user agents and one proxy server

Figure 21.10 shows a connection between two endpoints through two proxy servers. As shown in Figure 21.10, the endpoint A invites endpoint B for connection through the proxy server A. The proxy server A does not have the IP address of endpoint B. Therefore, a query is sent to the DNS server to obtain the IP address of endpoint B. The complete connection process is shown in Figure 21.10. Figure 21.11 shows the connection process of two PSTN phones via the Internet.

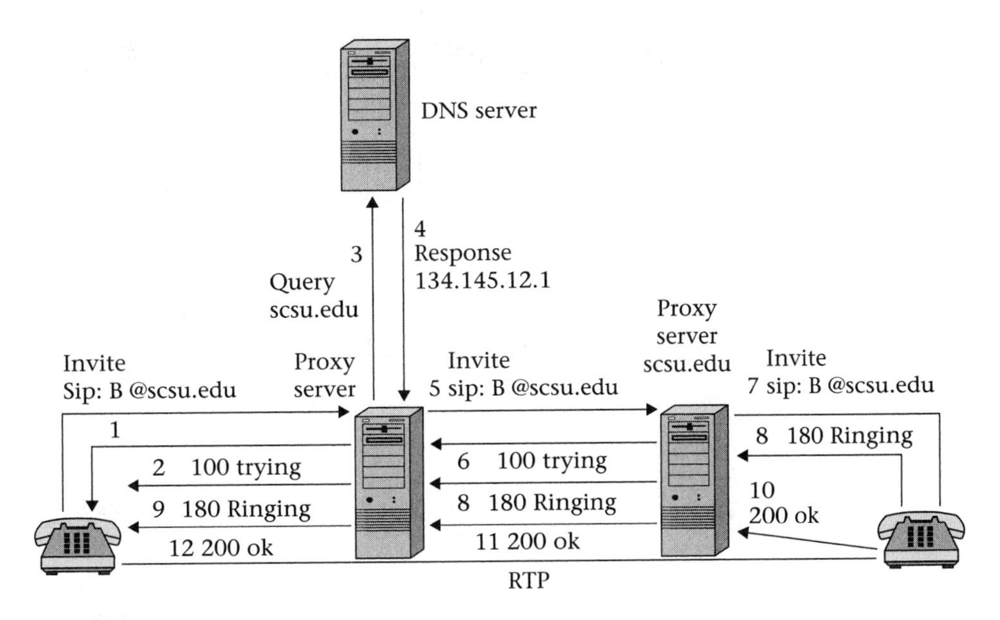


**FIGURE 21.10**
Connection between two user agents through more than one proxy server

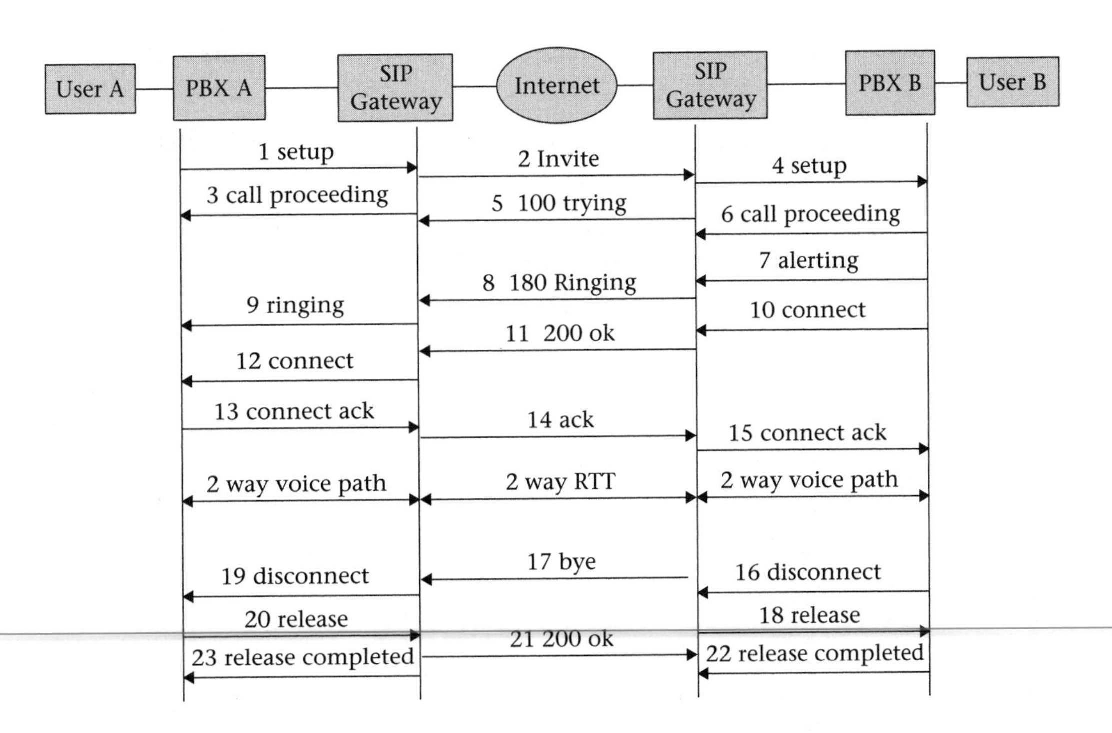


**FIGURE 21.11**
Connection of two PSTN phones using the Internet

## 21.13 VoIP Bandwidth Calculation

Voice packets must be received at a constant bit rate. The bit rate depends on the types of codec selection. The audio frame is made of 10 ms and the G.711 codec data rate is 64,000 bps; therefore, 10 ms of this 64,000 bps is 640 bits or 80 bytes.

The G.792 data rate is 8000 bps and 10 ms of this data rate is 80 bits or 10 bytes. The voice packet goes through RTP, UDP, and IP protocols and these protocols add their header to the voice frame, as shown in Figure 21.12.

**FIGURE 21.12**
Voice payload

| IP header 20 bytes | UDP header 8 bytes | RTP header 12 bytes | Voice payload Variable |
|---|---|---|---|

The voice packet goes through the network card (Ethernet, PPP, or Frame Relay) and the network card also adds its header and trailer, assuming an Ethernet NIC is used.

Ethernet has a 22-byte header and a 4-byte trailer, which gives a total of 26 bytes added to the voice packet. Therefore, the total voice packet for a 1-ms payload using G.711 codec is calculated as follows:

Voice packet = 80 bytes payload + 40 bytes (RTP, UDP, and IP) header + 26 Ethernet header = 146 bytes

The voice packet must reach its destination at 64,000 bits per second or 8,000 bytes per second. The 8,000 bytes per second is equal to 100 voice packets per second. Therefore, the bandwidth of channel should not be less than:

Bandwidth of a channel = 146 bytes * 100 = 116800 bps

## Summary

- VoIP reduces the cost of long distance communication.
- Voice quality plays an important role in VoIP.
- Factors that impact voice quality are transmission delay, jitter, and packet loss.
- Voice is converted by a PCM to a digital signal.
- Digital signals are converted to analog signals by an A/D converter.
- Protocols for VoIP are SIP and H.323.
- The components of H.323 and H.323 endpoints are H.323 gateway, H.323 gatekeeper, and Multi-Point Control Protocol (MCU).
- H.323 protocols are H.225 call signaling, H.245 call control, and T.125.
- Audio codecs for H.323 are G.711,G.722, and G.723.
- Video codecs for H.323 protocol are H.262 and H.263.
- RTP is used for transporting audio and video packets over UDP.
- SIP is an IETF protocol for VoIP.
- SIP components are SIP user agent, gateway, and SIP proxy server.

## Key Terms

Audio Codec
Call Signaling
H.323
H.225
H.323 Gateway
H.323 Gatekeeper
H.323 Terminal
Jitter Delay
Multi-Point Conferencing Unit (MCU)
Public Switch Telephone Network (PSTN)
Real-Time Protocol (RTP)
Real-Time Control Protocol (RTCP)
Session Initiation Protocol (SIP)
SIP Proxy Server
SIP Redirector Server
SIP User Agent
Transmission Delay
Video Codec
Voice over Internet Protocol (VoIP)
Voice Quality

## Review Questions

### • Multiple Choice Questions

1. Advantages of VoIP are reduced _____.
   a. cost
   b. delay
   c. packet lost
   d. none of the above
2. One of the factors that plays an important role in successful VoIP is _____.
   a. cost
   b. quality of service
   c. speed
   d. delay
3. The protocols currently used for VoIP are _____.
   a. TCP and SIP
   b. SIP and H.323
   c. UDP and H.323
   d. TCP and UDP
4. The Internet Engineering Task Force approved _____ for VoIP.
   a. TCP
   b. SIP
   c. H.323
   d. RTP
5. ITU approved the _____ protocol for VoIP.
   a. TCP
   b. SIP
   c. H.323
   d. RTP
6. An IP phone is an _____.
   a. H.323 terminal
   b. H.323 gateway
   c. H.323 gatekeeper
   d. none of the above
7. Of the following standards, _____ are used for video codecs.
   a. G.711 and H.261
   b. G.722 and H.263
   c. H.261 and H.263
   d. G.722 and G.723

8. The function of H.225 (RAS) is ______.
   a. connection setup
   b. admission
   c. registering endpoints
   d. opening and closing a logical channel
9. The function of an audio codec at the transmitting side is to ______.
   a. convert digital signals to analog
   b. convert voice to digital signals and compress
   c. convert analog signals to digital
   d. none of the above
10. RTP is used for ______.
    a. transporting data
    b. transporting voice
    c. transporting audio and video packets
    d. transporting images
11. The SIP gateway is used to connect a ______ to the Internet.
    a. LAN
    b. WAN
    c. PSTN
    d. DSL
12. A ______ accepts a SIP user agent request and forwards it to another user agent.
    a. SIP endpoint
    b. SIP gateway
    c. SIP proxy server
    d. SIP redirector server
13. A ______ accepts requests and directs the client to contact the alternate user agent.
    a. SIP endpoint
    b. SIP gateway
    c. SIP proxy server
    d. SIP redirector server

## • Short Answer Questions

1. What does VoIP stand for?
2. Does VoIP reduce or increase the cost of voice communications?
3. Why does VoIP reduce the cost of long distance communications?
4. What is the most important factor to consider in VoIP?

5. What are three factors that impact voice quality over the Internet?
6. Define transmission delay and describe what causes it.
7. Define jitter and describe what causes it.
8. Define packet loss and describe what causes it.
9. List the applications for VoIP.
10. What does PCM stand for and what is the function of PCM?
11. Describe the layers involved for transmission of VoIP.
12. What are the two protocols used for VoIP?
13. What is function of an audio codec at the transmission side?
14. What is a voice compression device called?
15. List the H.323 protocols.
16. List the components of H.323.
17. List the components of SIP.
18. What is the difference between a stateful and a stateless SIP proxy server?
19. What form of addressing does SIP use?
20. Describe the SIP connection operation.
21. Find the minimum bandwidth requirement for a VoIP channel using a 50-ms frame and a G.711 codec.
22. Find the minimum bandwidth requirement of a VoIP channel using a 20-ms audio frame and a G.729 codec.

CHAPTER 22

# Asynchronous Transfer Mode (ATM)

OUTLINE



OBJECTIVES

After completing this chapter, you should be able to:

- List the components of the Asynchronous Transfer Mode (ATM)
- Describe the advantages of ATM
- Describe the applications of ATM
- Show ATM cell format
- Explain ATM connection types and connection identifiers
- Discuss ATM switch characteristics
- Show ATM switch architecture
- Explain blocking in ATM switches

- Show the ATM User Network Interface (UNI) cell header and understand the function of each field
- Show the ATM Network-to-Network Interface (NNI) cell header and understand the function of each field
- Show ATM end user and ATM switch protocol
- Explain the functions of ATM Adaptation layer, ATM layer, and Physical layer
- Describe the application of each ATM Adaptation layer
- Show AAL1, AAL¾, and AAL5 cell formats

## INTRODUCTION

With **Asynchronous Transfer Mode (ATM)** a much wider array of information can be transmitted, for example; voice, data, images, CAT scans, MRI images, and video conferencing. This technology can be used for both private and public networks. ATM delivers bandwidth on demand, is not dependent on applications, and works at a data rate between 1.5 Mbps to 2 Gbps. All types of networking, from LANs to WANs, and from backbone to desktop, can integrate ATM technology. In addition, ATM is a transfer protocol for B-ISDN. Figure 22.1 illustrates a typical ATM network, consisting of switches and end users. ATM switches offer two types of interfaces: switch to switch interface (or Network-to-Network Interface, called *NNI*) and Switch-to-User Interface called *UNI*.

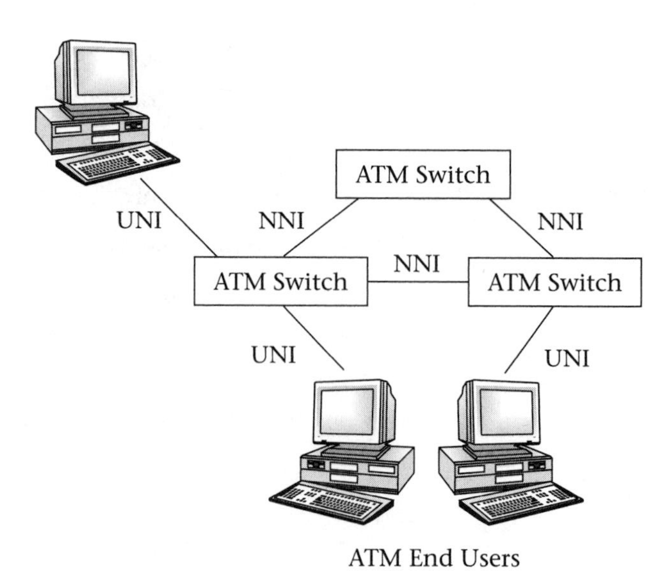


**FIGURE 22.1**
ATM network interface

## 22.1 ATM Network Components and Characteristics

The ATM network components are an ATM NIC, an ATM switch to relay each cell to its destination, and software. The most significant advantages of using ATM are as follows:

- Bandwidth on demand to users
- Connection-oriented service
- ATM switches use statistical multiplexing
- Higher quality of service
- Varied transmission media such as optical cable or twisted-pair wire
- Wider array of information can be handled
- Works with current LAN and WAN technologies and supports current protocols, such as TCP/IP

The characteristics of ATM are as follows:

**Cell Switching:** A cell carries information about routing; the switch reads the information and selects the route.

**Connection-Oriented Transmission:** A connection must be established between two stations before data can be transferred between them.

**Cell Size:** Fixed at 53 bytes, with a 5-byte header and a 48-byte payload, as shown in Figure 22.2.

**Delay:** Delay between transmissions is very small.

**Switching:** Occurs at very high speeds.

**FIGURE 22.2**
ATM cell

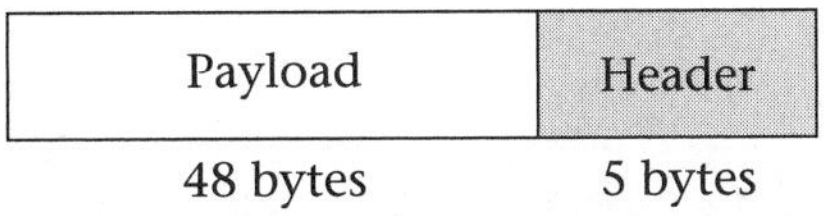

| Payload | Header |
|---|---|
| 48 bytes | 5 bytes |

## 22.2 ATM Forum

The **ATM forum** started in 1991, with four computer and networking vendors. Today, the forum has over 1,000 members. The forum writes specifications and definitions for ATM technology, and these specifications are submitted to the ITU for approval. The ITU-T standards organization works closely with the ATM forum.

## 22.3 Types of ATM Connections

ATM offers two types of connections:

1. **Permanent Virtual Connection (PVC):** A PVC is set up and taken down manually by a network manager. A set of network switches between the ATM source and destination are programmed with predefined values for the Virtual Channel Identifier and Virtual Path Identifier VCI/VPI. Transmission is more reliable with this type of connection than with SVC.
2. **Switched Virtual Connection (SVC):** SVC is a connection that is set up automatically by a signaling protocol. SVCs are more widely used because they do not require manual set up; however, they are not as reliable as PVCs.

### Connection Identifiers

There are two **connection identifiers** in an ATM cell header: the **Virtual Path Identifier (VPI)** and the **Virtual Channel Identifier (VCI).** These identifiers are used for routing and identification of cells. VPI and VCI do not represent destination addresses; instead they represent a connection that leads to the intended destination. One VPI may contain several VCIs, as shown in Figure 22.3.

Figure 22.4 shows a representation of the VPI and VCI. The VPI is the railroad number and the VCI is the wagon number. Each rail can

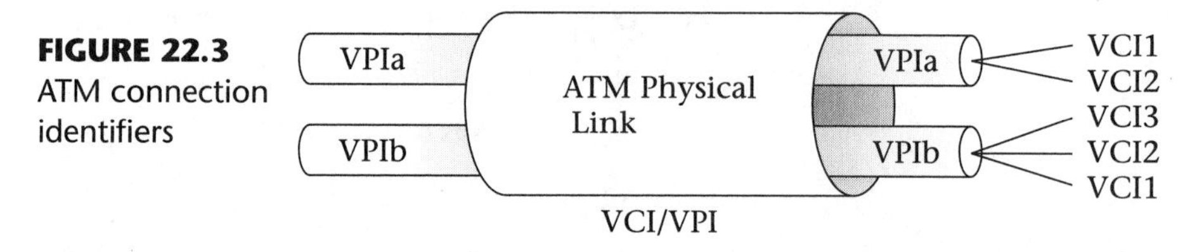


**FIGURE 22.3**
ATM connection identifiers

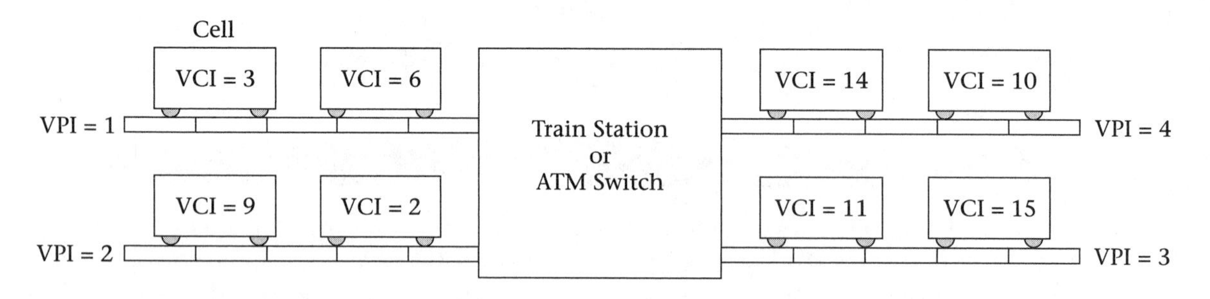


**FIGURE 22.4**
Representation of VPI and VCI using a train station example

transport several wagons. Each wagon has a unique number that is represented by its VCI. When a cell enters the switch, the switch can assign a new VCI number to the cell.

## 22.4 ATM Switch Operation

An ATM switch can process cells at an extremely high rate of speed. The ATM switch operates by performing the following functions:

1. A cell is received on an input port. Its header VPI/VCI is examined to determine the output port to which the cell should be forwarded.
2. VPI/VCI fields are modified to a new value based on the output port.
3. The Header Error Control (HEC) is used for error detection and correction in the header field of each cell. If the HEC cannot correct the error, the ATM switch will discard the cell.
4. The switch has a control unit that can modify the routing table.
5. The switch must support cell switching at a rate of at least one million cells per second.

Figure 22.5 shows an ATM switch with three ports. The cells enter the switch from port 1, and go out from ports 2 and 3 according to the routing table. The ATM switch has a routing table as defined by the ATM signaling protocol. Assume that Table 22.1 represents the routing table for Figure 22.5. Any cell from port 1 with VPI/VCI = 1/26 entering the switch will go out through port 2 with VPI/VCI = 2/45. Any cell entering from port 1 with VPI/VCI = 1/45 will go out through port 3 with VPI/ VCI = 3/39.

**TABLE 22.1** Routing Table for Figure 22.5

| Input | | Output | |
|---|---|---|---|
| Port | VPI/VCI | Port | VPI/VCI |
| 1 | 1/26 | 2 | 2/45 |
| 1 | 1/45 | 3 | 3/39 |

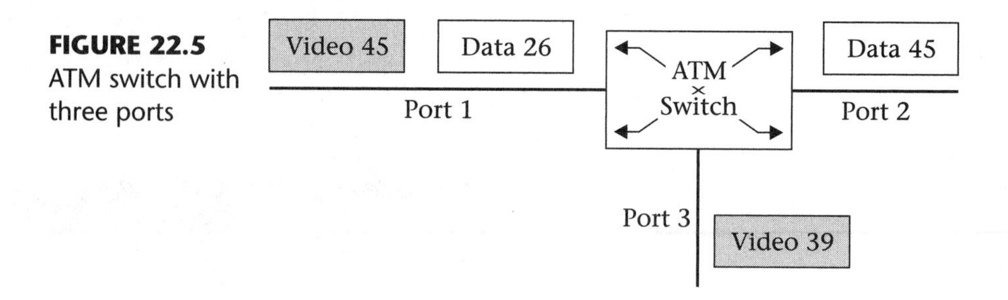


**FIGURE 22.5** ATM switch with three ports

### ATM Switch Characteristics

ATM switches should have the following characteristics:

- Have a large buffer capacity to store incoming cells and the routing table
- Be fast enough to transfer incoming cells to the output ports
- Have from 16 to 32 input and output ports
- Support all AAL types
- Support PVC and SVC connections
- Support point-to-multi-point connections
- Support congestion control (able to overcome congestion)

## 22.5 ATM Switch Architecture

The most important component of an ATM network is the switch, which must be capable of processing millions of cells per second. Figure 22.6 shows the general architecture of an ATM switch using Statistical Packet Multiplexing (SPM). SPM dynamically allocates bandwidth to the active input channels. The function of a control processor is to control the input/output buffers and update the routing table of the switch. There are several different architectures used for ATM switches, such as the Delta switch matrix and the Banyan switch matrix.

### ATM Switch Blocking

An ATM switch will generally experience two types of blocking or traffic congestion: fabric blocking and head of the line blocking.

**Fabric blocking** occurs when fabric capacity is less than the sum of its input data rate. In this case the switch must drop some of the cells. Some ATM switches are limited to 16 or 32 OC-3 input ports. **Head of the line blocking** occurs when an output port is congested and a cell is waiting in the input port. The switch must drop some of the cells in the output port. Some switches randomly discard the cells, and all stations must

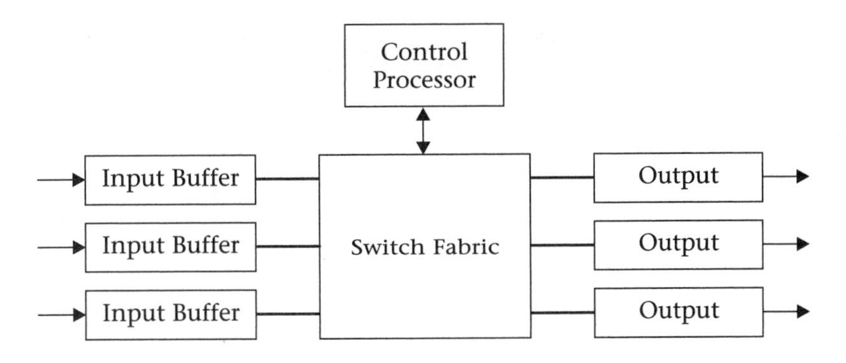


**FIGURE 22.6**
General architecture of an ATM switch

retransmit all the cells. Also, some switches have intelligence systems that drop cells belonging to one source.

## 22.6 ATM Connection Setup Through Signaling

ATM signaling is initiated by an end user who wants to establish a connection through an ATM network to a destination. The signaling packet is sent through the network from switch to switch with VPI = 0 and VCI = 5, and connection identifications are set up for each switch until the packet reaches its destination point. Figure 22.7 shows the first ATM user making a connection with an ATM switch or ATM network. Then the ATM network makes a connection with the ATM end user.

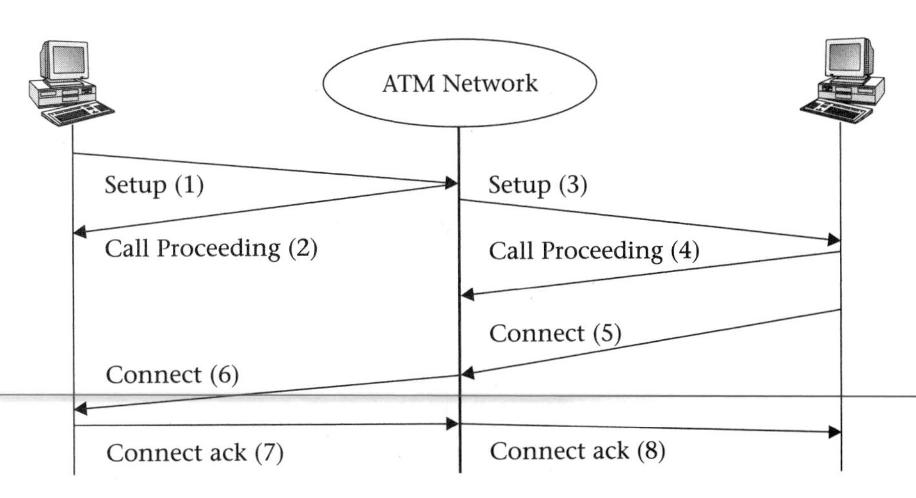


**FIGURE 22.7**
Connection set up through ATM signaling

## 22.7 ATM Cell Format

**ATM cell format** uses VLSI technology to segment data at high speeds to the cell. Each cell consists of 53 bytes, in which there is a 5-byte header and a 48-byte payload, as shown in Figure 22.8. ATM defines two header formats: one for the User Network Interface (UNI) cell header and another for the Network-to-Network Interface (NNI) cell header.

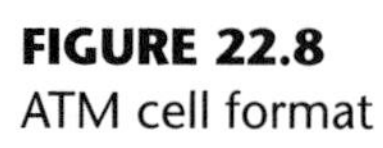

**FIGURE 22.8**
ATM cell format

| Payload | Header |
|---|---|
| 48 bytes | 5 bytes |

### UNI Cell Header

The **User Network Interface (UNI)** header cell defines communication between ATM end stations (workstation or router) and ATM switches. Figure 22.9 illustrates the UNI header cell format.

| Field length in bits | 4 | 8 | 16 | 3 | 1 | 8 |
|---|---|---|---|---|---|---|
| | GFC | VPI | VCI | PT | CLP | HEC |

**FIGURE 22.9**
UNI ATM cell header

**GFC:** 4-bit Generic Flow control can be used to provide local functions, such as identifying multiple stations sharing a single ATM interface. This is currently not used.

**VPI:** 8-bit Virtual Path Identifier used with VCI to identify the next destination of a cell as it passes through a series of ATM switches.

**VCI:** 16-bit Virtual Channel Identifier used with VPI to identify the next destination of a cell as it passes through a series of ATM switches.

**PT:** 3-bit Payload Type. The first bit indicates whether the payload is data or control data. The second bit indicates congestion, and the third bit indicates whether the cell is the last cell in the series that represents the AAL5 frame.

**CLP:** 1-bit Congestion Loss Priority, which indicates whether a cell should be discarded if it encounters extreme congestion. This bit is used for quality of service (QoS).

**HEC:** 8-bit Header Error Control. These checksums calculate only for the header. This is an 8-bit CRC that can detect all single errors and certain multiple-bit errors and can correct single-bit errors.

**NNI Cell Header**

The **Network-to-Network Interface (NNI)** defines communication between ATM switches. The format of an NNI header cell is shown in Figure 22.10.

**FIGURE 22.10**
NNI cell header format

| Field length in bits | 12 | 16 | 3 | 1 | 8 |
|---|---|---|---|---|---|
| | VPI | VCI | PT | CLP | HEC |

The VPI field is 12 bits, allowing ATM switches to assign a larger value to the VPI.

## 22.8 ATM Protocol

Figure 22.11 shows the ATM endpoint operational model and the ATM switch operational model. The ATM endpoints consist of three layers: the

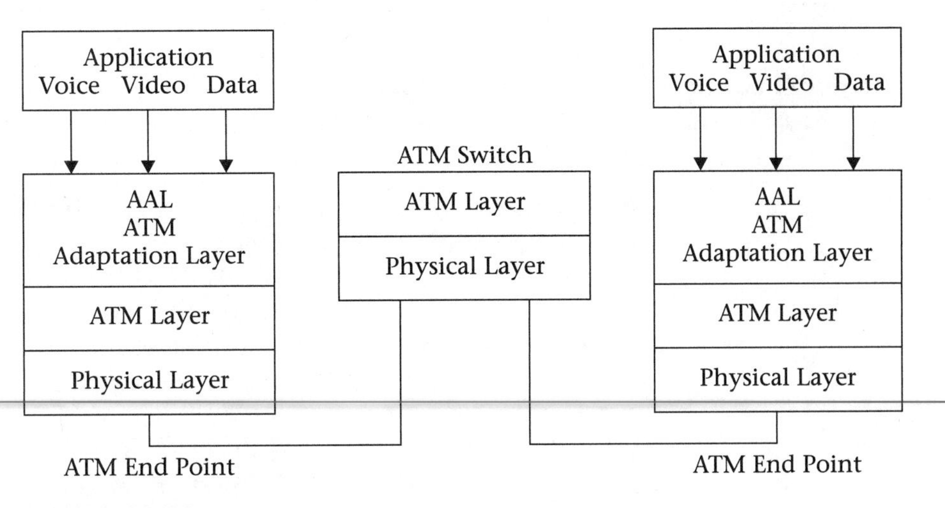


**FIGURE 22.11**
ATM endpoints connected to an ATM switch

ATM Adaptation layer (AAL), the ATM layer, and the Physical layer. An ATM switch consists of an ATM layer and a Physical layer. The ATM Adaptation layer, ATM layer, and Physical layer are divided into sublayers, as shown in Table 22.2.

**TABLE 22.2** ATM Sublayers

| Layer | Sublayer | Functions |
|---|---|---|
| ATM Adaptation Layer | Convergence Sublayer (CS) | Provides service to the lower and upper layers; dependent on the type of AAL. |
| | Segmentation and Reassembly (SAR) | Segmentation of information to 48 bytes. |
| ATM Layer | | Cell header generation and extraction.<br>Cell demultiplexing and multiplexing.<br>VPI/VCI translation. |
| Physical Layer | Transmission Convergence (TC) | Cell delineation, HEC generation, and cell verification.<br>Add idle cell if necessary. |
| | Physical Medium Dependent (PMD) | Bit timing and physical medium dependent. |

## Physical Layer Functions

As shown in Table 22.2, the Physical layer is divided into two sublayers, the Physical Medium Dependent (PMD) layer and the Transmission Convergence (TC) layer.

***Physical Medium Dependent (PMD) Functions.*** The PMD sublayer provides bit transmission, coding, electrical and optical conversion, and bit timing (the generation of signals suitable for transmission media).

Many LAN technologies, such as Ethernet and Token Ring, specify certain transmission media. Optical cable, coaxial cable, and twisted-pair cable can all be used as ATM transmission media.

The ATM forum has defined different types of transmission media, such as SONET, which uses Multi-Mode Fiber (MMF) cable and Single Mode Fiber (SMF) cable. Multi-mode fiber cable uses the same 4B/5B encoding as FDDI and twisted-pair cable. Table 22.3 gives examples of physical transmission media for ATM.

**TABLE 22.3** Transmission Media for ATM Network

| Layer | Rates (Mbps) | Media Type |
|---|---|---|
| OC-48 | 2488 | SMF |
| OC-24 | 1244 | SMF |
| OC-12 | 622.08 | SMF |
| STS-12C | 662.08 | UTP-5 |
| STS-3 | 155.52 | MMF and UTP-5 |
| STS-1 | 51.84 | UTP-3 |
| DS-3 | 45 | Coaxial Cable |

SMF = Single Mode Fiber
MMF = Multi-Mode Fiber
UTP-3 = Unshielded Twisted-Pair Cat-3
OC-48 = Optical Carrier
STS-1 = Synchronous Transport Signal-1
DS-3 = Digital Signal

***Transmission Convergence Sublayer (TC).*** The functions of the TC sublayer are as follows:

- Extracting the cell from the Physical layer (extracting cells from the SONET envelope).
- Cell delineation—scrambling the cell before transmission and descrambling the cell after transmission.
- The TC receives a cell from the Physical layer, calculates the HEC header, and then compares it with the cell's HEC header. The TC uses the results of the comparison for error correction in the cell header. If the error cannot be corrected, the cell will be discarded.
- The Transmission Convergence sublayer receives the cell from the ATM layer and generates the HEC; it then adds the HEC to the cell.

**ATM Layer** The ATM layer performs the following functions:

- **Cell Header Extraction and Generation**: Add the ATM cell header (except HEC) and remove the cell headers of incoming cells in an end station.
- **VPI/VCI translation**: Done in the ATM switch. The VPI/VCI value of an incoming cell is translated to a new VPI/VCI value for the outgoing cell.

- **Generic Flow Control:** Done on the user side (User-to-Network Interface) to determine the destination of the receiving cell.
- **Multiplexes and Demultiplexes cells**: Can be done from several different connections at the ATM switch.

### ATM Adaptation Layer (AAL)

The **ATM Adaptation layer (AAL)** converts the large Service Data Unit (SDU) of the upper layer to 48 bytes for the ATM cell payload. The AAL is designed so that ATM can become more flexible, and be able to handle all types of traffic. The ATM Adaptation layer is divided into two sublayers: the Convergence sublayer (CS) and Segmentation and Reassembly sublayer (SAR).

After a connection is set up by the ATM signaling protocol, the Convergence sublayer accepts higher layer traffic for transmission. The SAR segments each packet received from the CS into smaller units and adds a header or trailer depending on the type of AAL, to form the 48 bytes of payload.

ATM can be used for various applications. Therefore, different types of AALs are needed to provide service to upper layer applications. The AALs are divided into four classes of traffic and the ATM layer offers four types of AAL protocol. Each AAL protocol is used for a specific application, from class A through class D, as shown in Table 22.4.

**TABLE 22.4** AAL Types and Their Application

| Application Type | Class A | Class B | Class C | Class D |
|---|---|---|---|---|
| AAL Protocol | AAL1 | AAL2 | AAL¾ and AAL5 | |
| Timing Relation | Required | Required | Not Required | |
| Bite Rate | Constant Bit Rate | Variable Bit Rate | | |
| Connection Type | Connection Oriented | Connection Oriented | | Connectionless |
| Application | Audio | Compressed Video | Data | |

**Class A:** Constant Bit Rate (CBR). Connection oriented, with a timing relation between source and destination required. Applied to uncompressed voice and video.

**Class B:** Variable Bit Rate (VBR). Connection oriented, with a timing relation between source and destination required. Applied to compressed video and audio.

**Class C:** Variable Bit Rate (VBR). Connection oriented, with a timing relation between source and destination not required. Applied to data.

**Class D:** Variable Bit Rate (VBR). Connectionless, with a timing relation between source and destination not required. Applied to connectionless data transfer.

## 22.9 Types of Adaptation Layers

ATM networks offers four different types of adaptation layers, in order to handle different types of traffic: AAL1, AAL2, AAL¾, and AAL5.

**Adaptation Layer Type 1** Adaptation Layer Type 1 (AAL1) is designed to carry Class A traffic with constant bit rates (CBR), such as circuit emulation (which provides the same kind of services as a traditional leased-line or a time-division multiplexer) and is used for uncompressed video and telephone traffic applications. It is important not to disorder cells for voice communication; therefore a sequence number is added in the SAR sublayer. Figure 22.12 shows the AAL1 cell format.

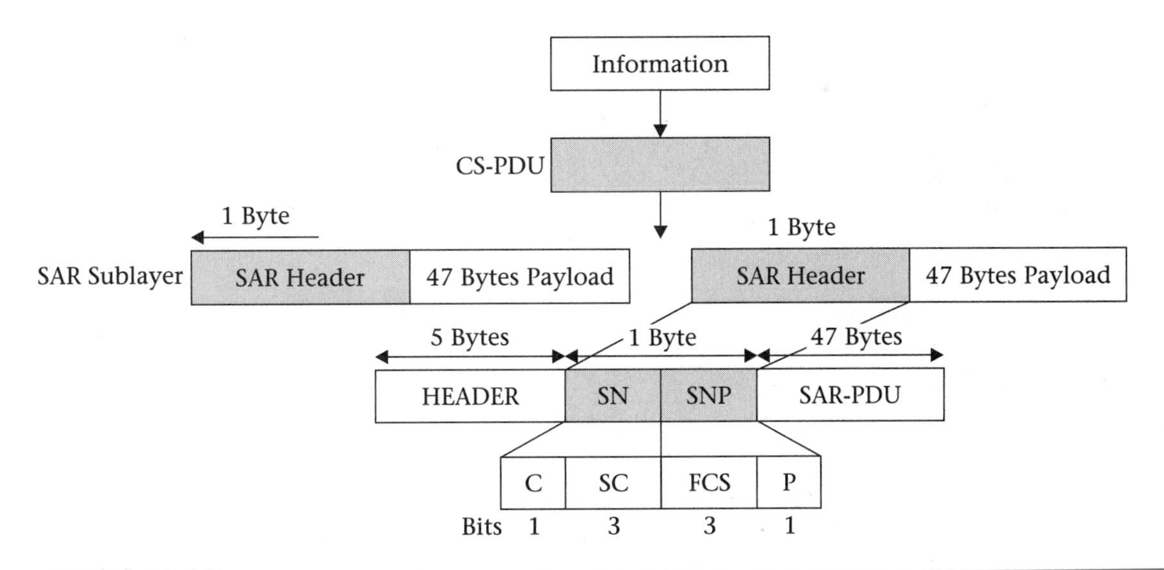


**FIGURE 22.12**
AAL1 cell format

The following describes the functions of the SAR header in Figure 22.12:

**SN:** Sequence Number is four bits. The SC field is used for cell sequence numbers to detect any missing cell or disordered cell. The C bit is used for timing information.

**SNP:** Sequence Number Protection field. The FCS bit is used for error detection on an SN field using CRC polynomial $X^3 + X^2 + 1$. The P bit is used as a parity bit for the FCS field.

## Adaptation Layer Type 2

Adaptation Layer Type 2 (AAL2) is used for variable bit rate applications such as compressed voice and audio. Figure 22.13 shows the AAL2 cell format. Information from the Application layer is passed to the SAR sublayer. The SAR sublayer segments the receiving information into 45 bytes of payload and adds a 1-byte header and a 2-byte trailer to each payload. Then SAR passes them to the ATM layer.

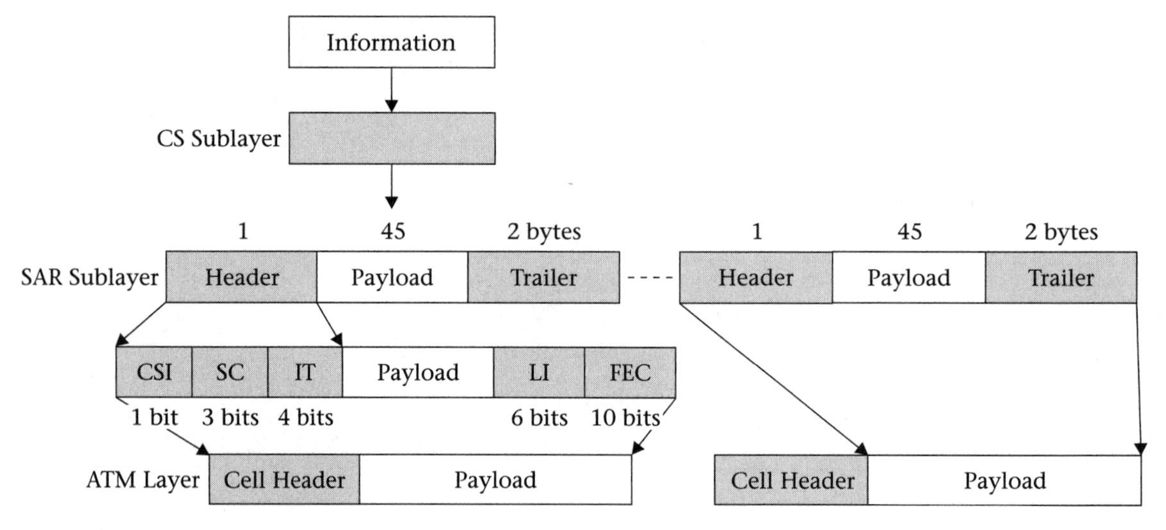


**FIGURE 22.13**
AAL2 protocol data unit

The following describe the functions of the SAR header in Figure 22:13:

**CSI:** Convergence sublayer identification is not defined yet.

**SC:** Sequence Count is used to detect lost cells.

**IT:** Information Type is used to indicate the position of payload, relative to the entire message. The IT indicates the beginning of the message, continuation of the message, and end of the message.

**LI:** Length Indicator is used to indicate the length of the message.

**FEC:** Forward Error Correction is used to correct some errors.

## Adaptation Layer Type ¾

Adaptation Layer Type ¾ (AAL¾) started out as two separate Adaptation layers (AAL3 and AAL4). The specification of two AALs merged together and the new type is called *AAL¾*. AAL¾ is designed to take variable length frames up to 64-k bytes and segment them into cells.

ITU recommends AAL¾ for data that is sensitive to loss but not to delay. AAL¾ can be used in connection-oriented and connectionless-oriented transmissions. Figure 22.14 shows the AAL¾ format. The Convergence sublayer accepts information from the upper layer and adds four bytes to the CS header and four bytes to the CS trailer (this is called *CS-PDU*). The CS transfers CS-PDU to the SAR sublayer, and SAR sublayer segments the CS-PDU into a 44-byte payload. The SAR adds a 2-byte SAR header and a 2-byte SAR trailer to each payload.

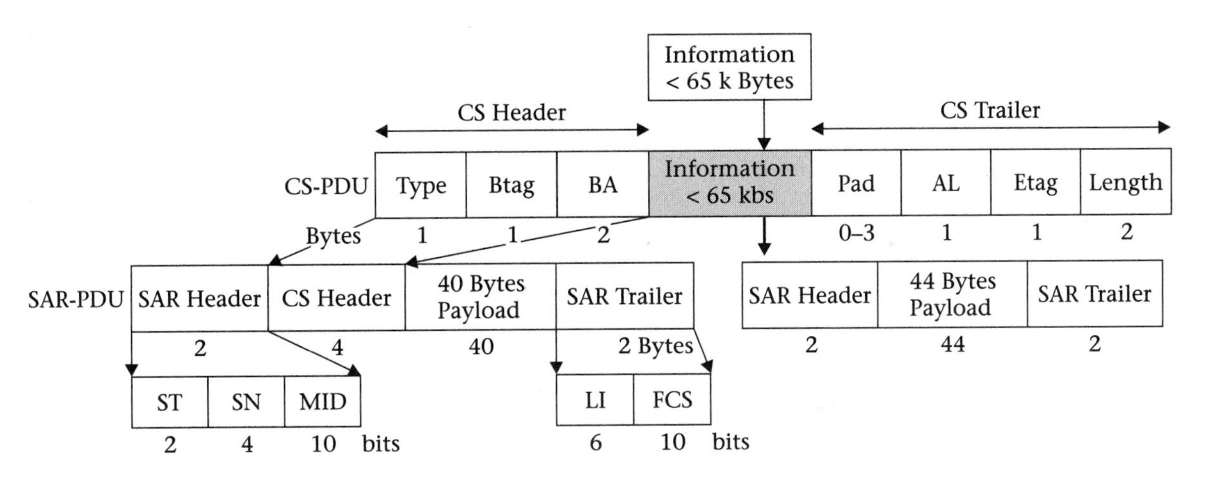


**FIGURE 22.14**
AAL¾ CS sublayer PDU and SAR sublayer PDU

The function of each field in CS-DPU is as follows:

**Type Field:** Set to zero.

**BEtag and Etag:** These two fields must have the same number to indicate the start and end of CS-PDU, CS-PDU to the receiver. The transmitter changes the Etag and BEtag for each successive CS-PDU.

**BA:** Buffer allocation, used by the destination to allocate buffer space for incoming cells.

**PAD:** This field is 4 bytes and is used to make the total number of bytes in CS-PDU multiples of 4 bytes.

**Length:** Indicates the size of a complete CS-PDU.

**SAR:** The SAR sublayer segments the CS-PDU to 44 bytes and adds its header and tailer to each segmented CS-PDU (as shown in Figure 20.14), to form the 48 bytes of an ATM cell.

**ST:** Segment Type indicates if the cell is first, last, or a continuation.

**AL:** Alignment byte is a dummy byte to make the trailer 4 bytes.

**SN:** Sequence number for detecting missing cells.

**MID:** Multiplexer Identifier.

**LI:** Indicates the number of useful bytes in the SAR-PDU.

**FCS:** Frame Check Sequence is used for error detection.

These 48 bytes are passed to the ATM layer, which adds a 5-byte header to it.

**Adaptation Layer Type 5**

Adaptation Layer Type 5 (AAL5) is an efficient way to transfer information in ATM, because all control information is in the last cell. The AAL5 format can be used for LAN emulation, and can handle very large Ethernet and Token Ring frame formats. AAL5 is used for variable lengths of information up to 65-K bytes. These 65-K bytes are segmented into a series of cells, each with the same header. The last cell arrives at the destination carrying all the control information needed to handle the packet. Figure 22.15 shows the AAL5 cell format. AAL5 provides services similar to those that AAL¾ provides, but uses fewer control bytes.

Information from upper layers passes to the CS. The CS adds a trailer to the information to generate a CS-PDU. The CS passes the CS-PDU to the SAR for segmentation of up to 48 bytes of cells. The segmented cells are passed to the ATM layer. The ATM layer adds a 5-byte header to each cell and passes the cells to the Physical layer for transmission.

The following describes the function of each field of the Convergence sublayer (CS) trailer:

**UU:** AAL layer User-to-User Identification.

**CPI:** Common Port Identification (it is not defined).

**Length:** Indicates number of bytes in payload or data field.

**FCS:** Used for error detection in CSC-PDU.

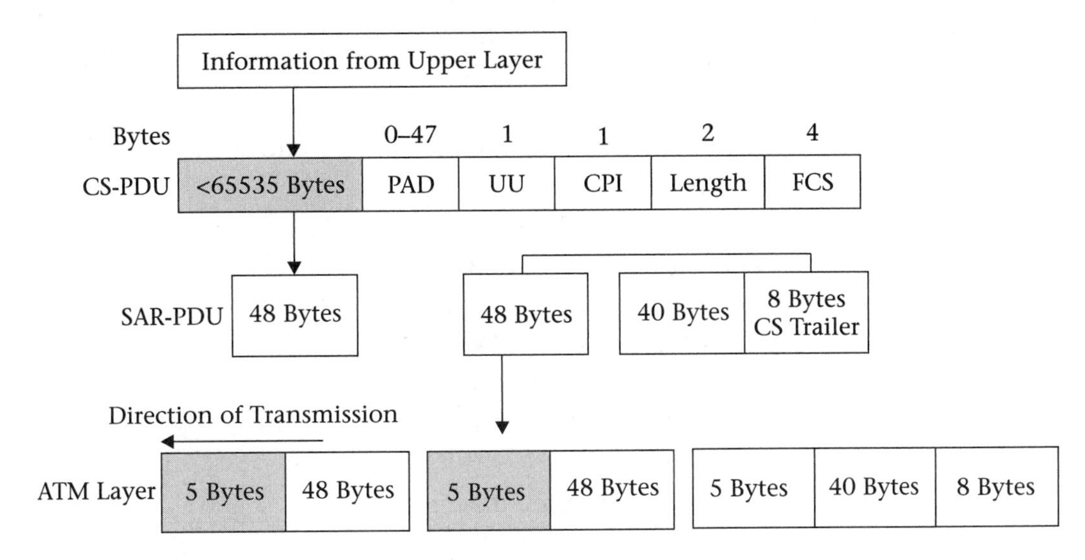


**FIGURE 22.15**
AAL5 cell format

## 22.10 Comparing ATM and Gigabit Ethernet

ATM can be used for both LANs and WANs, though it is not popular for LANs because it is expensive and complex. Most LAN designers prefer Gigabit Ethernet to ATM. However, unlike Gigabit Ethernet, ATM supports a constant bit rate for video and image transmission, provides real-time communication, and offers quality of service (QoS). Adapting the QoS protocol (IEEE 802.1Q) in the upper layer of Gigabit Ethernet enables it to prioritize packets and assign high priority to packets used in real-time communication and assign low priority to packets that are not time sensitive.

Gigabit Ethernet is growing so fast that it has become one of the most popular LANs, and ATM is more widely used for WANs.

## Summary

- Asynchronous Transfer Mode (ATM) is the next generation of networking technology that can handle all types of information such as data, voice, video, and images.
- An ATM network consists of an ATM switch and an ATM end user.
- The advantages of ATM networks are variable bandwidth, Quality of Service (QoS), the fact that it handles all types of traffic, high bandwidth, and small cell size.
- An ATM cell is made up of 48 bytes of payload and 5 bytes of header.
- ATM offers two types of connections: Permanent Virtual Connection (PVC) and Switched Virtual Connection (SVC).
- The connection identifiers in an ATM cell are Virtual Path Identifier (VPI) and Virtual Channel Identifier (VCI).
- PVC involves manual setup and disconnection by a network manager.
- SVC involves automatic setup and disconnection by a signaling protocol.
- An ATM switch is an electronic device with several input and output ports. It can process millions of cells per second.
- The ATM end user protocol consists of the ATM Adaptation layer, ATM layer, and Physical layer.
- The ATM switch protocol consists of the ATM layer and Physical layer.
- ATM Adaptation Layer Type1 (AAL1) is designed to carry traffic with a constant bit rate.
- AAL¾ is used to transfer variable length frames with up to 65-k bytes.
- AAL5 is an efficient way to transfer large frames, such as LAN emulation.

## Key Terms

Asynchronous Transfer Mode (ATM)

ATM Adaptation Layer (AAL)

ATM Cell Format

ATM Forum

ATM Layer

ATM Protocol

Connection Identifiers

Fabric Blocking

Head of the Line Blocking

Network-to-Network Interface (NNI)
Permanent Virtual Connection (PVC)
Switched Virtual Connection (SVC)
User Network Interface (UNI)
Virtual Channel Identifier (VCI)
Virtual Path Identifier (VPI)

## Review Questions

### • Multiple Choice Questions

1. ATM is a transfer protocol for ______.
   a. ISDN
   b. N-ISDN
   c. B-ISDN
   d. FDDI
2. ______ is not a component of ATM.
   a. Software
   b. NIC
   c. An ATM switch
   d. A video card
3. ATM has a fixed cell size of ______ bytes.
   a. 10
   b. 12
   c. 53
   d. 64
4. ATM has a ______ byte header.
   a. 1-
   b. 2-
   c. 5-
   d. 10-
5. ATM has a ______ byte payload.
   a. 1-
   b. 48-
   c. 5-
   d. 58-
6. ATM offers ______ type (s) of connections.
   a. 6
   b. 2
   c. 3
   d. 1

7. An ATM switch will be involved with ______ type(s) of congestion.
   a. 1
   b. 2
   c. 3
   d. 4
8. ATM offers ______.
   a. bandwidth on demand
   b. variable bandwidth
   c. constant bandwidth
   d. none of the above
9. The organization that ratified the standard for ATM is ______.
   a. ITU
   b. IEEE
   c. EIA
   d. IBM
10. ATM can be used for ______.
    a. LANs
    b. WANs
    c. LANs and WANs
    d. none of the above
11. The signaling protocol for ATM is used to set up a ______.
    a. permanent virtual connection
    b. switched virtual connection
    c. virtual connection
    d. none of the above
12. The type of ATM Adaptation layer used for real-time video and voice transmission is ______.
    a. ALL¾
    b. AAL2
    c. AAL1
    d. AAL5
13. The type of AAL used for transferring data efficiently is ______.
    a. AAL2
    b. AAL1
    c. AAL5
    d. AAL¾
14. ATM class A service is used for ______.
    a. applications that require variable bit rate and are connection oriented

b. applications that require constant bit rate and are connection oriented
c. applications that require variable bit rate and are connection oriented
d. none of the above

15. ATM class B service is used for ______.
a. applications that require variable bit rate and are connection oriented
b. applications that require constant bit rate and are connection oriented
c. applications that require variable bit rate and are connection oriented
d. none of the above

16. The type of AAL used for class A service is ______.
a. AAL2
b. AAL1
c. AAL5
d. AAL¾

17. The connection identifiers in an ATM cell header are ______.
a. VPI
b. VCI
c. VPI and VCI
d. none of the above

18. The ATM protocol at the end user consists of ______.
a. 3 layers
b. 2 layers
c. 1 layer
d. 4 layers

19. An ATM switch protocol consists of ______.
a. 3 layers
b. 2 layers
c. 1 layer
d. 4 layers

20. The ATM layer architecture from top to bottom is ______.
a. ATM Adaptation layer, Physical layer, and ATM layer
b. ATM Adaptation layer, ATM layer, and Physical layer
c. ATM layer, ATM Adaptation layer, and Physical layer
d. none of the above

21. The function of the ATM Adaptation layer is ______.
    a. segmentation and reassembly of the data unit
    b. cell header generation
    c. HEC generation
    d. to transmit information to destination
22. The function of an ATM layer is ______.
    a. segmentation and reassembly of the data unit
    b. cell header generation
    c. HEC generation
    d. b and c
23. An ATM switch consists of ______
    a. ATM Adaptation layer and Physical layer
    b. ATM layer and Physical layer
    c. ATM Adaptation layer and Physical layer
    d. b and c

## • Short Answer Questions

1. What does ATM stand for?
2. What is the data rate of ATM?
3. What are the characteristics of ATM?
4. List the ATM switching types.
5. What is the data unit of ATM?
6. What are the advantages of ATM?
7. Show an ATM cell.
8. Explain cell switching.
9. How many bytes are in an ATM cell header?
10. How many bytes is an ATM payload?
11. Explain permanent virtual connection.
12. How many bytes is an ATM cell?
13. What are connection identifiers?
14. Explain switched virtual connection.
15. What causes blocking in an ATM switch?
16. What are the functions of an ATM switch?
17. Show the ATM protocol.
18. What are the types of blocking in ATM switch?
19. What is the function of the ATM Adaptation layer?

CHAPTER 23

# Network Security

OUTLINE



OBJECTIVES

After completing this chapter, you should be able to:

- Explain different types of network attacks
- Define the elements of network security
- Understand the basics of cryptography
- Describe different types of encryptions
- Understand the application of digital certificates and signatures
- Discuss the applications of the Secure Socket Layer (SSL)
- Explain the operation of EAP
- Discuss firewall applications

INTRODUCTION

The transfer of information across networks and the Internet is increasing exponentially in e-commerce and business transactions. Network security plays an important role in successful e-commerce. Currently, many people are accessing their bank accounts, buying and selling stocks, and paying bills over the Internet. People using these services need to rely on secure transactions. This means that the information transmitted should not be able to be accessed or modified by anyone other than the authorized user. Network security is implemented to protect information in transit and also to protect a system from an attack. Attacks are categorized as direct attacks and indirect attacks.

- **Direct attack:** The attacker is able to disrupt the system by breaking passwords and accessing the system to modify information. For example, a person breaks the security of a bank server and then alters account information.
- **Indirect attack:** The attacker obtains the information and data in a system (such as a name, address, social security number, or credit card number). For example, a person using a cable modem can use a packet sniffer to capture packets transmitted over the modem and obtain such information. Packet sniffer is a piece of software that captures packets going out of and coming in to a network. The packet sniffer is legitimately used for network monitoring and analyzing.

## 23.1 Elements of Network Security

In order to have a secure network, the following network security components are necessary:

**Secrecy or confidentiality:** Secrecy provides privacy and protects information from being intercepted. If a person intercepts the information, it would be compromised.

**Authentication:** Authentication methods verify the identity of a person or computer accessing the network.

**Integrity:** Integrity maintains data consistency and prevents tampering with information.

**Non-repudiation:** Non-repudiation provides proof of origin to the recipient.

## 23.2 Introduction to Cryptography

**Cryptography** is the analysis and deciphering of codes and ciphers. In computer networking, cryptography is the science of keeping information secure. The process of encoding and decoding information is called

**encryption** and **decryption,** respectively. Figure 23.1 shows a cryptographic model.

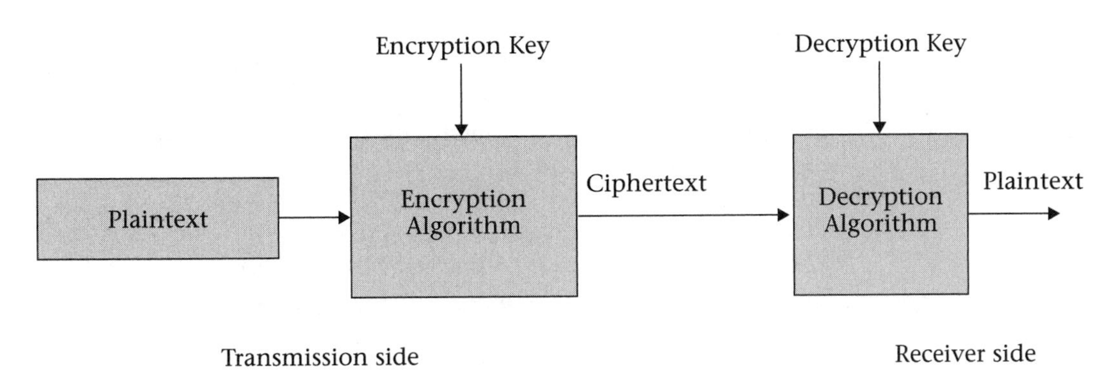


**FIGURE 23.1**
Cryptographic model

In this model, the message is referred to as **plaintext** or **cleartext** and the encrypted text is called **ciphertext.** The plaintext is encrypted by an encryption algorithm and an encryption key. The ciphertext is then transmitted over the communication channel. At the receiving side, the ciphertext is decrypted by a decryption algorithm and a decryption key, resulting in plaintext. The encryption and decryption algorithms are called **ciphers.** Cryptanalysis is the art of breaking ciphers or decrypting information without having the key. Cryptography can be divided into two classes: classical cryptography and modern cryptography. Classical cryptography was used historically for non-computing applications such as character substitution. Modern cryptography is now used for data communication. The types of modern cryptography are: **symmetric key cryptography** (private key or secret key) and **public key cryptography.**

## Symmetric Key Cryptography

In symmetric cryptography the transmitter and receiver of a message share a similar key for encryption and decryption, as shown in Figure 23.2.

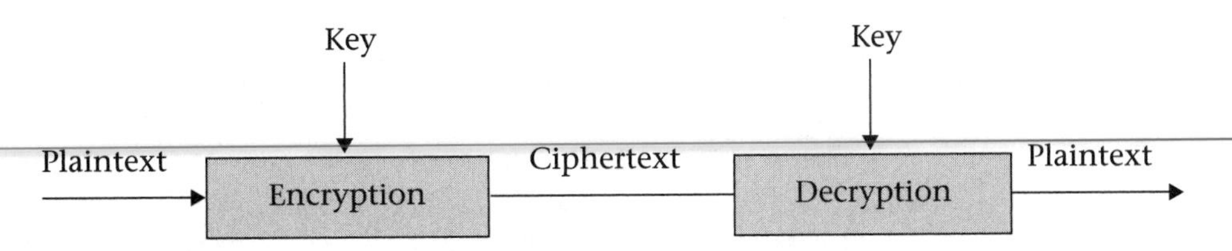


**FIGURE 23.2**
Block diagram of symmetric cryptography

Plaintext is encrypted using an encryption algorithm and a key, resulting in ciphertext, which is then transmitted to the receiver. The receiver uses the same key as the transmitter to decrypt the ciphertext into plain text.

The advantage of symmetric cryptography is that it is simple and fast. The disadvantage is that the transmitter and the receiver must exchange keys. The symmetric algorithm can be divided into **stream ciphers** and **block ciphers.**

***Stream Ciphers.*** The plaintext is encrypted one bit at a time; the text is converted to binary; and then the binary string is XORed with another binary string (the key) which results ciphertext.

### Example 23.1

| | | | | |
|---|---|---|---|---|
| The word "BOOK" in binary form is: | 1000010 | 1001111 | 1001111 | 1001011 |
| The key is: | 0111111 | 0111110 | 0111010 | 1010101 |
| The result of XOR is ciphertext: | 1111101 | 1110001 | 1110101 | 0011110 |

The ciphertext is transmitted to the receiver, which uses the same key to decrypt the text.

***Block Ciphers.*** A block cipher is a group of bits that are encrypted such as the **Data Encryption Standard (DES)** that takes 64 bits per block. Some of the classical block cipher encryptions are substitution and transposition.

***Substitution Ciphers.*** Substitution replaces each character in the text with a different character. This is an early cryptography method that was developed by Caesar and is known as the *Caesar Cipher*. Caesar Cipher substitutes the original letter with a different letter that comes later in the order of the alphabet. For example, the letter "A" is replaced with the fourth letter in the alphabet, substituting "A" with "E." Another method of substitution is to add a number to the ASCII value of the letter. The ASCII value of "A" is 41h. Adding 4 to 41h results in 45h (the letter "E"). The number 4 is the value of the key.

**Example: 23.2.** Encrypt the word "BOOK" assuming the encryption key is 4. The word "BOOK" in ASCII code is 41, 4F, 4F, 4B. Adding 4 to each character results in 45, 53, 53, 4F, which stands for "ESSK."

***Transposition Ciphers.*** In a transposition cipher, the plaintext is divided into groups of characters where the number of groups defines the key. For example, the phrase "WELCOME TO CLASS" is divided into groups of four characters as follows:

| 1 | 2 | 3 | 4 |
|---|---|---|---|
| W | E | L | C |
| O | M | E | – |
| T | O | – | C |
| L | A | S | S |

Reading from columns 1 to 4, the ciphertext is "WOLTEMOALE-SC-CS." In this example, the key is 1234. Also note that the order of the key can be changed to 2341. As encryption methods become more sophisticated, they are harder to break. In 1997, the National Institute of Standards and Technology (NITS) recommended the Data Encryption Standard for federal use. The algorithm used for DES is called the *Data Encryption Algorithm.*

## Data Encryption Standard

Figure 23.3 shows a block diagram of DES. The message is divided into 64 bits of $b_{63}\ b_{62}\ \ldots\ b_0$ plaintext; the initial key is 56 bits. The function of the permutation box is to change the order of bits in the plaintext.

**FIGURE 23.3** Block diagram of DES

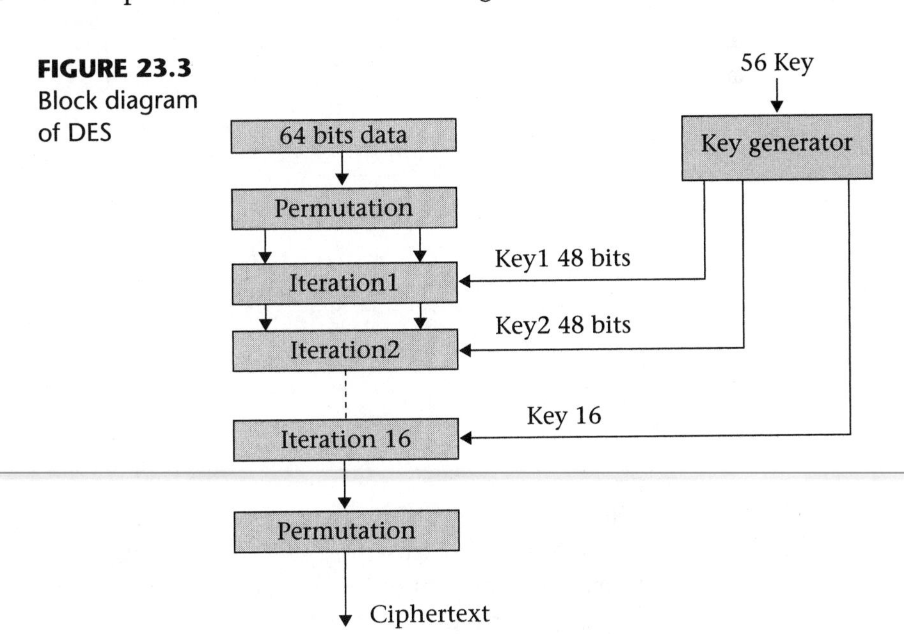

For example, bit $b_0$ becomes $b_{58}$, $b_1$ becomes $b_{50}$, etc. The function of the key generator is to generate 16 different 48-bit keys from a 56-bit key. As shown in Figure 23.3, the plaintext goes through 16 iterations using different keys and the results are permutated to generate the ciphertext.

Figure 23.4 shows a diagram of a single iteration of DES. The output of the permutation box is divided into two groups of 32 bits, called $L_0$ and $R_0$. $R_0$ is changed to 48 bits by using the expansion/permutation table. The output of the expansion/permutation box is XORed with the 48-bit key resulting in 48 bits. The 48 bits are converted by the substitution box (S-BOX is a table used for converting 48 bits to 32 bits) to 32 bits. These 32 bits are XORed with $L_0$ yielding the result $R_1$ for the next iteration. The output of $R_0$ becomes $L_1$ for the next iteration.

Key generation is done by dividing 56 bits into two groups of 28 bits. Both groups of 28 bits are shifted by one or two bits and the output becomes the input to the permutation table where 48 bits are generated for the first key. This method is repeated 16 times to produce 16 different keys.

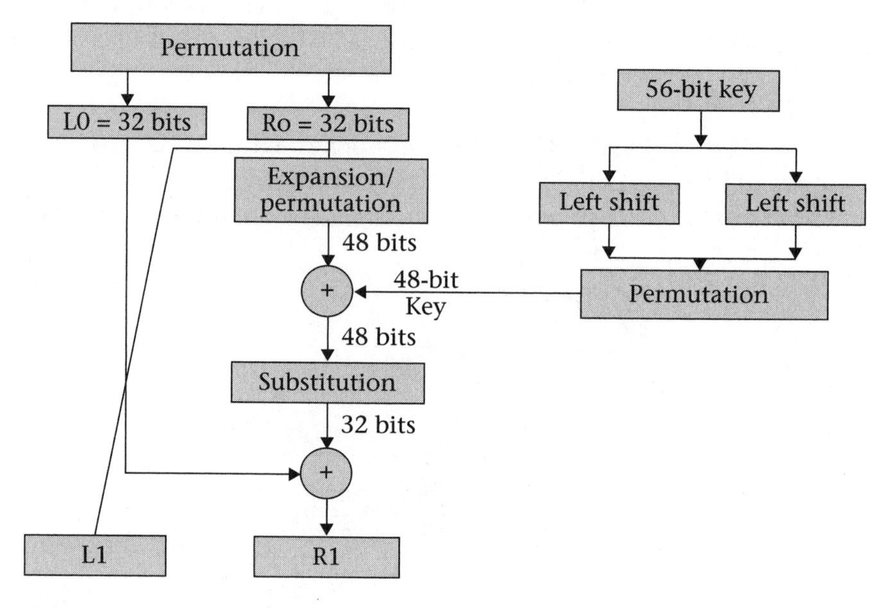


**FIGURE 23.4**
Block diagram of one iteration of DES

## Public Key Cryptography

Public key cryptography, which is also called *asymmetric cryptography,* uses a different key for encryption and decryption. Public key cryptography uses two keys: a public key for encrypting plaintext and a private key for decrypting ciphertext, as shown in Figure 23.5. The key generator produces the public key and the private key. The key generator transmits the public key to the transmitter. Then, the transmitter uses the public key and an encryption algorithm to produce the ciphertext. The transmitter then transmits the ciphertext to the receiver and the receiver uses the private key and a decryption algorithm to decrypt the ciphertext.

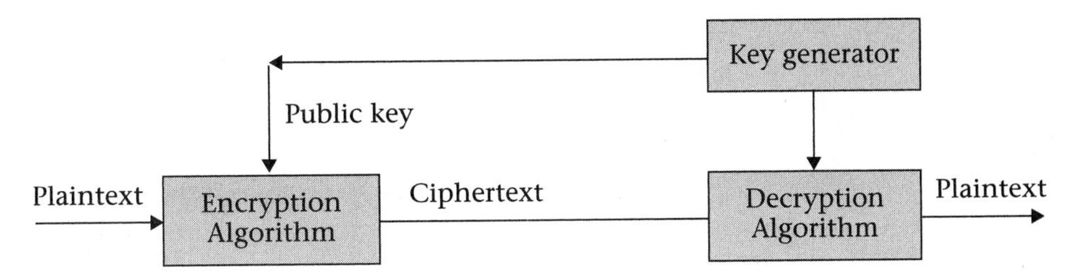


**FIGURE 23.5**
Public key encryption

The application of pubic key cryptography can be illustrated as follows: assume that a stockbroker serves 100 customers, as shown in Figure 23.6. The server generates a public key $K_p$ for each customer, private key $K_s$, a number N, and the private keys kept by the server. When any customer connects to the server, the server passes the customer $K_p$ and N. The customer uses these two numbers to encrypt information and transmit ciphertext to the broker's server. The server then uses the private key of the specific customer to decrypt information.

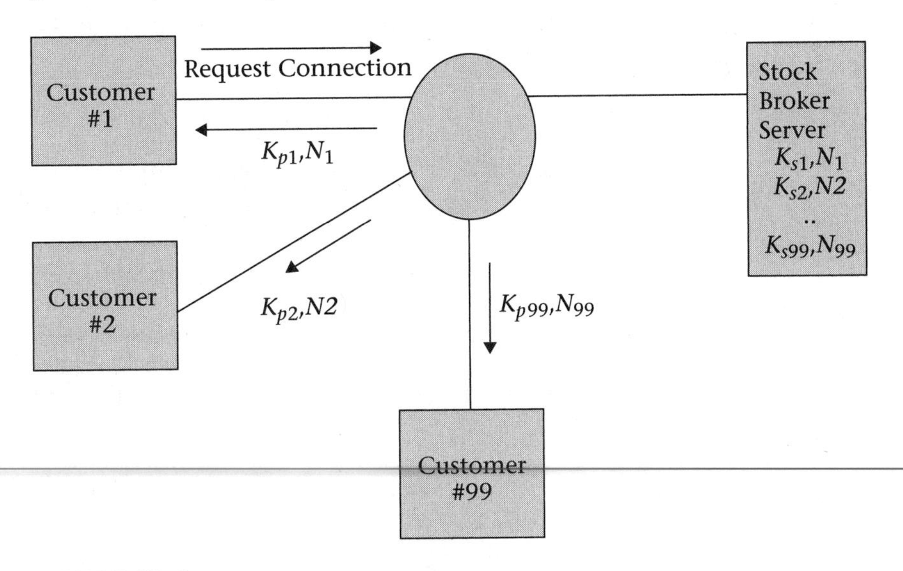


**FIGURE 23.6**
Application of public key cryptography

**RSA Algorithm** One of the most useful public key cryptography algorithms is the **RSA algorithm,** which is named after its inventors Ron Rivest, Adi Shamir, and Leonard Adleman. RSA is based on number theory and works as follows:

1. Select two prime numbers $p$ and $q$ (a prime number is a number divisible only by 1 and itself). The larger the number, the harder it is to break the RSA algorithm.
2. Let $N = p * q$, where $N$ is the product of two prime numbers, $p$ and $q$.
3. Let $Z = (p - 1) * (q - 1)$, where $Z$ is the product of $p - 1$ and $q - 1$.
4. Select $K_p$ (public key) such that $K_p$ is less than $N$ and has no common factors of $Z$.
5. Find $K_s$ (secret key) such that $K_s * K_p - 1$ is divisible by $Z$ or $K_s = (1 + n * Z)/K_p$; Where $n = 0,1,2, \ldots$
6. The transmitter uses $K_p$ and $N$ to encrypt message $M$ using equation 23.1:

$$EM = M^{K_p} \text{ modulo } N \tag{23.1}$$

   where:

   $EM$ is ciphertext

   The transmitter then transmits $EM$ to the receiver.

7. The receiver decrypts $EM$ using $K_s$ and $N$ according to equation 23.2:

$$M = EM^{K_s} \text{ Modulo } N \tag{23.2}$$

**Example 23.3.** Find $K_s$ and $K_p$ assuming $p = 3$ and $q = 5$. Also encrypt message $M = 3$ at the transmitted side; then decrypt ciphertext at the receiving side (in this example the prime numbers are purposely selected very small to give an understanding of the procedure).

$$N = 3 * 5 = 15 \text{ and } Z = (3 - 1)\,(5 - 1) = 8$$

Select a number that is less than $N$ and is not a factor of $Z$ (for this example, $K_p = 3$).

$$K_s = (1 + n * 8)/3 \text{ and } n = 4, \text{ thus } K_s = 11.$$

The transmitter uses $K_p$ to encrypt $M = 3$.

$$EM = 3^3 \text{ modulo } 15 = 12.$$

The transmitter transmits ciphertext of 12 to the receiver; then the receiver uses the secret key with $N$ to decrypt the ciphertext of the value 12.

$$M = EM^{K_s} \text{ modulo } N$$

or

$$M = 12^{11} \text{ modulo } 15 = 3$$

## 23.3 Digital Signatures

In business transactions, we sign various documents such as checks and contracts. The reason we sign documents is to indicate our understanding and acceptance of the contents of the document. The objective of a **digital signature** is to sign an electronic document rather than a paper document. Public cryptography is one of the methods used for digital signatures. An example of this is shown in Figure 23.7. In Figure 23.7, User A sends a document to User B. User B encrypts the document with its private key and sends it back to User A. User A stores the encrypted document in case User B denies the signature. User A decrypts the document for his or her use.

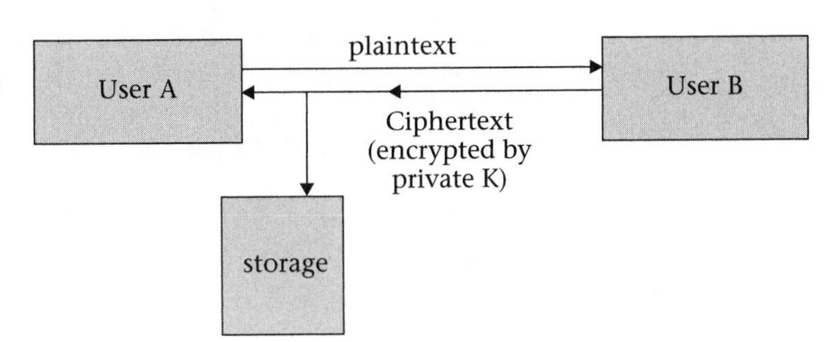


**FIGURE 23.7** Digital signature of a document using a public key

The disadvantage of this method is that the entire document must be encrypted and stored at the receiver side, thus requiring a large amount of memory. Another method is to sign a message digest (that is, a summary of the document contents) rather than the entire document. A message digest is a summary of the message, such as a frame check sequence.

Consider Figure 23.8. User A sends a document to User B for a signature. User B generates a message digest from the document and then encrypts the message digest and sends it back with the document to User A. User A stores the message digest as the signature of the document. The function that is used to generate a message digest is called a **hash function**.

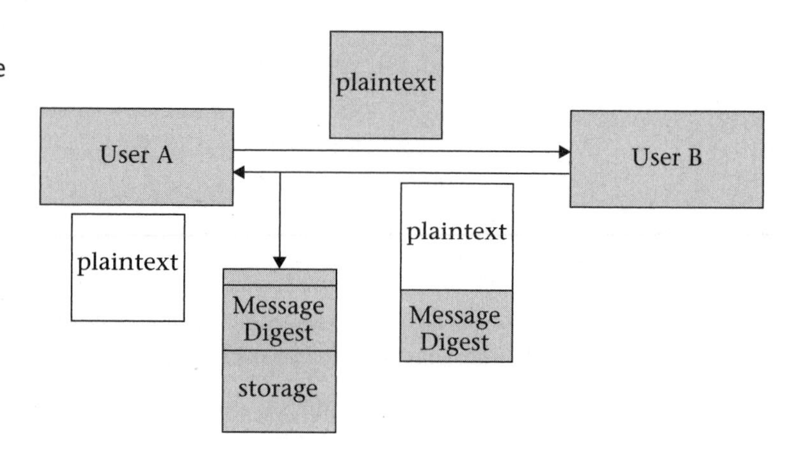


**FIGURE 23.8** Digital signature of a message digest

## Hash Function

Typically, a hash function generates a **hash value** (message digest) from a given message. One of the characteristics of a hash value is that it is impossible to use the hash value to generate the original message. Some of the methods that generate hash values from a plaintext are Frame Check Sequence (FCS) and checksum.

***Frame Check Sequence (FCS) as a Hash Value.*** As shown in Figure 23.8, User B generates the FCS of the information, encrypts the FCS, and sends it to User A. If one alters the document, the new FCS would differ from the one generated by User B.

***Checksum (Generating Checksum from a Message).*** Information can be broken into blocks of characters and arranged in columns. The message $6578100 can be broken into four ASCII characters per block as follows, and the checksum of each column can be calculated:

| $657 | 44 | 36 | 35 | 37 |
|---|---|---|---|---|
| 8100 | 38 | 31 | 30 | 30 |
| Checksum | 7C | 67 | 65 | 67 |

If the message changes to $5578200, it will generate the same checksum; therefore, the checksum method is not a strong message digest.

The simplest form of a hash function is to break the message into m blocks of n bits as follows:

| Block0 | $b_{01}$ | $b_{02}$ | $b_{03}$ | ... | $b_{0n}$ |
|---|---|---|---|---|---|
| Block1 | $b_{11}$ | $b_{12}$ | $b_{13}$ | ... | $b_{1n}$ |
| ... | ... | ... | ... | ... | ... |
| Blockm | $b_{m1}$ | $b_{m2}$ | $b_{m3}$ | ... | $b_{mn}$ |

Hash bits are represented by:

$$H = H_1H_2H_3 \dots Hn$$

where:

$$H_1 = b_{01} \text{ XORed } b_{11} \text{ XORed } b_{21} \text{ XORed} \dots \text{XORed bm1}$$

Or in general:

$$Hi = b_{i1} \text{ XORed } b_{i2} \text{ XORed } b_{i3} \text{ XORed } b_{i4} \text{ XORed} \dots \text{XORed } b_{in}$$

**Example 23.4.** Find the hash code for the word "WELCOME."

| W | 1010111 |
|---|---|
| E | 1000101 |
| L | 1001100 |
| C | 1000011 |
| O | 1001111 |
| M | 1001101 |
| E | 1000101 |
| Hash Code | 1011010 |

Some of the most popular message digests are as follows:

- MD4 is a message digest type 4 that produces 128-bit hash values or message digests.
- MD5 is an improved version of MD4 that produces 128-bit hash values.
- SHA (Secure Hash Algorithm) generates 160-bit hash values.

## 23.4 Kerberos

**Kerberos** is an authentication system that was developed at the Massachusetts Institute of Technology (MIT). It is used for the authentication of two systems over an unsecured network. The components of Kerberos are shown in Figure 23.9. For User A to access Server B, the following steps should take place:

1. User A logs into the Kerberos (KS) and requests access to Server B.
2. The KS asks User A for a password, a user ID, and a server ID.
3. KS checks the user's password and ID and also checks if User A has permission to access Server B.
4. If User A passes step 3, then KS creates a ticket that contains a user ID and a server ID. KS encrypts the ticket and transmits it to User A and Server B. Only Server B can decrypt the ticket.
5. User A sends the ticket to Server B; Server B decrypts the ticket and compares it with the ticket that was sent by KS. If both tickets are the same, then User A is allowed to access Server B.

**FIGURE 23.9**
Kerberos authentication system

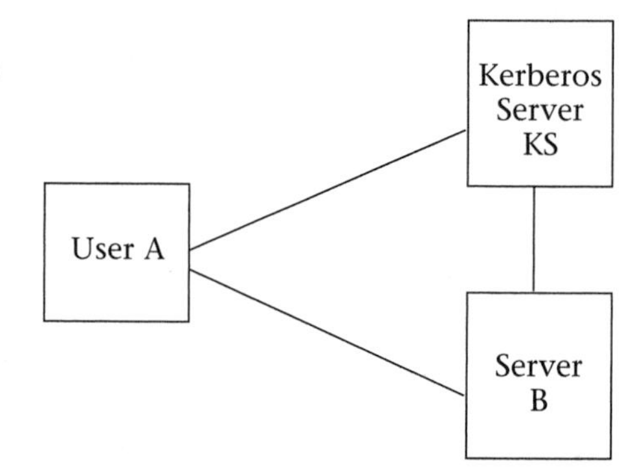


## 23.5 Certificates

In Figure 23.10, the customer requests a web page from a stockbroker by logging into the broker's server. The server requests an account number and a password for verification. If the information submitted by the customer is correct, then the server sends its public key to the customer for encryption.

Suppose an intruder installs a system between a customer's and a broker's server, as shown in Figure 23.10. When the customer requests the broker's web page, the intruder system responds with a fake web page. The customer logs into the fake web page and the intruder sends a fake public key to the customer to use for encryption. To check the authenticity of the public key, the public key must be certified to identify itself.

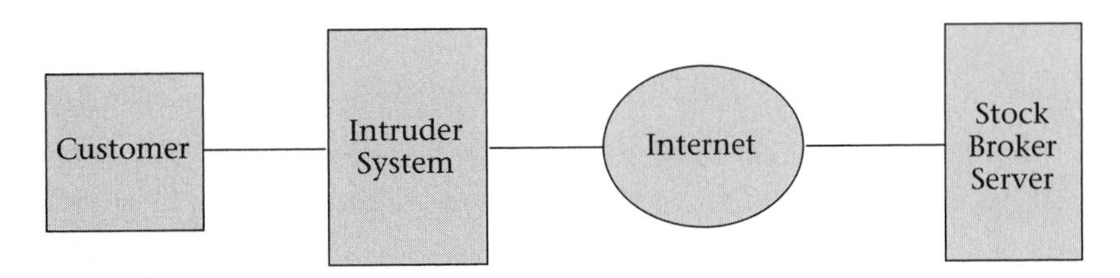


**FIGURE 23.10**
Application of certificates

**Certificates** verify the identity of a server, a program, or personal identification information such as a pictured driver license and number. A driver's license is issued by the Department of Motor Vehicles. Digital certificates are issued in a similar fashion by the Certificate Authority (CA). Some of the certification authorities are the U.S Postal Service, Commercenet, and Versign.

A digital certificate contains a public key, the full name of the certificate holder, the name of the certification authority, and a digital signature of the CA. All certificates use the X.509 standard, which has been recommended by the ITU. The X.509 standard defines the information in a certificate. This information is usually divided into two sections: the data section and signature section.

The data section of X.509 contains the following information:

- **Version Number:** Identifies the X.509 version such as V1, V2, and V3.
- **Certificate Serial Number:** This number is assigned to a certificate by the CA.
- **Signature Algorithm:** The type of algorithm that is used for the signature of a certificate.
- **Name of the Certificate Authority:** The name of the certificate authority.
- **Duration of the Certificate:** The period in which a certificate is valid.
- **Public Key Algorithm:** The algorithm that is used to generate a public key.

The signature section contains the following information:

- The cryptographic algorithm used by the CA to create its digital signatures.
- The digital signature of all data in the certificates.

For a user to check the validity of a certificate, he or she must perform the following functions:

1. Check the expiration date.
2. Check if the certified authority name is on the trusted list.
3. Find the digital certificates of all data and compare against those attached to the certificates.

When all of these tests are passed, then the certificate is valid. Figure 23.11 shows a certificate.

**FIGURE 23.11** Screenshot of a certificate

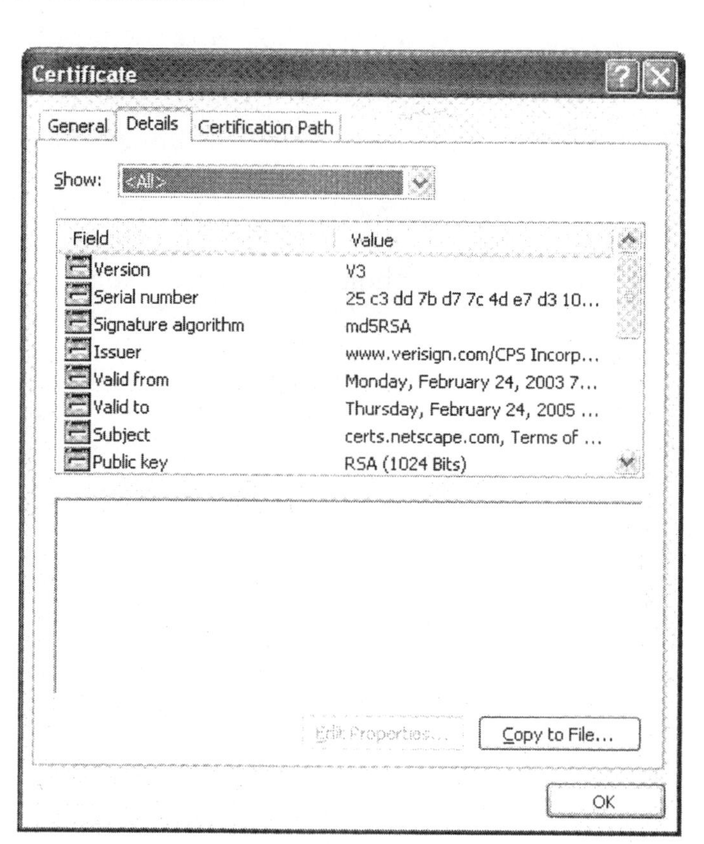


## 23.6 Secure Socket Layer (SSL) Protocol

The **Secure Socket Layer (SSL)** protocol was developed by Netscape Communication Corporation in 1995 for secure communication between a web browser and a web server. Version 3 is the latest version of SSL. Both Netscape Navigator and Internet Explorer support SSL. SSL provides data encryption, server authentication, message integrity, and client authentication (optional) for both ends. The SSL protocol is located between the Transport level and the Application level, as shown in Figure 23.12.

**FIGURE 23.12**
TCP/IP model with SSL

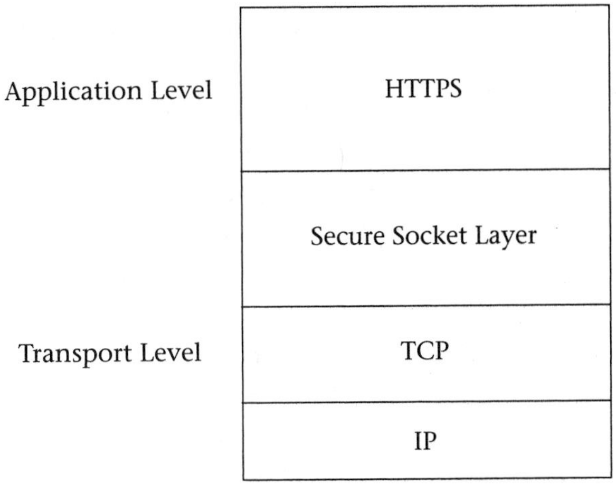

| Level | Layer |
|---|---|
| Application Level | HTTPS |
| | Secure Socket Layer |
| Transport Level | TCP |
| | IP |

HTTPS is Hypertext Transfer Protocol over SSL and it uses Netscape's SSL as a sublayer to replace HTTP. HTTPS uses port 443, whereas HTTP uses port 80 for interaction with TCP. SSL is composed of two layers: SSL Handshake protocol and SSL Record protocol.

**SSL Handshake Protocol**

The SSL handshake protocol performs a series of exchanging messages between SSL clients and servers to authenticate each other and negotiate encryption algorithms established via a secure connection. These messages are as follows:

1. The client requests a secure connection from the server by sending the following information to the server: SSL version, list of key sizes, encryption algorithms, and compression algorithms.
2. The server responds to the client's request with its SSL version, encryption algorithm, compression algorithm, and key size which is selected from the client's list.
3. The server transmits its X.509 certificate to the client, which includes the server's public key. The client then uses the certificate to identify the server.
4. The server requests the client's certificate (optional).
5. The client submits his or her certificate.

6. The sever sends the message "done" to the client.
7. The client chooses a 48-byte random number called the *pre-master key* and encrypts this number using the server public key (information from step 3), and then transmits it to the server.
8. Both the client and the server use random numbers to generate the secret key for encryption and decryption.
9. The client informs the server that all messages will be encrypted by a secret key.
10. The server sends a similar message and an SSL handshake is completed.

**SSL Record Protocol** At the transmitting side, the record protocol accepts the data from the upper layer and fragments the data for compression. The compressed data is then passed to the TCP layer. At the receiving side, the record protocol decrypts the information, decompresses it, and sends it to the application.

## 23.7 IEEE 802.1X Extensible Authentication Protocol (EAP)

One of the most important issues in the development of WLAN is security. In a wired LAN, users have a direct connection to the network whereas WLAN users do not. WLAN users must verify their identity before accessing the network. Authentication is a process used by a wireless station or a wired station to identify itself on the network. A password is used for authentication when accessing a network's resources.

Typically, the communication channel between a user and a WLAN is not secure. An attacker can monitor the communication channel and collect user passwords. The network administrator operates the access point (where a user accesses the network). The attacker can set up a rogue access point thus making a potential user access the hacker's access point. Therefore, a method is required such that the user can verify the authenticity of the access point. This method of authentication is called **mutual authentication.**

The IEEE 802.1X is a standard for port-based network access control and it is an open standard for authentication of both wireless and wired stations using an authentication server. Figure 23.13 shows the components of a network employing the IEEE 802.1X standard. These components are described next.

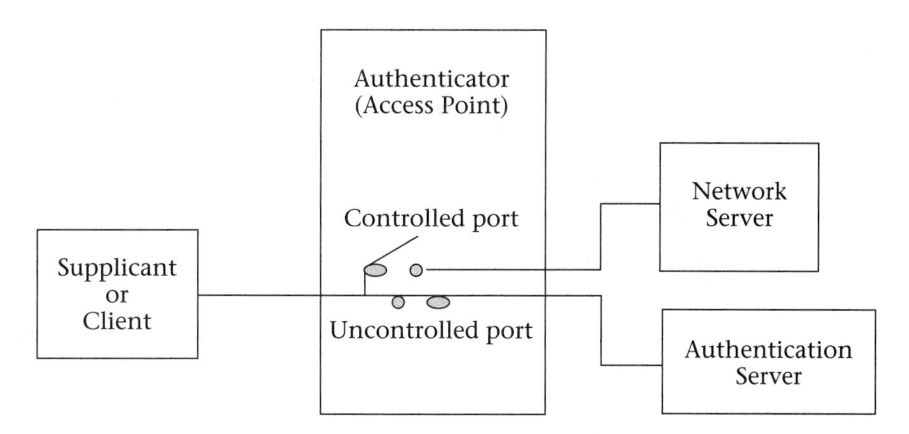


**FIGURE 23.13**
Components of IEEE 802.1X EAP

**Supplicant:** A supplicant can be any device using IEEE 802.11 protocol for networking or any PC connected to the network.

**Authenticator:** The authenticator can be an access point for 802.11 LAN or a switch for a wired LAN. The authenticator uses controlled and uncontrolled ports for authentication of a supplicant.

**Authentication Server:** The authentication server performs an authentication process for a supplicant. One type of authentication server is called a *RADIUS (Remote Authentication Dial-In User Service)*. A RADIUS is both an authentication server and an accounting server and together they are used to authorize a station on the network.

**Authentication Protocol:** An authentication protocol is a procedure that is used by the client and the authentication server for the authentication process. The IEEE 802.1X uses EAP for exchanging messages during the authentication process. IEEE 802.1X does not define the authentication protocol. Some of the most popular authentication protocols are as follows:

1. **EAP-MD5 (Message Digest 5):** EAP-MD5 is a password-based authentication protocol.
2. **EAP-TLS (EAP-Transport Layer Security):** EAP-TLS is based on mutual authentication of the client to the server and the server to the client. Both the client and the server must be assigned a digital certificate.
3. **LEAP:** (EAP Cisco): Used by CISCO.

4. **EAP-TTLS:** (Tunneled TLS).
5. **PEAP:** (Protected EAP).

**802.1X Operation** Figure 23.13 shows the components of an 802.1X. The authenticator (that is, the access point) contains both the logical controlled and the uncontrolled ports for authentication. When using the uncontrolled logical port, the client can communicate with the authentication server but he or she does not have access to the network services. The following steps describe the authentication process of a client by an authentication server:

1. The client requests an association with the AP (access point).
2. The AP responds to the client's request.
3. The client sends an EAP start message to the AP.
4. The AP requests the identity of the client (such as the user name).
5. The client sends an EAP packet containing his or her identity to the authentication server.
6. The authentication server identifies the client.
7. The authentication server can accept or reject the client request.
8. Upon accepting the client's request, the AP sends an EAP success packet to the client. Then, the AP authorizes the controlled port so that the client can access the network services.

## 23.8 Firewalls

A **firewall** is a system that is used for preventing unauthorized users or others to access private networks. Firewalls are located between private networks and the Internet (un-trusted network), as shown in Figure 23.14. Firewalls can be implemented by a combination of software and hardware.

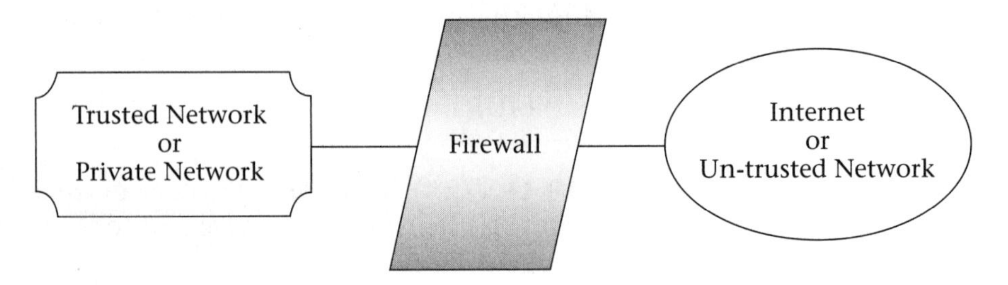


**FIGURE 23.14**
Location of a firewall

Firewalls examine all packets leaving and entering the private network, blocking packets that do not meet security criteria. Firewall technologies can be classified into the following types:

1. packet filtering firewall
2. application proxy server or network address translation firewall
3. stateful firewall

### Packet Filtering

A firewall can examine each incoming packet's headers, such as the IP header, TCP header, and UDP header; and, based on security criteria set by the network administration, accept or reject any packet. These criteria are as follows:

1. **Source IP Address and Destination IP Address:** A firewall can be programmed to block packets based on the IP address of the packet. This task is done at the Network level of the TCP/IP model.
2. **Protocol Type:** The firewall can block packets based on protocols such as TCP, UDP, and ICMP protocols. For example, the protocol filtering can filter any packet intended for ICMP. The protocol filtering is done at the Network level of the TCP/IP model.
3. **Source Port and Destination Port Filtering:** The firewall is able to block packets based on the port number of the incoming packets. The port number is used to define application protocols such as HTTP, SMTP, Telnet, and TFTP. For example, a firewall may block any incoming mail by blocking SMPT or port 25. Port filtering is done at the Transport level of the TCP/IP model.
4. **Packet Payload (Information on the Payload):** A firewall is able to block packets based on the information contained in the packets' payloads such as containing a "dirty word" or a specific sentence. The information filtering is done at the Application level of the TCP/IP model.

### Proxy Application Server or Network Address Translation (NAT)

The proxy server application is located between the trusted network and the untrusted network. Assume that a client of the private network wants to access an untrusted network, such as the Internet. First, it sends a request to the proxy server and the proxy server uses its IP address to send the request, on behalf of the client, to the destination server. Therefore, the destination server sees only the proxy server's IP address. The IP address of the clients in the private network are not exposed to the Internet. A proxy server should contain a software module for each application protocol such as HTTP, Telnet, SMTP, and TFTP.

**Stateful Firewall** A stateful firewall keeps track of its connection and it can determine whether the incoming packets belong to the current connections. This type of firewall can be configured such that it will deny any connection from an untrusted network to the trusted server. The untrusted network can send packets only to the trusted server if the trusted server requests them; otherwise the non-requested packets will be discarded by the firewall.

## Summary

- In a direct attack, the attacker breaks the security of the system.
- In an indirect attack, the attacker gets information and data (such as a name, address, social security number, or credit card number).
- Elements of network security are secrecy, authentication, integrity, and non-repudiation.
- When plaintext is encrypted by an encryption algorithm, it is called *ciphertext.*
- The algorithm that is used for encryption and decryption is called a *cipher.*
- Cryptography is the science of keeping information secret.
- Modern cryptography is classified into two types: symmetric cryptography (private key) and asymmetric cryptography (public key).
- In symmetric cryptography, the transmitter and receiver use the same key for encryption and decryption.
- In asymmetric cryptography (public key), the transmitter and receiver use different keys for encryption and decryption.
- The Data Encryption Algorithm (DEA) is an example of symmetric cryptography.
- The RSA Algorithm is an example of public key cryptography.
- A certificate is used to identify a server to a client.
- X.509 is a standard that is used in a certificate.
- A digital signature is a process that is used by a client to sign an electronic document.
- The Secure Socket Layer (SSL) is used for secure communication between a web browser and a web server.
- Authentication is a process to identify a client to a server, and a server to a client.

## Key Terms

- Authentication
- Authenticator
- Block Cipher
- Certificates
- Cipher
- Ciphertext
- Cleartext
- Cryptography
- Data Encryption Standard (DES)
- Decryption
- Digital Signature
- Direct Attack
- Encryption
- Extensible Authentication Protocol (EAP)
- Firewall
- Hash Function
- Hash Value
- Indirect Attack
- Integrity
- Kerberos
- Mutual Authentication
- Plaintext
- Public Key Cryptography
- RSA Algorithm
- Secrecy
- Secure Socket Layer (SSL)
- Stream Cipher
- Symmetric Key Cryptography

## Review Questions

### • Multiple Choice Questions

1. ______ provides privacy and protects information from an attacker.
   a. Authentication
   b. Secrecy
   c. Integrity
   d. Non-repudiation
2. Plaintext encrypted is called a/an ______.
   a. cipher
   b. encryption
   c. ciphertext
   d. hash value
3. When a transmitter and a receiver use the same key, the key is called a ______.
   a. public key
   b. private key
   c. secret key
   d. hash key
4. The science of encryption and decryption is called ______.
   a. authentication
   b. cipher
   c. cryptography
   d. hash value

5. The algorithm used for encryption and decryption is called a ______.
   a. cleartext
   b. cipher
   c. ciphertext
   d. digital signature
6. Stream ciphers encrypt information ______.
   a. one bit at a time
   b. one byte at a time
   c. one block at time
   d. all at the same time
7. Of the following algorithms, ______ uses a private key.
   a. DES
   b. RAS
   c. hash function
   d. digital signature
8. Computer A uses a public key for encrypting its information and sends it to station B; then ______.
   a. Computer B uses its public key for decryption
   b. Computer B uses its private key for decryption
   c. Computer B uses Computer A's public key for decryption
   d. Computer B randomly chooses a key for decryption
9. The objective of a digital signature is ______.
   a. to verify the identity of a user or a client
   b. to provide privacy for a document
   c. to inform the user of the acceptance of the document
   d. for the user to copy the document
10. A hash function is used for ______.
    a. a digital signature
    b. digital certificates
    c. encryption
    d. authentication
11. Keberos is used for ______.
    a. encryption of a document
    b. digital signatures
    c. digital certificates
    d. authentication purposes
12. A certificate is used to identify a ______.
    a. client
    b. server
    c. client and a server
    d. none of the above
13. A certificate is issued by a/the ______.
    a. client
    b. server
    c. certificate authority
    d. Internet
14. X.509 is a standard for ______.
    a. digital signatures
    b. encryption
    c. certificates
    d. hash functions
15. SSL provides ______.
    a. encryption
    b. authentication
    c. integrity
    d. all of the above

16. SSL is located between the ______.
    a. TCP and IP protocols
    b. TCP and HTTP protocols
    c. TCP and HTTPS protocols
    d. TCP and DNS protocols
17. A password is a type of ______.
    a. digital signature
    b. authentication
    c. certificate
    d. all above
18. IEEE 802.1X is a standard for ______.
    a. digital signatures
    b. authentication
    c. certificates
    d. all above

## • Short Answer Questions

1. List the types of attacks and explain each type.
2. List the elements of network security and explain the function of each.
3. What is the definition of cryptography?
4. Show the cryptography model.
5. What is a cipher?
6. List the classes of cryptography.
7. Explain symmetric cryptography.
8. Explain asymmetric cryptography.
9. Distinguish between stream ciphers and block ciphers.
10. Use the substitution method to encrypt the word "NETWORK" assuming the encryption key is 5.
11. Find the encrypted text for "COMPUTER SCIENCE DEPT" using a transposition cipher assuming each row is made of five characters and the encryption key is 43215.
12 What is the function of a hash function?
13. What are some popular hash functions?
14. Explain digital signatures.
15. Explain the function of Kerberos.
16. Find the public key, the private key, and N using the RSA algorithm for two prime numbers, 7 and 11.
17. Use the information in problem 16:
    a. Encrypt message 6 using the public key.
    b. Decrypt the encrypted message using the private key.
18. What does SSL stand for?
19. Show the TCP/IP model for SSL.

20. What does https stand for?
21. What is the function of a certificate authority?
22. What does EAP stand for?
23. What is the application of the RADIUS server?
24. List four authentication protocols.
25. What is the function of a firewall?
26. What are the different types of firewalls?
27. List the functions of a packet filtering firewall.
28. Explain the operation of a stateful firewall.

CHAPTER 24

# Universal Serial Bus (USB) and PCI-Express

**OUTLINE**



**OBJECTIVES**

After completing this chapter, you should be able to:

- Discuss the components of USB
- List the functions of the USB host controller
- Explain the functions of the USB token packet and USB token types
- Show the packet format of the USB token packet and USB data packet
- Explain how the USB host controller configures a USB device

**INTRODUCTION**

The **Universal Serial Bus (USB)** is a computer serial bus that enables users to connect peripherals such as a mouse, keyboard, modem, CD-ROM, scanner, and printer to a computer without any need for configuration. A personal computer equipped with USB will allow the computer be configured automatically as devices get connected to it. This means that USB has the capability to detect when a device has been added or removed from a PC.

USB is a true plug-and-play bus. Up to 127 peripherals can be connected to a PC with a USB. USB version 1.1 (USB 1.1) was released in 1998 and it supports data rates of 12 Mbps (full speed) and 1.5 Mbps (low speed). The low speed is used for devices such as a mouse, keyboard, and joystick. The USB version 2.0 (USB 2.0) supports data rates of 480 Mbps (high speed) and is compatible with USB 1.1. Specifications for USB 2.0 were developed by seven leading computer manufacturers and were published in 1999. The maximum cable length for a USB connection is five meters.

## 24.1 USB Architecture

Figure 24.1 shows USB architecture. The USB system is logically a tree topology but physically it is a star topology, because each USB device communicates directly with the root hub. There is only one host in any USB system.

A USB system consists of the USB host controller, USB root hub, USB hub, USB cable, USB device, client software, and host controller software.

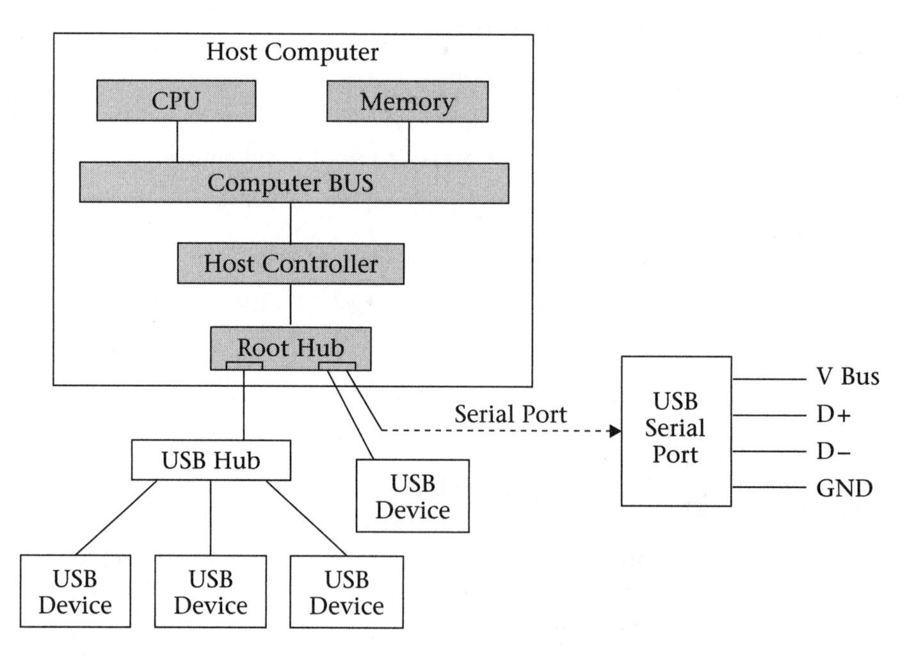


**FIGURE 24.1**
USB Architecture

**Host Controller** The **host controller** initiates all data transfers, and the **root hub** provides the connection between devices and the host controller. The root hub receives any transaction generated by the host controller and transmits it to USB devices. The host controller uses polling to detect if a new device is connected to or disconnected from the bus. The USB host controller performs the following functions:

1. Setting up the device for operation (device configuration)
2. Packet generation
3. Serializer/deserializer
4. Processing requests from the device and host
5. Managing USB protocol
6. Managing flow between the host and USB devices
7. Assigning address to devices
8. Executing client software
9. Collecting status bits from USB ports

**Root Hub** The root hub performs power distribution to devices, enables and disables ports, and reports the status of each port to the host controller. The root hub provides the connection between the host controller and USB ports.

**Hub** The hub is used to expand the number of devices connected to a USB system. The hub can detect when a device is attached or removed from a port. Figure 24.2 shows the architecture of a hub. The host is connected to an *upstream port* and the USB devices are connected to a *downstream port.* The transaction function is used to convert high-speed (480 Mbps) packets coming from the upstream port to the low- or full- speed packets for transmission to the low- and full-speed devices or vice versa. When an upstream port is the same speed as the downstream ports, then the repeater in the hub transfers the packet to the ports.

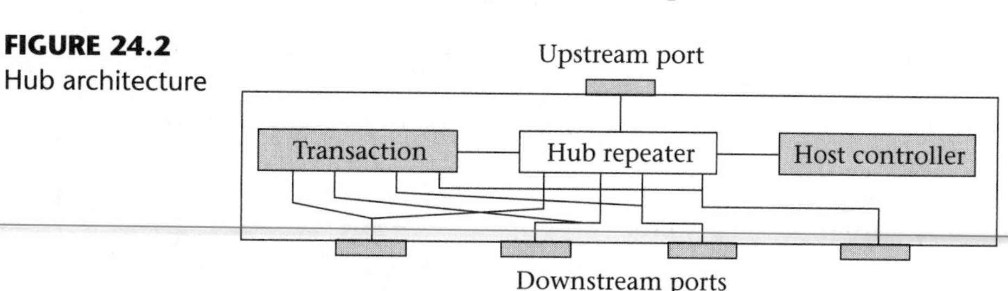


**FIGURE 24.2**
Hub architecture

In downstream transmission all devices that are connected to the hub will receive the packet. However, the only device that accepts the packet is the one whose address matches the address in the token. In upstream transmission the device sends the packet to the hub and the hub transmits the packet to its upstream port only.

**USB Cable** Figure 24.1 shows a USB port with four pins, which consist of four wires. V-bus is used to power devices, and $D^+$ and $D^-$ pins are used for signal transmission.

**USB Device** USB devices are divided into classes such as hub, printer, or mass storage. The USB device has information about its configuration such as class, type, manufacturer's ID, and data rate. The host controller uses this information to load the device software from a hard disk. A USB device might have multiple functions, such as volume control in a speaker. Each function in a USB device is specified by the end point address.

## 24.2 USB Packet Format

The communication between the host controller and the device is called a *transaction*. A transaction usually consists of three packets, as shown in Figure 24.3:

1. **Token packets** are transmitted by the host controller to the USB device. Token packets determine types of transactions, which are IN, OUT, and SETUP.
2. **Handshake packets** are used for acknowledgment.
3. **Data packets** are used for transmitting data.

| Token Packet | Data Packet | Handshake Packet |
|---|---|---|

**FIGURE 24.3**
Transaction packets

**Token Packet** Figure 24.4 shows the format of a token packet. The token defines the type of transaction such as data, setup, ACK, and NAK that would take place after transmission of a token packet.

| 8 or 32 bits | 4 bits | 4 bits | 7 bits | 4 bits | 5 bits | 2 bits |
|---|---|---|---|---|---|---|
| Sync | PID0 PID1 PID2 PID3 | $\overline{PID0}$ $\overline{PID1}$ $\overline{PID2}$ $\overline{PID3}$ | Address | ENDP | FCS | EOP |

**FIGURE 24.4**
Format of a token packet

The following describes the function of each field:

**Sync:** This field is used to synchronize the clock of the host controller with the clock of the USB device. The sync field is 8 bits for low/full-speed and 32 bits for high speed.

**PID (Packet ID):** PID is 4 bits and determines the type of packet to be transmitted by the host controller or USB device after the token packet. The complement of PID bits is used for error detection.

**PID = 0000** means OUT packet, which informs the USB device that the next packet is sent by the host to the USB.

**PID = 0001** means IN packet, which informs the USB device that the next packet is from the USB device to the host.

**PID = 1101** means **SETUP transaction**.

**Address:** This field is 7 bits and the host controller assigns a 7-bit unique address to the device that is attached to the USB bus. Address zero is used as the default address for the device; therefore 127 devices can be connected to a USB port.

**ENDP (End Point):** This field is 4 bits and it is used to identify the function at the end point, such as the volume control in a speaker.

**FCS (Frame Check Sequence):** This field is 5 bits and is used for error detection and covers address and ENDP fields.

**EOP (End of packet):** This field is 2 bits.

## Handshake Packet

A **handshake packet** consists of only the PID field, as shown in Figure 24.5. The handshake packets are used to report the status of transactions. The PID field defines the type of handshake packet, and they are as follows:

**PID = 0010** means ACK, indicating that the packet is received without error.

**PID = 1010** means NAK, indicating there was an error in the packet.

**PID = 1110** means stall. This is a response by the function to an IN or OUT token from the host and means that the function is unable to receive or transmit data.

| 8 or 32 bits | 4 bits | 4 bits | |
|---|---|---|---|
| Sync | PID0 PID1 PID2 PID3 | $\overline{\text{PID0}}$ $\overline{\text{PID1}}$ $\overline{\text{PID2}}$ $\overline{\text{PID3}}$ | EOP |

**FIGURE 24.5**
Format of handshake packet

**Data Packet** Figure 24.6 shows the format of a data packet. The maximum data payload size for a low-speed device is 8 bytes and for a full- or high-speed device is 1023 bytes. The FCS field covers the data field only.

| 8 or 32 bits | 4 bits | 4 bits | | 16 bits | 2 bits |
|---|---|---|---|---|---|
| Sync | PID0 PID1 PID2 PID3 | $\overline{PID0}$ $\overline{PID1}$ $\overline{PID2}$ $\overline{PID3}$ | Data | FCS | EOP |

**FIGURE 24.6**
Format of a data packet

The USB architecture supports four types of data flow: control data transfer, bulk data transfer, isochronous data transfer, and interrupt data transfer. The amount of data that is exchanged between the host controller and the device is determined by the setup stage.

**Bulk Data Transfer:** Used for transferring large blocks of data to the devices that do not have any specific bandwidth requirement, such as transferring a file to the printer. The host controller assigns lowest priority to the bulk data transfer.

**Interrupt Data Transfer:** Used for timely and reliable delivery of data, such as a keyboard control. When the user presses any key, a hardware interrupt is generated. The USB does not support hardware interrupt, therefore USB uses a polling interval to read the keys of the keyboard.

**Isochronous Data Transfer:** Used for real-time communication that requires fixed bandwidth and bandwidth is pre-assigned before transferring the data. For example, the transmission of information to a speaker or reading audio from a CD requires a constant bit rate. Isochronous transactions use the maximum data payload of 1023 bytes, but other types use the maximum of 64 bytes.

**Control Data Transfer:** Used for setup and configuring any USB device that is attached to the USB port. The control data transfer consists of setup, data, and status stages.

**Setup Stage:** The host controller initiates the setup stage. A setup packet consists of a token packet, data packet, and handshake packet. The host controller sends requests to the device to obtain a device description.

**Data Stage:** In the data stage, the host controller reads or writes to the device using data IN or data OUT.

**Status Stage:** This is used to confirm the status of the control transfer, such as ACK, NACK, and stall.

## 24.3 USB Device Configuration Process

All USB devices attach to the USB port or USB hub. When a device is connected to a USB port, the status bit of the port is set to one and the host controller polling the port indicates that a new device is connected. Upon removal of any device from the port, the status bit of the port changes back to zero, meaning no device is connected to this port. When the host controller detects a new device, the following processes would take place in configuring the device:

1. The host controller enables the port where a device is detected in order to give power to the device.
2. The host controller resets the device. By resetting the device, the default address of the device becomes zero.
3. The host controller assigns a 7-bit address to the device.
4. The host controller reads the device descriptors.
5. The host controller uses device descriptors to download and execute the device software for operation.

When USB devices are attached to or removed from a USB hub, the host uses a process known as **enumeration** to identify the device and assigns a unique 7-bit address to the device.

## 24.4 USB Transaction

The USB transactions consist of IN, OUT, and SETUP, and each transaction consists of a token packet, data packet, and status packet.

**IN Transaction**

IN transactions occur when the host wants to read data from a device. In the **IN transaction** the host sends an IN token to the USB device. The following are three cases that may occur in an IN transaction:

1. The device accepts the **IN token** and transmits its data packet to the host. The host controller checks the data packet and if the data does not have any errors the host sends an ACK frame (see Figure 24.7). If the host detects any errors then it does not respond with handshaking.
2. When the device detects errors in the token, then it is unable to accept the token and the device responds by sending a NACK packet, as shown in Figure 24.8.
3. If the device is unable to transmit data, the device transmits a **stall packet,** as shown in Figure 24.9.

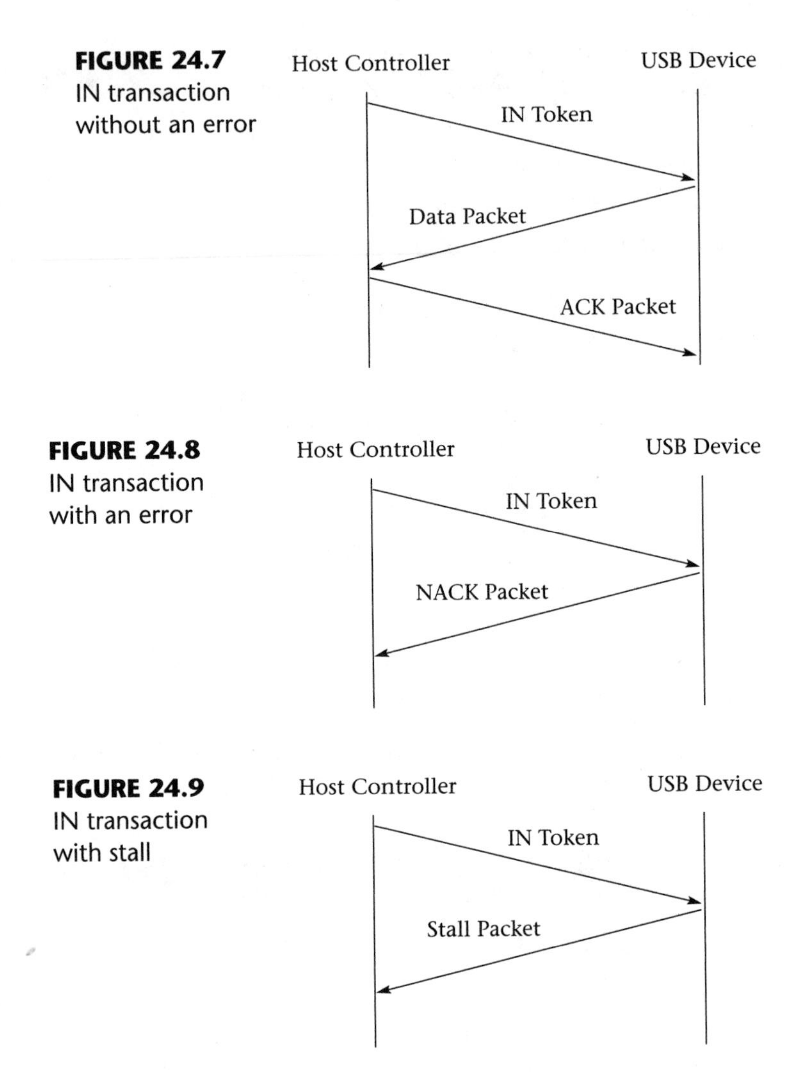


**FIGURE 24.7** IN transaction without an error

**FIGURE 24.8** IN transaction with an error

**FIGURE 24.9** IN transaction with stall

**OUT Transaction** **OUT transactions** are used to transmit data from the host to a USB device. The host controller sends an **OUT token,** which is followed by a data packet. If the data packet is accepted by the device, then the device transmits an ACK packet to the host controller. If the device detects any errors in the data, it responds with a NACK. If the device is unable to send a data packet, it responds with a stall packet.

**Setup Transaction** The setup transaction is used to setup the USB device for operation, and the following process takes place.

1. The host controller sends a Setup token followed by a data packet, which requests the device descriptor from the USB device. If the

data packet is accepted by the USB device, then the device transmits an ACK packet to the host controller.

2. The host controller sends an IN token to the USB device; then the USB device transmits the first eight bytes of its descriptor as a data packet to the host controller. Then the host controller transmits an ACK packet to the USB device.
3. The host controller sends an IN token to the USB device and the USB device transmits the last four bytes of its descriptor to the host controller. The host controller transmits an ACK packet to the USB device.

The host controller uses the device descriptor to download and to execute the device software for operation. Figure 24.10A shows the Setup Transaction process.

**FIGURE 24.10A** Setup Transaction Process

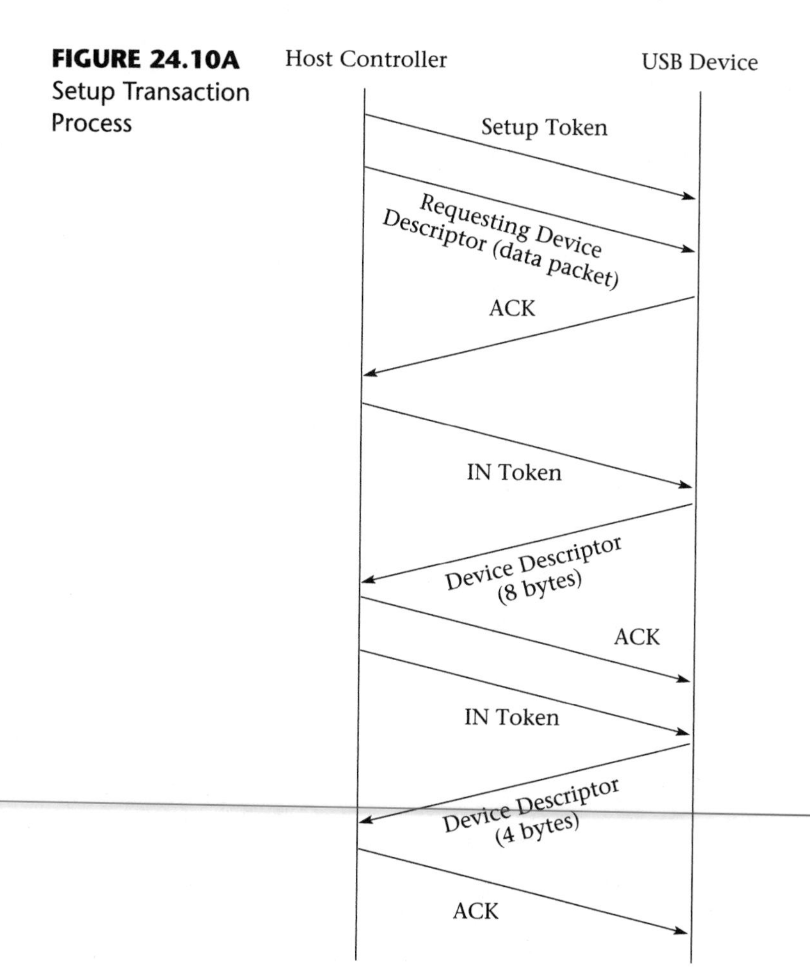

## 24.5 Split Transaction

A split transaction is used between the hub and the host controller when low- and full-speed devices are connected to the high-speed hub, as shown in Figure 24.10B. A split transaction requires a split token. The split token informs the hub about the devices that require full- or low-speed transmission. The split transmission uses two split tokens for transmission. One is the Start Split Token (SS Token), which is used for the start of the split transaction. The second one is called *Complete Split Token (CS Token)* and is used to indicate completion of the split transaction. Figure 24.11 shows the split token packet format.

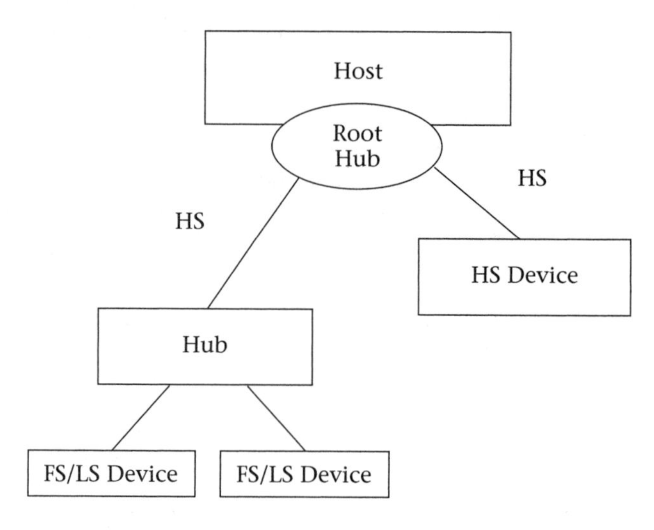


**FIGURE 24.10B**
Connection of low/full-speed devices to high-speed hub

HS: High Speed
FL: Full Speed
LS: Low Speed

| | 8 bits | 7 bits | 1 bit | 7 bits | 1 bit | 2 bits | 5 bits |
|---|---|---|---|---|---|---|---|
| Sync | PID | Hub Address | SC | Port Address | S | ET | FCS |

**FIGURE 24.11**
Split token format

The following describes the functions of each field:

**Hub Address (7 bits):** Identifies the hub.

**SC (Start/Complete):** When the SC bit is zero, then the token is a Start Split Token (SS token) and when the SC bit is 1, then the token is Complete Split Token (CS token).

**Port Address:** Address of the port that the USB device is connected to.

**S (Speed):** S = 0 indicates that the device is low speed; S = 1 indicates that the device is full speed.

**ET (Endpoint Type):** ET defines the type of data flow. ET = 00 means control data; ET = 01 means isochronous data; ET = 10 means bulk data; and ET = 11 means interrupt data.

Figure 24.12 shows an IN split transaction between a host and a device and the sequence of operations are as follows:

1. The host controller transmits the SS token and the IN token to the hub.
2. The hub grabs the tokens and transmits an IN token to the device.
3. The device sends the data packet to the hub.
4. The hub sends the data packet to the host controller.
5. The host controller sends a CS token to the hub.
6. The hub sends an ACK packet to the device.

**FIGURE 24.12** Split operation of an IN transaction

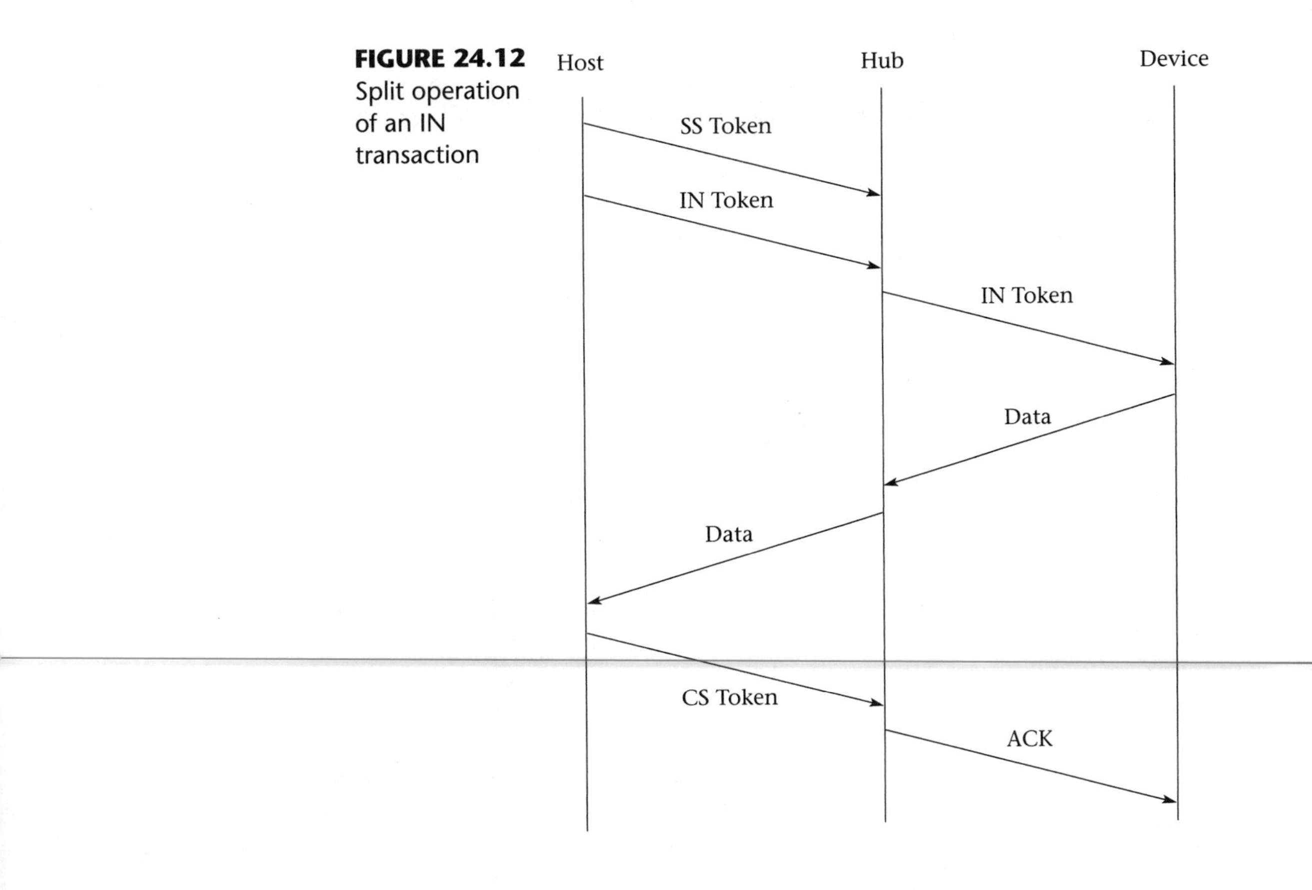

## 24.6 PCI-Express Bus

PCI was introduced in the mid 90s with 33-MHz frequency and the speed of the bus was subsequently increased to 66 MHz. Due to new development in networking technology such as Gigabit Ethernet and I/O devices that demand more bandwidth, there was a need for a new bus technology with higher bandwidth. PIC-Express was approved by a Special Interest Group in 2002 and chipsets were available in the market in 2004. PCI-Express has the following features:

- Provides point-to-point connection between devices
- Is a serial bus
- Uses pocket and layer architecture
- Compatible with PCI bus through software
- End-to-end link data integrity (error detection)
- Isochronous data transfer
- Selectable bandwidth

## 24.7 PCI-Express Architecture

Figure 24.13 shows PCI-Express architecture. The function of the host bridge is to interface the CPU bus with memory and the PCI-Express switch. The switch is used to increase the number of PCI–Express ports.

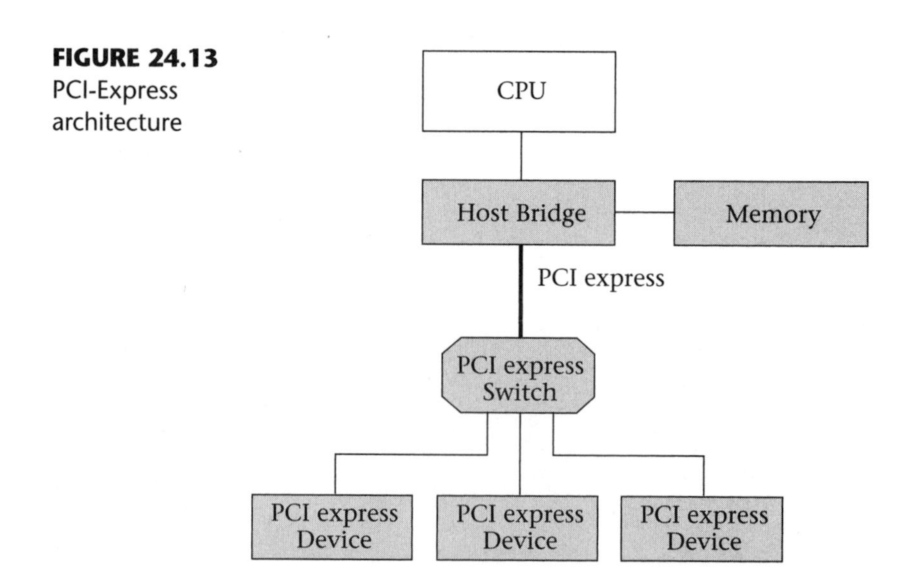


**FIGURE 24.13** PCI-Express architecture

## 24.8 PCI-Express Protocol Architecture

Figure 24.14 shows PCI-Express protocol architecture. The protocol consists of a Software layer, Transaction layer, Data Link layer, and Physical layer.

**FIGURE 24.14**
PCI-Express protocol architecture

| Software Layer |
| --- |
| Transaction Layer |
| Data Link Layer |
| Physical Layer |

### Software Layer

The Software layer is used for compatibility with PCI, initialization, and enumeration of the devices connected to the PCI-Express.

### PCI-Express Physical Layer

Figure 24.15 shows two devices that are connected through PCI-Express link (lane). Each lane is made of four wires and each PCI-Express lane consists of two simplex connections, one for transmitting packets and the other for receiving packets. The PCI-Express lane has a transfer rate of 2.5 Gigabytes per second in each direction.

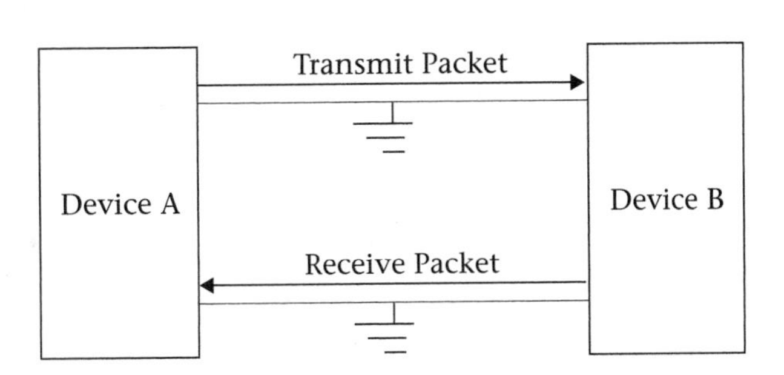


**FIGURE 24.15**
PCI-Express connections PC X1

PCI-Express can be configured as X1 (one lane), X2 (2 lanes), X4 (4 lanes), X16 (16 lanes), and X32 (32 lanes) where each lane consists of four wires. A serial link clock is embedded into the data by using 8B/10B encoding and packet format at the Physical layer is shown in Figure 24.16.

### Transaction Layer

The Transaction layer responds to services such as read or write requests from the Software layer, assembles or disassembles the packets, performs link configuration, creates a request packet and passes it to the Data Link layer. The Transaction layer also receives responses to requests from the destination. The packet header is used to identify the destination address (32-bit or 64-bit address). Packet format at the Transaction layer is shown in Figure 24.16.

**Data Link Layer** The function of the Data Link layer is to provide reliable delivery of the packet to the destination. The Data Link layer adds FCS for error detection and sequence number, and ACK and NACK for flow control. Packet format at the Data Link layer is shown in Figure 24.16.

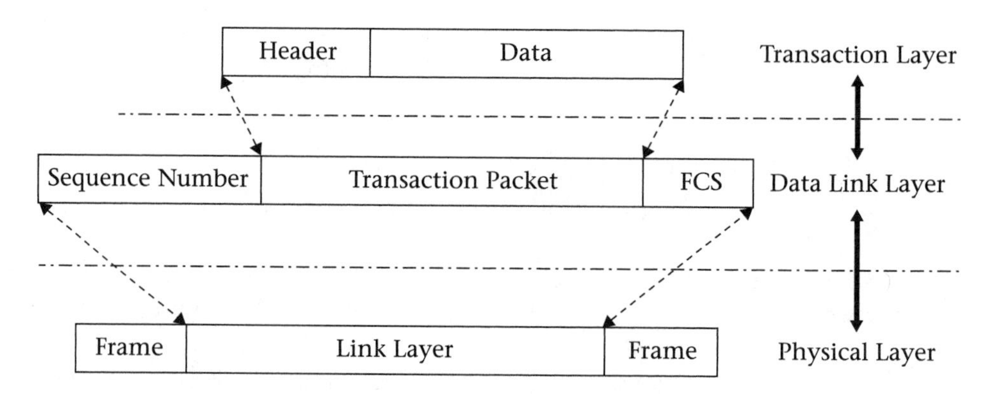


**FIGURE 24.16**
Packet format at each layer of a PCI-Express bus

## Summary

- The Universal Serial Bus (USB) is a computer serial bus that enables users to connect peripherals such as the mouse, keyboard, modem, CD-ROM, scanner, and printer to the outside of a computer without any need for configuration.
- Components of USB are the host controller, USB root hub, USH hub, client software, host controller software, and USB cable.

- USB version 2.0 data rate is 480 Mbps (high speed) and USB version 1.1 data rates are 12 Mbps (full speed) and 1.5 Mbps (high speed).
- The host controller initiates all transactions.
- The host controller uses IN token packets to transfer data from a device that is connected to a USB device.
- The host controller uses OUT tokens to transfer data from the host to a USB device.
- USB offers four types of data transfer: control data transfer, bulk data transfer, isochronous data transfer, and interrupt data transfer.
- The maximum number of devices that can be connected to a USB bus is 127.
- Split transaction is used for converting high-speed packets to the low/full-speed packets.
- PCI-Express is a serial bus that is compatible with the PCI bus at the Software layer.
- PCI-Express protocol architecture consists of the Software layer, Transaction layer, Data Link layer, and Physical layer.
- PCI-Express uses PCI-Express switch to expand the number of ports.
- PCI-Express uses two half-duplex connections that are called *lanes*.
- The PCI-Express lane data rate is 2.5 Gbps.

## Key Terms

Bulk Data Transfer
Control Data Transfer
Enumeration
Handshake Packet
Host Controller
Interrupted Data Transfer
IN Token
IN Transaction
Isochronous Data Transfer
OUT Token
OUT Transaction
Root Hub
SETUP Transaction
Stall Packet
Universal Serial Bus (USB)

## Review Questions

### • Multiple Choice Questions

1. What is the data rate of USB version 2.0? ______
   a. 1.5 Mbps
   b. 12 Mbps
   c. 480 Mbps
   d. 500 Mbps
2. What is the data rate of low-speed USB? ______
   a. 1.5 Mbps
   b. 12 Mbps
   c. 480 Mbps
   d. 500 Mbps
3. What is the data rate of full-speed USB? ______
   a. 1.5 Mbps
   b. 12 Mbps
   c. 480 Mbps
   d. 500 Mbps
4. What is the maximum size of the cable for a USB bus? ______
   a. 5 meters
   b. 10 meters
   c. 1 meter
   d. 100 meters
5. What is the default address for a device connected to a USB system? ______
   a. FF Hex
   b. 00 Hex
   c. 20 Hex
   d. 01 Hex
6. What is the maximum number of devices that can be connected to a USB system? ______
   a. 200
   b. 20
   c. 10
   d. 127
7. What is the function of an IN token? ______
   a. Inform the device that the next packet is from the host controller to the device.
   b. Inform the device that the next packet is from the device to the host controller.
   c. Used for setup of a USB device.
   d. Used for setup of a root hub.
8. Which of the following devices initiates a transition? ______
   a. USB device
   b. hub
   c. root hub
   d. host controller
9. Which of the following devices generates a token? ______
   a. USB device
   b. hub
   c. root hub
   d. host controller

## • Short Answer Questions

1. What is the high-speed USB data rate?
2. What is the full-speed USB data rate?
3. What is the function of a token in a USB system?
4. What is the topology of a USB system?
5. How many devices can be connected to a USB system?
6. What are the functions of the host controller?
7. What is the function of the root hub?
8. List the transaction packets.
9. List the types of token packets.
10. How many bits are in a USB device address?
11. List the handshake packets.
12. What is the application of bulk transfer in a USB system?
13. What is the application of isochronous data transfer?
14. Which device initiates transactions in a USB system?
15. What is the enumeration process?
16. List the USB transactions.
17. Why does USB version 2.0 support split transactions?

# Appendix A: Computers and Network Connectors

## Data Connectors

Computer manufacturers use five different shell sizes for connectors—shell sizes 1 through 5 as shown in Figure A.1. Table A.1 shows the size of each shell. Each shell size offers two types of connectors: standard DB and HD (high density). The DB connectors use two rows of pins and HD connectors use three rows of pins. HD-15 is used for video connectors.

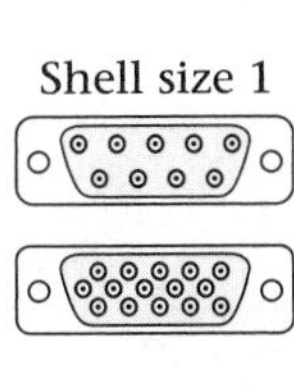

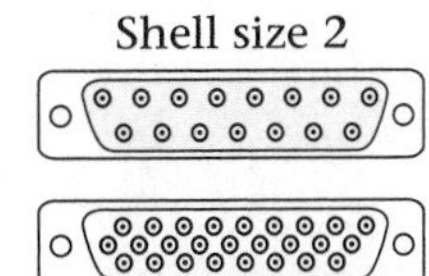

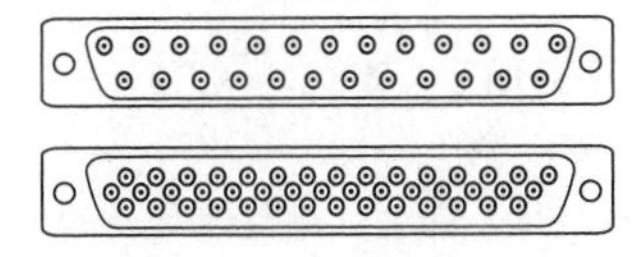

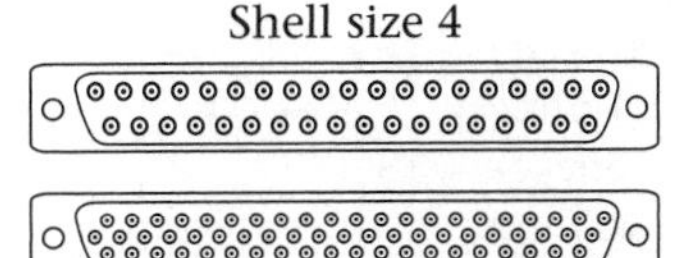

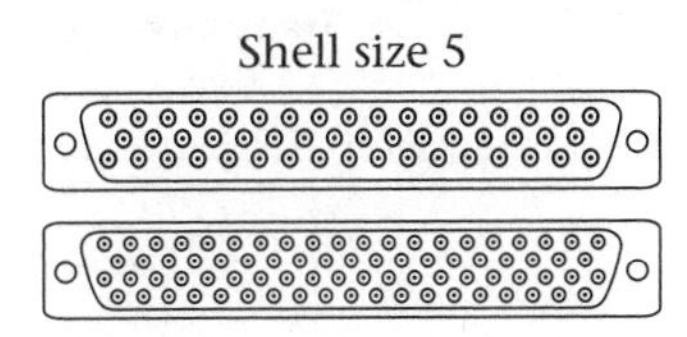


**FIGURE A1** Five different shells

**TABLE A1** Shell Sizes

| Shell Type | Size in Inches |
|---|---|
| Size 1 | 1.2 * 0.5 |
| Size 2 | 1.55 * 0.5 |
| Size 3 | 2.0 * 0.5 |
| Size 4 | 2.7 * 0.5 |
| Size 5 | 2.6 * 0.5 |

## Serial Data Transmission

Serial transmission is one of the methods of transmitting data from one device to another. The following describes the serial interfaces, which are used for serial transmission.

**EIA-232 or RS-232C Standard**

The RS-232 interface defines the electrical function of the pins and the mechanical function of the connector.

The EIA revised the RS-232C standard in 1989, and called the revision *RS 232D* (connector with 25 pins). RS232 is a standard connection for serial communication, which has been approved by the EIA.

All modems use RS-232 connections and all PCs have a RS-232 port. RS-232 supports two types of connectors: a 25-pin D-type connector (DB-25), as shown in Figure A2 and a 9-pin D-type connector (DB-9), as shown in Figure A3. The EIA has approved RS-422 and 423, which are successors to EIA 232. In RS232, zero is represented by voltage levels between +3 volts and +12 volts and one is represented by –3 volts to −12 volts. RS-232 is used for 20 Kbps up to 50 feet.

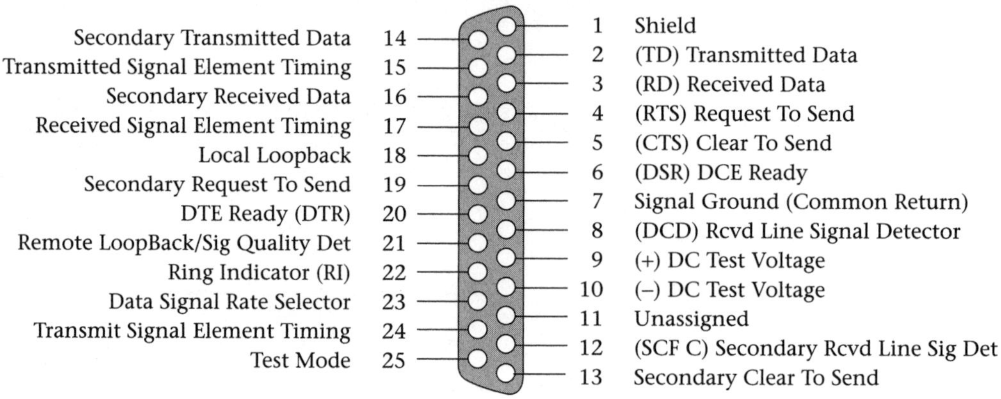


**FIGURE A2** EIA-232 interface using DB-25

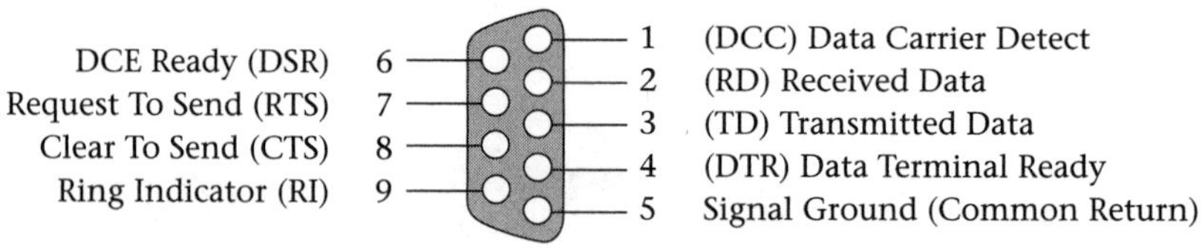


**FIGURE A3** EIA-232 interface using DB-9

## RS-449 Interface Standard

The biggest shortcoming of RS232 is its limited transmission and distance.

The EIA approved RS-499 as an improved version of RS-232 using DB-37 connectors for serial transmission, as shown in Figure A4. It is designed to increase the bandwidth and the distance of the cable. It is used to transmit data up to 10 Mbps over a 40-foot distance or to transmit data at a rate of 100 Kbps over 4000 feet.

RS-499 INTERFACE

| Signal Designation | Pin Number | Pin Number | Signal Designation |
|---|---|---|---|
| | | 1 | Shield |
| Receive Common | 20 | 2 | Signaling Rate Indicator |
| | 21 | 3 | |
| Send Data | 22 | 4 | Send Data |
| Send Timing | 23 | 5 | Send Timing |
| Receive Data | 24 | 6 | Receive Data |
| Request to Send | 25 | 7 | Request to Send |
| Receive Timing | 26 | 8 | Receive Timing |
| Clear to Send | 27 | 9 | Clear to Send |
| Terminal in Service | 28 | 10 | Local Loopback |
| Data Mode | 29 | 11 | Data Mode |
| Terminal Ready | 30 | 12 | Terminal Ready |
| Receiver Ready | 31 | 13 | Receiver Ready |
| Select Standby | 32 | 14 | Remote Loopback |
| Signal Quality | 33 | 15 | Incoming Call |
| New Signal | 34 | 16 | Select Frequency |
| Terminal Timing (B) | 35 | 17 | Terminal Timing |
| Standby/Indicator | 36 | 18 | Test Mode |
| Send Common | 37 | 19 | Signal Ground |

**FIGURE A4**
RS-499 interface

## V.35 Standard

The V.35 standard was developed by the ITU for interfacing DTE or DEC to a high-speed digital carrier. The most common application of V.35 is for interfacing routers or DSU to a T-1 link. Figure A5 shows the V.35 interface.

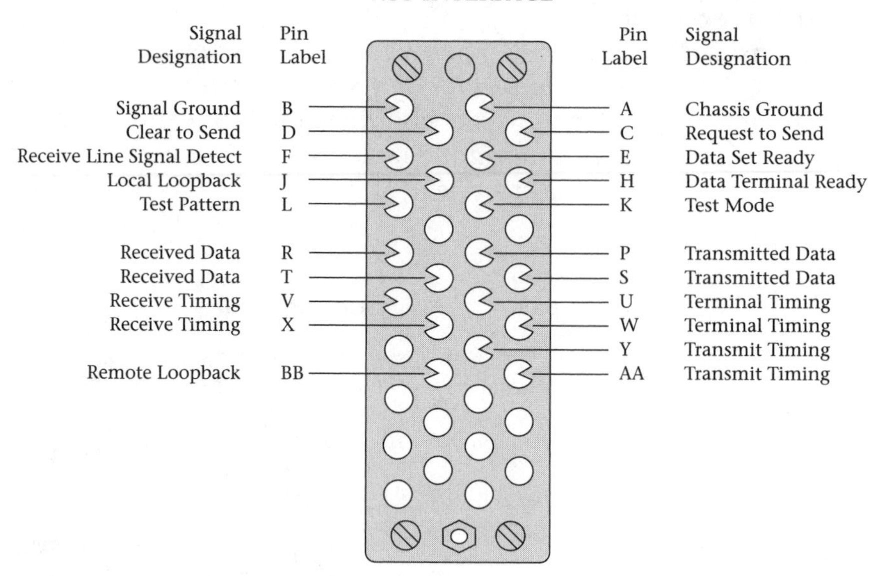


**FIGURE A5**
V.35 interface

## Null Modems

The application of a null modem is to connect two PC serial ports together in order to transfer information between two PCs. Figure A6 shows the connection between two RS232 DB-9 connectors.

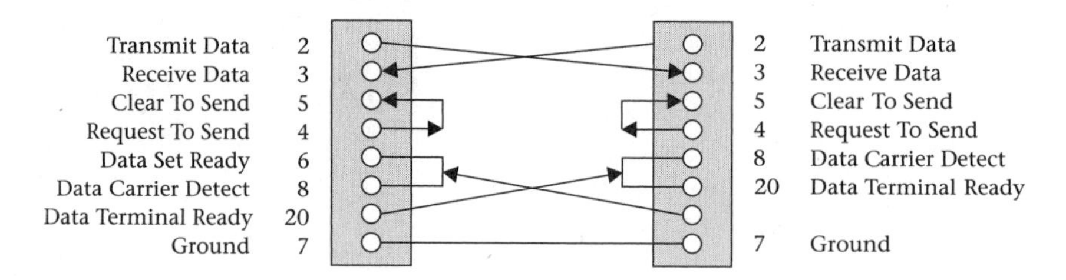


**FIGURE A6**
Null modem connection

## X.21

The X.21 interface was defined by ITU and uses a balanced circuit (two lines used to transmit each signal; each line uses opposite polarity). The X.21 interface uses 15 pins. X.21 is used to connect DTE to public data networks, such as ISDN.

**Small Computer System Interface (SCSI)**

The SCSI standard was defined by the American National Standard Institute (ANSI) for connecting daisy chained, multiple I/O devices such as scanners, hard disks, and CD-ROMs to a microcomputer, as shown in Figure A7. Figure A8 shows SCSI connectors.

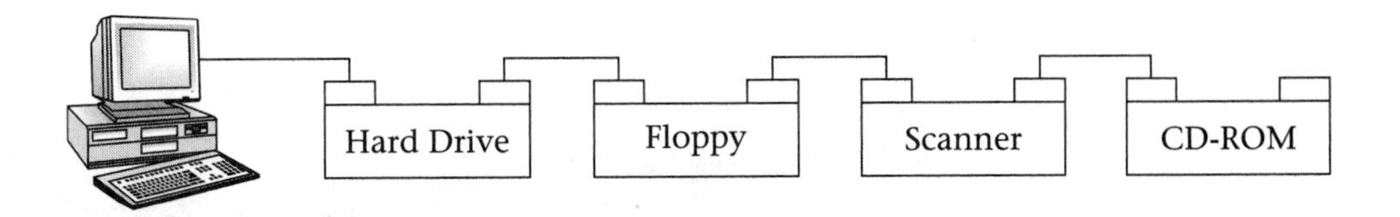


**FIGURE A7**
Several devices connected by daisy chain

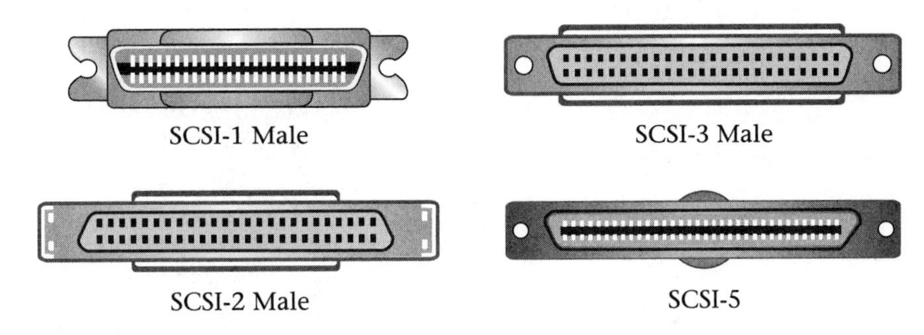


**FIGURE A8**
SCSI-1, SCSI-2, SCSI-3, and SCSI-5 connectors

**Universal Serial Bus (USB)**

Figure A9 shows USB male and female connectors.

**FIGURE A9**
Male and female USB connectors

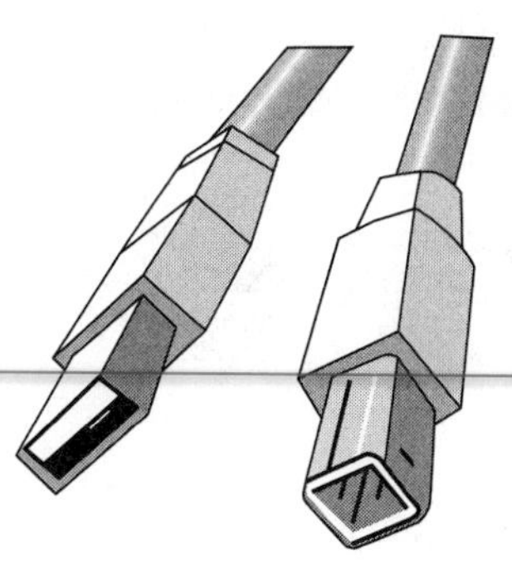

**FireWire** Figure A10 shows male and female connectors for FireWire.

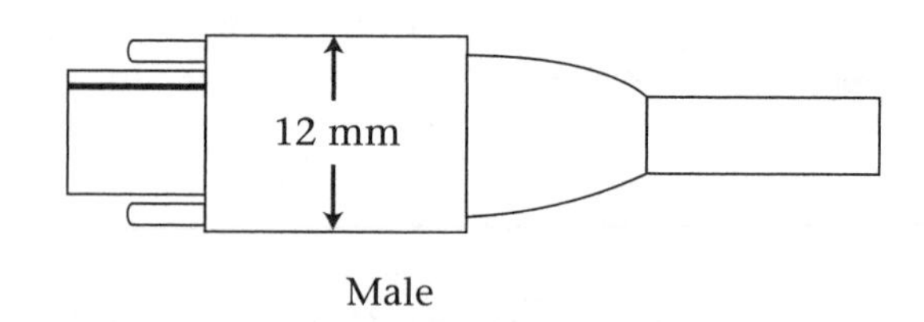


**FIGURE A10**
Female and male FireWire connectors

## Parallel Connectors

Parallel transmission is used for the high-speed transmission of data. Figure A11 shows a DB-25 female parallel connector for a PC port. Parallel cable is limited to 10 feet. IEEE-1284 enhanced parallel cable is used for high-speed printing. Figure A12 shows Centronics connectors.

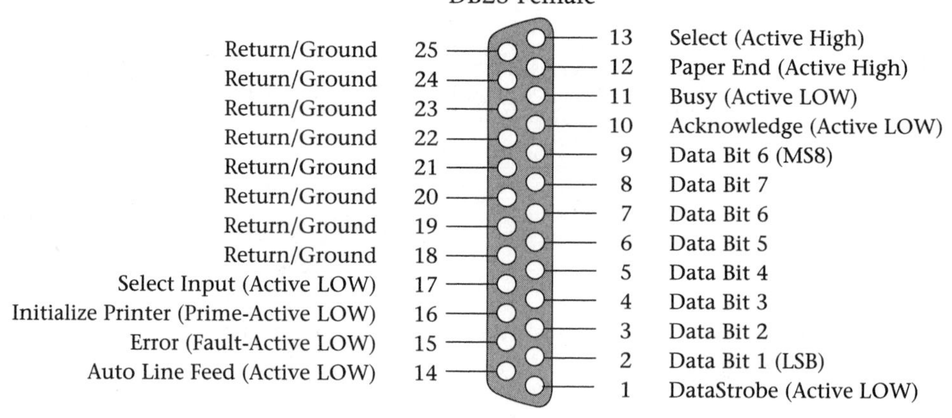


**FIGURE A11**
Parallel interface using DB-25

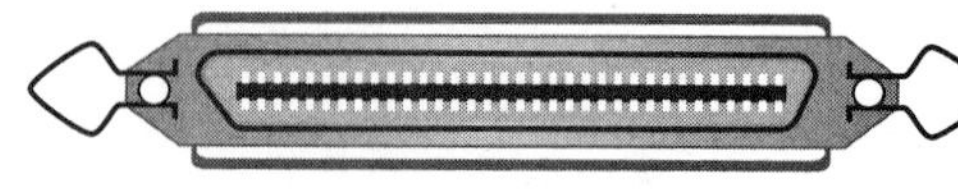


**FIGURE A12**
Centronics connectors

## Modular Connectors

RJ-11, RJ-12, and RJ-45 connectors are used with unshielded twisted-pair cable (UTP). RJ-45 uses four pairs of wires, RJ-12 uses three pairs of wires, and RJ-11 uses two pairs of wires. Figure A13 shows RJ connectors.

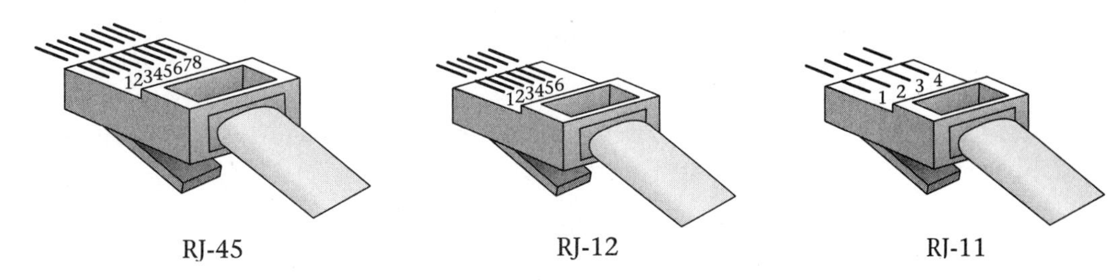


**FIGURE A13**
RJ connectors

**DIN-Type Connectors** DIN connectors are used for keyboards and mice and come with four different types as shown in Figure A14:

1. 4-pin mini DIN
2. 5-pin mini DIN is used for an AT style keyboard
3. 6-pin mini DIN is used for PS/2 style keyboards and mice
4. 8-pin mini DIN is used for Sun keyboards

**FIGURE A14**
DIN connectors

**Coaxial Cable Connectors** There are three types of connectors used in coaxial cable, as shown in Figure A15:

1. RCA type connector
2. F-type connector
3. BNC connector

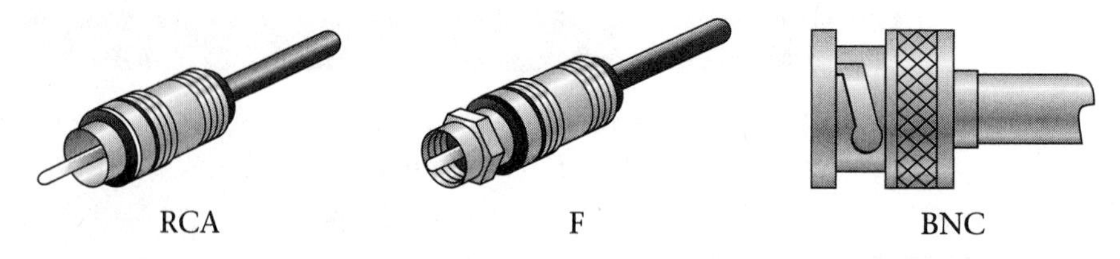


**FIGURE A15**
Coaxial cable connectors

**Fiber-Optics Connectors**

There are three type of connectors commonly used for fiber-optics cable, as shown in Figure A16:

1. MTR-J connector
2. SC connector
3. ST connector

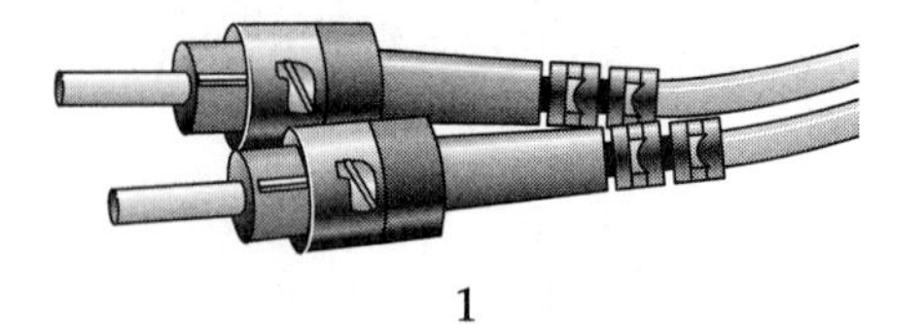
1

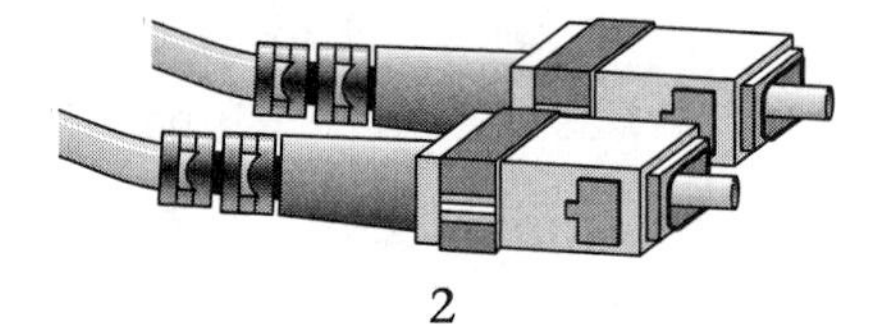
2

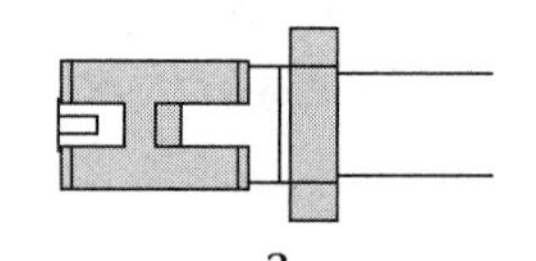
3
*MTRJ Connectors
*Dual Fiber/Single Jacket

**FIGURE A16**
Fiber-optic connectors

**FDDI Connector**

Figure A17 shows an FDDI connector.

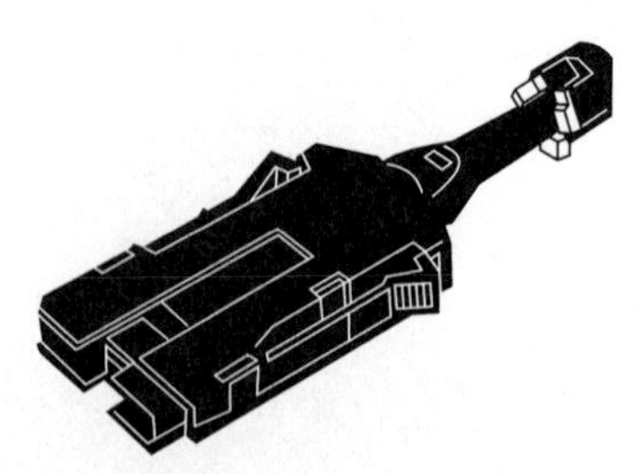

*FDDI (FSD) Connectors
*Ceramic Ferrules

**FIGURE A17**
FDDI connector

# Appendix B: Answers to Odd Questions

## Chapter 1 Introduction to Communications Network

### Multiple Choice Questions

**1.** a, **3.** a, **5.** b, **7.** b, **9.** c, **11.** b, **13.** d, **15.** a

### Short Answer Questions

1. **What are the components of a communication model?**
   transmitter, commutation link, and receiver
3. **What is the function of the client in a file server model?**
   submitting tasks to the server for processing
5. **What are the advantages of a client/server mode?**
   Less information travels through the network compared to the file/server model.
7. **A Network Operating System runs on a** ___*server*___.
9. **What is the disadvantage of a fully connected topology?**
   In fully connected topology, when stations are far away from each other, it is impossible to connect them because they require too much wiring.
11. **What are the types of networks?**
    LAN, MAN, WAN, and Internet
13. **Explain Internet operation.**
    The Internet is a collection of networks connected by gateways.
15. **What are the advantages of bus topology?**
    Simple, low cost, and easy to expand the network.

## Chapter 2 Introduction to Data Communications

### Multiple Choice Questions

**1.** a, **3.** a, **5.** a, **7.** c, **9.** c, **11.** a, **13.** b, **15.** a, **17.** a

## Short Answer Questions

1. **Sketch an analog signal.**

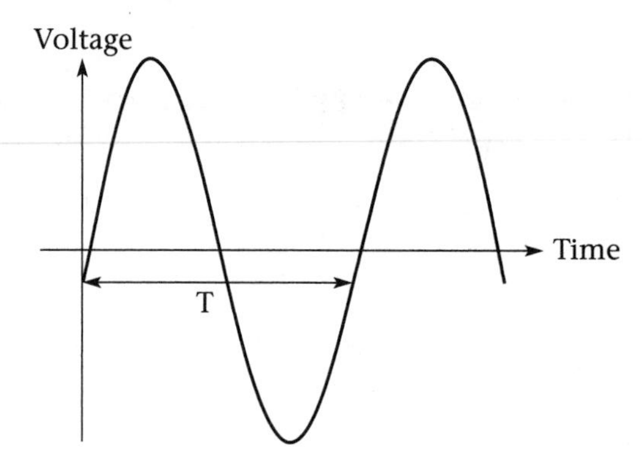


3. **What is the unit of frequency?**
   Hertz
5. **Explain the amplitude of an analog signal.**
   The amplitude of an analog signal is its voltage at any given time.

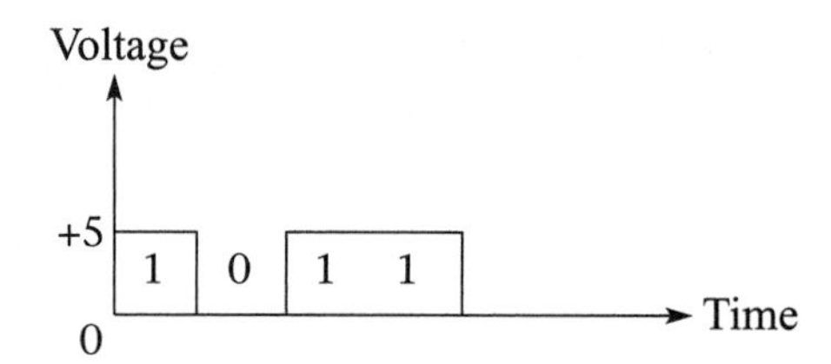


7. **What is a bit?**
   One digit in binary is called a *bit,* represented as a 0 or 1.
9. **What is a word?**
   Two or more bytes are called a *word.*
11. **Convert the following binary number to decimal.**

$$(11111111)_2 = 1*2^0+1*2^1+1*2^2+1*2^3+1*2^4+1*2^5+1*2^6+1*2^7 = 255$$

$$(10110001)_2 = 181$$

13. **Convert the word DIGITAL to binary using the ASCII table.**

| 1000100 | 1001001 | 1000111 | 1001001 | 1010100 | 1000001 | 1001100 |
|---|---|---|---|---|---|---|
| D | I | G | I | T | A | L |

15. **Write your name in ASCII, then change the result to hexadecimal.**
17. **What is parallel transmission?**
    In parallel transmission multiple bits are transmitted simultaneously.
19. **Explain the following terms:**
    **a. simplex.**
    **b. half duplex.**
    **c. full duplex.**

    a. The transmission of information goes in one direction.
    b. Two devices transmit information to each other, but one at a time.
    c. Both devices can receive and send information simultaneously.
21. **Why is a clock pulse needed for transmission of a digital signal?**
    A clock pulse is needed in order for the receiver to recognize the speed of incoming data.
23. **Show a clock pulse.**
    The clock pulse is alternating ones and zeros.

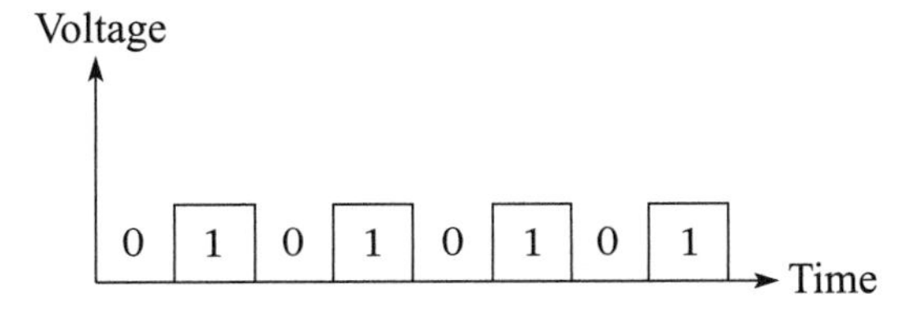


25. **List methods of error detection.**
    Parity bit, Block Check Character (BCC), Cyclic Redundancy Check (CRC), and one's complement of sums.
27. **Represent binary 110101 with a polynomial.**

$$X^5 + X^4 + X^2 + 1$$

29. **Show the CRC circuit for 1011.**

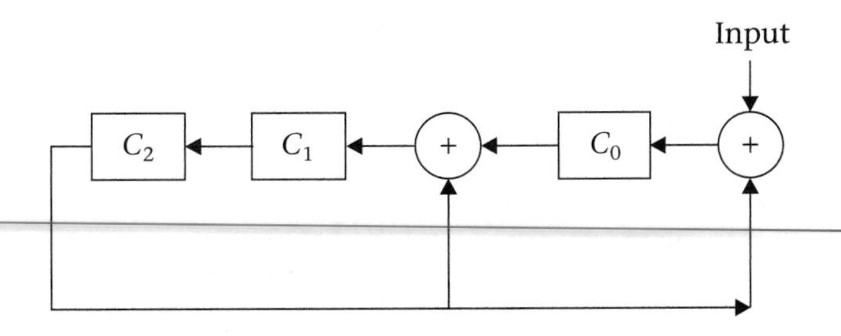

**31. Find the one's complement of sums for the word "NETWORK".**

| Character | ASCII |
|---|---|
| N | 1001110 |
| E | 1000101 |
| T | 1010100 |
| W | 1010111 |
| O | 1001111 |
| R | 1010010 |
| K | 1001011 |
| | 0101010 and the one's complement is 1010101 |

**33. Draw Manchester encoding and differential Manchester encoding for binary 010110110.**

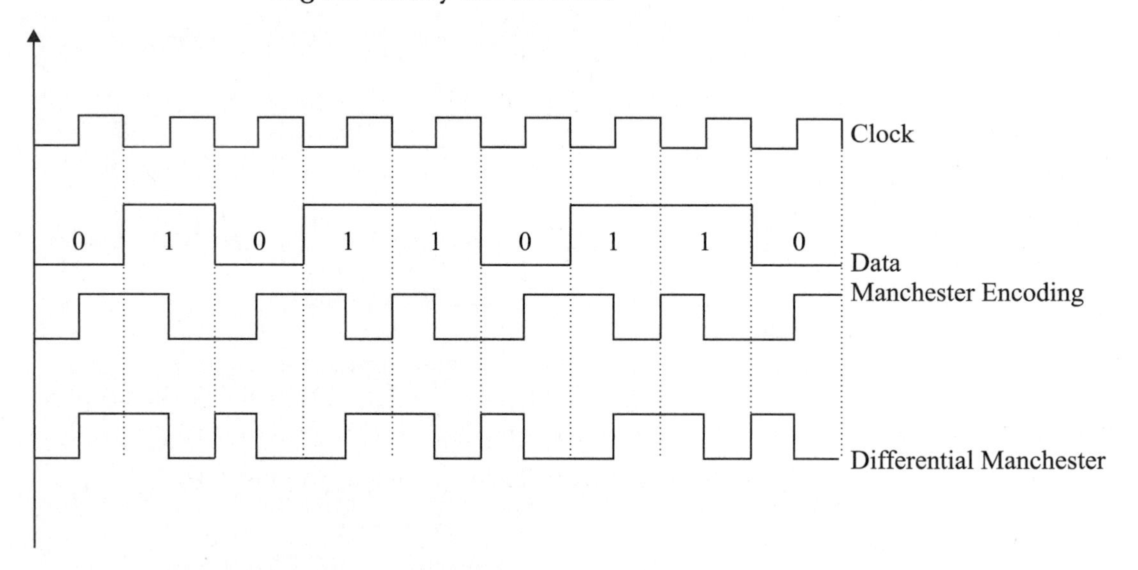


**35. Find the FCS for data unit 111011 with divisor 1011.**

```
        110001
1011 | 111011000
       1011
       01011
        1011
         00001000
             1011
             0011
```

FCS is 0011

**37. Find the check bit for 10001101.**

$$C1 = M1 \; xor \; M2 \; xor \; M4 \; xor \; M5 \; xor \; M7$$
$$= 1 \; xor \; 0 \; xor \; 1 \; xor \; 0 \; xor \; 0 = 0$$
$$C2 = M1 \; xor \; M3 \; xor \; M4 \; xor \; M6 \; xor \; M7$$
$$= 1 \; xor \; 1 \; xor \; 1 \; xor \; 0 \; xor \; 0 = 1$$
$$C4 = M2 \; xor \; M3 \; xor \; M4 \; xor \; M8$$
$$= 0 \; xor \; 1 \; xor \; 1 \; xor \; 1 = 1$$
$$C8 = M5 \; xor \; M6 \; xor \; M7 \; xor \; M8$$
$$= 0 \; xor \; 0 \; xor \; 0 \; xor \; 1 = 1$$

Then
Check bits are 1110:

# Chapter 3 Introduction to Computer Architecture

## Multiple Choice Questions

**1.** c, **3.** b, **5.** c, **7.** a, **9.** b and c

## Short Answer Questions

1. **List the components of a microcomputer.**
   microprocessor, memory, programmable interrupt, programmable parallel input/output, direct memory access, and bus.
3. **List the functions of an ALU.**
   Arithmetic Logic Unit performs arithmetic operations (addition, subtraction) and logic operations (AND, OR, and NOT).
5. **Distinguish between a CPU and a microprocessor.**
   The components of a CPU are ALU, registers, and control unit. If these three units are in more than one integrated circuit, it is a CPU. If these three components are in one IC, it is a microprocessor.
7. **What is SRAM and its application?**
   Static Random Access Memory (SRAM) is used for cache memory.
9. **Explain the function of the address bus and data bus.**
   The address bus carries the address of information and the data bus carries information.
11. **What is the capacity of a memory IC with 10 address lines and 8 data lines?**

$$2^{10} * 8 = 1024 * 8 \text{ bits or } 1024 \text{ bytes}$$

13. **What does EEPROM stand for? What is its application?**
    Electrically Erasable Read Only Memory (EEPROM) is a type of ROM that can be erased electrically.
15. **What is SIMM?**
    SIMM stands for Single-In-Line Memory Module and it is a type of DRAM packaging.
17. **List the types of memory used for main memory.**
    DRAM and SDRAM
19. **Explain the function of DMA.**
    DMA stands for Direct Memory Access, and is used for transferring blocks of data from memory to an I/O device or vice versa.
21. **What is the application of a serial port?**
    A serial port is used for asynchronous transmission such as with a modem.
23. **Explain plug and play.**
    When adding new hardware to the computer, the Interrupt (IRQ), I/O address, and DMA address must be set. Some computers with PCI and EISA bus have the capability to set IRQ and I/O addresses automatically.
25. **What is the application of FireWire?**
    It is a high-speed serial bus used for connecting digital devices such as video and camcorder to a computer.
27. **What does IDE stand for?**
    Integrated Device Electronics, which is a type of disk controller
29. **Explain the differences between CISC processors and RISC processors.**
    The characteristics of a Complex Instruction Set Computer (CISC) are as follows:

    - Control unit is micro-code
    - Uses long instruction set
    - Uses indexed and indirect addressing

    The characteristics of a Reduced Instruction Set Computer (RISC) are as follows:

    - Control unit is hardware
    - Uses small instruction sets
    - Uses a large number of registers

# Chapter 4 Communication Channels and Media

## Multiple Choice Questions

**1.** a, **3.** b, **5.** c, **7.** c, **9.** a

## Short Answer Questions

1. **List the communications media.**
   conductors, optical cable, and wireless communication
3. **What does STP stand for?**
   Shielded Twisted-Pair cable
5. **What is the performance of Cat-5 UTP?**
   Cat-5 UTP can handle signals with speeds up to 100 MHz.
7. **What are the advantages of fiber-optics cable over conductor?**
   It can transmit information a longer distance.
   Immune to external noise.
   Hard to tap the cable (more secure than conductor).
9. **Explain SMF cable.**
   Single Mode Fiber cable is used for long distance transmission, and only one ray of light can travel through it.
11. **What is the application of Single Mode fiber?**
    Long distance communication
13. **What are the types of microwave communication?**
    Terrestrial and satellite
15. **What are the signal sources for optical cable?**
    Laser Diode and Light Emitted Diode (LED)
17. **What are the advantages of STP over UTP?**
    In STP, the cable signal is more immune to noise.
19. **Explain latency and the causes of latency.**
    Latency is the time that it takes to transmit a packet from the source to the destination. This time consists of transmission time, propagation delay, and buffering time.
21. **For the following case, what should the transmitter voltage be for 100 meters of Cat-5 cable with a receiver voltage of 500 mV?**
    a. **Transmitted signal at 20 MHz**
    b. **Transmitted signal at 100 MHz**

a. From Table 4.3, attenuation for 100 meters cat-5 at frequency 20 MHz is 9.3.

$$Av = 10\log_{10}\frac{V_t}{V_r}, \qquad 9.3 = 20\log_{10}\frac{V_t}{0.5} \quad \text{or}$$

$$Vt = 0.5 * 10^{0.465} = 2.9 \text{ volts}$$

b. From Table 4.3, attenuation for cat-5 at 100 MHz is 22.

$$22 = 20\log_{10}\frac{V_t}{0.5} \quad \text{or} \quad VT = .5 * 10^{1.1} = 12.5 \text{ volts}$$

**23. A packet of 10000 bytes is transmitted over a 100-Km cable with bandwidth of 100 Mbps. Calculate the following:**

**a. Propagation delay of the link**

**b. Transmission time**

**c. Latency of the packet**

**d. RTT**

a. $T_x = 1000 * 8/100 * 10^6 = 0.08$ ms
b. $T_p = 100 * 1000$ meters$/2 * 10^8 = 0.5$ ms
c. $T = 0.08+0.5 = 0.58$ ms
d. $RTT = 2 * 0.58$

**25. Find the time that it takes to transfer a 1000-K byte of information from a server that is located 4000 Km away from a host computer. Assume you are using a modem with a data rate of 50 Kbps and the size of each packet is 1000 bytes.**

$T_x$ for one packet $= 1000 * 8/50000 = 1.6$ second
$T_p = 4000 * 1000$ meters$/2 * 10^8 = 0.02$ second

Latency for one packet $T = 1.6 + 0.02 = 1.62$ sec
Latency for 1000 packet $= 1.62 * 1000 = 1620$ second
$= 27$ minutes

**27. 1500 bytes of data are transmitted over 200 Km of fiber-optic cable.**

**a. Find the data rate such that the transmission time becomes equal to the propagation time.**

**b. What is the throughput of this communication link?**

a.

$$T_x = T_p$$

$$1500 * 8/\text{channel bandwidth} = 200 * 1000/2 * 10^8$$

$$\text{Channel bandwidth} = 12 \text{ Mbps}$$

$$T_x = 1500 * 8/12 * 10^6 = 1 \text{ ms}$$
$$T_p = 200 * 1000/2 * 10^8 = 1 \text{ ms}$$
$$T = 1+1 = 2 \text{ ms}$$

b. Throughput = Packet size/Latency = $1500 * 8/2 * 10^{-3} = 6$ Mbps

**29. Find the bandwidth of a communication channel in order to transfer the data at a rate of 100 Mbps; assume $\frac{S}{N}$ ratio is (50 $dB$).**

$$\text{Decibels} = 10 * \log_{10}\left(\frac{S}{N}\right) \text{ or } \quad 50 = 10\log_{10}\left(\frac{S}{N}\right)$$
$$\frac{S}{N} = 10^5$$

$$\text{Max Data Rate (MDR)} = W * \log_2\left(1 + \frac{S}{N}\right) bps$$
$$100 * 10^6 = w \log_2 (1 + 100000)\ bps$$
$$100 * 10^6 = w \ln (100001)/\ln 2$$
$$W = 6 \text{ Mhz}$$

# Chapter 5 Multiplexer and Switching Concepts

## Multiple Choice Questions

**1.** a, **3.** c, **5.** b, **7.** a, **9.** b, **11.** b

## Short Answer Questions

**1. Show an 8-to-1 MUX and 1-to-8 DMUX.**

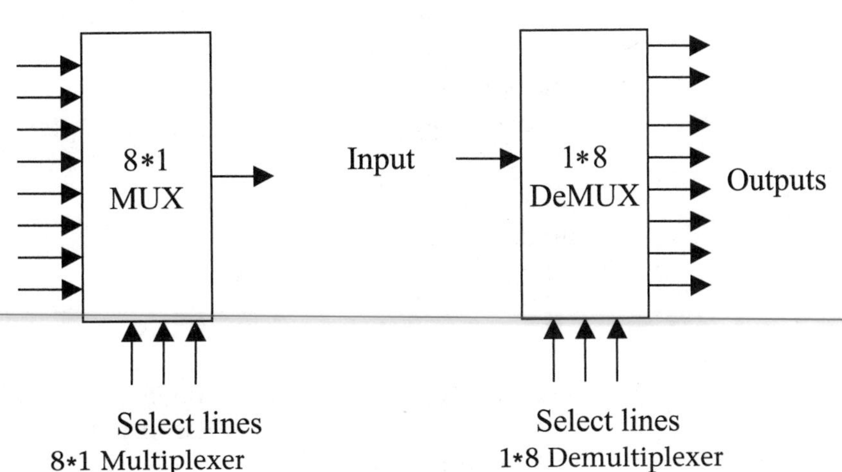


8*1 Multiplexer

1*8 Demultiplexer

3. **Explain TDM operation.**
   In Time Division Multiplexer (TDM), each input is assigned equal time to transmit its information.
5. **What is the difference between CDM and TDM?**
   In CDM all users transmit simultaneously but in TDM users transmit information one at a time.
7. **What are the types of WDM?**
   DWDM and CWDM
9. **What is the type of signal used between two central switches of a telephone system?**
   digital signal
11. **What does PCM stand for and what is its application?**
    PCM stands for Pulse Code Modulation, and it is a method to convert human voice to a digital signal.
13. **How many voice channels does a T1 link carry?**
    24 voice channels
15. **What is the data rate of the human voice and why?**
    The data rate of the human voice is 64 Kbps: the human voice is sampled 8000 times per second, with each sample represented by 8 bits. 8 * 8000 equals 64 Kbps.
17. **What is the difference between DS-1 and a T1 link?**
    DS-1 is the frame format of a T1 link.
19. **How many voice channels can be carried by a T3 link?**
    672 voice channels
21. **Show the frame format of a T1 link.**

| 1 | Byte #24 | Byte | | Byte #1 |
|---|---|---|---|---|

23. **The following inputs are connected to a 4 * 1 statistical multiplexer. Show the outputs of this multiplexer.**

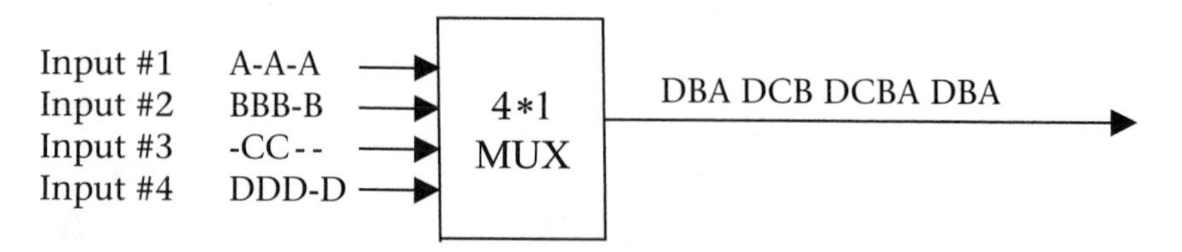


25. **Use the chip sequence on Table 5.3 to find data transmitted for node C at the receiver side. Assume the nodes A, B, and C transmit the following data:**

| Node A | 111 |
|---|---|
| Node B | 010 |
| Node C | 001 |

| | | | |
|---|---|---|---|
| node A | −1−1−1−1 | −1−1−1−1 | −1−1−1−1 |
| node B | +1−1+1−1 | −1+1−1+1 | +1−1+1−1 |
| node C | −1−1+1+1 | −1−1+1+1 | +1+1−1−1 |
| Sum | −1−3+1−1 | −3−1−1+1 | +1−1−1−3 |
| node C's chip bits | +1+1−1−1 | +1+1−1−1 | +1+1−1−1 |
| Inner product of node C's chip bits and Sum | (−1−3−1+1)/4 | (−3−1+1−1)/4 | (+1−1+1+3)/4 |
| Result of inner products | −1 | −1 | +1 |
| Data for node C is | 0 | 0 | 1 |

## Chapter 6 Standard Organization and OSI Model

### Multiple Choice Questions

**1.** a, **3.** c, **5.** c, **7.** a, **9.** a, **11.** b, **13.** a, **15.** c, **17.** b

### Short Answer Questions

1. **Explain why it is necessary to have standards for networking products.**
   Developing standards for computers enables hardware and software products to be compatible.
3. **Define an open system.**
   An open system is a set of protocols that allows two computers to communicate with each other, regardless of the manufacturer, CPU, or operating system.
5. **Explain the function of the Application layer.**
   The Application layer enables users to access networks through tools such as Telnet, FTP, and Network Neighborhood.
7. **Explain the function of the Session layer.**
   to set up a session between two applications, such as login in Telnet
9. **List the functions of the Network layer.**
   setting up a physical connection, disconnecting a connection, and routing information

11. **Define the functions of the Physical layer.**
    electrical and mechanical interface with communications media, converting electrical and optical signals to bits and *vice versa*
13. **Show the frame format of IEEE 802.2.**

| 1 byte | 1 byte | 1 or 2 bytes | |
|---|---|---|---|
| DSAP | SSAP | Control | Information |

15. **What is IEEE 802.3?**
    a standard for the MAC and Physical layers in Ethernet networks
17. **What are the frame transmission methods?**
    asynchronous and synchronous
19. **What does SDLC stand for?**
    Synchronous Data Link Control
21. **What is the application of HDLC?**
    High Level Data Link Control is used to define the frame format of the Data Link layer.
23. **Explain bit insertion in bit-oriented synchronization.**
    01111110 represents the start and end flags. If the information field contains 01111110, the receiver will detect it as the end or start of the frame. To avoid this problem, the transmitter inserts an extra zero after five ones are repeated in the information field. The receiver will discard this extra zero.
25. **Name the organization that sets standards for cable and connectors.**
    Electronic Industry Association (EIA)
27. **List four communication protocols.**
    TCP/IP, NetBEUI, IPX/SPX, and Nwlink
29. **Which layer performs error detection?**
    Data Link layer
31. **Which layer is responsible for the physical connection between the source and the destination?**
    Network layer
33. **Which layer performs login?**
    Session layer
35. **Show the frame format of bit synchronization.**

| 01111110 | Information | 01111110 |
|---|---|---|

37. **Show the ASCII code representation of the SYN character, STX, and ETX.**
SYN = 16h, STX = 02h, and ETX = 03h

39. **Show the transmitted frame, after bit insertion, for the following frame: 01111111000000011111011111110**
After bit insertion, the result is
01111101100000001111100111110110.

# Chapter 7 Modem, Digital Subscriber Line (DSL), Cable Modem, and ISDN

## Multiple Choice Questions

**1.** c, **3.** a, **5.** b, **7.** c, **9.** b and c, **11.** c, **13.** c

## Short Answer Questions

1. **What does modem stand for?**
Modulation and Demodulation

3. **How is a data rate determined?**
the number of bits per second

5. **Explain ASK modulation.**
Amplitude shift keying, or amplitude modulation, represents changes in amplitude of the signal by binary ones and zeros.

7. **Explain PSK modulation.**
Phase Shift Keying uses changes in the phase of the signal to represent ones and zeros.

9. **What is the speed of the current telephone modems being produced?**
56 Kbps

11. **Distinguish between data rate and baud rate.**
Data rate is the number of bits per second and baud rate is the number of signals per second.

13. **What does DSL stand for?**
Digital Subscriber Line

15. **Explain ADSL operation.**
ADSL uses current telephone wiring. These wires can handle signals up to 1 MHz, which is divided into 4-kHz channels. The first channel is used for telephone conversation and 249 channels are used for transmitting and receiving information.

17. **Explain xDSL.**
DSL is implemented in several different technologies, called *xDSL*. Types of xDSL are ADSL, HDSL, SDSL, and VDSL.

19. **Is ADSL dependent on cable length?**
Yes, the data rate of ADSL is dependent on the length of the cable between the subscriber and the telephone central switch.

21. **What are the components of a cable TV system?**
head end, trunk cable, feeder cable, amplifier, and drop cable

23. **What is the bandwidth of a TV channel?**
6 MHz

25. **What type of modulation is used in cable TV modems for downstream transmission?**
64 QAM (Qaudrature Amplitude Modulation) or 256 QAM

27. **List the devices that can be connected to a cable modem.**
a 10 Base-T repeater, or a PC with 10-Base-T NIC

29. **What is the data rate of a modem using frequency shift keying, with a baud rate of 300 signals per second?**
In FSK, each signal is represented by one bit; therefore the data rate is 300 bps.

31. **Calculate the number of bits represented by each signal, for a modem using PSK modulation, with a data rate of 2400 bps and a baud rate of 600.**
2400/600 = 4 bits

33. **Calculate the baud rate of a 32-QAM signal with a data rate of 25 Kbps.**
$2^5 = 32$, 25 Kbps/5 = 5000 signals per second

35. **List the types of ISDN.**
narrowband ISDN and broadband ISDN

37. **List the data rate of the B and D channels for BRI.**
B channel is 64 Kbps, and D channel is 16 Kbps.

39. **How many devices can be connected to BRI ISDN?**
two telephones and one computer

# Chapter 8 Ethernet and IEEE 802.3 Networking Technology

## Multiple Choice Questions

**1.** a, **3.** a, **5.** a, **7.** a, **9.** b, **11.** b

## Short Answer Questions

1. **Define the following terms:**
   **A. 10 Base-T**
   **B. 10 Base-5**
   **C. 10 Base-2**

   A. 10 Base-T: 10 Mbps, baseband, using twisted-pair cable as the transmission media.
   B. 10 Base-5 or Thick Net: 10 Mbps, baseband, using thick coaxial cable as the transmission media. The length of a network segment is 5 * 100 meters.
   C. 10 Base-2 or Thin Net: 10 Mbps, baseband, using thin coaxial cable as transmission media. The length of a network segment is 185 meters.

3. **What do UTP and STP stand for?**
   Unshielded Twisted-Pair Cable and Shielded Twisted-Pair Cable

5. **What is a network segment?**
   the maximum length of the network without a repeater

7. **Explain the function of a repeater or hub.**
   accepting information from one port and retransmitting it to other ports

9. **Describe the access method for Ethernet.**
   Any station requiring access to the network performs the following tasks:
   a. Listens to the network. If the network is busy, the station keeps listening. If the network is not busy, it transmits information and listens for a collision.
   b. Transmits a jam signal on collision detection.

11. **What is IEEE 802.2?**
    a standard for Logical Link Control (LLC)

13. **What is the MAC Address?**
    Also called the *physical address,* the MAC address is the 48 bits assigned to the NIC by its manufacturer.

15. **Explain collision in Ethernet.**
    When two stations transmit information on the Ethernet network at the same time, a collision will occur.

17. **Explain broadcast address.**
    A station uses the broadcast address in the destination field of the frame to send the frame to all stations on the network. The destination address field for broadcast is set to all ones.

19. **What is the size of an NIC card's address?**
    48 bits
21. **What is the function of a transceiver?**
    receiving information from network media and transmitting information to the network media
23. **Determine the maximum size of a network using three repeaters for following networks:**
    **A. 10 Base-T**
    **B. 10 Base-5**
    **C. 10 Base-2**
    A. 400 meters, B. 2000 meters, C. 740 meters
25. **What is the maximum size of a frame determined by IEEE 802.3?**
    1512 bytes
27. **How do computers distinguish one another on an Ethernet network?**
    from their hardware, or MAC address
29. **What is the function of the FCS field in the Ethernet frame format?**
    error detection
31. **What is the function of the length field in an Ethernet frame?**
    identifying the number of bytes in an information field
33. **List the IEEE sublayers of the Data Link layer.**
    LLC and MAC

## Chapter 9 Token Ring and IEEE 802.5 Networking Technology

### Multiple Choice Questions

**1.** c, **3.** b, **5.** d, **7.** c, **9.** a, **11.** b

### Short Answer Questions

1. **What is the function of MAU?**
   Computers are connected to Multiple Access Units (MAU) to form a ring connection.
3. **How many stations can be connected to each MAU?**
   8 stations
5. **How many stations can be connected to a Token Ring network when:**
   **a. Stations are connected to MAU by UTP.**
   **b. Stations are connected to MAU by STP.**

a. 72 stations
b. 260 stations

7. **Explain the functions of an active monitor.**
An active monitor performs the following functions:
a. Generates a token.
b. Removes unwanted frames from the ring.
c. Purges the ring in case of ring failure.
d. Broadcasts a MAC frame every seven seconds to inform stations that there is an active monitor.
e. Detects the loss of tokens and frames.

9. **Explain the process that takes place by inserting a station into the ring?**
a. Physical insertion of station.
b. Station sends MAC frames to the MAU to test the connection.
c. The station verifies its unique address.
d. Finds out the address of its upstream neighbor.

11. **What type of signaling is used in a Token Ring network?**
Differential Manchester Encoding

13. **How many bits is a token made up of?**
24 bits

15. **What is purpose of J and K bits in Token Ring frame format?**
J and K bits are violations of Differential Manchester Encoding (J–K bits do not convert to Differential Manchester Encoding); therefore the SD and ED fields do not appear in the information field of the frame.

# Chapter 10 Fast Ethernet Networking Technology

## Multiple Choice Questions

**1.** c, **3.** c, **5.** a, **7.** b, **9.** a, **11.** a

## Short Answer Questions

1. **Explain the following terms:**
a. **100 BASE-4T:** 100 Mbps, baseband, using 4 pairs of Cat-3 UTP as transmission media
b. **100 BASE-TX:** 100 Mbps, baseband, using 2 pairs UTP as transmission media
c. **100 BASE-FX:** 100 Mbps, baseband, using fiber-optical cable as transmission media

3. **What is the difference between 100 Base-TX and 100 Base-4T?**
   100 Base-TX uses two pairs of Cat-5 UTP, while 100 Base-4T uses four pairs of Cat-3 UTP.
5. **What is the application of a Class II repeater in Fast Ethernet?**
   connecting stations with the same type of Fast Ethernet cards
7. **Name the IEEE committee that developed the standard for Fast Ethernet.**
   IEEE 802.3u
9. **What is the function of a Convergence sublayer?**
   The Convergence sublayer is used to interface the MAC layer with the Physical layer.
11. **What type of encoding is used for 100 Base-4T?**
    8B/6T (8 bits to 6 bit ternary)
13. **Convert 64 hex to 5-bit symbols and then show the corresponding MLT digital signal using Table 10.2.**
    **64 5 01110 01010**

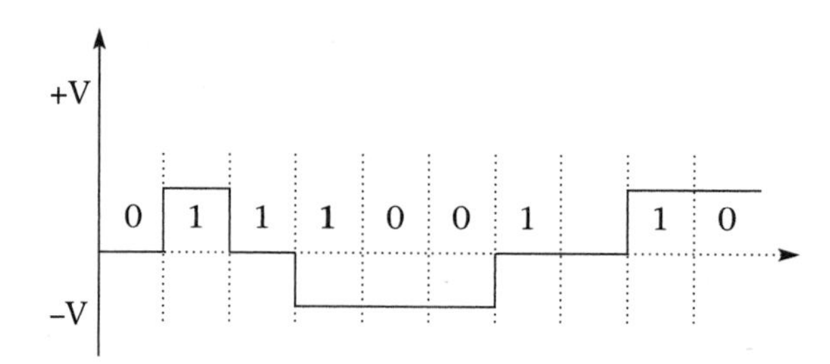


## Chapter 11 Local Area Networks Interconnection Devices

### Multiple Choice Questions

**1.** d, **3.** a, **5.** b, **7.** b, **9.** c, **11.** b, **13.** d, **15.** d, **17.** a, **19.** b, **21.** a

### Short Answer Questions

1. **List the LAN interconnection devices.**
   repeaters
   bridges
   routers
   switches
   gateways

3. **Describe the function of a bridge.**
   connecting different segments of a network
5. **Explain the operation of a transparent bridge.**
   The transparent bridge, or learning bridge, does not require any programming. The bridge learns the location of a station and records the NIC address of that station, as well as the port number by which the station is connected to the bridge.
7. **List the functions of a router.**
   A router accepts a frame from its port and retransmits the frame according to its routing table. A router can also change the frame format, for example: IEEE 802.3 to IEEE 803.5.
9. **What is different between a router and L2 Switch?**
   L2 Switch operates at Data Link layer and Router operates at Network layer.
11. **Explain dynamic routers.**
    A dynamic router automatically generates a routing table.
13. **A gateway operates in which layer(s) of the OSI model?**
    Application layer
15. **Explain switch operation.**
    The switch is used to connect different segments of a LAN, in order to increase network throughput.
17. **What is the application of an asymmetric switch?**
    It provides switching between segments having different bandwidths: 10 Mbps to 100 Mbps or 100 Mbps to 1000 Mbps.
19. **Explain the operation of a store and forward switch.**
    A store and forward switch stores the packet, then checks for errors. If there is no error in the packet, it retransmits the packet to the destination port. If the packet contains an error, it discards the packet.
21. **What is the difference between a router and an L3 switch?**
    An L3 switch is a type of router that uses hardware rather than software.
23. **Suppose a company has two working groups, A and B. Group A has four computers and group B has three computers, all connected to an eight-port Ethernet switch. Both groups need to access a common file server: FS1. There is an in-house requirement that group A's computers should not be able to see Group B's computers on the network.**
    a. **Draw a diagram showing an Ethernet switch with seven computers and a file server.**

**b. Show the VLAN connectivity matrix for the above requirements.**

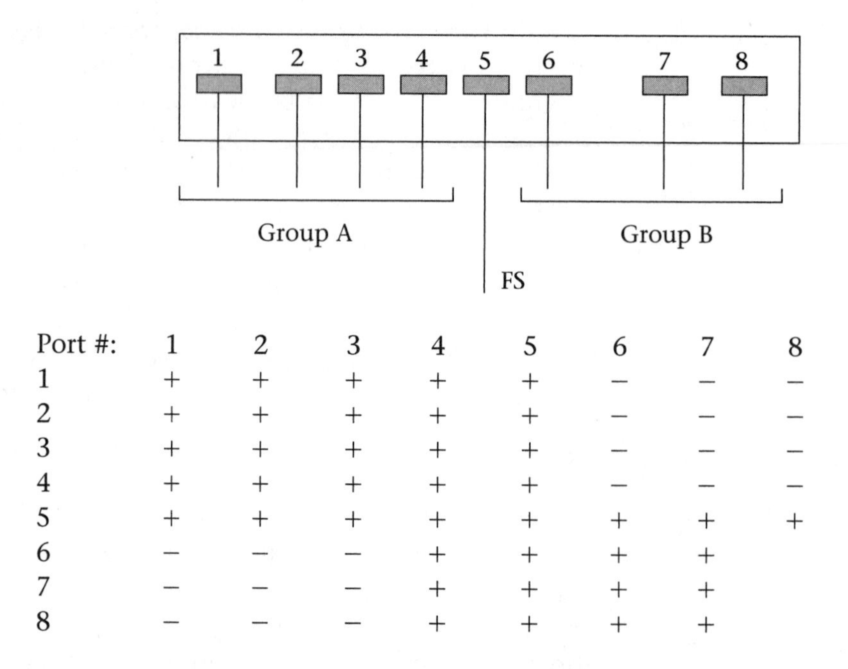


| Port #: | 1 | 2 | 3 | 4 | 5 | 6 | 7 | 8 |
|---|---|---|---|---|---|---|---|---|
| 1 | + | + | + | + | + | − | − | − |
| 2 | + | + | + | + | + | − | − | − |
| 3 | + | + | + | + | + | − | − | − |
| 4 | + | + | + | + | + | − | − | − |
| 5 | + | + | + | + | + | + | + | + |
| 6 | − | − | − | + | + | + | + | |
| 7 | − | − | − | + | + | + | + | |
| 8 | − | − | − | + | + | + | + | |

# Chapter 12 Gigabit and 10 Gigabit Ethernet Technology

## Multiple Choice Questions

**1.** b, **3.** a, **5.** b, **7.** c, **9.** a, **11.** c, **13.** c, **15.** d, **17.** b

## Short Answer Questions

**1. What is the IEEE number for Gigabit Ethernet?**
IEEE 802.3z

**3. What type of frame is used by Gigabit Ethernet?**
Ethernet frame format

**5. List the transmission media used in Gigabit Ethernet.**
multi-mode fiber, single mode fiber, UTP, and shielded balanced twisted pair

**7. What are the hardware components of Gigabit Ethernet?**
(1) Gigabit Ethernet interface card
(2) switches that can handle 100-Mbps Fast Ethernet and 1000-Mbps Ethernet

(3) Gigabit Ethernet switch
(4) transmission media that handles 1000 Mbps

9. **Explain the following terms:**
   a. **10 GBASE-R:** 10 Gbps, baseband, short wavelength, used for LAN
   b. **10 GBASE-SW:** 10 Gbps, baseband, short wavelength, used for WAN
   c. **10 GBASE-LR:** 10 Gbps, baseband, long wavelength, used for LAN
   d. **10 GBASE-LW:** 10 Gbps, baseband, short wavelength, used for WAN
   e. **10 GBASE-ER:** 10 Gbps, baseband, extended long wavelength, used for LAN
   f. **10 GBASE-EW:** 10 Gbps, baseband, extended long wavelength, used for WAN
   g. **10 GBASE-LX4:** 10 Gbps, baseband, WWDM, used for WAN

## Chapter 13 Fiber Distributed Data Interface (FDDI)

### Multiple Choice Questions

**1.** a, **3.** a, **5.** c, **7.** a, **9.** b

### Short Answer Questions

1. **What does FDDI stand for?**
   Fiber Distributed Data Interface

3. **What is the data rate for FDDI?**
   100 Mbps

5. **Explain the function of an FDDI concentrator.**
   An FDDI concentrator is used as a ring and stations are connected to the concentrator.

7. **Explain the function of SAS.**
   SAS is a Single Attachment Station, which attaches to only the primary FDDI ring.

9. **Explain the function of a bypass relay switch.**
   The optical bypass switch is used to isolate stations from the ring.

11. **What type of encoding is used in FDDI?**
    4B/5B (4-bit binary to 5-bit symbol)

13. **List the advantages of a Fast Ethernet switch over FDDI.**
    A Fast Ethernet NIC is less expensive than an FDDI NIC.
    Fast Ethernet can use UTP as a transmission media, which is less expensive than fiber cable.

## Chapter 14 Frame Relay Networks

### Multiple Choice Questions

**1.** c, **3.** b, **5.** b, **7.** c, **9.** a

### Short Answer Questions

1. **What layers of the OSI model does frame relay operate on?**
   both the Data Link layer and Physical layer of the OSI model
3. **What is the function of FRAD?**
   The function of FRAD (Frame Relay Assembler and Disassembler) is to convert the frame format from a site network to the frame format of the frame relay network and vice versa.
5. **What are the components of frame relay networks?**
   a. frame relay access equipment:
      1. customer premises equipment, such as a LAN
      2. a Frame Relay Assembler and Disassembler (FRAD), such as a router

   b. frame relay switches
   c. frame relay service: public carriers provide frame relay services

## Chapter 15 Synchronous Optical Network (SONET)

### Multiple Choice Questions

**1.** b, **3.** a, **5.** a, **7.** a, **9.** c, **11.** b, **13.** b

### Short Answer Questions

1. **What does SONET stands for?**
   Synchronous Optical Network
3. **What is the application of SONET?**
   SONET is a high-speed optical data carrier.
5. **What is the transmission media for SONET?**
   fiber-optic cable

7. **List the SONET components.**
   (a) STS multiplexer
   (b) STS demultiplexer
   (c) ADD/DROP multiplexer
   (d) regenerator
9. **What is OC-1?**
   Optical Carrier type 1 (OC-1)
   SONET converts STS-1 to OC-1 (optical signal) with a data rate of 51.84 Mbps.
11. **How many bytes is STS-1?**
   810 bytes
13. **How many STS-1s must be multiplexed to generate a STS-3?**
   three
15. **Show the frame format of SONET.**

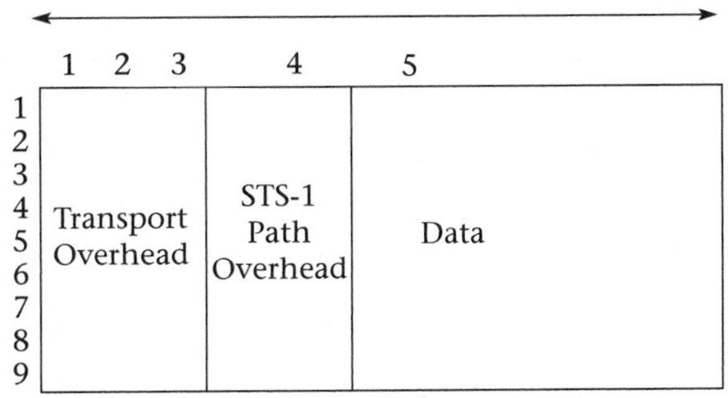

| | 1 2 3 | 4 | 5 |
|---|---|---|---|
| 1 2 3 4 5 6 7 8 9 | Transport Overhead | STS-1 Path Overhead | Data |

17. **Calculate the data rates of VT3 and VT6.**
   A VT3 frame is 54 bytes transmitted at 8000 frames per second; therefore the data rate of
   VT3 is 54 * 8 * 8000 bps = 3.152 Mbps.
   A VT6 frame is 108 bytes, and its data rate is 108 * 8 * 8000 bps.
19. **What is STS-n?**
   Synchronous Transport Signal Type n

# Chapter 16 Internet Protocols (Part I)

## Multiple Choice Questions

**1.** c, **3.** d, **5.** a, **7.** c, **9.** b, **11.** c, **13.** b, **15.** d, **17.** a, **19.** a, **21.** b, **23.** a

## Short Answer Questions

1. **What is the language used to create files for the web?**
   HTML (Hyper Text Markup Language) is used to create files for the WWW (World Wide Web).
3. **What protocol is used to access the Internet by a modem?**
   Point-to-Point Protocol (PPP) or Slip
5. **List the protocols in the Transport level of TCP/IP.**
   TCP (Transmission Control Protocol) and UDP (User Datagram Protocol)
7. **What is the size of an IP address?**
   32 bits
9. **Explain the function of IP.**
   delivering packets to a destination
11. **What is the size of a TCP header?**
    20 bytes
13. **What is the size of a UDP header?**
    8 bytes
15. **If an organization uses the third byte of the IP address for a subnet, how many subnets can be assigned to the class B address?**
    $2^8 = 256$
17. **What is the function of ARP?**
    requesting physical addresses from a destination
19. **What is the current version of IP?**
    IPv6
21. **What is the size of IPv6?**
    128 bits
23. **List the Internet applications.**
    e-mail, telnet, WWW, and FTP
25. **List two applications for UDP.**
    TFTP (Trivial File Transfer Protocol) and RCP (Remote Call Procedure)
27. **TCP/IP was developed by ____________.**
    University of California Berkeley
29. **What are the functions of ICMP?**
    Internet Control Message Protocol is used for handling error messages and control messages.

31. **Explain the function of Telnet.**
    Telnet is one of the application protocols of TCP/IP, and is used for remote connection of one computer to another computer.

33. **What is an MTU?**
    Maximum Transfer Unit (MTU) is the largest frame length that can be sent through the Internet, and it depends on the Network Interface Card of the station.

35. **How many bits are in an IPV6 address?**
    128 bits

37. **There are two computers, A and B, with IP addresses of 174.20.45.37 and 174.20.67.45. These computers have a subnet mask ID of 255.255.0.0. Can you determine if these two computers are located in the same network?**
    Yes, both networks are using class B addresses.

# Chapter 17 Internet Protocols (Part II)

## Multiple Choice Questions

**1.** a, **3.** c, **5.** c, **7.** c, **9.** c, **11.** c

## Short Answer Questions

1. **What is the application of DHCP?**
   DHCP dynamically assigns an IP address to the client.

3. **List the DHCP IP address allocation methods.**
   automatic allocation, dynamic allocation, and manual allocation

5. **What does PPP stand for?**
   Point-to-Point Protocol

7. **Show PPP layered architecture.**
   Refer to Figure 17.8.

9. **What are the functions of Link Control Protocol?**
   establish connection, configure connection parameters, and negotiate link parameters such as authentication method and MRU

11. **Explain Challenge Handshakes Authentication (CHAP)**
    The remote server sends a random number to the client station, the client generates a hash value from a random number, then the client sends its name, password, and hash value to the server. The server generates a hash value from the random number that was

sent to the client; if both hash values are the same, then the client is authenticated.

13. **In question #12 determine type of the authentication that is used.**
PAP
15. **Describe tunneling and its application.**
Tunneling is a method that places one packet inside another packet for transmission over a public network. It is used for secure transmission of information in corporations having multiple sites.
17. **What is the function of a router?**
The function of a router is to determine the path for transporting information through internetworking (routing).
19. **Explain static routing and dynamic routing.**
A static routing table is configured manually but a dynamic routing table is configured automatically.

# Chapter 18 Wireless Local Area Networks (WLAN) or IEEE 802.11

## Multiple Choice Questions

**1.** a, **3.** b, **5.** c, **7.** b, **9.** b, **11.** a, **13.** b, **15.** d

## Short Answer Questions

1. **What does WLAN stand for?**
Wireless LAN
3. **What are the WLAN topologies?**
managed wireless network (BSS and ESS) and wireless unmanaged network (Ad-Hoc)
5. **What is an Ad-Hoc topology?**
Ad-Hoc topology is made of wireless clients without access points.
7. **What is the function of an access point device in WLAN?**
An access point acts like a bridge, receiving information and re-transmitting through the air and its wired port.
9. **What are the IEEE standards for WLAN and their data rates?**
802.11 with data rate of 1 and 2 Mbps
802.11b with data rate 1, 2, 5.5, and 11 Mbps
802.11a with data rate 6, 12, 18, 24, 36, 45, and 54 Mbps
802.11g with data rate 6, 12, 18, 24, 36, 45, and 54 Mbps

11. **Explain ODFM operation.**
    Each channel for 802.11a is 20 MHz, and OFDM divides this channel into 52 subchannels of 300 KHZ each. OFDM uses 48 channels for transmission of information in parallel form.
13. **Explain DSSS operation.**
    In DSSS each bit is broken down to a pattern of bits and is called *chip bits,* then these chip bits are modulated and transmitted.
15. **How many non-overlapping channels are offered by IEEE 802.11 and what are they?**
    three and they are 1, 6, and 11
17. **What are the ranges of frequency for U-NII band?**
    5.15–5.25 GHz low band
    5.25–5.35 GHz middle band
    5.725–5.825 high band
19. **What are the types of frame formats for WLAN?**
    management frame, control frame, and data frame
21. **Which devices transmit beacon frames?**
    Beacon frames are transmitted periodically by access point.
23. **What does CSMA/CA stand for?**
    Carrier Sense Multiple Access with Collision Avoidance (CSMA/CA)
25. **What is the function of Wi-Fi?**
    to offer certifications for wireless LAN equipment interoperability
27. **Explain multi-path fading.**
    When RF signals are transmitted through air, it takes multiple paths to reach the destination due to obstacles on the way: The receiver receives multiple signals with different delays and amplitudes.
29. **What is the maximum transmitter power for WLAN?**
    1 watt

# Chapter 19 Bluetooth Technology

## Multiple Choice Questions

**1.** c, **3.** d, **5.** b, **7.** b, **9.** b, **11.** a, **13.** b, **15.** a

## Short Answer Questions

1. **What are the applications of Bluetooth technology?**
   connecting keyboard, mouse, and any peripheral to PC

3. **Show Piconet topology.**
   M: master station, S: slave station, P: parked station

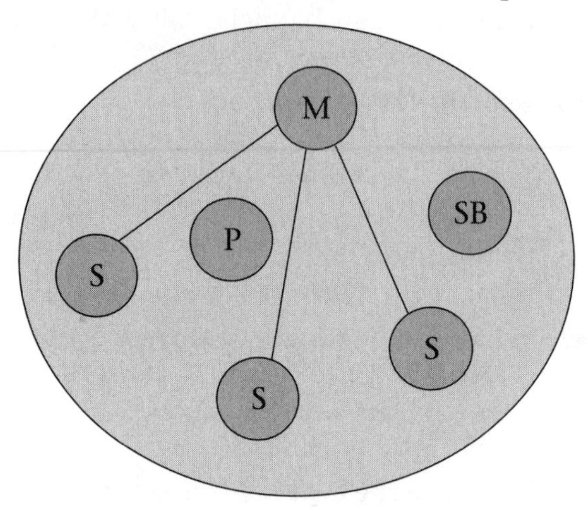


5. **How many devices can be in one Piconet?**
   7 devices
7. **How many slaves are in a Piconet?**
   6 slaves
9. **Show Bluetooth protocol architecture.**
   Refer to Figure 19.3.
11. **List the power transmissions for Bluetooth.**

| Power Class | Max Power | Min power | Distance |
|---|---|---|---|
| Class 1 | 100 mW | 1 mW | 100 meters |
| Class 2 | 2.5 | 0.25 mW | 10 meters |
| Class 3 | N/A | 1 mW | N/A |

13. **What are the types of physical connections supported by Bluetooth?**
    The physical connections supported by Bluetooth are synchronous connection oriented and asynchronous connectionless links.
15. **How many SCO links are supported by the Physical layer?**
    In the Physical layer, up to three SCO links can be supported simultaneously.
17. **List the types of error correcting methods used by Bluetooth.**
    Error correction methods used by Bluetooth are FEC (1/3 or 2/3) and ARQ.

# Chapter 20 Wireless MAN

## Multiple Choice Questions

**1.** b, **3.** c, **5.** b, **7.** d

## Short Answer Questions

1. **What is the IEEE number for Wireless MAN?**
   IEEE 802.16
3. **What is the topology for IEEE 802.16?**
   IEEE 802.16 uses PMP and optional mesh topology.
5. **What is the frequency spectrum of IEEE 802.16?**
   10–66 GHz
7. **What is difference between IEEE 802.16a and the 802.16 Physical layer?**
   802.16 operates in line-of-sight but 802.16a operates in non-line-of-sight.
9. **What are the types of connections that are used by IEEE 802.16?**
   basic connection, primary management connection, and secondary management connection
11. **List the four different management frames.**
    Uplink Channel Descriptor (UCD), Downlink Channel Descriptor (DCD), DL-Map, and UL-Map (see Table 20.1)
13. **Show the FDD frame format.**

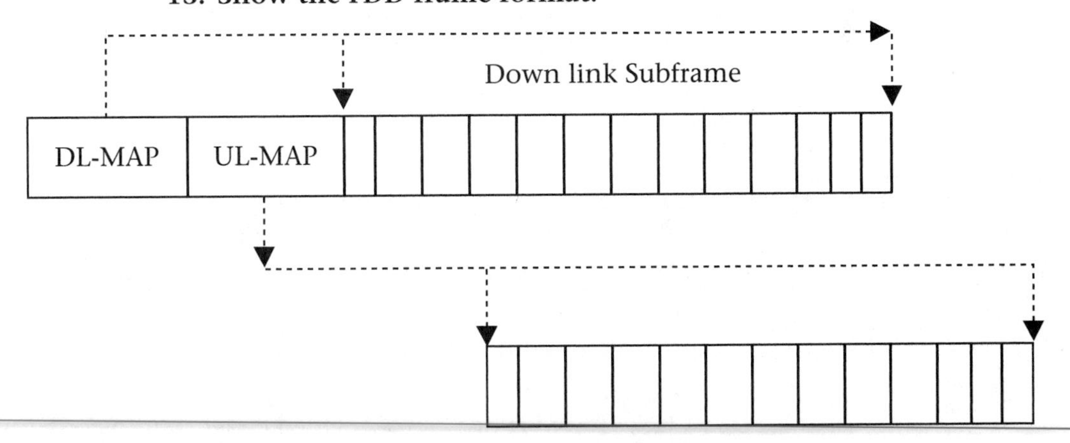


15. **What is the function of a DCD frame?**
    Data Channel Descriptor broadcasts by BS to SS periodically for informing subscriber stations of BS transmitter power, FDD, or TDD frame duration.

17. **What is the function of a UL-Map frame?**
    The uplink message is transmitted by BS to SSs to allocate uplink channels to them. It contains an uplink Channel ID (8 bits), the number of UL-Map information elements, and allocation start time for uplink transmission that is defined by the UL-Map in the mini-slots.

## Chapter 21 Voice over Internet Protocols (VoIP)

### Multiple Choice Questions

**1.** a, **3.** b, **5.** c, **7.** c, **9.** b, **11.** c, **13.** d

### Short Answer Questions

1. **What does VoIP stand for?**
   Voice over Internet Protocol
3. **Why does VoIP reduce the cost of long distance communications?**
   The link can be shared by multiple users at the same time.
5. **What are three factors that impact voice quality over the Internet?**
   transmission delay, jitter delay, and packet lost
7. **Define jitter and describe what causes it.**
   It is the difference in arrival time between packets caused by delay due to network congestion.
9. **List the applications for VoIP.**
   computer–computer voice communication, corporations with multiple sites, and VoIP by service provider
11. **Describe the layers involved for transmission of VoIP.**
    At the transmitting side: PCM, Compression, RTP, UDP, IP, and NIC
    At the receiving side: NIC, IP, UDP, RTP, decompression, and D/A
13. **What is function of an audio codec at the transmission side?**
    converting audio signal to digital, then compression of digital bits
15. **List H.323 protocols.**
    H.245, H.225, H.261, H.263, video codec, audio codec, and TCP/IP protocols
17. **List the components of SIP.**
    user agent, server, gateway

19. **What form of addressing does SIP use?**
    SIP uses URLs for addressing such as
    Sip:elahia1@southernct.edu

21. **Find the minimum bandwidth requirement for a VoIP channel using a 50-ms frame and a G.711 codec.**
    G.711 codec data rate is 64000 bps, therefore 50 ms of 64000 bps is 3200 bits or 400 bytes. Voice packet size = 400 bytes payload + 40 bytes header (RTP, UDP, and IP)+26 bytes Ethernet Header = 466 bytes.

    This vice packet must receive at destination 64 kbps or 8000 bytes per second.
    8000/466 = 17 voice packets per second
    or 17 * 8 * 466 = 63376 bps

## Chapter 22 Asynchronous Transfer Mode (ATM)

### Multiple Choice Questions

**1.** c, **3.** c, **5.** b, **7.** b, **9.** a, **11.** b, **13.** c, **15.** c, **17.** c, **19.** b, **21.** a, **23.** b

### Short Answer Questions

1. **What does ATM stand for?**
   Asynchronous Transfer Mode

3. **What are the characteristics of ATM?**
   cell switching, connection oriented transmission, fixed cell size of 53 bytes, and switching occurs at very high speed

5. **What is the data unit of ATM?**
   a cell, and it is 53 bytes

7. **Show an ATM cell.**

| 48 bytes payload | 5 bytes |
|---|---|

9. **How many bytes are in an ATM cell header?**
   5 bytes

11. **Explain Permanent Virtual Circuits.**
    Permanent Virtual Circuits (PVC) are set up and taken down manually by the network manager.

13. **What are connection identifiers?**
    virtual path identifier and virtual circuit identifier

15. **What causes blocking in an ATM switch?**
An ATM switch must deal with two types of congestion: fabric blocking and head end blocking. Fabric blocking happens when the switch capacity is less than the sum of the input's data rate. Head blocking happens when an output port is congested.

17. **Show the ATM protocol.**
Please refer to Figure 22.11.

19. **What is the function of the ATM Adaptation layer?**
The ATM Adaptation layer converts the large Service Data Unit (SDU) to the 48 bytes required for an ATM cell payload.

# Chapter 23 Network Security

## Multiple Choice Questions

**1.** b, **3.** b and c, **5.** b, **7.** a, **9.** c, **11.** d, **13.** c, **15.** d, **17.** b

## Short Answer Questions

1. **List the types of attacks and explain each type.**
direct attack and indirect attack
In a direct attack, the attacker is able to access the system by breaking a password.
In an indirect attack, the attacker obtains information such as social security and credit card numbers.

3. **What is the definition of cryptography?**
process that encrypts and decrypts information

5. **What is a cipher?**
An encryption and decryption algorithm is called a *cipher.*

7. **Explain symmetric cryptography.**
In symmetric cryptography, the transmitter and receiver use the same key for encryption and decryption.

9. **Distinguish between stream ciphers and block ciphers.**
A stream cipher encrypts information bit by bit and a block cipher encrypts information using a block of bits.

11. **Find the encrypted text for "COMPUTER SCIENCE DEPT" using a transposition cipher, assuming each row is made of five characters and the encryption key is 43215.**
    1 2 3 4 5
    COMPU
    TE R-S
    CIENC
    EDEP-
    The result of encryption is P NEMREEOEIDUSC
13. **What are some popular hash functions?**
    MD4, MD5, and SHA
15. **Explain the function of Kerberos.**
    Kerberos is an authentication method that authenticates both end users.
17. **Use the information in problem 16:**
    **a. Encrypt message 6 using the public key.**
    **b. Decrypt the encrypted message using the private key.**
19. **Show the TCP/IP model for SSL.**
    Refer to Figure 23.12.
21. **What is the function of a certificate authority?**
    The certificate authority issues digital certificates.
23. **What is the application of the RADIUS server?**
    It is an authentication server.
25. **What is the function of a firewall?**
    A firewall is a system for preventing unauthorized users access to private networks.
27. **List the functions of a packet filtering firewall.**
    filter IP address, protocol type, packet payload, and source port and destination port

# Chapter 24 Universal Serial BUS (USB) and PCI-Express

## Multiple Choice Questions

**1.** c, **3.** b, **5.** b, **7.** b, **9.** d

## Short Answer Questions

1. **What is the high speed USB data rate?**
   460 Mbps
3. **What is the function of a token in a USB system?**
   A token is a transaction packet transmitted by the host controller to a USB device and it determines the types of transactions.
5. **How many devices can be connected to a USB system?**
   127 devices
7. **What is the function of the root hub?**
   A root hub provides connections between devices and the host controller.
9. **List the types of token packets.**
   IN, OUT, and SETUP
11. **List the handshake packets.**
    Handshake packets are ACK, NACK, and stall packets.
13. **What is the application of isochronous data transfer?**
    It is used for real-time communication.
15. **What is the enumeration process?**
    a process used by the host controller to identify a device when attached and assign it a unique 7-bit address
17. **Why does USB version 2.0 support split transactions?**
    so it is able to handle low- and full-speed devices

# Appendix C: Acronyms

## A

**AAL** ATM Adaptation Layer

**ACK** Acknowledgment

**ACL** Asynchronous Connectionless

**ADSL** Asymmetrical Digital Subscriber Line

**ALU** Arithmetic Logic Unit

**AM** Amplitude Modulation

**ANSI** American National Standards Institute

**API** Application Program Interface

**ARP** Address Resolution Protocol

**ARPA** Advanced Research Project Agency

**ARPANET** Advanced Research Project Agency Network

**ASCII** American Standard Code for Information Interchange

**ASIC** Application Specific Integrated Circuit

**ASK** Amplitude Shift Keying

**ATM** Asynchronous Transfer Mode

**AUI** Attachment Unit Interface

## B

**B Channel** Bearer Channel

**BCD** Binary Coded Decimal

**B-ISDN** Broadband Integrated Services Digital Network

**BBN** Bolt Beranek & Newman

**BENC** Backward Explicit Congestion Notification

**BIOS** Basic Input / Output System

**BNC** British Naval Connector

**BRI** Basic Rate Interface

**BSD** Berkeley Software Distribution

**BSS** Basic Service Set

## C

**CA** Certificate Authority

**CAD** Computer Aided Design

**CAT** Categories

**CBR** Constant Bit Rate

**CD-ROM** Compact Disk Read Only Memory

**CISC** Complex Instruction Set Computer

**CLP** Congestion Loss Priority

**CPI** Common Port Identification

**CPU** Central Processing Unit

**C/R** Command/Response

**CRC** Cyclic Redundancy Check

**CS** Convergence Sublayer

**CSMA/CD** Carrier Sense with Multiple Access and Collision Detection

**CSU/DSU** Channel Service Unit/Data Service Unit

## D

**DA** Destination Address

**DARPA** Defense Advanced Research Projects Agency

**DAS** Dual Attachment Station

**DCD** Down Link Channel Descriptor

**DES** Data Encryption Standard

**DE** Discard Eligibility

**DEMUX** Demultiplexer

**DF** Do not Fragment

**DHCP** Dynamic Host Configuration Protocol

**DIMM** Dual Inline Memory Module

**DLC** Data Link Control

**DLCI** Data Link Connection Identifier

**DMA** Direct Memory Access

**DMT** Discrete Multi-Tone

**DNS** Domain Name System

**DOD** Department of Defense

**DOS** Disk Operating System

**DPAM** Demand Priority Access Method

**DRAM** Dynamic Random Access Memory

**DSAP** Destination Service Access Point

**DSL** Digital Subscriber Line

**DS-n** Digital Signal level n

**DSS** Direct Sequence Spread Spectrum

**DTE** Data Terminal Equipment

## E

**EAP** Extensible Authentication Protocol

**EPROM** Erasable Programmable Read Only Memory

**EEPROM** Electrically Erasable Programmable Read Only Memory

**EHF** Extreme High Frequency

**EIA** Electronics Industries Association

**EIDE** Extended Integrated Disk Electronics

**EISA** Extension Industry Standard Architecture

**EPROM** Erasable Programmable Read Only Memory

**ESC** Escape

**ESD** End of Stream Delimiters

**ESP** Encapsulation Security Protocol

**ESS** Extended Service Set

**ETX** End of Text

## F

**FCC** Federal Communication Commission

**FCS** Frame Check Sequence

**FDD** Frequency Division Duplex

**FDDI** Fiber Distributed Data Interface

**FDM** Frequency Division Multiplexing

**FECN** Forward Explicit Congestion Notification

**FHSS** Frequency Hopping Spread Spectrum

**FM** Frequency Modulation

**FRAD** Frame Relay Assembler and Disassembler

**FSK** Frequency Shift Keying

**FTP** File Transfer Protocol

## G

**GFC** Generic Flow Control

## H

**HDLC** High Level Data Link Control

**HDSL** High bit rate Digital Subscriber Line

**HEC** Header Error Control

**HELN** Hardware Address Length

**HFC** Hybrid Fiber Coax

**HTML** Hypertext Markup Language

**HTTP** Hypertext Transfer Protocol

## I

**IAB** Internet Architecture Board

**IANA** Internet Assigned Numbers Authority

**IBM** International Business Machine

**ICs** Integrated Circuits

**ICMP** Internet Control Message Protocol

**IDE** Integrated Disk Electronics

**IEEE** Institute of Electrical and Electronic Engineers

**IESG** Internet Engineering Steering Group

**IETF** Internet Engineering Task Force

**IF** Information Field

**INIC** Internet Network Information Center

**I/O** Input / Output

**IP** Internet Protocol

**IPv6** Internet Protocol Version 6

**IPX/SPX** Internet Packet Exchange/Sequence Packet Exchange

**IR** Infrared

**IRTF** Internet Research Task Force

**ISA** Industry Standard Architecture

**ISDN** Integrated Services Digital Network

**ISM** Industrial, Scientific, and Medical Band

**ISO** International Organization for Standardization

**ITU** International Telecommunications Union

## L

**LAA** Locally Administered Address

**LANs** Local Area Networks

**LAP** Link Access Protocol

**LAPB** Link Access Procedure Balanced

**RAS** Registration Admission and Status

**LED** Light Emitting Diode

**LD** Laser Diode

**LLC** Logical Link Control

**L3 Switch** Layer 3 Switch

**LW** Long-Wave

## M

**MAC** Media Access Control

**MAN** Metropolitan Area Network

**MAU** Multiple Access Unit or Multistation Access Unit

**MCA** Micro-Channel Architecture

**MCU** Multipoint Conference Unit

**MD4** Message Digest type 4

**MD5** Message Digest type 5

**MF** More Fragment

**MII** Media Independent Interface

**MMF** Multi-Mode Fiber

**MTU** Maximum Transfer Unit

**MUX** Multiplexer

**MW** Medium Wave

## N

**NCP** Network Control Protocol

**NDIS** Network Device Interface Specification

**NetBEUI** NetBIOS Extended User Interface

**NetBIOS** Network Basic Input Output System

**NFS** Network File System

**NIC** Network Interface Card

**NISDN** Narrowband Integrated Services Digital Network

**NLM** Network Loadable Modules

**NNI** Network to Network Interface

**NOS** Network Operating System

**NRZ** Non-Return to Zero

**NRZ-I** Non-return to Zero Inverted

**NRZ-L** Non-Return Zero Level

**NSF** National Science Foundation

**NT** New Technology

**NT1** Network Termination Device

## O

**OAM** Operation, Administration and Maintenance

**OC** Optical Carrier

**ODI** Open Data Link Interface

**OSI** Open System Interconnection

## P

**PAM** Pulse Amplitude Modulation

**PC** Personal Computer

**PCI** Peripheral Component Interconnection

**PCM** Pulse Code Modulation

**PCMCIA** Personal Computer Memory Card International Association

**PDM** Physical Medium Dependent

**PDU** Protocol Data Unit

**PLEN** Length of IP Address

**PMP** Point-to-Multipoint

**PnP** Plug and Play

**POTS** Plain Old Telephone Service

**PPP** Point-to-Point Protocol

**PRI** Primary Rate Interface

**PSK** Phase Shift Keying

**PSTN** Packet Switch Telephone Network

**PT** Payload Type

**PVC** Polyvinyl Chloride

**PVC** Permanent Virtual Connection

## Q

**QAM** Quadrature Amplitude Modulation

## R

**RADIUS** Remote Authentication Dial-In User Service

**RADSL** Rate Adaptive Asymmetric Digital Subscriber Line

**RAM** Random Access Memory

**RARP** Reverse Address Resolution Protocol

**RF** Radio Frequency

**RG** Radio Government

**RI** Routing Information

**RISC** Reduced Instruction Set Computer

**ROM** Read Only Memory

**RPC** Remote Call Procedure

**RTCP** Real Time Control Protocol

**RTP** Real Time Protocol

**RZ** Return to Zero

## S

**SA** Source Address

**SAR** Segmentation and Re-assembly

**SAS** Single Attachment Station

**SB** Standby Station

**SCO** Synchronous Connection Oriented

**SCSI** Small Computer Systems Interface

**SDH** Synchronous Digital Hierarchy

**SDLC** Synchronous Data Link Control

**SDP** Service Discovery Protocol

**SDRAM** Synchronous DRAM

**SDSL** Symmetrical Digital Subscriber Line

**SFD** Start Frame Delimiter

**SHA** Secure Hash Algorithm

**SHF** Super High Frequency

**SIE** Serial Engine Interface

**SIG** Special Interest Group

**SIMD** Single Instruction Multiple Data

**SIMM** Single In-Line Memory Module

**SIP** Session Initiation Protocol

**SMF** Single Mode Fiber

**SMTP** Simple Mail Transfer Protocol

**SMT** Station Management

**SNA** System Network Architecture

**SNMP** Simple Network Management Protocol

**SONET** Synchronous Optical Network

**SPE** Synchronous Payload Envelope

**SPID** Service Profile Identifiers

**SPM** Statistical Packet Multiplexing

**SQE** Signal Quality Error

**SRAM** Static Random Access Memory

**SSAP** Source Service Access Point

**SSL** Secure Socket Layer

**STM** Synchronous Transport Module

**STP** Shielded Twisted Pair

**STS** Synchronous Transport Signal

**STX** Start of Text

**SVC** Switched Virtual Connection

**SW** Short Wave

## T

**TA** Terminal Adapter

**TC** Transmission Convergence

**TCP** Transmission Control Protocol

**TCP/IP** Transmission Control Protocol/Internet Protocol

**TDD** Time Division Duplexing

**TDM** Time Division Multiplexing

**TFTP** Trivial File Transfer Protocol

**TOS** Type Of Service

**TTL** Time To Live

## U

**UAA** Universal Administered Address

**UCD** Uplink Channel Descriptor

**UDP** User Datagram Protocol

**UHF** Ultra High Frequency

**UNI** User Network Interface

**U-NII** Unlicensed National Information Infrastructure Band

**URL** Uniform Resource Locator

**UTP** Unshielded Twisted Pair

## V

**VBR** Variable Bit Rate

**VCI** Virtual Channel Identifier

**VDSL** Very high speed Digital Subscriber Line

**VESA** Video Electronics Standards Association

**VHF** Very High Frequency

**VLAN** Virtual Local Area Network

**VLF** Very Low Frequency

**VoIP** Voice over IP

**VPI** Virtual Path Identifier

**VPN** Virtual Private Network

## W

**WAN** Wide Area Network

**Windows NT** Windows New Technology

**WLAN** Wireless Local Area Network

**WWW** World Wide Web

# Appendix D: Glossary

**10 Base-2** 10 Mpbs baseband Ethernet network using thin coaxial cable for transmission media and maximum length of network 200 meters without repeater

**10 Base-5** 10 Mpbs baseband Ethernet network using thick coaxial cable for transmission media and maximum length of network 500 meters without repeater

**10Base-T** 10 Mpbs baseband Ethernet network using Cat-5 UTP cable for transmission media

**100 Base T4** 100 Mbps baseband Fast Ethernet using four pairs Cat-3 UTP cable

**100 Base TX** 100 Mbps baseband Fast Ethernet using Cat-5 UTP cable

**100 Base FX** 100 Mbps baseband Ethernet using fiber-optic cable for transmission media

**1000 Base T** Gigabit Ethernet, 1000 Mbps baseband Ethernet using Cat-5 cable

**1000 Base-LX** Gigabit Ethernet, 1000 Mbps Ethernet using long wavelength laser (1300 nm)

**1000 Base-SX** Gigabit Ethernet, 1000 Mbps Ethernet using short wavelength laser (850 nm)

**1000 Base-CX** Gigabit Ethernet, 1000 Mbps Ethernet using twinax cable

**10 Gigabit Ethernet** 10000 Mbps Ethernet technology

**10GBase-ER** 10 Gigabit Ethernet that uses extended wavelength laser (1550 nm) for LAN

**10GBase-LR** 10 Gigabit Ethernet that uses long wavelength laser (1300 nm) for LAN

**10GBase-SW** 10 Gigabit Ethernet that uses short wavelength laser (850 nm) for WAN

**10GBase-LX4** 10 Gigabit Ethernet that uses four wavelength lasers and WDM

**4B/5B Encoding** Converting 4 bits to 5-bit symbol

**8B/10B Encoding** Converting 8 bits to 10-bit symbol

**8B/6T** Converting 8 bits to 6 bits ternary

**Access Point (AP)** A device that acts like a bridge in Wireless LAN; it bridges a wired LAN with wireless devices

**Active Monitor** A station in Token Ring network that manages the network

**Address Resolution Protocol** An Internet protocol that maps IP address to physical address

**Add/Drop Multiplexer** A multiplexer that has the capability to drop and add low rate signals from a high-rate multiplexed signal without demultiplexing the signal

**Ad-Hoc** A Wireless LAN topology that does not use any access point

**Advanced Research Project Agency (ARPA)** Agency funded by the Department of Defense for the purpose of research, it created the ARPANET in 1968

**Audio Codec** A device that converts voice signal to digital signal and vice versa

**American National Standards Institute (ANSI)** Organization for developing standards related to computers in the United States

**American Standard Code for Information Interchange (ASCII)** Assigns a 7-bit code to characters in the keyboard and also includes several nonprintable control characters

**Amplitude Shift Keying (ASK)** A type of modulation that uses different amplitudes to represent binary one and zero

**Analog Bandwidth** The difference between the highest and lowest frequencies of an analog signal that a communication channel can carry

**Analog Signal** A signal whose amplitude is a function of time and that changes gradually as time changes

**Analog to Digital Converter** A device that converts an analog signal to a digital signal

**Application Layer** Layer 7 of OSI model, it is a protocol that enables users to access the network

**Arithmetic Logic Unit (ALU)** Part of the CPU that performs arithmetic and logic operation

**ARP Command** Displays host IP address with its MAC address

**Asymmetric DSL (Asymmetric Digital Subscriber Line)** A type of modem that provides higher data rates downstream than upstream

**Asynchronous Transmission** Data is transmitted one frame at a time. The frame consists of a character to be sent as well as synchronization information

**ATM (Asynchronous Transfer Mode)** A network standard that uses cell switching with a cell size of 53 bytes

**ATM Adaptation Layer** Used to convert an application data unit to ATM data unit and vice versa

**ATM Cell** A unit of information used for Asynchronous Transfer Mode. Its size is 53 bytes

**Attenuation** The measurement of the reduction in strength of a signal

**Authentication** A method used to verify the identity of a computer or a user that is attempting to access a system

**Automatic Repeat Request (ARQ)** A method of flow control used in networking

**B-ISDN (Broadband Integrated Services Digital Network)** An ISDN network similar to narrowband ISDN except it uses fiber-optic cable

**Bandwidth** Amount of information that can be transmitted by a communication channel

**Baseband Mode** The entire bandwidth of cable is used to transmit only one signal

**Basic Service Set** The simplest variation of the IEEE 802.11 WLAN standard, sometimes called an *ad hoc network,* requires no network infrastructure

**Baud Rate** Number of signals per second

**Beacon Frame** A frame periodically sent by the access point to the wireless device

**Binary Number** Also called *base 2,* it can take on values of only 0 or 1

**Bipolar Encoding** The signal voltage is represented by three voltage levels: positive, zero, and negative voltage

**Bit Oriented Transmission** A method of transferring information bit by bit

**Block Check Character (BCC)** Method that is used for error detection using a block of characters

**Block Cipher** A group of encrypted bits

**Bluetooth** A wireless technology the enables peripherals to be connected to a PC without cable

**Bridge** A networking device that operates in the Data Link layer of the OSI model, it is used to connect the LANs together and filter frames based on MAC address

**Broadband Transmission** When the bandwidth of a cable is used to carry more than one signal

**Broadband Wireless** Wireless technology that employs broadband transmission methods to efficiently use bandwidth

**Broadcast Address** Address value to transmit information to all hosts on the network

**Burst Error** Occurs when two or more consecutive bits of a frame have changed

**Bus Topology** Network topology where the stations are connected together through a single cable

**Byte Oriented Transmission** Method where information is transmitted byte by byte

**Cable Modem** A device that connects a user's computer to the Internet using a medium provided by a cable television provider

**Carrier Sense Multiple Access with Collision Avoidance (CSMA/CA)** A method used by a wireless device to access Wireless LAN. It listens to the media, and if the media is free it sends a short packet to the access point in order to get permission for transmission

**Carrier Sense Multiple Access with Collision Detection (CSMA/CD)** A protocol in which a station listens to the bus, and transmits only if the bus is free. After transmission the station listens for collision. If a collision occurs, the station retransmits the packet if media is free to transmit

**Cat-5 Cable** Type of UTP cable capable of carrying up to 100-MHz signal

**Cat-6 Cable** Type of UTP cable can carrying up to 250-MHz signal

**Cell** The area covered by an access point

**Certificate Authority (CA)** A network authority that issues and manages security credentials and public keys for message encryption and decryption.

**Chip Bit** Method used to represent one bit with several bits

**Cipher** Algorithm that is used for encryption and decryption

**Circuit Switching** A physical connection established between two users for transmission of information

**Cladding** The material surrounding the core of an optical cable

**Class A Address** Range of IP addresses that encompasses 0.0.0.0 to 127.255.255.255

**Class B Address** Range of IP addresses that encompasses 128.0.0.0 to 191.255.255.255

**Class C Address** Range of IP addresses that encompasses 192.0.0.0 to 223.255.255.255

**Class D Address** Range of IP addresses that encompasses 224.0.0.0 to 239.255.255.255

**Class E Address** Range of IP addresses that encompasses 240.0.0.0 to 247.255.255.255

**Client/Server Model** Model in which the client submits a task to the server for processing

**Coarse Wave Division Multiplexing (CWDM)** Combines up to 16 wavelengths onto a single fiber using the ITU standard of 20-nm spacing between the wavelengths ranging from 1310 nm to 1610 nm

**Codec** A device that converts an analog signal to digital and vice versa

**Code Division Multiplexing (CDM)** Allows multiple users to transmit simultaneously over one transmission line

**Connection Oriented Transmission** Transmission in which a connection is established between source and destination before data transmission

**Connectionless Transmission** Data communication method in which communication occurs between hosts without setting up a connection.

**Control Characters** Non-printable ASCII code characters that are used for control

**CPU (Central Processing Unit)** Part of computer that processes the information

**Crosstalk** Type of noise that is generated by a transmission line that carries a strong signal coupled with a transmission line carrying a weak signal

**Cryptography** Science of keeping information secret

**Cut-Through Switch** A type of switch that reads the first few bytes of a packet to determine the destination address in order to forward the packet to the destination

**Cyclic Redundancy Check (CRC)** Type of error detection that can detect single, multiple, and burst errors

**Data Encryption Standard** A method for data encryption and decryption that uses a private key

**Data Link Layer** Layer 2 of OSI model that performs framing, error detection, and retransmission

**Data Link Protocol** Protocol that is used in the Data Link layer of the OSI model

**Data Rate** Number of bits in one second

**D-Channel** A data channel is 16 Kpbs for BRI and 64 Kbps for PRI and is used in ISDN to carry a control signal

**Decibel** Logarithmic unit describing the power ratio

**Decryption** Process that uses encrypted information to recover the information

**Demodulation** Converting the modulated signal to the original signal

**Demultiplexer** Device that accepts a single multiplexed signal and recovers the original signals that were multiplexed

**Destination Address** Address of recipient frame

**Differential Manchester Encoding** Encoding scheme where the data and the synchronizing clock signal are held contained in a single signal. One of the two bits is represented by no transition at the beginning of a pulse and the other is represented by a transition at the beginning of a pulse period

**Digital Signal** Digital signal represented by two voltages; one voltage represents binary zero and other voltage represents binary one

**Digital Signature** An electronic rather than a written signature that can be used by someone to authenticate the identity of the sender of a message or the signer of a document

**Digital Subscriber Line (DSL)** A type of modem that uses a telephone line for transmission of information

**Direct Attack** Method of attack where the attacker accesses the system directly through an action such as breaking the password

**Directional Antenna** An antenna that transmits in a specific direction

**Direct Sequence Spread Spectrum (DSSS)** Used in WAN transmissions where a data signal at the sending station is combined with a higher data rate bit sequence that is capable of dividing the user data according to a spreading ratio

**Domain Name System (DNS)** System used on the Internet for translating names of the network to IP address

**Dynamic Host Configuration Protocol (DHCP)** A protocol used to automatically assign IP addresses to the hosts on the network

**Dynamic RAM (DRAM)** A type of semiconductor memory that is used for the main memory of a computer

**Electrical Industrial Association (EIA)** Specifies electrical specifications for cables and connectors

**Electromagnetic Wave** A wave of energy having a frequency that is propagated as a periodic disturbance of the electromagnetic field when an electric charge oscillates or accelerates

**Email** An application protocol for TCP, used for electronic mail

**Encryption** A process of scrambling information that only the receiver is able to recover

**Encryption Key** A code that is used for encrypting of information

**Error Correction** A method of error detection with the capability to correct the error present in transmitted information

**Error Detection** A method used to determine if errors exist in transmitted information

**Ethernet** A LAN network that uses bus topology

**Extended Service Set (ESS)** WLAN topology with multiple access points

**Fast Ethernet** Ethernet technology with data rates of 100 Mbps

**Federal Communication Commission (FCC)** An independent commission created by the US government charged with regulating interstate and international communications by radio, television, wire, satellite, and cable

**Fiber Distributed Interface (FDDI)** A high-speed LAN with data rate of 100 Mbps that is primarily used to interconnect LANs together using ring topology and fiber-optic cable as media

**File Transfer Protocol (FTP)** Internet application protocol that is used to transfer files from a remote computer without directly logging onto it

**Firewall** Combination of hardware and software that separates internal networks from the Internet in order to prevent unauthorized users, access to the internal network

**Firewire** High-speed serial bus with bandwidth capabilities up to 400 Mbps

**Forward Error Correction** By using additional information transmitted along with the data, receiver is able to correct errors itself without reference to the transmitter

**Frame** Unit of information in networking

**Frame Relay** A network offered by telecommunications carriers to corporations for connection of their separated LANs located in different geographic regions

**Frame Relay Assembler/Disassember (FRAD)** Converts frame format of LAN to frame format of frame relay or vice versa

**Frame Check Sequence (FCS)** Result generated by CRC for error detection

**Frequency** The number of cycles in one second

**Frequency Division Duplex (FDD)** Multiplexing scheme in which numerous signals are combined for transmission on a single communication line or channel

**Frequency Division Multiplexing (FDM)** Divides the bandwidth of a communication channel into smaller channels that are used to transmit simultaneously

**Frequency Hopping Spread Spectrum (FHSS)** In FHSS, the frequency band is divided into 79 channels and the transmitter transmits each part of its information in a different channel

**Frequency Shift Keying (FSK)** A method of modulation that uses change of frequency to represent zero and one

**Full Duplex** Mode in which both transmitter and receiver have the capability to transmit simultaneously

**Fully Connected Topology** Topology in which each station connects directly to other stations

**Gateway** A device that operates on all levels of the OSI model and can connect the networks with different protocols

**Gigabit Ethernet** Ethernet technology with data rate of 1000 Mbps

**Go-Back-N** A method for flow control in computer networking

**Graded Index Multimode Fiber** Fiber-optic cable that possess a varying index of refraction, which enables longer transmission distances

**H.323** Set of standards developed by ITU for Voice over Internet

**Hacker** Someone who tries to break into computer systems

**Half Duplex** Communication between two devices that uses one line for transmission in both directions, but only one at a time

**Hamming Code** A type of error correction that can detect and correct one error

**Hash Function** Used to generate encrypted messages and transform a string of characters usually into a shorter fixed-length value or a key that represents the original string of information

**High Level Data Link Protocol (HDLC)** A synchronous, bit oriented protocol for use in the Data Link layer

**HTTPS** Hyper Transport Protocol over Secure Socket Layer (SSL) using Netscape SSL as sublayer

**Hub** Also called a *repeater,* it operates at the Physical layer of the OSI model

**Hyper Text Markup Language (HTML)** Language that is used to create Web documents

**Hypertext Transfer Protocol (HTTP)** A file retrieving program that can access distributed and linked documents on the Web

**IEEE 802.2** IEEE standard that defines the Logical Link Control

**IEEE 802.3** IEEE standard for Ethernet network

**IEEE 802.4** IEEE standard for Token Bus network

**IEEE 802.5** IEEE standard for Token Ring network

**IEEE 802.1q** IEEE standard for VLAN

**IEEE 802.3u** IEEE standard for Fast Ethernet

**IEEE 802.3z** IEEE standard for Gigabit Ethernet

**IEEE 802.11** IEEE standard for Wireless LAN

**IEEE 802.16** IEEE standard for Wireless MAN

**Indirect Attack** Method of attack where the attacker obtains an individual's information and data, such as address and credit card number

**Industrial Scientific Medical Band (ISM)** Frequency band that does not require permission from FCC with a range of frequencies of 902 MHz to 928 MHz for the Industrial band, 2.4 GHz to 2.48 GHz for the Scientific band, and 5.725 to 5.85 GHz for the Medical band

**Integrated Services Digital Network (ISDN)** A communication protocol offered by telephone companies that permits telephone networks to carry data, voice, and other source traffic

**International Telecommunication Union (ITU)** An organization established by the United Nations, has its membership in virtually every government in the world

**Institute of Electrical and Electronic Engineering (IEEE)** Professional organization that develops standards for LAN and WLAN

**Integrity** Maintains data consistency and prevents tampering of information

**Internet** Collection of networks around the world connected together by routers or gateways

**Internet Protocol (IP) Address** The 32-bit address assigned to each host, divided into classes A, B, C, D, or E

**Internet Control Message Protocol (ICMP)** Used to report messages during packet delivery by the Internet, such as network unreachable and packet is too large

**Internet Engineering Task Force (IETF)** An international committee that develops standards for the Internet, such as IPv6

**Internet Protocol (IP)** The function of IP protocol is to deliver the packet to the destination on the Internet

**Internet Protocol Version 6 (IPV6)** The 128-bit address destined to replace the existing 32-bit IPV4 address

**Internet Service Provider (ISP)** A company that provides Internet access to other companies or individuals

**IPconfig Command** An Internet command that displays network settings, physical address, IP address, and subnet mask

**IP Security Protocol** Developed by IETF to protect information from attacks on any VPN; supports integrity, authentication, and confidentiality and delivery of packets in a secure manner

**Kerberos** An authentication system developed at MIT used for authentication of both systems

**LAN (Local Area Network)** Used to connect office and building computers located close together

**Laser** Optical ray used for transmission over fiber-optic cable

**Light Emitted diode (LED)** Generates an optical ray for use in fiber-optic cable

**Link Access Procedure Balanced (LAPB)** Modification of HDL by ITU, a Data Link layer protocol

**Logical Link Control (LLC)** Sublayer of Data Link layer that defines how the data will be multiplexed

**Loop Back Address** The last IP address of each class used for testing: Class A 127.0.0.1, Class B 191.255.0.0, and Class C 223.2555.255.0

**MAC Address** The 48-bit hardware address of NIC

**Mail Server** Manages incoming email of an organization

**Manchester Encoding** A type of signal encoding in which the clock pulse is embedded in the data signal

**MAN (Metropolitan Area Network)** Type of network that covers a wider area than a LAN, such as a metropolis

**Maximum Transfer Unit (MTU)** Defines the largest size of packet that a router can accept for transmission from a network

**Medium Access Control (MAC)** Sublayer of Data Link layer, it defines the method that a station uses to access the network

**Message Switching** Method in which the entire message passes to the switch for delivery

**Modal Bandwidth** The maximum bandwidth of fiber cable per 1 km, represented by MHz/Km

**Modem** Device that enables a user's computer to access the Internet through a phone line or cable TV

**Modulation** The addition of information (or the signal) to an electronic or optical signal

**Multicast Address** A single address that is assigned to multiple hosts

**Multimode Fiber (MMF)** Type of fiber cable that can propagate multiple rays of light simultaneously

**Multi-Path Fading** When a communication signal transmitted through the air takes multiple paths to reach the destination

**Multiplexer** A device that allows multiple users to transmit over one channel by combining multiple signals into one

**Narrow Band Signal** Signal with narrow spectrum or single frequency

**Netstat Command** Displays information about the network that the host is connected to

**Network Layer** Layer 3 of OSI model, the Network layer provides the following functionality: connection setup, disconnection, and routing of information

**Network-to-Network Interface (NNI)** In ATM, this defines the connection between switches or nodes

**Network Topology** Describes the method that connects computers together

**Non-Periodic Signal** A signal that does not exhibit a repeated pattern

**Non-Return to Zero** Coding scheme where zero is represented by a positive voltage and one is represented by a negative voltage

**Nslookup Command** Displays the host domain name with its IP address

**Nyquist Theorem** Used to find the maximum data rate of a noisy channel

**Odd Parity** Parity method used to check for errors

**Omni-Directional Antenna** An antenna that transmits in all directions

**One's Complement of Sums** Error detection method that computes the one's complement of the sum of the bits to transmit on both ends of the transmission. If the values differ, there is an error

**Open System Interconnection (OSI)** A model for networking developed by ISO for interoperability between equipment designed for networks

**Optical Fiber Cable** Type of cable that uses rays of light for transmission of information

**Optical Transponder** Used to change the wavelength of optical signals

**Parallel Transmission** Transmits multiples of bits over multiple wires simultaneously

**Parity Bit** An extra bit added to information for error detection, capable of detecting only one error

**Password Authentication Protocol (PAP)** A protocol where the client transmits a name and password to a remote server for verification

**Periodic Signal** A signal that repeats a pattern within a measurable time period

**Permanent Virtual Connection (PVC)** A virtual connection between source and destination that is set up and taken down manually by a network administrator

**Phase Shift Keying (PSK)** A method of transmitting and receiving digital signals, which requires that the phase of a transmitted signal must vary in order to convey information

**Physical Layer** Layer 1 of OSI model, it converts bits to signals to transmit over a communication channel as well as accepts signals from communications channels for conversion to bits

**Piconet** A Bluetooth topology with one master and six slaves

**Ping Command** Used for checking whether the destination host is reachable

**Point-to-Multipoint Topology** Topology used in wireless MAN where a base station transmits to subscriber stations

**Point-to-Point Protocol (PPP)** Internet protocol that is used for serial connection between host and Internet using a device, such as a modem

**Polar Encoding** Encoding scheme where positive and negative voltages are used to represent binary one and zero

**Port Number** A field in TCP header that determines the application that generated the data and is used for transmission or receipt of data

**Presentation Layer** Layer 6 of OSI model, it defines how the data will be represented, such as in ASCII or encrypted format

**Private Branch Exchange (PBX)** A private telephone switch system

**Private Key Encryption** Uses the same key for encryption and decryption

**Propagation Delay** The amount of the time that it takes a signal to travel from source to destination

**Propagation Speed** Speed of a signal over a given medium

**Protocol** A set of rules and standards used so that computers can communicate

**Proxy Server** A server that is connected between a private network and a public network, the proxy server filters out information based on its content or IP address

**Public Key Encryption** Encryption that uses different keys for encryption and decryption

**Public Switch Telephone Network (PSTN)** A term that refers to a public telephone network

**Pulse Code Modulation** Method for converting an analog signal to a digital signal

**Quadrature Amplitude Modulation (QAM)** A combination of amplitude shift keying and phase shift keying modulation methods

**Real-Time Transport Protocol (RTP)** Internet protocol used for real-time data communications

**Relay Agent** When the DHP server is not located in a broadcast domain of the client station, the router that the client station is connected to requires special software (relay agent) in order to transmit a client request to the DHCP server

**Remote Login** A host that uses Telnet to connect to a remote server

**Repeater** A networking device with multiple ports, it accepts frames from one port and transmits them to other ports

**Ring Topology** Network topology where computers are connected in a cascading order through a MAU to make a ring

**RJ-45** Type of connector used to connect UTP cable especially in an Ethernet network

**Roaming** When a station in a wireless LAN moves from one cell to another cell without losing connection

**Router** A device that operates in the Network layer of the OSI model, it determines the path the information will take in order to reach the destination

**Routing Table** A table that the router uses to find the next hop for the packet

**RS-232** A standard for serial port and serial communication

**Sampling Rate** Defines the number of samples taken from an analog signal in order to convert it to a digital signal

**Scatternet** A Bluetooth topology that is capable of connecting multiple Piconets together

**Secure Socket Layer (SSL)** Developed by Netscape for secure communication between a Web browser and a Web server

**Serial Transmission** Transmission in which information is transmitted bit by bit over a single wire

**Service Set Identifier (SSI)** An ID that is assigned to an access point, which the client must know in order to communicate with it

**Session Initiation Protocol (SIP)** Protocol developed by IETF for Voice over IP

**Session Layer** Layer 5 of OSI model, it is used to set up a session between source and destination applications

**Shielded Twisted Pair (STP)** Pair of copper wires that are twisted together and shielded for electronic noise

**Signal-to-Noise Ratio (S/N)** Ratio of the power of the signal over the power of the noise, represented in decibels

**Simple Network Management Protocol (SNMP)** Internet protocol that is used for network management and monitoring

**Simplex Mode** Transmitter that transmits in only one direction

**Single-Mode Fiber (SMF)** Type of fiber cable that propagates only one ray of light at a time

**SIP Redirector Server** A server used by Voice over IP to redirect a call to a new location once the endpoint has moved

**Sliding Window** Used by source and destination to inform each other about the size of their buffers

**Spread Spectrum Signal (SSS)** In SSS, information is transmitted over multiple frequencies or channels

**Star Topology** Topology where all computers are connected to a central station or hub

**Statistical Packet Multiplexing** Type of multiplexer that dynamically allocates bandwidth only to those active inputs of the multiplexer

**Stop and Wait** A method of flow control in a network

**Store and Forward Switch** Type of switch that stores incoming packets and checks for errors. If there are no errors in the packet, it is forwarded to the destination; otherwise it is discarded

**Stream Cipher** Encrypts information bit by bit

**STS Multiplexer** Multiplexer used specifically in a SONET

**Subnet Mask** Separate host ID and network ID of a given IP address

**Supervisory Frame** Type of frame used by networks for error and flow control

**Switched Virtual Connection** A virtual connection in an ATM network set up on demand by the signaling control point

**Symmetric Cryptography** A cartography method where source and destination use a similar key for encryption and decryption

**Symmetric Digital Subscriber Line (SDSL)** Type of DSL modem that uses the same data rate for upstream and downstream transmission

**Symmetric Switch** Type of switch that connects the segment of the LANs with the same data rate

**SYN Character** An ASCII special character that is used for synchronization

**Synchronous Data Link Control (SDLC)** Developed by IBM for the data link protocol of system network architecture

**Synchronous Optical Network (SONET)** Extremely high-speed carrier that uses optical cable as transmission media

**Synchronous Payload Envelope (SPE)** Payload of STS-1 frame plus path overhead, used in SONET networks

**Synchronous Transmission** Transmission scheme where the transmitter uses its clock pulse to inform the receiver of the transmission speed

**Synchronous Transport Signal (STS)** The basic electrical signal for SONET

**T1 Link** Digital transmission link with a bandwidth of 1.54 Mbps

**TCP/IP Protocol** Protocols used for the Internet

**Telnet** An application protocol for the Internet that enables clients to connect to a remote server

**ThickNet** A thick coaxial cable of 10 mm used for 10Base-5

**ThinNet** A thin coaxial cable of 5 mm used for 10Base-2

**Time Division Multiplexing** Multiplexing method where each input is allocated a discrete time quanta

**Time to Live** Amount of time that a packet can travel in the Internet

**Token** A three-byte frame used as the access method in Token Ring network

**Token Bus** Network topology that is physically a bus topology but logically is a Ring topology

**Token Ring Network** Network topology that use Ring topology and token as an access method

**Tracert Command** Command that displays the path of the packet from the source to the destination

**Transceiver** A device capable of transmitting and receiving information

**Transmission Control Protocol (TCP)** Transport level protocol for the Internet, used to reliably transmit information

**Transmission Media** Media that is used for transmitting information such as a conductor, fiber-optic cable, or wireless MAN

**Transmission Time** Time that it takes to transmit one bit on a transmission channel

**Transport Layer** Layer 4 of OSI model, used for reliable communication between source and destination

**Tunneling** A method used for secure connection between LANs over a public network (Internet)

**Twisted-Pair Wire** A pair of wires wrapped around each other

**Unicast Address** Address that specifies a single network device

**Unicode** 16-bit code used for encoding characters in all languages

**Unipolar Encoding** Encoding method where only positive voltage or negative voltage is used to represent zero and one

**Universal Serial Bus** Computer serial bus that enables users to a externally connect a peripheral

**Unlicensed National Information Infrastructure Band (U-NII)** 300 MHz of spectrum that the FCC has made available for devices that provide short-range, high-speed wireless digital communications without licensing

**Unshielded Twisted Pair (UTP)** One the most popular networking cables, it uses four pairs of twisted wires

**User Datagram Protocol (UDP)** A connectionless protocol for transmission of information without reliability

**User to Network Interface (UNI)** An interface between a user and an ATM switch

**Video Codec** A device that converts video signal to digital and vice versa

**Virtual Circuit** A logical connection between the source and the destination using a specific path for packet transmission

**Virtual LAN (VLAN)** Group of hosts on one or more LANs that are configured by software

**Virtual Path (VP)** A bundle of virtual channels

**Virtual Path Identifier (VPI)** A connection identifier that is used in ATM networks

**Virtual Private Network (VPN)** A method that uses public networks, which act as private networks

**Virtual Tributary** Mapping lower data rate frames into a higher data rate STS-1 payload

**Voice over IP (VoIP)** Enables users to transfer voice packets over the Internet

**Wave Division Multiplexing (WDM)** Enables single fiber-optic cable to transmit multiple rays of lights at different wavelengths

**Wavelength** Distance traveled by a wave over one period

**Wide Area Network (WAN)** Network that covers a large geographic area, such as cities or continents

**Wired Equivalent Privacy (WEP)** A method to protect wireless LAN from intruders

**Wireless Fidelity (Wi-Fi)** Wireless Ethernet alliance that offers certification for wireless equipment interoperability

**Wireless LAN (WLAN)** A type of local area network that uses air as transmission media

**Wireless MAN** A type of WAN that uses air as the transmission media

**World Wide Web** Collection of documents on the Internet

**X.509** A standard by IITU to define information used in a certificate

# Bibliography


Beasley, J. *Networking,* Prentice-Hall, 2004.

Bisallion, T. and Werner, B. *TCP/IP with Windows Illustrated,* McGraw-Hill, 1998.

Black, U. *ATM: Foundation for Broadband Networks,* Prentice-Hall, 1995.

Black, U. *Frame Relay Networks,* McGraw-Hill, 1996.

Comer, E.D. *Computer Networks and Internets,* 2nd edition, Prentice-Hall, 1999.

Feit, S. *TCP/IP Architecture and Implementation with IPv6 and IP Security,* 2nd edition, McGraw-Hill, 1997.

Garcia, A. and Widjaja, J. *Communication Networks,* McGraw-Hill, 2004.

Halsall, F. *Multimedia Communications,* Addison-Wesley, 2001.

Held, G. and Sarch, R. *Data Communications,* 3rd edition, McGraw-Hill, 1995.

Hura, G. and Singhal, M. *Data and Computer Communications,* CRC, 2000.

Mel, H. and Baker, D. *Cryptography,* Addison-Wesley, 2001.

Panko, R. *Business Data Communications and Networking,* 2nd edition, Prentice-Hall, 1999.

Shaeda, N.K. *Multimedia Information Networking,* Prentice-Hall, 1999.

Shay, W.A. *Understanding Data Communications & Networks,* 3rd edition, Thomson, 2004.

Stalling, W. *Data and Computer Communications,* 6th edition, Prentice-Hall, 2000.

Stalling, W. *ISDN and Broadband ISDN with Frame Relay and ATM,* 3rd edition, Prentice-Hall, 1995.

Stalling, W. *Network Security Essential,* Prentice-Hall, 2000.

Stalling, W. *Wireless Communications and Networks,* Prentice-Hall, 2002.

Stevens, W.R. *TCP/IP Illustrated,* Vol. 1, Addison-Wesley, 1994.

Tanenbum, A.S. *Computer Networks,* 4th edition, Prentice-Hall, 2003.

Thomas, S.A. *IPng and the TCP/IP Protocols: Implementing the Next Generation Internet,* Wiley, 1996.

Wu, C.H. and Irvin, J.D. *Emerging Multimedia Computer Communication Technologies,* Prentice-Hall, 1999.

# Web Links

**ATM**

http://www.atmforum.com

**Cable Modem**

http://www.cabledatacomnews.com
http://www.cablelabs.com
http://www.cablemodem.com

**DHCP**

http://www.dhcp.org/

**DSL**

http://www.adsl.com
http://www.westell.com/adslWpr.html

**Firewall**

http://www.oms.co.za
http://www.3com.com

**Frame Relay**

http://www.ascend.com
http://www.frame-relay.indiana.edu

**Gigabit Ethernet**

http://www.gigabit-ethrnet.org

**Home Networking**

http://www.commsdesign.com

**IEEE standards**

http://standards.ieee.org/getieee802/

**Internet**

http://www.vbns.com
http://www.internetvalley.com
http://rs.internic.net
http://www.broadwatch.com

### Network Security

http://wp.netscape.com
http://www.cisco.com
http://www.iacr.org
http://www.ssh.fi

### Point-to-Point Protocol

http://www.ietf.org

### SONET

http://www.atis.org
http://www.niuf.nist.gov

### 10 Gigabit Ethernet

http://grouper.ieee.org

### Universal Serial BUS

http://www.usb.org

### Voice over IP

http://www.h323forum.org/papers/
http://www.iec.org
http://voip.internet2.edu

### Wireless Networking

http://www.bluetooth.com
http://grouper.ieee.org
http://www.computerhope.com
http://www.mobilian.com
http://www.palowireless.com
http://www.proxim.com
http://www.3com.com/wireless
http://www.wimaxforum.org
http://www.wlana.com
http://www.xilinx.com

# Index

## A

## B

## J



## K



## L

## Q



## R



## S

## T